**Voltage-Gated Ion Channels
as Drug Targets**

*Edited by
David J. Triggle,
Murali Gopalakrishnan,
David Rampe, and Wei Zheng*

Methods and Principles in Medicinal Chemistry

Edited by R. Mannhold, H. Kubinyi, G. Folkers

Editorial Board
H.-D. Höltje, H. Timmerman, J. Vacca, H. van de Waterbeemd, T. Wieland

Previous Volumes of this Series:

H.-J. Böhm, G. Schneider (eds.)

Protein-Ligand Interactions

Vol. 19

2003
ISBN 3-527-30521-1

R.E. Babine S. S. Abdel-Meguid (eds.)

Protein Crystallography in Drug Discovery

Vol. 20

2004
ISBN 3-527-30678-1

Th. Dingermann, D. Steinhilber, G. Folkers (eds.)

Molecular Biology in Medicinal Chemistry

Vol. 21

2004
ISBN 3-527-30431-2

H. Kubinyi, G. Müller (ed.)

Chemogenomics in Drug Discovery

Vol. 22

2004
ISBN 3-527-30987-X

T. I. Oprea (ed.)

Chemoinformatics in Drug Discovery

Vol. 23

2005
ISBN 3-527-30753-2

R. Seifert, T. Wieland (eds.)

G-Protein Coupled Receptors as Drug Targets

Vol. 24

2005
ISBN 3-527-30819-9

O. Kappe, A. Stadler

Microwaves in Organic and Medicinal Chemistry

Vol. 25

2005
ISBN 3-527-31210-2

W. Bannwarth, B. Hinzen (eds.)

Combinatorial Chemistry

Vol. 26

2005
ISBN 3-527-30693-5

G. Cruciani (ed.)

Molecular Interaction Fields

Vol. 27

2005
ISBN 3-527-31087-8

M. Hamacher, K. Marcus, K. Stühler, A. van Hall, B. Warscheid, H. E. Meyer (eds.)

Proteomics in Drug Design

Vol. 28

2005
ISBN 3-527-31226-9

Voltage-Gated Ion Channels as Drug Targets

Edited by
David J. Triggle, Murali Gopalakrishnan, David Rampe,
and Wei Zheng

WILEY-VCH

WILEY-VCH Verlag GmbH & Co. KGaA

Series Editors

Prof. Dr. Raimund Mannhold
Biomedical Research Center
Molecular Drug Research Group
Heinrich-Heine-Universität
Universitätsstraße 1
40225 Düsseldorf
Germany
Raimund.mannhold@uni-duesseldorf.de

Prof. Dr. Hugo Kubinyi
Donnersbergstrasse 9
67256 Weisenheim am Sand
Germany
kubinyi@t-online.de

Prof. Dr. Gerd Folkers
Collegium Helveticum
STW/ETH Zürich
8092 Zürich
Switzerland
folkers@collegium-ethz.ch

The Volume Editors

Prof. David J. Triggle
126 Cooke Hall
State University of New York at Buffalo
Buffalo
New York 14260
USA

Dr. Murali Gopalakrishnan
Abbott Laboratories
Neuroscience Research
Department R47W, Bldg AP9A-3
100 Abbott Park Road
Chicago
Ilinois 60064-6125
USA

Dr. David Rampe
sanofi-aventis
Route 202-206
PO Box 6800
Bridgwater NJ 08807-0800
USA

Dr. Wei Zheng
NIH Chemical Genomics Center
National Human Genome Research Institute
National Institutes of Health
9800 Medical Center Drive
Bethesda, MD 20892
USA

Library of Congress Card No.: Applied for
British Library Cataloguing-in-Publication Data:
A catalogue record for this book is available from the British Library.

**Bibliographic information published by
Die Deutsche Bibliothek**
Die Deutsche Bibliothek lists this publication in the Deutsche Nationalbibliografie; detailed bibliographic data is available in the Internet at <http://dnb.ddb.de>.

© 2006 WILEY-VCH
Verlag GmbH & Co. KGaA, Weinheim

Typesetting Hagedorn Kommunikation, Viernheim
Printing Strauss GmbH, Mörlenbach
Bookbinding Litges & Dopf, Heppenheim
Cover Design Grafik-Design Schulz, Fußgönnheim

Printed in the Federal Republic of Germany.
Printed on acid-free paper.

ISBN-13: 978-3-527-31258-0
ISBN-10: 3-527-31258-7

Contents

Voltage-Gated Ion Channels as Drug Targets. Edited by D. Triggle
Copyright © 2006 WILEY-VCH Verlag GmbH & Co. KGaA, Weinheim
ISBN 3-527-31258-7

X | *Contents*

Preface

The present volume of our series "Methods and Principles in Medicinal Chemistry" is dedicated to "Voltage-gated Ion Channels" and their impact as targets for drug design.

Ion channels are ubiquitously distributed throughout cellular life; they represent integral membrane proteins that both produce and transduce the electrical signals crucial to the maintenance and function of cells. Ion channels gate or regulate the ion flow between the cytoplasmic compartment and the extracellular space and between subcellular compartments. They open and close in response to changes in membrane potential, changes in ion concentrations on either side of the membrane, and agonist binding to the channel or closely associated regulatory proteins. Under pathological conditions ion channels contribute to or drive a variety of disease processes from achalasia and arrhythmias to xerostomia and vertigo.

Ion channels can be classified in several distinct ways. The most common classification refers to the ions for which they are selective. Their primary mode of stimulus allows a classification into ligand- and voltage-gated channels. Alternatively, they can be classified by their electrophysiological properties and by their pharmacological sensitivity to toxins and synthetic drugs. Increasingly, they are classified according to their sequences, demonstrating that ion channels exist as super-families with considerable structural homology between the members, despite very different electrophysiological and pharmacological properties.

Whereas our initial understanding of ion channel structure and function was largely due to electrophysiological data, our knowledge was remarkably advanced by the work of Roderick MacKinnon and his coworkers on the three-dimensional structures of potassium channels. We are now at a stage where it becomes increasingly possible to start the integration of structural and functional data to provide a detailed understanding of both channel function and how drugs interact with and modulate such channel function.

A main characteristic of ion channels is their remarkable sensitivity to chemical modulation. Ion channels are excellent targets for drug design because: a) they are loci for integrated cellular communication: many inputs control the level of cellular membrane potential and thus the degree of excitation or inhibition; b) they are highly efficient molecular machines that enable the selective permeation of certain ions; c) there exists a multiplicity of channel types and subtypes; d) each

Voltage-Gated Ion Channels as Drug Targets. Edited by D. Triggle
Copyright © 2006 WILEY-VCH Verlag GmbH & Co. KGaA, Weinheim
ISBN 3-527-31258-7

channel type and subtype typically has a multiplicity of discrete ligand binding sites that are allosterically coupled to the gating and permeation machinery of the channel; and e) the binding characteristics of the ligand can be modulated, both quantitatively and qualitatively, by factors such as membrane potential or channel phosphorylation.

The present volume comprises eight sections, the first four of which deal with basic background information. An introduction by the volume editors is followed by in-depth overviews dedicated to STRUCTURE AND FUNCTION OF ION CHANNELS (by William A. Catterall), DRUG INTERACTIONS AT ION CHANNELS (by Bruce Bean and Stefan McDonough) and to ASSAY TECHNOLOGIES (by Derek Leishman and Gareth Waldron).

The fifth section refers to CALCIUM CHANNELS. Clinton Doering and Gerald Zamponi give a general overview on calcium channels, whereas the most important calcium channel subtypes including T-type channels (by Thomas Connolly and James Barrow), L-type channels (by David J. Triggle) as well as N-type channels (by Terry Snutch) are separately treated in follow-up chapters.

SODIUM CHANNELS represent the focus of the next section. An overview on this ion channel class by Doug Krafte, Ken McCormack and Mark Chapman is followed by a chapter on sodium channel subtype selectivity from Tito Gonzales.

A quite comprehensive section is dedicated to POTASSIUM CHANNELS starting with an overview by Murali Gopalakrishnan. Then five different potassium channel subtypes and their relevant modulators are described in adequate detail. They comprise Kv1.3 channels (reviewed by George Chandy, Heike Wulff, Christine Beeton; and Michael Pennington), Kv1.5 channels (by Stefan Peukert and Heinz Goegelein), Ca^{2+}-activated K^+ channels (by Sean C. Turner and Char-Chang Shieh), K_{ATP} channels (by William Carroll), and finally KCNQ channels (by Grant McNaughton Smith and Alan Wickenden).

The last section is dedicated to GENETIC AND ACQUIRED CHANNELOPATHIES. Introductory remarks to this important aspect of ion channel research are given by Dennis Wray. Structural and ligand-based models for HERG and their application in medicinal chemistry are then debated by Yi Li, Giovanni Cianchetta, and Roy Vaz. The concluding chapter of this volume, written by Armando Lagrutta and Joseph Salata, treats relevant safety issues in ion channel drug development.

The series editors believe that this book is unique in its topic and presentation and adds a fascinating facet to the series. We are indebted to all authors for their well-elaborated contributions and we would like to thank the volume editors David Triggle, Murali Gopalakrishnan, David Rampe, and Wei Zheng for their enthusiasm to organize this volume. We also want to express our gratitude to Renate Doetzer and Frank Weinreich from Wiley-VCH for their valuable contributions to this project.

January 2006
Raimund Mannhold, Düsseldorf
Hugo Kubinyi, Weisenheim am Sand
Gerd Folkers, Zürich

1
Introduction – On Ion Channels

Murali Gopalakrishnan, David Rampe, David Triggle, and Wei Zheng

Ion channels are membrane proteins that are pore forming to permit ion transit according to concentration and electrochemical gradients. They are ubiquitously distributed throughout cellular life and indeed some form of ion moving (and other solute moving) device must have developed very early in cellular evolution and certainly must have evolved simultaneously with the development of the cell membrane.

Ion channels are also one of the mechanisms by which cells respond to informational inputs. Under physiological conditions these channels permit an orderly movement of ions across cell membranes and contribute both to cellular signaling processes and to the maintenance of cellular homeostasis. Under pathological conditions ion channels contribute to or drive various diseases processes from achalasia and arrhythmias to xerostomia and vertigo.

Ion channels are allosteric proteins that undergo significant conformational transitions in response to various informational inputs, including mechanical tension, voltage gradients and endogenous and exogenous chemical signals. The latter inputs are of particular interest to the clinician, pharmacologist and the pharmaceutical industry since they represent opportunities, both real and potential, for drug intervention. However, as usual, Nature, the supreme medicinal chemist, long ago seized this opportunity and a remarkable number of toxins from venomous species are directed against both ligand- and voltage-gated ion channels. Indeed, one of the principal characteristics of ion channels is their remarkable sensitivity to chemical modulation.

The chemical species depicted in Fig. 1.1 provide adequate testimony to this point. Currently, there exists ion channel therapeutics for anesthesia, anxiety, epilepsy, hypertension, insomnia and pain and excellent opportunities for ion channel therapeutic modulation in, for example, affective disorders, allergic disorders, autoimmune diseases, contraception, incontinence, and stroke.

Ion channels are excellent targets for drug design because they are:

1. Loci for integrated cellular communication: many inputs control the level of cellular membrane potential and thus the degree of excitation or inhibition.

Voltage-Gated Ion Channels as Drug Targets. Edited by D. Triggle
Copyright © 2006 WILEY-VCH Verlag GmbH & Co. KGaA, Weinheim
ISBN 3-527-31258-7

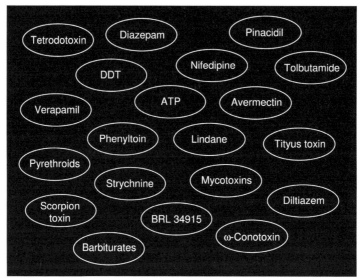

Fig. 1.1 Drugs and toxins that act at ion channels.

2. Highly efficient molecular machines that permeate ions selectively at rates that can approach diffusion-controlled.
3. There exists a multiplicity of channel types and subtypes.
4. Each channel type and subtype typically has a multiplicity of discrete ligand binding sites that are coupled allosterically to the gating and permeation machinery of the channel.
5. The binding characteristics of the ligand can be modulated, both quantitatively and qualitatively, by factors such as membrane potential or channel phosphorylation.

There are thus provided opportunities for multiple modes of interaction that can in principle generate a common pharmacological and therapeutic endpoint. Figure 1.2 depicts drug interaction at the L-type voltage-gated calcium channels, site of interaction several of cardiovascular drugs: three separate sites are indicated, but there are probably at least eight such sites, all linked allosterically to the guts of the calcium channel. Figure 1.3 provides a second example of multiple drug interaction where a neuroprotective strategy might be to prevent cellular depolarization and block calcium entry into neurons: this can be achieved through various discrete strategies from blockade of voltage-gated calcium channels to activation of ATP-dependent potassium channels. Parenthetically, Fig. 1.2 also presents a dilemma common to a number of therapeutic strategies, namely that for some diseases (in this case neuroprotection during ischemic stroke) various pathologic mechanisms are operative and modulation of only one may not be an effective strategy.

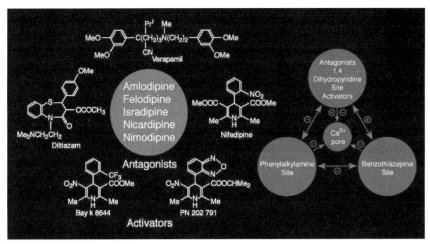

Fig. 1.2 Drug interactions at the L-type voltage-gated calcium channel.

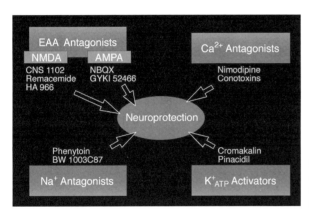

Fig. 1.3 Neuroprotective strategies implemented by drugs acting at diverse ion channels.

Until very recently our understanding of ion channel structure and function has been derived largely from electrophysiological data, leading to the representation depicted in Fig. 1.4. Despite the cartoon-like characteristics of Fig. 1.4, the electrophysiology underlying it has been astonishingly successful in explaining channel function and how drugs interact with channels. The past decade has seen major advances in our knowledge of channel structure and function, starting with the remarkable work of Roderick Mackinnon and his colleagues and their success in providing solid-state structures of the potassium channels. We are now at a stage where it becomes increasingly possible to start the integration of structural and functional data to provide for a detailed understanding of both channel function and of how drugs interact with and modulate such channel function.

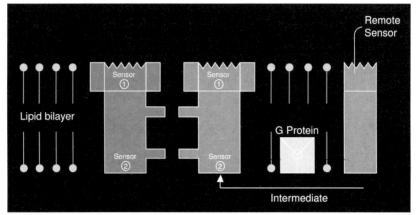

Fig. 1.4 Two-dimensional representation of an ion channel, depicting the major structural features.

It is thus believed that this book will appear at an appropriate time in our understanding of ion channels. All four editors have worked extensively for several years on the medicinal chemistry and pharmacology of ion channels and recognize that for a successful approach to the development of new drugs active at ion channels it is increasingly necessary to follow an integrated approach. A medicinal chemistry approach in the absence of an understanding of ion channel function and behavior and without recognizing what assay technologies are telling us is not likely to be successful. Far greater integration of chemical, biochemical, biophysical, pharmacological and structural approaches are necessary. Additionally, there are drug discovery programs where it is necessary that certain types of channel-modulating behavior not be found, notably activity at HERG channels. This has resulted in the expenditure of millions of research dollars annually and supports a growing "cottage industry" of small companies specializing in ion channel safety assays. Thus, even non-ion channel programs need to be aware of at least some aspects of ion channel structure and function.

This book is organized to optimize such an interdisciplinary approach. Although the primary emphasis is on drugs active at voltage-gated calcium, potassium and sodium channels the chemical pharmacology of these drugs is set against a background of channel classification, function and structure. Accordingly, the initial chapters deal with, respectively, channel structure and function, state-dependent interactions of drugs with channels and the assay technologies for drug screening. These three initial chapters provide the necessary background for the more detailed understanding of drug actions at calcium, potassium and sodium channels. These channels are discussed in three separate sections, each of which starts with an overview of the respective channel class. The volume concludes with a three-chapter section on ion channel diseases or "Channelopathies" and discussions of channel safety, with particular reference to HERG.

The next decade is almost certain to see continuing major advances in our knowledge of channel structure as more ion channels provide three-dimensional views. The challenge will be to link that structural knowledge to the definition of channel function and to our knowledge of drug action at those channels. That should lead to therapeutic advances for arrhythmias, neurodegenerative disorders, pain and stroke, all of which are unmet or underserved medical needs and where ion channels are significant contributors to the underlying pathologies.

We thank all of our contributors to this volume. They have all put aside other activities to contribute their specific knowledge and expertise and the success of the book will be entirely due to them.

2
The Voltage-gated Ion Channel Superfamily

William A. Catterall

2.1
Introduction

Electrical signals control contraction of muscle, secretion of hormones, sensation of the environment, processing of information in the brain, and output from the brain to peripheral tissues. In excitable cells, electrical signals also have an important influence on intracellular metabolism and signal transduction, gene expression, protein synthesis and targeting, and protein degradation. In all of these contexts, electrical signals are conducted by members of the ion channel protein superfamily, a set of more than 140 structurally related pore-forming proteins. In addition, members of this protein superfamily are crucial in maintaining ion homeostasis in the kidney and in many different cell types and participate in calcium signaling pathways in nonexcitable cells. This introductory chapter describes the voltage-gated ion channel families and gives an overview of the common features of their structure and function.

2.2
Voltage-gated Sodium Channels

The founding member of the ion channel superfamily in terms of its discovery as a protein is the voltage-gated sodium channel. These channels are responsible for the rapid influx of sodium ions that underlies the rising phase of the action potential in nerve, muscle, and endocrine cells. Neurotoxin labeling, purification and functional reconstitution showed that sodium channels from mammalian brain contain voltage-sensing and pore-forming elements in a single protein complex of one principal α subunit of 220 to 260 kDa and one or two auxiliary β subunits of approximately 33 to 36 kDa (Catterall, 1984; Catterall, 2000a). The α subunits of sodium channels contain four homologous domains that each contain six hydrophobic, probable transmembrane segments (Fig. 2.1). A membrane-reentrant loop between the fifth and sixth (S5 and S6) transmembrane segments

Voltage-Gated Ion Channels as Drug Targets. Edited by D. Triggle
Copyright © 2006 WILEY-VCH Verlag GmbH & Co. KGaA, Weinheim
ISBN 3-527-31258-7

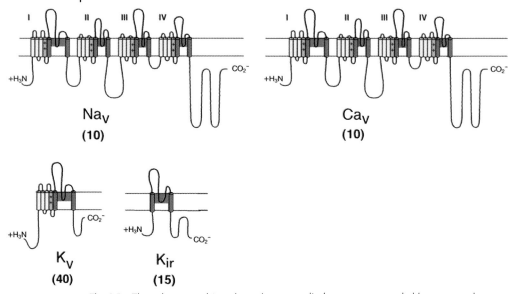

Fig. 2.1 The voltage-gated ion channel protein family. The different members of the ion channel protein family structurally related to the voltage-gated ion channels are illustrated as transmembrane folding diagrams in which cylinders represent probable transmembrane alpha helices. Red, S5-S6 pore-forming segments; green, S4 voltage sensor; yellow, S1-S3 transmembrane segments.

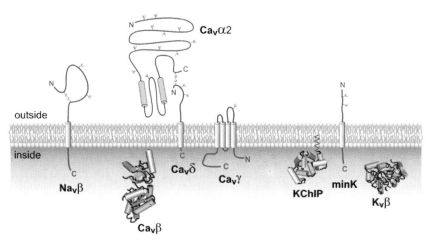

Fig. 2.2 Auxiliary subunits of the voltage-gated ion channels. Auxiliary subunits of Na_v, Ca_v, and K_v channels are illustrated, with cylinders representing predicted alpha helices of the transmembrane subunits. N-linked carbohydrate chains are indicated by Ψ. Intracellular auxiliary subunits are illustrated by their predicted three-dimensional structures.

forms the narrow extracellular end of the pore while the S6 segments form the intracellular end (Fig. 2.1, red). The pore is formed in the center of a pseudosymmetric array of the four domains, and a single α subunit containing four domains is able to receive voltage signals and activate its intrinsic pore. The channel responds to voltage by virtue of its S4 segments (Fig. 2.1, green), which contain repeated motifs of a positively charged amino acid residue followed by two hydrophobic residues, and move outward under the influence of the membrane electric field to initiate a conformational change that opens the pore. Drugs that block the pore of sodium channels are important as local anesthetics, antiarrhythmic drugs, and antiepileptic drugs. Their receptor sites are formed by amino acid residues in the S6 segments of domains I, III, and IV (Catterall, 2000a; Ragsdale et al., 1994). Nine voltage-gated sodium channel α subunits, designated $Na_v1.1$ to $Na_v1.9$ have been functionally characterized. They comprise a single subfamily of proteins with greater than 70% amino acid sequence identity in their transmembrane segments (Goldin et al., 2000). One additional related α subunit, which defines a second subfamily, $Na_v2.1$, is known but is apparently not voltage-gated.

Sodium channel auxiliary subunits, $Na_vβ1$ to $Na_vβ4$, interact with the different α subunits and alter their physiological properties and subcellular localization. These proteins have a single transmembrane segment, a large N-terminal extracellular domain that is homologous in structure to a variable chain (V-type) immunoglobulin-like fold, and a short C-terminal intracellular segment (Fig. 2.2) (Isom et al., 1992; Isom et al., 1995; Morgan et al., 2000; Yu et al., 2003). The $Na_vβ$ subunits interact with α subunits through their extracellular Ig-fold domains, modulate α subunit function, and enhance their cell surface expression (McCormick et al., 1999). Like other proteins with an extracellular Ig-fold, they also serve as cell adhesion molecules by interacting with extracellular matrix proteins, cell adhesion molecules, and cytoskeletal linker proteins (Ratcliffe et al., 2000; Srinivasan et al., 1998; Ratcliffe et al., 2001; Kazarinova-Noyes et al., 2001; Malhotra et al., 2002). A mutation in a conserved cysteine in the Ig-fold of the $Na_vβ1$ subunit causes familial epilepsy (Wallace et al., 1998). The $Na_vβ$ subunits are a recent evolutionary addition to the family of ion channel associated proteins, as they have only been identified in vertebrates.

2.3
Voltage-gated Calcium Channels

Voltage-gated calcium channels are the key signal transducers of electrical signaling, converting depolarization of the cell membrane into an influx of calcium ions that initiates contraction, secretion, neurotransmission, and other intracellular regulatory events (Catterall, 2000b). Skeletal muscle calcium channels, first identified by drug labeling, purification, and functional reconstitution, have a principal α1 subunit of 212 to 250 kDa, which is similar to the sodium channel α subunit (Curtis and Catterall, 1984; Takahashi et al., 1987). cDNA cloning and sequencing showed that the α1 subunit of calcium channels is analogous to the sodium channel α subunits in structural organization (Fig. 2.1) and is approximately 25% identical in amino acid sequence in the transmembrane regions (Tanabe et al., 1981). The α1 subunit is associated with auxiliary α2δ, β, and γ subunits that are unrelated to the sodium channel auxiliary subunits (Fig. 2.2). As for the sodium channel α subunit, the calcium channel α1 subunit is sufficient to form a voltage-gated calcium-selective pore by itself. Ten functional calcium channel α1 subunits are known in vertebrates, and they fall into three subfamilies that differ in function and regulation (Ertel et al., 2000). The Ca_v1 subfamily (Ca_v1.1 to Ca_v1.4) conduct L-type calcium currents that initiate contraction, endocrine secretion, and synaptic transmission at the specialized ribbon synapses involved in sensory input in the eye and ear (Hofmann et al., 1994; Striessnig, 1999). L-type calcium currents also are important regulators of gene expression and other intracellular processes. Blockers of Ca_v1.2 channels, including phenylalkylamines, dihydropyridines, and benzothiazepines, are important in the therapy of cardiovascular diseases such as hypertension, cardiac arrhythmia, and angina pectoris (Triggle, 1999). These drugs bind to three distinct receptor sites formed primarily by the S5 and S6 segments in domain III and the S6 segment in domain IV (Hockerman et al., 1997; Striessnig, 1999). The Ca_v2 subfamily of calcium channels (Ca_v2.1 to Ca_v2.3) conduct N-, P/Q- and R-type calcium currents that initiate fast synaptic transmission at synapses in the central and peripheral nervous systems and are blocked specifically by peptide neurotoxins from spider and cone snail venoms (Snutch and Reiner, 1992; Dunlap et al., 1995; Catterall, 2000b; Olivera et al., 1994). The Ca_v3 subfamily of calcium channels (Ca_v3.1 to Ca_v3.3) conduct T-type calcium currents that are important for repetitive action potential firing of neurons in the brain and in the pacemaker cells of the sinoatrial node in the heart (Perez-Reyes, 2003). The functional and regulatory properties and protein–protein interactions of different subfamilies of ion channels these channels are adapted to their different roles in electrical signaling and cellular signal transduction.

Ca$_v$1 and Ca$_v$2 channels have four distinct auxiliary subunits, $Ca_v\alpha$2, $Ca_v\beta$, $Ca_v\gamma$, and $Ca_v\delta$ (Fig. 2.2) (Takahashi et al., 1987; Catterall, 2000b), which each comprise a small protein family. The $Ca_v\alpha$2 and $Ca_v\delta$ subunits are encoded by the same gene (Ellis et al., 1988), whose translation product is proteolytically cleaved and disulfide linked to yield the mature extracellular α2 subunit glycoprotein of 140 kDa and transmembrane disulfide-linked δ subunit glycoprotein of 27 kDa (De Jongh et al., 1990). Four $Ca_v\alpha$2δ genes are known (Arikkath and Campbell, 2003). The four $Ca_v\beta$ subunits are all intracellular proteins with a common pattern of alpha helical and unstructured segments (Ruth et al., 1989; Arikkath and Campbell, 2003). They have important regulatory effects on cell surface expression and they also modulate the gating of calcium channels, causing enhanced activation upon depolarization and altered rate and voltage dependence of inactivation (Arikkath and Campbell, 2003). Recent structural modeling and X-ray crystallography studies have revealed that these subunits contain conserved, interacting SH3 and guanylate kinase domains like the MAGUK family of scaffolding proteins (Van Petegem et al., 2004; Chen et al., 2004; Opatowsky et al.,

2004; McGee et al., 2004; Takahashi et al., 2004), and therefore may interact with other intracellular proteins. Eight Ca$_v$γ subunit genes encode glycoproteins with four transmembrane segments (Jay et al., 1990; Arikkath and Campbell, 2003). Although the Ca$_v$γ1 subunit is associated specifically with skeletal muscle Ca$_v$1.1 channels, other Ca$_v$γ subunits interact with other calcium channels, glutamate receptors and possibly with other membrane signaling proteins (Arikkath and Campbell, 2003). Thus, the γ subunits discovered as components of calcium channels apparently have a more widespread role in assembly and cell surface expression of other membrane signaling proteins.

2.4
Voltage-gated Potassium Channels

Voltage-gated potassium channels are activated by depolarization, and the outward movement of potassium ions through them repolarizes the membrane potential to end action potentials, hyperpolarizes the membrane potential immediately following action potentials, and plays a key role in setting the resting membrane potential. In this way, potassium channels control electrical signaling and regulate ion flux and calcium transients in nonexcitable cells. The first voltage-gated potassium channels were cloned from *Drosophila* based on a mutation that causes the *Shaker* phenotype (Jan and Jan, 1997). They are composed of four transmembrane subunits that each is analogous to a single domain of the principal subunits of sodium or calcium channels (Fig. 2.1). The voltage-gated potassium channels are remarkable for their diversity. They include 40 different channels that are classified into 12 distinct subfamilies based on their amino acid sequence homology (K$_v$1 to K$_v$12) (Gutman et al., 2003). These α subunits can assemble into homo- and heterotetramers, leading to a wide diversity of different channel complexes. The diversity of potassium channels allows neurons and other excitable cells to precisely tune their electrical signaling properties by expression of different combinations of potassium channel subunits. The voltage-gated potassium channels are important drug targets for treatment of cardiac arrhythmia (Anderson et al., 2002) and are also the molecular targets of many drugs that cause long QT syndrome as a side effect (Sanguinetti and Mitcheson, 2005). Voltage-gated potassium channels are also potential therapeutic targets for immune modulation (Chandy et al., 2004) and other indications.

K$_v$1 channels are often associated with an intracellular K$_v$β subunit (Fig. 2.2; K$_v$β1-3) (Scott et al., 1993; Rettig et al., 1994; Pongs et al., 1999), which interacts with the N-terminal T1 domain and forms a symmetric tetramer on the intracellular surface of the channels (Gulbis et al., 2000). The K$_v$β subunits are superficially similar to the Ca$_v$β subunits in their cytoplasmic location, but are not related in amino acid sequence or structure. The N-terminus of K$_v$β subunits of vertebrates serves as an inactivation gate for K$_v$1 α subunits (Pongs et al., 1999), and is thought to enter the pore and block it during sustained channel opening (Zhou et al., 2001). This is a unique example of a direct physical role for an auxiliary subunit in channel gating, rather than modulating the gating process of its associated pore-forming α subunit. K_v4 channels interact with the K channel interacting proteins KChIp1-4, which are members of the neuronal calcium sensor family of calmodulin-like calcium regulatory proteins and have four EF hand motifs (Fig. 2.2) (An et al., 2000). The KChIps enhance expression of K_v4 channels and modify their functional properties by binding to a site in the intracellular T1 domain, similar to the interaction of $K_v\beta$ subunits with K_v1 channels. The K_v7, K_v10, and K_v11 channels associate with a different type of auxiliary subunit – the minKlike subunits. These five closely related proteins have a single transmembrane segment and small extracellular and intracellular domains (Takumi et al., 1988; Abbott et al., 2001b). Although these subunits are topologically similar to $Na_v\beta$ and $Ca_v\delta$ subunits, they do not have significant amino acid sequence similarity. The minK-like subunits are important regulators of K_v7 channel function (Sanguinetti et al., 1996; Barhanin et al., 1996), and mutations in one of these auxiliary subunits causes a form of familial long QT syndrome, which predisposes to dangerous cardiac arrhythmias (Abbott et al., 2001a). In addition, recent work indicates that these subunits also associate with K_v3 and K_v4 channels and are responsible for a form of inherited periodic paralysis (Abbott et al., 2001a; Zhang et al., 2001). In light of this work, it is possible that all K_v channels associate with minK-related subunits. If this hypothesis is true, the K_v channels would then resemble the NaV and Ca_v channels in having an associated subunit with a single transmembrane segment and short intracellular and extracellular domains. It will be intriguing to learn if there is a common function for these similar auxiliary ion channel subunits.

2.5
Inwardly Rectifying Potassium Channels

The inwardly rectifying potassium channels (K_{ir}) are the structurally simplest ion channels, containing four subunits that each have two transmembrane segments and a membrane-reentrant pore loop between them (Jan and Jan, 1997) (Fig. 2.1). As their name implies, they conduct potassium more effectively inward than outward, but their outward conductance of potassium is physiologically important to repolarize cells near the resting membrane potential. They are structurally related to the S5 and S6 segments of the voltage-gated ion channels, and their pore loop forms the extracellular end of the pore as in that family of proteins. The two transmembrane segments and the pore loop of a related bacterial potassium channel have been shown directly by X-ray crystallography to form an ion-selective pore (Fig. 2.3) (Doyle et al., 1998). Many inwardly rectifying potassium channels are gated in a voltage-dependent manner by intracellular divalent cations that bind in their pore, including Mg, spermine, and spermidine. Other Kir channels are gated by direct binding of ATP, G protein βγ subunits, and inositol phospholipid messengers.

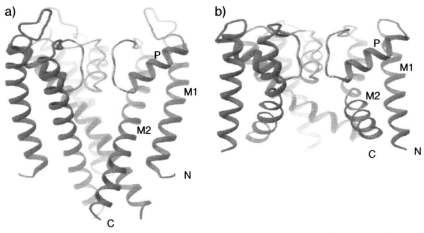

Fig. 2.3 Structure of the pores of two bacterial potassium channels in open and closed states. The structures of two bacterial potassium channels, KcsA (A, closed) and MthK (B, open) related to the inwardly rectifying potassium channels are illustrated. Yellow marks the backbone carbonyls that form the narrow ion selectivity filter. Orange indicates the hinge glycine residue at which the M2/S6 segments of ion channels bend upon opening.

2.6
Common Aspects of Ion Channel Structure and Function

The functions of the voltage-gated ion channel proteins can be divided into three complementary aspects: ion conductance, pore gating, and regulation. All members of the family share a common pore motif, with variations appropriate to determine their different ion selectivity. It is likely that this pore motif first evolved in the bacterial ion channels that resemble inward rectifiers. The X-ray crystal structure of one of these channels has shown that the narrow outer mouth of the pore is formed by the P loops between the M1 and M2 segments that are analogous to the S5 and S6 transmembrane segments of the 6-TM family members while the length of the pore is formed by a tilted bundle, "inverted teepee" arrangement of the M2 segments (Doyle et al., 1998) (Fig. 2.3). This structure suggests that the pore is closed at its intracellular end and discriminates ions at the narrow ion selectivity filter at its extracellular end. This model of pore formation fits well with a wealth of structure–function data on the voltage-gated ion channels and the cyclic nucleotide-gated channels. The closure of the pore at the intracellular end suggests that this is the site of pore gating. Support for this idea comes from recent work. The structure of a calcium-activated bacterial potassium channel with calcium bound shows that the crossing of the M2 segments at the intracellular end of the pore is opened by bending of the helix at a highly conserved hinge glycine residue, consistent with the hypothesis that this is the mechanism of pore opening in two-transmembrane-segment bacterial potassium channels (Fig. 2.3) (Jiang et al., 2002b; Jiang et al., 2002a). This conclusion is further supported by mutagenesis studies of six-transmembrane bacterial sodium channels in which substitution of proline for the hinge glycine to favor bending in that position greatly stabilizes the activated state of the channel (Zhao et al., 2004).

The opening and closing of the voltage-gated sodium, calcium, and potassium channels are primarily gated by changes in membrane potential (Armstrong, 1975), which cause movement of gating charges across the membrane and drive conformational changes that open and close the pore (Armstrong, 1981). Their voltage-sensing and voltage-dependent gating depends on the S1 to S4 segments of these channels, which can be viewed as an evolutionary addition to the pore-forming segments. The detailed mechanism of voltage-dependent gating remains unknown, but extensive structure–function studies and recent X-ray crystallography provide a progressively clearer picture. The positively charged S4 segments are thought to undergo outward and rotational movement through the protein structure during the gating process, as proposed in the sliding helix and helical screw models of gating (Catterall, 1986; Guy and Seetharamulu, 1986). This movement has been detected in fluorescent labeling experiments (Bezanilla, 2000), and the S4 voltage sensor is oriented across the membrane with its gating charges in an outward position in the structure of a K_v channel with its pore open (Long et al., 2005). These results nicely define the structure of the open channel with its activated voltage sensor. The structure of the resting state of the voltage sensors and the mechanism of coupling of the outward movement of the S4 voltage sensors to the bending of the S6 segments and opening of the pore are important aspects that must be defined in future work.

2.7
Conclusions

The voltage-gated ion channel superfamily is one of the largest families of signaling proteins, following the G protein-coupled receptors and the protein kinases in the number of family members. The family is likely to have evolved from a 2-transmembrane-segment ancestor like the modern inwardly rectifying potassium channels and bacterial potassium channels. Modular additions of intracellular regulatory domains for ligand binding and a 4-transmembrane-domain for voltage gating have produced extraordinarily versatile signaling molecules with capacity to respond to voltage signals and intracellular effectors and to integrate information coming from these two types of inputs. The resulting signaling mechanisms control most aspects of cell physiology and underlie complex integrative processes like learning and memory in the brain and coordinated movements of muscles. The evolutionary appearance and refinement of these signaling mechanisms is one of the hallmark events, allowing the development of complex multicellular organisms. The many functions of these signaling proteins have made them ideal targets for drug discovery, and many families of therapeutically important drugs act by altering ion channel function.

References

Abbott, G.W., Butler, M.H., Bendahhou, S., Dalakas, M.C., Ptacek, L.J., Goldstein, S.A. (**2001a**). MiRP2 forms potassium channels in skeletal muscle with $K_v3.4$ and is associated with periodic paralysis. *Cell 104*, 217–231.

Abbott, G.W., Goldstein, S.A., Sesti, F. (**2001b**). Do all voltage-gated potassium channels use MiRPs? *Circ. Res. 88*, 981–983.

An, W.F., Bowlby, M.R., Betty, M., Cao, J., Ling, H.P., Mendoza, G., Hinson, J.W., Mattsson, K.I., Strassle, B.W., Trimmer, J.S., Rhodes, K.J. (**2000**). Modulation of A-type potassium channels by a family of calcium sensors. *Nature 403*, 553–556.

Anderson, M.E., Al-Khatib, S.M., Roden, D.M., Califf, R.M. (**2002**). Cardiac repolarization: current knowledge, critical gaps, and new approaches to drug development and patient management. *Am. Heart J. 144*, 769–781.

Arikkath, J., Campbell, K.P. (**2003**). Auxiliary subunits: essential components of the voltage-gated calcium channel complex. *Curr. Opin. Neurobiol. 13*, 298–307.

Armstrong, C.M. (**1975**). Ionic pores, gates, and gating currents. *Q. Rev. Biophys. 7*, 179–210.

Armstrong, C.M. (**1981**). Sodium channels and gating currents. *Physiol. Rev. 61*, 644–682.

Barhanin, J., Lesage, F., Guillemare, E., Fink, M., Lazdunski, M., Romey, G. (**1996**). K_vLQT1 and lsK (minK) proteins associate to form the I_{Ks} cardiac potassium current. *Nature 384*, 78–80.

Bezanilla, F. (**2000**). The voltage sensor in voltage-dependent ion channels. *Physiol. Rev. 80*, 555–592.

Catterall, W.A. (**1984**). The molecular basis of neuronal excitability. *Science 223*, 653–661.

Catterall, W.A. (**1986**). Molecular properties of voltage-sensitive sodium channels. *Annu. Rev. Biochem. 55*, 953–985.

Catterall, W.A. (**2000a**). From ionic currents to molecular mechanisms: The structure and function of voltage-gated sodium channels. *Neuron 26*, 13–25.

Catterall, W.A. (**2000b**). Structure and regulation of voltage-gated calcium channels. *Annu. Rev. Cell Dev. Biol. 16*, 521–555.

Chandy, K.G., Wulff, H., Beeton, C., Pennington, M., Gutman, G.A., Cahalan, M.D. (**2004**). K^+ channels as targets for specific immunomodulation. *Trends Pharmacol. Sci. 25*, 280–289.

Chen, Y.H., Li, M.H., Zhang, Y., He, L.L., Yamada, Y., Fitzmaurice, A., Shen, Y., Zhang, H., Tong, L., Yang, J. (**2004**). Structural basis of the alpha1-beta subunit interaction of voltage-gated calcium channels. *Nature 429*, 675–680.

De Jongh, K.S., Warner, C., Catterall, W.A. (**1990**). Subunits of purified calcium channels. α2 and δ are encoded by the same gene. *J. Biol. Chem. 265*, 14738–14741.

Doyle, D.A., Cabral, J.M., Pfuetzner, R.A., Kuo, A.L., Gulbis, J.M., Cohen, S.L., Chait, B.T., MacKinnon, R. (**1998**). The structure of the potassium channel: Molecular basis of potassium conduction and selectivity. *Science 280*, 69–77.

Dunlap, K., Luebke, J.I., Turner, T.J. (**1995**). Exocytotic Ca^{2+} channels in mammalian central neurons. *Trends Neurosci. 18*, 89–98.

Ellis, S.B., Williams, M.E., Ways, N.R., Brenner, R., Sharp, A.H., Leung, A.T., Campbell, K.P., McKenna, E., Koch, W.J., Hui, A., Schwartz, A., Harpold, M.M. (**1988**). Sequence and expression of mRNAs encoding the alpha 1 and alpha 2 subunits of a DHP-sensitive calcium channel. *Science 241*, 1661–1664.

Ertel, E.A., Campbell, K.P., Harpold, M.M., Hofmann, F., Mori, Y., Perez-Reyes, E., Schwartz, A., Snutch, T.P., Tanabe, T., Birnbaumer, L., Tsien, R.W., Catterall, W.A. (**2000**). Nomenclature of voltagegated calcium channels. *Neuron 25*, 533–535.

Goldin, A.L., Barchi, R.L., Caldwell, J.H., Hofmann, F., Howe, J.R., Hunter, J.C., Kallen, R.G., Mandel, G., Meisler, M.H., Berwald Netter, Y., Noda, M., Tamkun, M.M., Waxman, S.G., Wood, J.N., Catterall, W.A. (**2000**). Nomenclature of voltage-gated sodium channels. *Neuron 28*, 365–368.

Gulbis, J.M., Zhou, M., Mann, S., MacKinnon, R. (**2000**). Structure of the cytoplasmic beta subunit-T1 assembly of voltage-dependent potassium channels. *Science 289*, 123–127.

Gutman, G.A., Chandy, K.G., Adelman, J.P., Aiyar, J., Bayliss, D.A., Clapham, D.E., Covarriubias, M., Desir, G.V., Furuichi, K., Ganetzky, B., Garcia, M.L., Grissmer, S., Jan, L.Y., Karschin, A., Kim, D., Kuperschmidt, S., Kurachi, Y., Lazdunski, M., Lesage, F., Lester, H.A., McKinnon, D., Nichols, C.G., O'Kelly, I., Robbins, J., Robertson, G.A., Rudy, B., Sanguinetti, M., Seino, S., Stuehmer, W., Tamkun, M.M., Vandenberg, C.A., Wei, A., Wulff, H., Wymore, R.S. (**2003**). International Union of Pharmacology. XLI. Compendium of voltage-gated ion channels: potassium channels. *Pharmacol. Rev. 55*, 583–586.

Guy, H.R., Seetharamulu, P. (**1986**). Molecular model of the action potential sodium channel. *Proc. Natl. Acad. Sci. U.S.A. 508*, 508–512.

Hockerman, G.H., Peterson, B.Z., Johnson, B.D., Catterall, W.A. (**1997**). Molecular determinants of drug binding and action on L-type calcium channels. *Annu. Rev. Pharmacol. Toxicol. 37*, 361–396.

Hofmann, F., Biel, M., Flockerzi, V. (**1994**). Molecular basis for Ca^{2+} channel diversity. *Annu. Rev. Neurosci. 17*, 399–418.

Isom, L.L., De Jongh, K.S., Patton, D.E., Reber, B.F.X., Offord, J., Charbonneau, H., Walsh, K., Goldin, A.L., Catterall, W.A. (1992). Primary structure and functional expression of the β1 subunit of the rat brain sodium channel. *Science 256*, 839–842.

Isom, L.L., Ragsdale, D.S., De Jongh, K.S., Westenbroek, R.E., Reber, B.F.X., Scheuer, T., Catterall, W.A. (**1995**). Structure and function of the β2 subunit of brain sodium channels, a transmembrane glycoprotein with a CAM-motif. *Cell 83*, 433–442.

Jan, L.Y., Jan, Y.N. (**1997**). Cloned potassium channels from eukaryotes and prokaryotes. *Annu. Rev. Neurosci. 20*, 91–123.

Jay, S.D., Ellis, S.B., McCue, A.F., Williams, M.E., Vedvick, T.S., Harpold, M.M., Campbell, K.P. (**1990**). Primary structure of the gamma subunit of the DHP-sensitive calcium channel from skeletal muscle. *Science 248*, 490–492.

Jiang, Y., Lee, A., Chen, J., Cadene, M., Chait, B.T., MacKinnon (**2002a**). Crystal structure and mechanism of a calcium-gated potassium channel. *Nature 417*, 515–522.

Jiang, Y., Lee, A., Chen, J., Cadene, M., Chait, B.T., MacKinnon, R. (**2002b**). The open pore conformation of potassium channels. *Nature 417*, 523–526.

Kazarinova-Noyes, K., Malhotra, J.D., McEwen, D.P., Mattei, L.N., Berglund, E.O., Ranscht, B., Levinson, S.R., Schachner, M., Shrager, P., Isom, L.L., Xiao, Z.C. (2001). Contactin associates with sodium channels and increases their functional expression. *J. Neurosci. 21*, 7517–7525.

Long, S.B., Campbell, E.B., Mackinnon, R. (2005). Crystal structure of a mammalian voltage-dependent Shaker family potassium channel. *Science 309*, 897–903.

Malhotra, J.D., Koopmann, M.C., Kazen-Gillespie, K.A., Fettman, N., Hortsch, M., Isom, L.L. (**2002**). Structural requirements for interaction of sodium channel beta 1 subunits with ankyrin. *J. Biol. Chem. 277*, 26 681–26 688.

McCormick, K.A., Srinivasan, J., White, K., Scheuer, T., Catterall, W.A. (1999). The extracellular domain of the β1 subunit is both necessary and sufficient for β1-like modulation of sodium channel gating. *J. Biol. Chem. 274*, 32 638–32 646.

McGee, A.W., Nunziato, D.A., Maltez, J.M., Prehoda, K.E., Pitt, G.S., Bredt, D.S. (**2004**). Calcium channel function regulated by the SH3-GK module in beta subunits. *Neuron 42*, 89–99.

Morgan, K., Stevens, E.B., Shah, B., Cox, P.J., Dixon, A.K., Lee, K., Pinnock, R.D., Hughes, J., Richardson, P.J., Mizuguchi, K., Jackson, A.P. (**2000**). β3: An additional auxiliary subunit of the voltage-sensitive sodium channel that modulates channel gating with distinct kinetics. *Proc. Natl. Acad. Sci. U.S.A. 97*, 2308–2313.

Olivera, B.M., Miljanich, G.P., Ramachandran, J., Adams, M.E. (**1994**). Calcium channel diversity and neurotransmitter release: The omega-conotoxins and omega-agatoxins. *Annu. Rev. Biochem. 63*, 823–867.

Opatowsky, Y., Chen, C.C., Campbell, K.P., Hirsch, J.A. (**2004**). Structural analysis of the voltage-dependent calcium channel beta subunit functional core and its complex with the alpha 1 interaction domain. *Neuron 42*, 387–399.

Perez-Reyes, E. (**2003**). Molecular physiology of low-voltage-activated t-type calcium channels. *Physiol. Rev. 83*, 117–161.

Pongs, O., Leicher, T., Berger, M., Roeper, J., Bahring, R., Wray, D., Giese, K.P., Silva, A.J., Storm, J.F. (1999). Functional and molecular aspects of voltage-gated potassium channel beta subunits. *Ann. New York Acad. Sci.* 868, 344–355.

Ragsdale, D.S., McPhee, J.C., Scheuer, T., Catterall, W.A. (1994). Molecular determinants of state-dependent block of sodium channels by local anesthetics. *Science 265*, 1724–1728.

Ratcliffe, C.F., Qu, Y., McCormick, K.A., Tibbs, V.C., Dixon, J.E., Scheuer, T., Catterall, W.A. (2000). A sodium channel signaling complex: Modulation by associated receptor protein tyrosine phosphatase β. *Nat. Neurosci.* 3, 437–444.

Ratcliffe, C.F., Westenbroek, R.E., Curtis, R., Catterall, W.A. (2001). Sodium channel β1 and β3 subunits associate with neurofascin through their extracellular immunoglobulin-like domain. *J. Cell Biol.* 154, 427–434.

Rettig, J., Heinemann, S.H., Wunder, F., Lorra, C., Parcej, D.N., Dolly, J.O., Pongs, O. (1994). Inactivation properties of voltage-gated K⁺ channels altered by presence of β-subunit. *Nature 369*, 289–294.

Ruth, P., Röhrkasten, A., Biel, M., Bosse, E., Regulla, S., Meyer, H.E., Flockerzi, V., Hofmann, F. (1989). Primary structure of the beta subunit of the DHP-sensitive calcium channel from skeletal muscle. *Science 245*, 1115–1118.

Sanguinetti, M.C., Curran, M.E., Zou, A., Shen, J., Spector, P.S., Atkinson, D.L., Keating, M.T. (1996). Coassembly of K$_v$LQT1 and minK (IsK) proteins to form cardiac I_{Ks} potassium channel. *Nature 384*, 80–83.

Sanguinetti, M.C., Mitcheson, J.S. (2005). Predicting drug-hERG channel interactions that cause acquired long QT syndrome. *Trends Pharmacol. Sci.* 26, 119–124.

Scott, V.E.S., Rettig, J., Parcej, D.N., Keen, J.N., Findlay, J.B.C., Pongs, O., Dolly, J.O. (1993). Primary structure of a β subunit of α-dendrotoxin-sensitive K⁺ channels from bovine brain. *Proc. Natl. Acad. Sci. U.S.A.* 91, 1637–1641.

Snutch, T.P., Reiner, P.B. (1992). Ca²⁺ channels: diversity of form and function. *Curr. Opin. Neurobiol. 2*, 247–253.

Srinivasan, J., Schachner, M., Catterall, W.A. (1998). Interaction of voltage-gated sodium channels with the extracellular matrix molecules tenascin-C and tenascin-R. *Proc. Natl. Acad. Sci. U.S.A. 95*, 15 753–15 757.

Striessnig, J. (1999). Pharmacology, structure and function of cardiac L-type calcium channels. *Cell Physiol. Biochem. 9*, 242–269.

Takahashi, M., Seagar, M.J., Jones, J.F., Reber, B.F., Catterall, W.A. (1987). Subunit structure of dihydropyridine-sensitive calcium channels from skeletal muscle. *Proc. Natl. Acad. Sci. U.S.A. 84*, 5478–5482.

Takahashi, S.X., Miriyala, J., Colecraft, H.M. (2004). Membrane-associated guanylate kinase-like properties of beta-subunits required for modulation of voltage-dependent calcium channels. *Proc. Natl. Acad. Sci. U.S.A. 101*, 7193–7198.

Takumi, T., Ohkubo, H., Nakanishi, S. (1988). Cloning of a membrane protein that induces a slow voltage-gated potassium current. *Science 242*, 1042–1045.

Tanabe, T., Takeshima, H., Mikami, A., Flockerzi, V., Takahashi, H., Kangawa, K., Kojima, M., Matsuo, H., Hirose, T., Numa, S. (1987). Primary structure of the receptor for calcium channel blockers from skeletal muscle. *Nature 328*, 313–318.

Triggle, D.J. (1999). The pharmacology of ion channels: with particular reference to voltage-gated Ca²⁺ channels. *Eur. J. Pharmacol. 375*, 311–325.

Van Petegem, F., Clark, K.A., Chatelain, F.C., Minor, D.L. Jr (2004). Structure of a complex between voltage-gated calcium channel beta-subunit and an alpha-subunit domain. *Nature 429*, 671–675.

Wallace, R.H., Wang, D.W., Singh, R., Scheffer, I.E., George, Jr. A.L., Phillips, H.A., Saar, K., Reis, A., Johnson, E.W., Sutherland, G.R., Berkovic, S.F., Mulley, J.C. (1998). Febrile seizures and generalized epilepsy associated with a mutation in the sodium channel β1 subunit gene *SCN1B*. *Nat. Genet. 19*, 366–370.

Yu, F.H., Westenbroek, R.E., Silos-Santiago, I., Scheuer, T., Catterall, W.A., Curtis, R. (2003). Sodium channel β4: A disulfide-linked auxiliary subunit structurally and functionally similar to β2. *J. Neurosci. 23*, 7577–7585.

Zhang, M., Jiang, M., Tseng, G.N. (2001). minK-related peptide 1 associates with K$_v$4.2 and modulates its gating function: potential role as beta subunit of cardiac transient outward channel? *Circ. Res. 88*, 1012–1019.

Zhao, Y., Yarov-Yarovoy, V., Scheuer, T., Catterall, W.A. (**2004**). A gating hinge in sodium channels: a molecular switch for electrical signaling. *Neuron 41*, 859–865.

Zhou, M., Morais-Cabral, J.H., Mann, S., MacKinnon, R. (**2001**). Potassium channel receptor site for the inactivation gate and quaternary amine inhibitors. *Nature 411*, 657–661.

3
State-dependent Drug Interactions with Ion Channels

Stefan I. McDonough and Bruce P. Bean

3.1
Introduction

Voltage-gated sodium, calcium, and potassium ion channels are currently among the most promising classes of targets for drug discovery. Although the general functional roles of voltage-dependent ion channels have been known for many decades, only within the last several years has it been possible to correlate the ion channels present in particular cell types with specific gene products of known amino acid sequences and, in some cases, partially-defined structures. In parallel, there have been rapid advances in the technology for heterologous expression of ion channels and rapid development of screening technologies for identification of compounds affecting ion channel function. The result is a recent surge of interest in ion channels as drug targets. As detailed in subsequent chapters, individual channels have excellent validation as drug targets for untreated disorders in many therapeutic areas, including chronic pain, epilepsy, cardiac arrhythmia, inflammation, and diabetes. Channels related to but distinct from the target channels are often responsible for off-target ion channel liabilities: cardiac, CNS, and metabolic. Accordingly, with the biology and the tools recently at hand, pharmaceutical companies large and small are increasingly focusing on developing compounds that bind to voltage-gated channels in a subtype-specific manner.

A fundamental feature of ion channels is that, considered as binding sites for drug molecules, they are moving targets. The normal operation of voltage-dependent ion channels involves conformational changes large enough to switch between states with completely open or completely closed water-filled pores. These changes in *gating state*, which typically occur on a millisecond time scale in response to changes in membrane voltage or the presence of neurotransmitters, are often accompanied by dramatic changes in drug binding affinity. In the case of voltage-gated sodium and calcium channels, many drugs that inhibit the channels bind with higher affinity to gating states that are reached when the membrane is depolarized (open states or so-called "inactivated" states) and more weakly to the gating states ("closed" or "resting" states) that are typical of the rest-

Voltage-Gated Ion Channels as Drug Targets. Edited by D. Triggle
Copyright © 2006 WILEY-VCH Verlag GmbH & Co. KGaA, Weinheim
ISBN 3-527-31258-7

ing potential. Because changes in the transmembrane voltage of the cell drive the changes in gating state, this *state-dependent* inhibition is often manifested as *voltage-dependent* inhibition. For neurons and cardiac muscle cells that form action potentials, the state dependence of drug binding often results in a phenomenon whereby a higher frequency of action potentials results in a higher degree of block. This phenomenon is termed *use-dependent* inhibition, since inhibition increases as channels are "used" by cycling through various gating states during the action potentials. Voltage dependence and use dependence are often desirable features for drug action, because with these properties the degree of inhibition can be graded with the degree of channel activity. Several disorders involving excitable cells, including epilepsy, many cardiac arrhythmias, and some forms of neuropathic pain, originate from hyperexcitability of the cells. Thus, state-dependent drugs that target preferentially the channel states reached during activity can have little or no effect on normal, moderate activity but strongly inhibit pathologically high levels of activity. Such properties are almost a necessity in drugs targeting electrical activity of heart and brain, where block of normal activity would be disastrous.

Lacking high-resolution structures (or in most cases, even general knowledge) of the drug-binding epitopes of the channel protein, drug development against ion channels is currently best done with cell-based functional assays. In running such assays and comparing different assay platforms, it is cardinal to take state dependence into account. The main goal of this chapter is to summarize our current understanding of the mechanisms underlying state dependence of drug action and to discuss the practical consequences of the phenomenon in the context of the modern drug discovery process. Here we consider primarily voltage-gated sodium, calcium, and potassium channels, for which these concepts were first introduced. However, the concept of state dependence is by no means specific to voltage-gated ion channels. The same basic concepts are also important for drugs targeting ion channels gated by external ligands or by intracellular messenger pathways such as nicotinic acetylcholine receptors, glutamate receptors, TRP family channels, and P2X receptor families, many of which have outstanding validation as drug targets.

3.2
Ion Channels as Drug Receptors

The concept that the essential electrical function of neurons and muscle is due to ion channels – pores through the bilayer membrane – dates only to the 1960s and 1970s. Ironically, many of the most widely used clinical drugs that target ion channels were developed long before this. Typically, the drugs were discovered with whole animal or *ex vivo* whole-tissue assays, and only long afterward was it realized that their targets are ion channels. For example, phenytoin was synthesized in 1908 and discovered to have anticonvulsant activity in 1938, long before the existence of voltage-dependent sodium channels was inferred following the

Hodgkin–Huxley understanding of action potentials in neurons in 1952 [1]. Only during the 1960s and 1970s did it become clear that the primary mode of action of phenytoin involves use-dependent block of sodium channels in neurons. Similarly, the dihydropyridine class of antihypertensive drugs (e.g., nifedipine and amlodipine) was originally developed by screening for inhibitors of high-potassium induced contraction of smooth muscle strips. Only after several such drugs had already been introduced to clinical use was it realized that the primary mode of dihydropyridine action is inhibition of voltage-activated calcium channels in vascular smooth muscle. With similar assays, various drugs with anti-epileptic, antihypertensive, or analgesic activity were discovered, including carbamazepine, mibefradil, diltiazem, and mexilitine. All of these were found subsequently to be inhibitors of members of the voltage-gated sodium and/or calcium ion channel families. Many of these first-generation drugs have relatively poor selectivity among members of the family or even between ion channel families.

In 1952, Hodgkin and Huxley elucidated the action potential in excitable membranes as being due to a sodium-selective conductance that activates when the membrane is depolarized [1]. However, it was only during the 1960s and 1970s that the concept of ion channels as pores through membrane (as opposed to carrier-like mechanisms, for example) developed, mainly through the efforts of Clay Armstrong and Bertil Hille and their colleagues. A series of seminal experiments performed by Armstrong on potassium channels in squid axons provided early experimental evidence for the channel nature of the voltage-dependent conductances in axons and also gave the first example of a blocker whose interaction with the channel depended on the gating state of the channel. These experiments found that quaternary ammonium ions acted only from the inside of the squid axon membrane and could block or unblock only when the channel was opened by membrane depolarization (reviewed in [2]). Most strikingly, if channels were closed after entry of a blocking drug molecule, it could be shown that the molecule was trapped in the closed channel until the channel was opened again by a subsequent depolarization. These were the first experiments in which the presence of a drug binding site in an ion channel was explicitly proposed.

3.3
Ion Channels Adopt Multiple Conformations

Figure 3.1 illustrates the electrical current that flows during operation of voltage-dependent sodium channels of the type present in nerve and muscle cells. Under resting conditions, the cell membrane of an excitable cell has a negative resting potential, typically in the range of −60 to −80 mV (inside relative to outside, which is by convention defined as electrical ground). Thus, at rest, the cell membrane is *polarized*. If the voltage across the membrane is caused to become less negative or even positive, the membrane is said to be *depolarized* from rest. At normal resting potentials, sodium channels are closed (non-conducting). If the membrane is depolarized sufficiently (positive to about −40 mV), the voltage-dependent

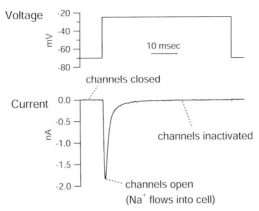

Voltage

Current

channels closed

channels inactivated

channels open
(Na$^+$ flows into cell)

Fig. 3.1 Time course of current through voltage-dependent sodium channels when channels are transiently opened by a voltage step from −70 to −25 mV. Data are from a voltage-clamp experiment performed on a dissociated neuron from the suprachiasmatic nucleus of a rat brain. Current carried by voltage-dependent sodium channels was defined and isolated from other currents by its sensitivity to tetrodotoxin.

sodium channels in the membrane open and carry sodium ions, which flow down their concentration gradient from outside to inside, thus carrying an inward current. A striking feature of the sodium current is that it is transient even if the depolarization is maintained. If the membrane is depolarized suddenly and the depolarization is maintained, which can be accomplished by the voltage-clamp technique, the sodium current is found to activate and reach a peak over a few hundred microseconds, and then decay with a time constant of about 1–2 ms. The decay of the current is termed *inactivation*.

This behavior suggests that the sodium channel can adopt at least three distinct conformations: a closed state, in which the channel is non-conducting, the predominant form when the membrane is at rest; an open state, occupied transiently during a depolarization; and an inactivated state, the non-conducting state reached late in the depolarization. Although the closed resting state and the inactivated state are both non-conducting, they can be distinguished because when the channels are in the closed resting state, they are available to be opened by a subsequent depolarization, while when the channels are in the inactivated state, they cannot be opened by a further depolarization.

Voltage-dependent calcium and potassium channels behave in a manner basically similar to sodium channels, except that inactivation is generally slower and less complete than for sodium channels. The main difference in channel gating between different subtypes of voltage-dependent channels of each class is in the rate and completeness of inactivation (and, to a lesser extent, in the voltage range over which the channels activate). The structural basis for conformational changes between closed, open and inactivated states of channels is still not known in detail. However, some of the general regions of the channel proteins involved in gating are known. At least for potassium channels, there is evidence that the "gate" of the channel that regulates whether it is non-conducting or conducting is on the intracellular face of the channel (reviewed in [3]). The simplest view of inactivation is that it might involve physical plugging of the channel by a tethered blocker that is part of the channel. For one type of potassium channel,

Shaker, there is strong evidence for this mechanism, and the blocker is known to correspond to the N-terminus of the channel molecule [4]. With the sodium channel, a particular cytoplasmic loop of the channel is believed to play a similar role [5]. However, there is also evidence that some slower forms of inactivation might be quite different, involving structural changes of the outer part of the pore.

Conversions between the different gating states are controlled by membrane voltage. At steady-state at any particular voltage, there is a dynamic equilibrium between channels in closed, open, and inactivated states. During large depolarizations from rest potentials, the normal sequence of channel states would be closed to open to inactivated (Fig. 3.1). However, with smaller depolarizations, there can be direct conversion of closed states into inactivated states without channel opening. The conversion of closed states into inactivated states is of particular interest for pharmacology, because it appears to be accompanied by a dramatic increase in binding affinity for many sodium and calcium channel blockers. The conversion of closed channels into inactivated channels is electrically silent, since both states are non-conducting. However, it can be detected as a reduction in the fraction of channels that are available to be opened by a larger depolarization (one large enough to cause channel opening).

A common experimental protocol for measuring the relative numbers of channels within a population that are closed or inactivated is shown in Fig. 3.2. A conditioning voltage is applied for long enough (usually about 1 or 2 s) to reach steady-state distribution between closed and inactivated channels, and the fraction of channels in the closed state at the conditioning voltage is assayed by a test depolarization to a voltage (–30 mV in this case) that opens channels from the closed state. As the conditioning membrane voltage ("holding voltage") is made more depolarized, the fraction of channels in the closed state decreases, the fraction of channels in the inactivated state increases, and the size of current evoked by the test step to –30 mV decreases. Plotting the size of the current evoked by the test voltage as a function of conditioning voltage (Fig. 3.2B) results in an availability curve, which essentially plots the fraction of channels in the resting closed state (available to be opened) as a function of voltage. Sometimes such plots are called "h-infinity" curves, since Hodgkin and Huxley used a parameter they called "*h*" to correspond to the fraction of channels in the closed, available state; in their mathematical model, "h-infinity" represents the steady-state value of this parameter at each voltage. Figure 3.2 shows an example for the case of T-type calcium channels studied in a cerebellar Purkinje neuron. The test current is maximal (about 1300 pA) following conditioning voltages negative to –90 mV. It is reduced to about 60% of maximal by a conditioning voltage of –80 mV, to about 40% by a conditioning voltage of –70 mV, and to near zero by a conditioning voltage of –50 mV. As originally shown by Hodgkin and Huxley [1], the voltage dependence of channel availability can be fit quite accurately by a curve drawn according to $1/(1 + \exp[(V - V_{1/2})/k])$, where $V_{1/2}$ is the voltage of half-maximal availability (the mid-point of the curve) and k is a slope factor that corresponds to the steepness of the curve (a steeper curve corresponding to a smaller slope factor). This form corresponds to the Boltzmann curve expected from the theory of thera)

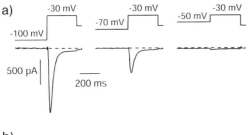

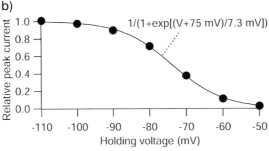

b)

Fig. 3.2 Effect of steady holding potential on calcium channel current elicited by voltage steps to −30 mV in a cerebellar Purkinje neuron (A). Voltage steps elicit a transient, inactivating current carried by barium ions flowing into the cell. External solution contained 5 mM Ba^{2+}; barium ions are efficient in carrying current through voltage-dependent calcium channels. With progressively depolarized holding potentials, the magnitude of the current elicited by the "test" step to −30 mV declines because an increasing fraction of the channels in the cell are in the inactivated state at the holding potential. The smooth curve in (B) was drawn according to $1/(1 + \exp[(V − V_{1/2})/k])$, where $V_{1/2} = −75$ mV and $k = 7.3$ mV. Current through "T-type" calcium channels was isolated by blocking "P-type" current by 10 μM ω-conotoxin MVIIC. (Modified from Ref. [32] and reprinted with permission.)

modynamics for the equilibrium between two states with a free energy difference, with the energy difference in this case corresponding to a charged particle moving across the membrane voltage.

3.4
Biophysics Meets Pharmacology:
State Dependence, Voltage Dependence, and the Modulated Receptor Model

The voltage dependence of the relative fraction of channels in closed or open states is of great interest for pharmacology because, for many channel-blocking drugs, the degree of inactivation at the resting potential affects dramatically the potency of the drug in blocking channels. Figure 3.3 shows an example of this phenomenon for block of T-type calcium channels by mibefradil, a drug developed for treating hypertension [6]. Channels were activated by depolarization to −30 mV, either from a holding voltage of −110 mV, where most channels would be in the closed, resting (non-inactivated) state, or from a holding potential of

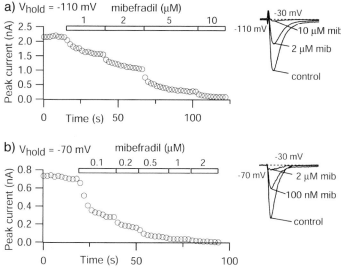

Fig. 3.3 Effects of holding potential on potency of mibefradil for blocking currents carried by T-type calcium channels in cerebellar Purkinje neurons. The effect of mibefradil was studied at a holding potential of –110 mV (A), where most channels are in the resting state, or a holding potential of –70 mV (B), where most channels are in the inactivated state. Note the difference in potency of block at the two different holding potentials. Right-hand panels: examples of calcium current elicited by steps to –30 mV, showing effects of different concentrations of mibefradil at the two different holding potentials. (Modified from Ref. [32] and reprinted with permission.)

–70 mV, where a majority (about 60%) of the channels would be in the inactivated state. When increasing concentrations of mibefradil were applied at a holding voltage of –110 mV (Fig. 3.3A), the concentration necessary for half-block of the test pulse current was about 2 μM. When drug was applied under identical conditions but at a holding voltage of –70 mV (Fig. 3.3B), block was much more potent, with a half-blocking concentration of about 100 nM.

Figure 3.4 shows how the voltage dependence of block by mibefradil can be understood quantitatively as arising from the voltage dependence of channel inactivation. Figure 3.4A illustrates dose–response curves for block of calcium current when the current was elicited by test steps delivered from various steady holding voltages from –110 to –70 mV. The IC$_{50}$ for block decreases from about 2 μM at a holding voltage of –110 mV to about 100 nM at a holding voltage of –70 mV. This behavior can be understood by hypothesizing that the drug binds relatively weakly to closed resting states and more tightly to inactivated states; thus, when a larger fraction of channels is in the inactivated state, the drug binds more tightly. Although the high-affinity drug binding to inactivated states involves only non-conducting, electrically silent states, it can be detected in the experiment because – at a particular steady holding voltage with a particular concentration of mibefradil present – there is an equilibrium between channels in four states: resting with

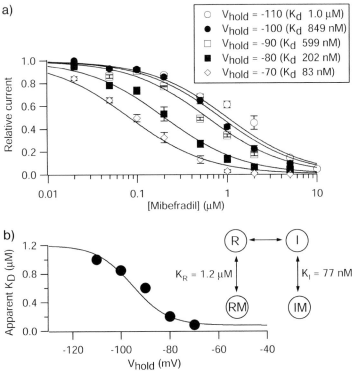

Fig. 3.4 Interpretation of the effect of holding potential on potency of mibefradil block. (A) Data points show mean ± standard error for determinations of block in 6–14 neurons for experiments performed as in Fig. 3.3. Smooth curves are drawn according to Langmuir curves for 1:1 binding, i.e., $1/(1 + [mibefradil]/K_{app})$, where K_{app} is the apparent dissociation constant. (B) Apparent dissociation constant as a function of holding voltage. The smooth curve is described by $K_{app} = 1/[h/K_R + (1 - h)/K_I]$, where h is the degree of inactivation at each voltage and is given by the smooth curve in Fig. 3.2, $h = 1/(1 + exp[(V - V_{1/2})/k])$, where $V_{1/2} = -75$ mV and $k = 7.3$ mV. K_R represents the dissociation constant for binding to the resting state of the channel and K_I is the dissociation constant for binding to the inactivated state. (Modified from Ref. [32] and reprinted with permission.)

no drug bound, resting with drug bound, inactivated with no drug bound, and inactivated with drug bound. Only channels in the resting state with no drug bound are available to be opened by the test depolarization, which then assays the fraction of channels in this state. Due to the dynamic equilibrium among the four states, binding of drug to the inactivated state reduces the fraction of channels in the (non-drug bound) resting state by siphoning channels into the inactivated, drug-bound state. Figure 3.4(B) plots the apparent dissociation constant for drug block of calcium current as a function of the holding voltage. The experimentally observed relationship can be fit well by the predictions of a simple state model in which channels equilibrate between resting and inactivated states (with the voltage dependence given by the Boltzmann function in Fig. 3.2), and where drug binds to inactivated channels with much higher affinity (dissociation constant 77 nM) than to resting channels (dissociation constant 1.2 µM). Drug binding is modeled to follow a conventional mass-action process, assuming an infinite reservoir of drug molecules and fixed number of channel molecules.

A crucial point from this analysis is that *the apparent voltage dependence of drug binding arises solely from the voltage dependence of channel gating and not from any direct voltage dependence of the interaction of drug molecules with the channel.* Thus, it is perfectly possible for an uncharged drug to have such voltage dependence in its blocking action. Voltage-dependent block due to channel inactivation does not imply that the drug molecule binds within the pore. A completely different mechanism of voltage-dependent block of a channel can arise from a charged blocker that actually enters the electrical field inside the pore of a channel as part of its blocking action. A well-known example is block of glutamate-activated channels of the NMDA receptor type by external magnesium ions: positively charged magnesium ions can enter the channel to a binding site within the pore but cannot permeate through the channel, and the resulting block is strongly voltage-dependent, tighter at more negative voltages, and relieved by depolarization.

As shown by the availability curve in Fig. 3.2, it is often an oversimplification to say that channels are entirely in the closed, non-inactivated, "resting" state at normal resting potentials. For the T-type calcium channels in cerebellar Purkinje neurons illustrated in Figs. 3.2–3.4, at a typical resting potential of –75 mV, the channels are roughly half-and-half in closed and inactivated states. Thus, the potency of block even in the absence of any additional depolarization reflects an intermediate position between the low potency block typical of closed channels and the high potency block typical of inactivated channels. This situation is also typical for voltage-dependent sodium channels in neurons, which are typically about 30–50% inactivated at a neuron's resting potential. However, in cardiac muscle, which typically has a more negative resting potential (–85 to –90 mV), resting inactivation of sodium channels is less, and most channels are in the low affinity closed state.

The situation illustrated in Figs. 3.2–3.4 for mibefradil block of T-type calcium channels is an instance of the "modulated receptor model" for ion channel block, originally introduced by Hille to account for the properties of block of sodium channels by local anesthetics like lidocaine (for review see [3]). The essential idea of the modulated receptor model is that the changes in channel conformation occurring during channel gating alter the binding sites for the drug molecules. For most blockers of sodium and calcium channels, binding affinity is higher for the inactivated channels than for closed channels. Lidocaine block of sodium channels, phenytoin block of sodium channels, nifedipine block of L-type Ca channels, and nitrendipine block of L-type calcium channels all behave in a manner very similar to the mibefradil block of T-type calcium channels illustrated in Figs. 3.2–3.4, and all can be understood as reflecting higher affinity binding to the inactivated state of the channels compared to the resting closed state [7–10]. Perhaps the easiest way to imagine the mechanism is if a single binding site is altered subtly by channel gating, so that binding is tighter for inactivated channels. However, in principle it is possible that the binding sites on closed and inactivated channels could be entirely different.

3.5
Use Dependence

Use dependence is the phenomenological term for inhibition that builds up with repetitive stimulation, either with repeated firing of action potentials under physiological conditions or with repeated applications of depolarizing pulses under experimental voltage-clamp conditions. Many drugs that interact with sodium channels and calcium channels have this property, which often (but not always) is seen with drugs that show voltage dependence in response to steady changes in holding potential. Figure 3.5 shows an example for lidocaine block of cardiac sodium channels. In this experiment, the voltage clamp protocol was to first establish a very negative voltage (−105 mV) in the steady-state, so that channels would be in the closed, non-inactivated state. Then, a train of repeated depolarizing pulses (500 ms-long pulses to −35 mV) was delivered at 1 Hz. The top panels show, superimposed, current elicited by the 1st and the 12th pulse in the train. With no drug present, the currents elicited by these pulses are almost identical. With 20 µM lidocaine, the 1st pulse of the train elicits a current almost unchanged from the control, but as the train continues, currents become progressively smaller. By the 12th pulse, the peak sodium current was reduced by about 40%. With 100 µM lidocaine, this use dependence was even more pronounced.

With each depolarization in the train, channels cycle through open and inactivated states. Between depolarizations, at the holding voltage, channels return from the inactivated state to the resting closed state. The simplest explanation of use dependence is if (1) binding of drug is tighter to either open or inactivated states than to the resting closed state, so that binding is enhanced during the depolarization; and (2) this "extra" binding is not completely removed between depolarizations. Since tight binding is likely to be accompanied by a slow unbinding rate, failure for the extra binding to be removed between pulses is a plausible correlate of tight binding to the inactivated state. This mechanism of use dependence can arise from interactions only with inactivated states and does not require drug interactions with open states.

Another mechanism of use dependence involves drug binding to open channels. For example, a molecule might require a channel to be open in order to enter the pore and bind to a site within the pore. In this case, binding can only occur during the short time that a channel is open during an action potential (or a depolarizing pulse if in voltage clamp). If binding does not reverse between pulses or action potentials (e.g., if the channel can close around the blocker and trap it in the channel), block will accumulate during a train of pulses or action potentials. Examples of such open channel-blockers include the antiarrhythmics diisopyramide, quinidine, flecainide, and encainide (reviewed in [11]).

a)

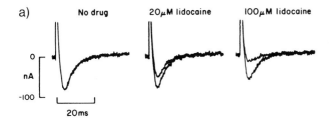

b)

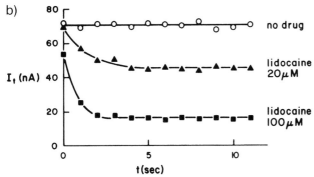

Fig. 3.5 Use-dependent effect of lidocaine on cardiac sodium currents. The sodium current was measured in short pieces of Purkinje fibers from rabbit heart during trains of 500-ms pulses from −105 to −35 mV delivered at 1.0 Hz. (A) Membrane currents measured on the 1st and 12th pulses in 0, 20, and 100 μM lidocaine. The rest period before each train was long enough for effects of previous trains to be removed. (B) Peak sodium current amplitudes for each of the pulses. The decrease in current magnitude has been fitted by an exponential curve, with a time constant of 1.3 s in 20 μM lidocaine and 0.7 s in 100 μM lidocaine. (Reproduced from *The J. Gen. Physiol.*, **1983**, *81*, 616 [7] by copyright permission of The Rockefeller University Press.)

Some drugs showing use dependence appear to have a mechanism that is a combination of open channel block and tight binding to inactivated states. These include the local anesthetics lidocaine, mexilitine, bupivacaine, and the antiarrhythmic amiodarone (reviewed in [11]). For many drugs that appear to use both pathways for drug binding, it is difficult to distinguish open channel block from a mechanistically different but functionally similar mechanism involving especially tight binding to the open conformation of the channel (but not necessarily within the pore). Thus for some cases it may be more precise to speak of "activated channel" binding rather than "open channel block" [11]. Open channel block is particularly important for the hERG cardiac potassium channel, since hERG is open for approximately half of the human lifetime.

In the case of lidocaine, Vedantham and Cannon [12] have shown that binding of lidocaine does not interfere with movement of the cytoplasmic loop believed to underlie inactivation of the sodium channel. Most significantly, bound lidocaine does not slow the recovery of the inactivation gate, showing that the failure of lidocaine-bound channels to "reprime" between depolarizations during use-dependent block does not represent slowing of recovery from inactivation but rather slow unbinding of lidocaine.

3.6
Physical Meaning of State Dependence

The major families of voltage-dependent channels have a fair degree of sequence homology. As detailed in later chapters, all consist of four subunits (or four homologous pseudo-subunit domains), each subunit or pseudo-subunit containing a particular transmembrane segment (the S4 segment) that contains positively charged lysine and arginine residues believed to serve as the voltage sensors for the channel. In the simplest view, energy derived from movement of these positively charged residues across the electric field of the membrane drives conformational changes in the rest of the channel to convert between closed and open states. It is oversimplified to view channel gating in terms of only closed, open, and inactivated conformations. The conversion of channels from the resting closed state into the open state occurs with a sigmoidal rather than exponential time course, implying that the channel moves through multiple closed states before opening; most likely, these multiple closed states correspond to states in which one, two, three, and four of the charged S4 regions have moved from the resting to the activated positions.

In the context of drug action, a particularly interesting question is how inactivation gets its voltage dependence. As shown in Fig. 3.2, the equilibrium between closed (resting) channels and inactivated channels is strongly voltage-dependent. The Hodgkin–Huxley model of voltage dependence postulated a strongly voltage-dependent process controlling inactivation separate from that controlling activation, which is also strongly voltage-dependent. However, a series of experiments by Armstrong and Bezanilla (reviewed in [13]) showed that inactivation derives its voltage dependence from the same charge movements that control the voltage dependence of activation, now identified with the movements of the S4 regions. One way this could happen is if inactivation can only occur if channels have opened. However, various experiments show that channels can inactivate without first opening. Therefore, current models of sodium channel gating (e.g., [14]) postulate that inactivation is coupled to activation in a looser way, so that inactivation becomes progressively more likely and more complete as successive S4 regions move from the resting to the activated positions, but without requiring actual channel opening. Physically, this mechanism could correspond to the cytoplasmic loop believed to act as a blocking particle binding progressively more tightly to the rest of the channel as successive S4 regions move.

This picture of channel gating suggests that substantial changes in channel structure occur before actual opening of the channel. If the changes in conformation of these partially activated states are large enough to greatly affect the binding of the inactivation particle, they may also affect the binding of drug molecules. Indeed, the state dependence of lidocaine binding to the sodium channel can be understood if the relevant conformational change that controls binding affinity is not opening or inactivation of the channel but rather activation in the sense of successive movement of S4 regions [12]. This version of the modulated receptor model suggests that the apparent correlation of tight drug binding to inactivation is because both are favored by the activation of S4 regions, not because the conformational change associated with inactivation itself changes drug binding affinity. Similar versions of the modulated receptor model have been developed by other investigators [15, 16]. At least for lidocaine binding to sodium channels, evidence is accumulating that supports this class of models [17].

The view that binding affinity depends on successive movements of S4 regions suggests that there is a continuum of binding sites with progressively tight binding, fitting better with the idea of a single binding site whose affinity changes with gating rather than distinct binding sites on resting, open, and inactivated channels. With lidocaine block of the sodium channel, site-directed mutagenesis shows the largest effects for changes in the middle of the S6 transmembrane segment, which experiments in other channel types suggest may be involved in forming the activation gate of the channel [18]. Thus a plausible model would be that movements of successive S4 regions produce changes in the position of S6 segments that then produce graded changes in a binding site formed in part by the S6 regions. However, since all parts of the channel are connected together in one way or another, changes in channel conformation resulting from site-directed mutagenesis at one region could well produce changes in binding sites at distant regions, through allosteric interactions of channel subunits and protein domains. Thus, it is difficult to define a binding site through site-directed mutagenesis with any confidence. Biochemical labeling of channel residues using crosslinking by drug molecules modified to have reactive elements provides more direct information about binding sites but has been carried out for only a few cases (e.g., [19]).

3.7
State Dependence in Drug Discovery

For compounds directed against voltage-gated channels, including several classes of clinical drugs, a preference for inactivated channels gives greater potency and a wider therapeutic window. For example, the dihydropyridine class of antihypertensives produces vasodilation via inhibition of Cav1.2 L-type calcium channels in vascular smooth muscle, but without excessive negative inotropy from inhibition of Cav1.2 in ventricular muscle. Although some of the difference in action of the two tissues may be due to different splice forms of Cav1.2 in the two types of muscle [20], the most important difference probably derives from the different resting potentials and different levels of resting inactivation in the two cases. Vascular smooth muscle typically has a less negative resting potential (−30 to −50 mV) than working heart muscle (−80 to −90 mV). Thus, calcium channels in smooth muscle have higher fractional inactivation at rest and there is more potent block in this tissue. Similarly, reentrant arrhythmias in ventricular cardiac muscle often arise from areas of diseased tissue that have slow conduction because depolarized resting potentials lead to partial inactivation of the sodium channels, thus slowing action potentials. By blocking conduction along the abnormal pathway while having minimal effects on normally polarized tissue, antiarrhythmics targeting sodium channels can disrupt the reentry and eliminate the arrhythmia [11].

Sodium channels are the primary targets of many of the most widely-used anticonvulsant agents, including phenytoin, carbamezepine, and lamotrigine. A simple view of their ability to disrupt abnormal firing during epilepsy without altering normal activity focuses on their use-dependent properties: if abnormal activity involves a higher frequency of firing, this would be preferentially inhibited. Yet, it is not necessarily clear that seizures always involve a higher rate *frequency* of firing of individual neurons as opposed to a pathological degree of *synchronization* of firing of many neurons. In this case, it may be that voltage dependence rather than use dependence is the key property of the drugs: synchrony of activity may lead to convergence of depolarizing excitatory postsynaptic potentials and consequent depolarization of the resting potential.

Chronic pain is probably the most prominent current target for development of new drugs targeted to ion channels. Here, voltage dependence or use dependence are likely to be highly desirable properties. The root cause of many forms of inflammatory or neuropathic chronic pain is likely to be hyperexcitability of either primary or secondary nociceptors, and channel blockers with strong use dependence should be highly advantageous in disrupting such hyperexcitability with minimal side effects.

By contrast, several drugs in the clinic are probably hampered by a lack of use dependence. Prialt® (ziconotide) delivered intrathecally to block Cav2.2 calcium channels is indeed highly effective against opiate-resistant pain but has a narrow therapeutic window and often causes CNS side effects [21]. Ziconotide probably acts by occluding the outer part of the pore region of the channel and probably has little state dependence. Therefore, ziconotide presumably blocks the Cav2.2 channels of non-affected neurons in the CNS equally well as the Cav2.2 channels in the presynaptic nerve terminals of hyperexcitable primary sensory afferents and spinal cord neurons that drive pain. A drug targeted to the same channels but with enhanced block of activated or inactivated channels might well have a wider therapeutic window. Likewise, the sodium channel blocker tetrodotoxin injected intra-muscularly is in phase II trials for relief of cancer pain. Since this sodium channel blocker has little use dependence or voltage dependence under normal conditions, hyperexcitable neurons are not targeted preferentially. This makes dosing difficult and systemic administration impossible to consider due to block of the nerves innervating the diaphragm muscle.

Some new methods and technologies for primary drug screening can be designed to bias hits towards those with strongest state dependence, and follow-up confirmation of hits and lead optimization can now take advantage of automated electrophysiology to build state dependence into a molecule. The several generations of FLIPR® machines from Molecular Devices are a common technology for cell-based functional assays. Although FLIPR® cannot control transmembrane voltage directly, investigators at Merck have used co-expression of an inward rectifier potassium channel and careful control of the extracellular potassium ion concentration to manipulate membrane potential and drive fractional inactivation of the calcium channel under study; this method can accurately detect state dependence of benchmark antagonists [22]. The method is in principle extendable to any voltage-gated channel. Perhaps the most exciting high-throughput screening tool for selecting use-dependent compounds is the E-VIPR® invented by Vertex (formerly Aurora Biosciences). This combines FRET-based dyes of unprecedented speed with external electric field stimulation of expressed channels in 384-well format to produce an optical readout of compound inhibition as a function of stimulation frequency [23]. Other technologies that manipulate stimulation frequencies or channel gating states directly are in development, but none is yet in widespread use.

Few other of the most commonly used primary screening technologies are well-suited to differentiate between strongly and weakly state-dependent compounds. For example, voltage-gated sodium channels inactivate so quickly that they are commonly screened in FLIPR® with compounds such as deltamethrin or veratridine that remove channel inactivation. Channels artificially held open produce a stronger and slower signal. However, such screens might miss detection of inhibitors that work by high-affinity binding to inactivated states or that require use dependence for optimal block. Careful design of the protocol can ameliorate such difficulties to some extent [24].

Automated electrophysiology is rapidly emerging as a practical tool for ion channel drug development. As detailed in later chapters, the most common current uses of automated electrophysiology are lead optimization, primary hit confirmation, and safety screening. The PatchXpress® developed by Axon Instruments (now part of Molecular Devices) and Sophion's QPatch® system each have voltage control sufficient to examine compounds for state dependence. The IonWorks® HT and Quattro™ machines from Molecular Devices do not provide continuous voltage clamp during the experiment and so are less reliable to test state dependence, but the throughput of the IonWorks® machines is sufficiently high to run even primary screens of focused libraries with electrophysiology rather than with indirect fluorescence or spectroscopy reporter assays.

3.8
Future Directions for Ion Channel Drug Discovery

Ongoing research in several areas promises even better drug discovery against ion channels. A completely new era will begin when it is possible to precisely define the three-dimensional structure of drug binding sites on channels of interest. Determination of three-dimensional structures of ion channels has begun, using cryo-electron microscopy [25], NMR [26], and high-resolution X-ray crystallography [27, 28]. So far, crystallographic structures have been obtained for only a handful of channels, and obtaining crystals of adequate quality has proven extremely difficult for eukaryotic channels. Nevertheless, it seems almost certain that, with enough effort, eventually investigators will succeed in obtaining structures of eukaryotic channels with drugs bound, opening the way to drug design by computational modeling. Another current frontier in ion channel research is the effort to elucidate the structural changes that are responsible for channel gating, and within the next 5 to 10 years it may be possible to define exactly how drug binding sites are changed during opening and inactivation of channels. The thought of designing ion channel inhibitors with the same computational sophistication as kinase inhibitors is tremendously appealing. For many channels that are the best targets for drug discovery, the most desirable drugs might be those that bind most tightly to inactivated channels (or channels with fully-activated S4 regions). This is fortunate, since it seems likely that, when channels are purified and crystallized, these are the forms the channels will take, since they are forms that would be present in the absence of a transmembrane voltage. In this case crystallization of channels with bound drug may naturally be in the form most relevant for drug design.

In the shorter term, new molecular, chemical, and screening methods are also likely to accelerate compound development. In theory, biotechnology and recombinant DNA techniques that make large molecules with many contact points could produce ion channel inhibitors with specificity matching those of peptide toxins. Antibodies that affect the function of voltage-gated [29, 30] and ligand-gated [31] ion channels have been described, although currently none engineered to a particular target have been reported. However, small molecule chemistry is producing ion channel-focused libraries for screening, some commercially available, based on structures of known ion channel inhibitors and on consensus information about drug-like small molecule properties, including absorption, metabolism, clearance, and CNS-penetrance. Screening and lead-optimization technologies designed for ion channels have arrived, and the capacity and sophistication of automated electrophysiology will only improve. Thus, the tools to exploit the biological information about the role of ion channels in disease and to make new drugs are increasingly at hand.

Oddly, in hindsight some classical methods of drug identification and development were biased to find compounds with strong state dependence, even before the concept of state dependence was clear. For example, the smooth muscle contraction assays used to assay antihypertensives used solution application of high potassium concentrations. The resulting depolarization, lasting many seconds, would bias channels strongly towards inactivation. It is perhaps no coincidence that the resulting calcium channel blockers were strongly state-dependent. Conversely, some drugs discovered with whole-animal pharmacology such as gabapentin (and now pregabalin) that may target ion channels are not strongly state dependent and may not even block channel current. Can the classical ion channel-targeted drugs be surpassed using drug discovery based on modern biological and chemical understanding together with new screening methods? The next five to ten years will tell.

Acknowledgments

We thank Stephen Hitchcock for helpful discussions and Alexander Choi Jackson for providing the data used in Fig. 3.1.

References

1. Hodgkin, A.L., A.F. Huxley. A quantitative description of membrane current and its application to conduction and excitation in nerve. *J. Physiol.* **1952**, *117*, 500–544.

2. Armstrong, C.M. Ionic pores, gates, and gating currents. *Q. Rev. Biophys.* **1974**, *7*, 179–210.

3. Hille, B., *Ion Channels of Excitable Membranes*, Sinauer Associates, Sunderland, MA, **2001**.

4. Zagotta, W.N., T. Hoshi, R.W. Aldrich. Restoration of inactivation in mutants of Shaker potassium channels by a peptide derived from ShB. *Science* **1990**, *250*, 568–571.

5. West, J.W., D.E. Patton, T. Scheuer, Y. Wang, A.L. Goldin, W.A. Catterall. A cluster of hydrophobic amino acid residues required for fast Na^+-channel inactivation. *Proc. Natl. Acad. Sci. U.S.A.* **1992**, *89*, 10 910–10 914.

6. Triggle, D.J. The physiological and pharmacological significance of cardiovascular T-type, voltage-gated calcium channels. *Am. J. Hypertens.* **1998**, *11*, 80S–87S.

7. Bean, B.P., C.J. Cohen, R.W. Tsien. Lidocaine block of cardiac sodium channels. *J. Gen. Physiol.* **1983**, *81*, 613–642.

8. Bean, B.P. Nitrendipine block of cardiac calcium channels: high-affinity binding to the inactivated state. *Proc. Natl. Acad. Sci. U.S.A.* **1984**, *81*, 6388–6392.

9. Sanguinetti, M.C., R.S. Kass. Voltage-dependent block of calcium channel current in the calf cardiac Purkinje fiber by dihydropyridine calcium channel antagonists. *Circ. Res.* **1984**, *55*, 336–348.

10. Kuo, C.C., B.P. Bean. Slow binding of phenytoin to inactivated sodium channels in rat hippocampal neurons. *Mol. Pharmacol.* **1994**, *46*, 716–725.

11. Carmeliet, E., K. Mubagwa. Antiarrhythmic drugs and cardiac ion channels: mechanisms of action. *Prog. Biophys. Mol. Biol.* **1998**, *70*, 1–72.

12. Vedantham, V., S.C. Cannon. The position of the fast-inactivation gate during lidocaine block of voltage-gated Na^+ channels. *J. Gen. Physiol.* **1999**, *113*, 7–16.

13. Armstrong, C.M. Sodium channels and gating currents. *Physiol. Rev.* **1981**, *61*, 644–683.

14. Kuo, C.C., B.P. Bean. Na^+ channels must deactivate to recover from inactivation. *Neuron* **1994**, *12*, 819–829.

15. Yeh, J.Z., J. Tanguy. Na channel activation gate modulates slow recovery from use-dependent block by local anesthetics in squid giant axons. *Biophys. J.* **1985**, *47*, 685–694.

16. Wang G.K., M.S. Brodwick, D.C. Eaton, G.R. Strichartz. Inhibition of sodium currents by local anesthetics in chloramines-T-treated squid axons. The role of channel activation. *J. Gen. Physiol.* **1987**, *89*, 645–667.

17. Wang S.Y., J. Mitchell, E. Moczydlowski, G.K. Wang. Block of inactivation-deficient Na^+ channels by local anesthetics in stably transfected mammalian cells: evidence for drug binding along the activation pathway. *J. Gen. Physiol.* **2004**, *124*, 691–701.

18. Liu, U., M. Holmgren, M.E. Jurman, G. Yellen. Gated access to the pore of a voltage-dependent K^+ channel. *Neuron* **1997**, *19*, 175–184.

19. Striessnig, J., H. Glossmann, W.A. Catterall. Identification of a phenylalkylamine binding region within the alpha 1 subunit of skeletal muscle Ca^{2+} channels. *Proc. Natl. Acad. Sci. U.S.A.* **1990**, *87*, 9108–9112.

20. Welling, A., A. Ludwig, S. Zimmer, N. Klugbauer, V. Flockerzi, F. Hofmann. Alternatively spliced IS6 segments of the α1C gene determine the tissue-specific dihydropyridine sensitivity of cardiac and vascular smooth muscle L-type Ca^{2+} channels. *Circ. Res.* **1997**, *81*, 526–532.

21. Staats, P.S., T. Yearwood, S.G. Charapata, R.W. Presley, M.S. Wallace, M. Byas-Smith, R. Fisher, D.A. Bryce et al. Intrathecal ziconotide in the treatment of refractory pain in patients with cancer or AIDS: a randomized controlled clinical trial. *J. Am. Med. Assoc.* **2004**, *291*, 63–70.

22. Xia, M., J.P. Imredy, K.S. Koblan, P. Bennett, T.M. Connolly. State-dependent inhibition of L-type calcium channels: cell-based assay in high-throughput format. *Anal. Biochem.* **2004**, *327*, 74–81.

23. Gonzalez, T. Measuring use-dependent compound inhibition of voltage-gated channels with electrical stimulation and FRET detection. *Abstract,* Aurora Biomed. Ion Channel Retreat, Vancouver, BC, **2004**.

24. Felix, J.P., B.S. Williams, B.T. Priest, R.M. Brochu, I.E. Dick, V.A. Warren, L. Yan, R.S. Slaughter, G.J. Kaczorowski, M.M. Smith, M.L. Garcia. Functional assay of voltage-gated sodium channels using membrane potential-sensitive dyes. *Assay Drug Dev. Technol.* **2004**, *2*, 260–268.

25. Unwin, N. Refined structure of the nicotinic acetylcholine receptor at 4 Å resolution. *J. Mol. Biol.* **2005**, *346*, 967–989.

26. Liu, Y.S., P. Sompornpisut, E. Perozo. Structure of the KcsA channel intracellular gate in the open state. *Nat. Struct. Biol.* **2001**, *8*, 883–887.

27. Doyle, D.A., J. Morais Cabral, R.A. Pfuetzner, A. Kuo, J.M. Gulbis, S.L. Cohen, B.T. Chait, R. MacKinnon. The structure of the potassium channel: molecular basis of K^+ conduction and selectivity. *Science* **1998**, *280*, 69–77.

28. MacKinnon, R. Potassium channels. *FEBS Lett.* **2003**, *555*, 62–65.

29. Vassilev P., T. Scheuer, W.A. Catterall. Inhibition of inactivation of single sodium channels by a site-directed antibody. *Proc. Natl. Acad. Sci. U.S.A.* **1989**, *86*, 8147–8151.

30. Engisch, K.L., M.M. Rich, N. Cook, M.C. Nowycky. Lambert-Eaton antibodies inhibit Ca^{2+} currents but paradoxically increase exocytosis during stimulus trains in bovine adrenal chromaffin cells. *J. Neurosci.* **1999**, *19*, 3384–3395.

31. Rogers, S.W., P.I. Andrews, L.C. Gahring, T. Whisenand, K. Cauley, B. Crain, T.E. Hughes, S.F. Heinemann, J.O. McNamara. Autoantibodies to glutamate receptor GluR3 in Rasmussen's encephalitis. *Science* **1994**, *265*, 648–651.

32. McDonough, S.I., B.P. Bean. Mibefradil inhibition of T-type calcium channels in cerebellar Purkinje neurons. *Mol. Pharmacol.* **1998**, *54*, 1080–1087.

4
Assay Technologies: Techniques Available for Quantifying Drug–Channel Interactions

Derek Leishman and Gareth Waldron

4.1
Introduction

> *"Electrophysiologists tend to prefer working alone in the corners of small rooms ... social interactions are inadmissible" –*
> The Axon Guide.

Quotes such as the above, despite having a grain of truth, potentiate a mythology that ion channels are "special" or "difficult". However, in many ways voltage-gated ion channels are "just" proteins in the same way that G-protein coupled receptors (GPCRs) and enzymes are "just" proteins. Ion channels and GPCRs are proteins that sense an alteration in their environment, undergo a conformational change and effect a change. They are membrane bound and thus transduce signals across the membrane. Some confusion can arise, however, as the natural ligand for voltage-gated ion channels is not a molecule but the ethereal membrane voltage, which does not have physical form.

As described earlier in this book, voltage-gated channels exist in physical conformations, closed (C), open (O) and inactivated (I).

$$C \xrightarrow{\Delta V} O \xrightarrow{\Delta V} I \tag{1}$$

These macroscopic gross states are approximations of multiple microscopic states that exist in dynamic equilibrium at *any* membrane potential – the proportion of time spent in each state is largely determined by the membrane potential. In this way voltage can be seen as a catalyst promoting the conformational change and it should be remembered that no state exists to the exclusivity of the others – even at extremes of voltage.

Parallels between closed, open and inactivated states can be drawn with models of GPCR activation where the receptor can exist in R, R* and endocytosed states. With respect to ion channels, techniques exist to directly measure the functional characteristics of *single* ion channel protein molecules in the open state in real

Voltage-Gated Ion Channels as Drug Targets. Edited by D. Triggle
Copyright © 2006 WILEY-VCH Verlag GmbH & Co. KGaA, Weinheim
ISBN 3-527-31258-7

time and these can be used to infer/calculate closed and inactivated state characteristics. This level of detail is not technically feasible as yet for other types of proteins (though some advances with fluorescent tagging are occurring, e.g. [1]). The availability of this resolution of data for ion channels has lead to the appreciation that

1. The ion channels naturally cycle through all three macroscopic states (especially in tissues where action potentials occur).
2. Compounds can have differing affinities for each of these states.

Thus, the experimental conditions needed to measure a compound's affinity for a voltage-gated ion channel need to take these points into account. In keeping with voltage-gated ion channels' best understood role of immediate and quick signaling, alterations in the balance of distribution between these three states can occur on timescale of milliseconds to seconds, making direct measurements of the effect of a compound on a single population of ion channels exclusively in a particular state difficult – all one can hope to do is bias the population to the desired state. The conformation that is easiest to measure is the open state, as this is the conducting state and generally the conduction of ions through the open state is the parameter that is directly, or indirectly, measured. Thus, any ion channel screening technique needs to either control the cycling of the protein through the different states or bias the state of the majority of the channels to a known conformation. The former tends to be technically difficult in higher throughput plate-based assays and it is also challenging to manipulate the channels into the desired state for the latter.

There are some needs for an assay for voltage-gated channels that are common with other targets that should not be overlooked. For example, the assays (or screen sequence at the very least) need to be sensitive enough to discriminate between agonist and antagonist effects at a channel. When setting up an assay the experimenter must take into account any translation – that is the rank order of potencies of known chemical entities should agree with previously acquired data, either in a similar system or, as typically occurs in drug discovery, in an *in vivo* or clinical setting. If using recombinant technology, the experimenter needs to decide upon the importance of knowing the exact molecular entity that it is necessary to screen against. For example, if the human clone is not available is an orthologue acceptable? Ion channels can consist of multiple subunits and deciphering the endogenous, pathophysiologically relevant constituent subunits is not trivial. For example, one of the most studied voltage-gated ion channels is the cardiac K^+ channel underlying the rapid component of the delayed rectifier K^+ current. Whether the endogenous channel consists of homotetrameric h-ERG-1a α-subunits, heterotetrameric h-ERG-1a/1b α-subunits, or in the presence or absence of the β-subunit MiRP1 is still not universally agreed [2, 3]. Similar issues exist for GPCRs and enzymes; thus, ion channels can show great similarities with any other drug targets.

4.2
Patch Clamp

The technique of using voltage clamp to control the membrane potential of the cell is common for patch clamp, planar patch clamp and two-electrode voltage clamp – hence the technique described here is also equally applicable to Sections 4.3 and 4.4.

4.2.1
Basic Description of Technique

The patch clamp technique is an extension of previous voltage clamp methodologies pioneered from the 1930s and 1950s by Cole, Hodgkin, Huxley and Katz (Fig. 4.1). The scope of the technique can be described by Ohm's law (Eq. 2).

$$V = IR \qquad\qquad (2)$$

The experimenter controls the voltage (V), records current (I), and the resistance (R) is a variable dependent upon the cell characteristics. These classical experiments used one electrode for controlling the voltage and a second for recording the current – hence the technique was limited to cells large enough to withstand impalement by two glass pipettes. A major breakthrough was the application of voltage control and current recording through a single pipette by Neher and Sakmann that allowed the application of this technique to cells smaller than the squid giant axon [4]. Thus, when recombinant technologies allowed, it was a natural extension to express ion channels in heterologous cell systems (*Xenopus* oocytes and mammalian cells) to study the ion channels in isolation from other proteins. This forms the mainstay of drug screening paradigms today.

The "patch" in patch clamp refers to the patch of cell membrane contained within the polished aperture (~1 µm diameter) of a glass pipette that is placed against the membrane of a single isolated cell and has suction applied to it form a high ($10^9\,\Omega$ – "gigaOhm") resistance seal. This physical interaction, with the subsequent rupture of the membrane within the patch, makes the connection between the cell and the amplifier. The quality of this seal is important as the current measured is the sum of the channel current and the leak current (escaping between the pipette and the cell); thus, if the seal is not of high resistance, the leak current dominates the current that the experimenter is interested in. In the most common variant of patch clamp – whole-cell patch clamp – this patch of membrane is ruptured by suction, allowing electrical access from the amplifier to the entire cell. The amplifier controls – "clamps" – the voltage across the cell and measures the current flowing through the ion channels in the cell membrane.

A pertinent analogy for the explanation of voltage clamp is that of a leaky tire. The cyclist needs use a pump (amplifier) to pump more air (current) into the tire (cell) to keep the pressure (voltage) where it is desired. Every so often the cyclist

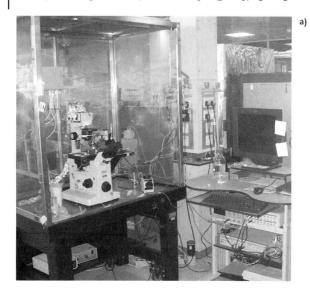

a)

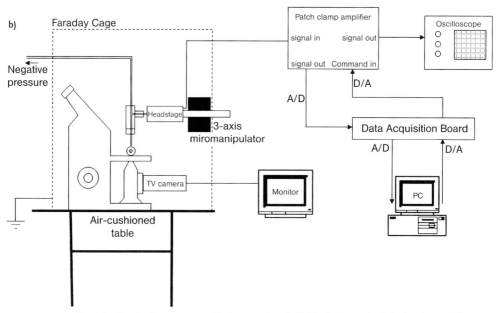

Fig. 4.1 Patch clamp set up. (A) An actual patch clamp rig, demonstrating the Heath-Robinson nature of the equipment. It is not possible to buy an entire set-up off the shelf and individual electrophysiologists have differing preferences, leading to no two rigs being identical. (B) Diagrammatic representation of a patch clamp rig.

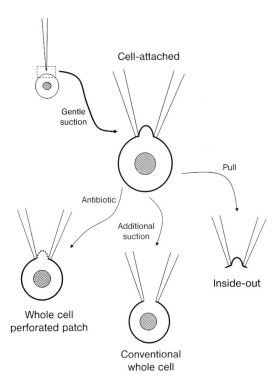

Fig. 4.2 Representation of different patch clamp configurations.

checks whether the pressure is at the desired level (feedback loop). The holes in the tire are analogous to the channels. If they are voltage-gated holes they would be opened, or shut, by sensing the pressure differential between the pressure inside and outside of the tire. Air passing through the holes affects the pressure inside the tire. One thing to note is that the passage of air (current carried by ions) is not an active process – the air moves passively down a pressure gradient. Where the analogy breaks down is the time scale over which this occurs. Today's voltage clamp amplifiers can accurately switch the voltage of the cell from a physiological minimum to a physiological maximum in less than a millisecond. The currents are recorded by repetitive voltage stimulation ("test pulse") and the amplitude of the elicited current (being the most frequently measured parameter) is plotted against time.

The patch clamp technique can be applied in various forms (whole-cell, cell-attached, excised patch etc.; see Fig. 4.2). The reader is directed to more in-depth reviews about the difference between these modes [5, 6]. For screening purposes the whole cell variant of patch clamp is the only viable option as other modes are much more time consuming.

The study of a single type of ion channel can be facilitated by the use of solutions of specific ionic composition. For example, replacement of K^+ in buffers with Cs^+ (which is impermeant to K^+ channels) allows for Na^+ channels to be studied in the absence of contaminating K^+ current. The use of selective channel blockers can either remove contaminating currents to isolate the channel of interest [e.g., 100 nM tetrodotoxin (TTX) will block TTX-sensitive Na_V channels, but leave TTX-insensitive Na_V channels largely untouched] or can be added at the end of an experiment to block all of the current of interest, thus defining the compound-sensitive current. This can often be digitally subtracted from recordings earlier in the experiment, thus isolating the current of interest. As different voltage-gated channels of similar type often have different voltage dependent characteristics, the use of voltage stimulation protocols that limit the voltage to that relevant to the current in question can also help. For example, L- and T-type Ca^{2+} channels have differing voltage ranges over which they are activated and inactivated. T-type channels are rapidly activated and inactivated at −30 mV whereas most L-type channels are not activated at this potential. By using test pulses to a potential depolarized enough to open channels (~0 mV) in combination with an interpulse holding potential of −80 mV, both currents can be recorded, whereas using an interpulse holding potential of −30 mV leads to only the L-type currents being recorded as the T-type channels are inactivated and not available for opening [7]. More complex voltage stimulation protocols are often used in diagnostic mode to tease out different components of a current, as in the identification of the two components of the cardiac delayed rectifier, IK_R and IK_S [8]. In whole-cell patch clamp, where the pipette has physical and electrical access to the interior of the cell, the volume of the pipette solution is in vast excess to that of the intracellular contents of the cell. This can be used advantageously to control the ionic composition of the intracellular solution, which can aid in isolating specific currents. However, many voltage-gated channels require, or have activities modulated by, intracellular diffusible second messengers (cAMP etc.). If the wash out of these factors is detrimental to the experiment they can be maintained in the cell by using antibiotics (amphotericin, nystatin; Fig. 4.2.) to perforate the patch of membrane under the pipette rather than using suction to completely remove it. Thus, small ions can reach equilibrium, without loss of larger intracellular components. However, this adds time and another layer of complexity to the experiment. The combination of these described techniques can allow for relatively selective recording of the activity of a single type of ion channel within a native cell. The use of null-background parental cell lines when constructing a heterologously expressing stable line, in combination with the tendency for over-expression of the channel of interest, immediately increases the proportion of channels of interest, relative to others, and simplifies the experiment.

4.2.2
Advantages and Disadvantages of Manual Patch Clamp

Patch clamp allows for the precise control of the channels measurements of channel activity can occur on a sub-millisecond time scale and with single digit pA resolution, resulting in what can be very detailed, high quality data. This absolute control over membrane potential is a key advantage that the voltage clamp technique has over other ion channel screening technologies – it allows for precise and repeated control of the conformation of the channel to either study one state in relative isolation or cycle the channel through the voltage cycle the channel would experience *in vivo*. For example, an action potential can be recorded and used as voltage stimulus for a single cloned channel [9].

Patch clamp offers a direct functional measurement of the effect of a compound on current flow – which is the effector mechanism of the channel. Other functional screening techniques tend to measure the effects of compounds indirectly. For example, voltage-sensing dyes report changes in membrane potential, which is a nonlinear reporter of alterations in the activity of ion channels. The high temporal resolution of patch clamp allows for exquisite characterization of a compound's effects on different phases of the current elicited by a voltage pulse or pulses (see Section 4.2.3). Utilization of different modes of patch clamp (Fig. 4.2) allows for detailed characterization of the mode of action of a particular compound, though these different modes of recording are probably more applicable to the needs of a development compound rather than primary screening due to the even further reduced throughput. Another advantage is the requirement in patch clamp for few cells in the experiment so that exactly the same technique can be used on clonal or native cells, which helps in the translation of an IC_{50} from clonal line to *in vivo*.

Despite these advantages, which lead patch clamp to be routinely touted as the "gold standard", critical downsides to the technique limit its usefulness. Primary among these is the extremely low throughput of the technique. This originates from several sources but has the effect of limiting a scientist to <10 data points per day. The technique involves manually guiding a fragile glass micropipette, viewed from below through microscope optics, to a single cell by movement of a 3-axis micromanipulator – and only one of the cells or the micropipettes will be in focus at one time. This process is time-consuming and physically demanding and thus requires a highly trained individual. Because the measurements of current occur from a single cell, unlike plate-based techniques (where the signal measured is the average or sum of many tens of thousands of cells) there is a requirement that each cell acts as its own control. The level of current measured from individual cells can vary widely (percentage coefficients of variation of current between cells of 50% are not unknown from a clonal cell line, even when normalized for cell size) and, consequently, in a screening scenario it is impractical to compare treated and untreated cells. Thus each cell needs to act as it own internal control measuring current before compound application and after. This can lead to extremely high confidence in the data, but has the downside of making the experiment a "time series" experiment and thus each cell can take up to an hour or longer to obtain data from. In this respect, patch clamp can be compared with tissue bath pharmacology. An advantage that tissue baths have, though, is that one operator can utilize 16–32 tissues in parallel. Until recently it has not been possible to patch more than one cell at a time (Section 4.3).

4.2.3
Use of Patch Clamp for Quantification of Drug–Channel Effects

Patch clamp allows for repetitive stimulation of the channels – something that occurs naturally in tissues that fire action potentials (cardiac muscle, nerves etc.). This is currently difficult to do in higher throughput fluorescent-based technologies. Many compounds exhibit preferential binding to one or more of these states and can be characterized as closed-, open- or inactivated-state blockers. Thus, to enable testing of a compound's ability to bind to a particular state of the channel, the channel must be placed in the appropriate state. This is most commonly done with the natural "ligand" of the voltage dependent channel – voltage itself, and voltage clamp is the most accurate method with which to induce changes between these states. Figure 4.3 shows how a simple voltage protocol elicits a current with distinct temporal components that are related to the state of the channel.

The high time- and amplitude of response-resolution of patch clamp allows for measurement of direct interaction of compounds with the channel. The generation of separate IC_{50}s for different states of the channel can be routinely achieved. Some examples are detailed below.

4.2.3.1   State-dependent Block
There exist therapeutic rationales for compounds that selectively block open K_V channels compared to closed channels and this can be determined in patch clamp. Figure 4.4 describes a protocol that discriminates between a compound

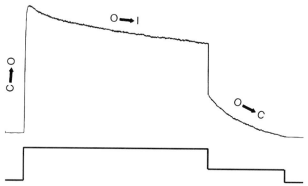

Fig. 4.3 A single depolarizing voltage pulse (bottom) elicits whole-cell current from a voltage-gated K^+ channel. The initial depolarization opens the channel (C→O). When held at a positive potential, a small degree of inactivation occurs (O→I), but a substantial fraction of the channels are still in the open state. Upon partial repolarization, the current amplitude drops (owing to the decrease in voltage driving K^+ through the channels) and moves the channel to a voltage that favors closing of the channel and, over time, the channels close (O→C), which is referred to as the "tail current". These voltage pulses are repeatedly elicited.

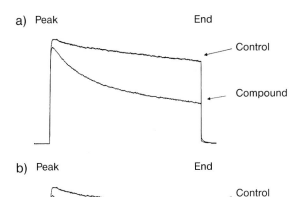

a) Peak   End

Control

Compound

b) Peak   End

Control

First Pulse

Steady state

Fig. 4.4 Example of a) open-channel block and b) combined open channel block with time-dependent accumulation of block of a K_V channel.

(or concentrations of the same compound) that demonstrates open and closed channel block. The depolarizing voltage pulse elicits a K_V channel current similar to that shown in Fig. 4.3. Figure 4.4(a) shows two overlaid traces of control and in the presence of compound. In the presence of compound, the magnitude of the peak current is barely affected, whereas the magnitude of the end pulse is dramatically inhibited, indicating that the channel is required to move to the open state before a blocking event can occur – open/inactivated state block. Practically, the differentiation between open state and inactivated state blockers is very difficult, especially in fast inactivating Na_V channels.

Figure 4.4(b) shows a similar phenomenon on the same type of channel but demonstrating the time-dependent accumulation of block. Over time (with a different compound) the peak current declines as well as the end pulse current. Whether this occurs is determined by the off-rate of the compound at the potential that the channel is held at in the inter-test-pulse period and the pulse frequency. Thus, such experiments can give an idea of the on/off kinetics.

4.2.3.2 Use/Frequency-dependent Block

The above analysis can be complemented to investigate state-dependence and use- (or frequency-) dependence of block. Quite often, compounds bind to the open and/or inactivated states of channels at lower concentrations than to the closed state but there are examples that reverse this trend or that show little discrimination. Figure 4.5 shows a paradigm whereby this can be quantified. Repetitive voltage test pulses are used to elicit current and establish a baseline. A compound is then washed on in the absence of any voltage pulsing when the channels are held

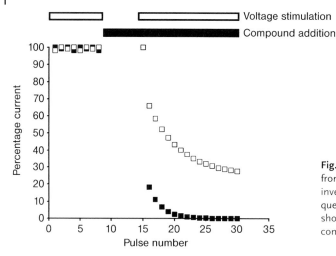

Fig. 4.5 Stylized results from an experiment to investigate use/frequency dependent block, showing two differing compounds.

in a closed state. Voltage pulsing is then resumed once the experimenter is sure that compound equilibrium has been reached in the bath. The effect of the compound on the current elicited by the first, and subsequent, pulses after exposure to the compound can be interpreted to give an idea of the state dependence of binding of the drug. Compound A (open symbols) shows no closed channel block as the first pulse after compound wash-in is similar to baseline whereas compound B (closed symbols) exhibits closed channel block as the first pulse elicits less current. Both compounds exhibit use dependence as the amount of block increases with further pulsing. Complementary experiments can investigate the frequency-dependence of block by comparing the rate of block with differing frequency rates of stimulation. An example where this would prove beneficial in patients is an anti-arrhythmic compound that preferentially affected faster heart rates. Unfortunately, current therapies often exhibit reverse-rate dependence in that they become less efficacious at higher rates of stimulation [10].

4.2.3.3 Inactivated State Block – From IC_{50}s

Many Na_V ligands can show differing IC_{50}s dependent upon the holding potential of the cell and show preferential binding to the inactivated state of the channel. This is referred to as the "Modulated Receptor Hypothesis" [11]. Briefly, when depolarization opens channels, a binding site is revealed so that the compound exhibits preferential inhibition of the channel in its depolarized state. The apparent affinity of a compound for the open state of a Na_V channel can be measured by an IC_{50} value. This can be calculated by measuring the decrease in current amplitude elicited by a test pulse when the cells are held during the interpulse interval at a (hyperpolarized) membrane potential so the channels are predominantly in the closed state (the test pulse elicits a closed to open transformation; Fig. 4.3).

The measurement of the affinity of a compound to the inactivated state of the channel (K_{inact}) is difficult to measure functionally as, by definition, the inactivated channels are not available for opening (and thus measurement). However, K_{inact} can be estimated by comparing two IC_{50} measurements – the resting state IC_{50} (K_{rest}) described above and the IC_{50} at a potential where a substantial proportion (but not all) of the channels are in the inactivated state (K_{app}) [12]. For an inactivated state blocker, K_{app} will be more potent than K_{rest} and K_{inact} can be calculated by rearrangement of Eq. (3), where h is the fraction of channels in the inactivated state at the more positive holding potential as measured from a steady-state inactivation curve.

$$K_{app} = \frac{1}{\left(\frac{h}{K_{rest}}\right) + \left(\frac{(1 - h)}{K_{inact}}\right)} \qquad (3)$$

4.2.3.4 Inactivated State Block – From Steady-state Inactivation Curves

Inactivated state block by the mathematics in Eq. (3) also predicts a leftward shift in the steady-state inactivation curve [13]. As the interpulse voltage (or, more correctly, the pre-pulse voltage) is held less hyperpolarized (closer to 0 mV), more channels are inactivated and thus fewer of them are available for opening in a subsequent test pulse. The fraction of available channels can be plotted against the membrane potential immediately before the test pulse. This curve, called the steady-state inactivation curve (as the pre-pulses need to be an order of magnitude of time longer than the test pulse), can be fitted with a Boltzmann curve and the membrane potential where half the channels are inactivated ($V_{0.5,inact}$) can be calculated. In the presence of an inactivated state blocker, $V_{0.5,inact}$ will be more negative than control. The change in the midpoint of the inactivation curve is related to the inactivated state affinity, K_{inact}, by Eq. (4), in which ΔV is the change in $V_{0.5,inact}$ and k is the slope of the steady-state inactivation curve.

$$\Delta V = k \ln \left(\frac{1 + \frac{[drug]}{K_{inact}}}{1 + \frac{[drug]}{K_{rest}}} \right) \qquad (4)$$

4.2.4
Caveats of Interpretations in Patch Clamp

In the ways described above, and others, $IC_{50}s$, K_is and K_Ds can be measured for voltage-gated ion channels. Although many of these calculations are interdependent care must be taken not to (directly) compare apples with oranges. As with all experimental techniques, the conditions used in the assay have a direct effect upon the results gained. For example, it is a relatively common in patch clamp to alter the ionic constituents in the assay (e.g., to aid in the isolation of specific currents). Raising the external concentration of K^+ from 4.5 to 140 mM has several biophysical effects but can make K_V currents easier to measure – though one may well expect pharmacological differences. Ba^{2+} is often used as a charge carrier for Ca^{2+} as the currents are larger and this, too, can lead to changes in pharmacology. Care must be taken in comparing numerical values obtained under differing experimental conditions.

4.3
Planar Patch Clamp

Until recently, patch clamp has been a manual, low throughput technique not suited for the rapid determination of structure–activity relationships that modern drug discovery requires. However, the glass pipette can now be replaced by a planar substrate [14]. In this paradigm, tens of thousands of cells of interest are acutely dissociated from the culture plate and added to the planar substrate and a single cell seals over a laser etched aperture in that substrate. The perforated patch technique or suction is used to achieve electrical access to the cell. Further to this, the technique is broadly the same as manual patch clamp. This advance has opened up the development of automated patch machines. Several automated patch clamp machines are now on the market (PatchXpress 7000A, Ionworks HT, Ionworks Quattro all from Molecular Devices; Qpatch from Sophion) and the sector is rapidly maturing.

Planar patch is, essentially, whole-cell, manual patch clamp with few differences, and thus has advantages and disadvantages similar to those described in Section 4.2.2. The main difference, illustrated in Fig. 4.6, is that instead of a manual manipulation of the glass pipette towards the cell as in manual patch clamp, the cell is attracted (by suction) to the aperture in the planar substrate that replaces the glass pipette. This has two effects: firstly it removes the physical technicalities of patch clamp and secondly it allows for parallel experimentation. Estimates of the increases in throughput vary from 5-fold to 800-fold dependent upon the machine and the mechanism of usage. Clearly, however, planar patch dramatically increases the capacity of the technique whilst having little or no effect upon the quality of the data.

One difference that distinguishes the machines (PatchXpress 7000A and QPatch vs. IonWorks and Quattro) is that the seal resistance (how "tightly" the cell sticks to the planar surface) may not be greater than the commonly accepted minimum for manual patch clamp of $1\,G\Omega$. Whether this is of functional consequence depends upon the nature of the experiment. The origin of high resistance seals in patch clamp originates from pre-clonal cell line days and was a dogma to distinguish ion channel currents from passive leak or time-independent currents. In a well-defined clonal cell line, where the ion channel current of interest vastly exceeds any native currents, the lowering of the "dignity" threshold for a seal resistance may well be acceptable. Early data comparing and contrasting planar patch data with $G\Omega$ seals compared to hundreds of $M\Omega$ would seem to suggest that lower seal resistance machines may be associated with right-shifted pharmacology in the region of 3-fold [15, 16]. Whether this is a direct consequence of the

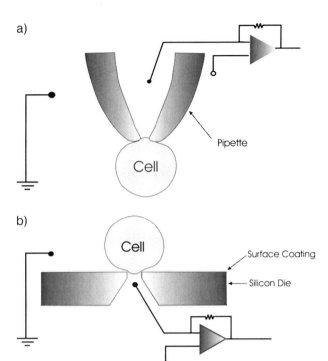

a)

Pipette

Cell

b)

Cell

Surface Coating

Silicon Die

Fig. 4.6 Diagrammatic representation of (a) a conventional and (b) a planar patch. (Figure courtesy of Sophion Bioscience A/S.)

lower seal resistance or due to some other mechanistic differences is unknown. For a hit-seeking program (as opposed to a safety pharmacology assay) this modest reduction in apparent potency is more than offset by the increase in throughput achieved by this technique.

Most of the present generation of automated patch clamp machines are static bath, whereas conventional patch has a flow of salt solution over the cells. There may be 100–1000 cells in the recording chamber in a conventional patch clamp rig, containing perhaps 1 mL of fluid, whereas >10 000 cells are placed into a well on a planar patch chip containing ~50 μL. This increase in density of cells, combined with the static nature of the fluid, may act as a sink for lipophilic compounds in planar patch, leading to an underestimation of potency. The magnitude of this effect will be most apparent with compounds of high affinity (as the sink:compound ratio will be high) and may vary between chips synthesized from different materials. The early evidence is that results with compounds of cLogP > 5 can be questioned [17]. How different this is from conventional patch, where compounds typically flow through thin bore tubing from compound reservoir to assay bath (high surface area/volume ratio), has not been publicly studied.

Despite these differences planar patch data appear to be directly comparable to manual data and goes some way to breaking the "information content $\propto$ 1/throughput" paradigm. They partially remove the throughput bottleneck that is the major downside to manual patch clamp and with future improvements will allow the utilization of patch clamp in a High-throughput Screening (HTS) environment.

4.4
Two-electrode Voltage Clamp (TEVC) of *Xenopus* Oocytes

Oocytes of the African clawed toad (*Xenopus laevis*) can be microinjected with cDNA or cRNA and use the endogenous translational machinery to produce high levels of functional ion channel expression at the cell surface. The size of *Xenopus* oocytes (~1 mm diameter) is both a boon and downside to their use. This large size allows the two-electrode voltage clamp method to be used (one electrode for current passing and one for voltage recording). This is technically much simpler to perform than patch clamping and can give precise voltage control and thus has all the advantages described for patch clamp in Section 4.2.2. However, due to the size of the cells, control of the intracellular cellular constituents (as is typically controlled in patch clamp) is not possible without using more advanced techniques such as cut-open oocyte patch [18] or excised macropatches [19].

A continual supply of cDNA or cRNA is required as each individual oocyte is one use only. However, the nucleic acids can be injected easily and the impalement of such large cells by two electrodes is technically trivial. For example, the need in patch clamp of a large vibration isolation table is not as great with such large cells as oocytes. Serial and parallel automated TEVC machines are now on the market and some pharma companies have custom built their own [20, 21]. The increase in oocytes used in the automated technique obviously requires more cRNA; however, the machines can also be used for automatic RNA injection and can be used in unattended mode. One technique of interest to pharma companies is site-directed mutagenesis to include alanine, cysteine or non-natural amino acids into ion channels [22]. The rapid and robust protein expression, coupled with the ease of measurement of current, in oocytes can be used to rapidly define the binding sites of compounds.

As with all parental cells, oocytes do express some endogenous channels (e.g., GIRK 5 and Cl_{Ca}) that can either be confounding to making conclusions or can be utilized as an aid in signal measurement. The correct identification of the molecular constituents of the IK_{Ach} current (now known to be Kir3.1/Kir3.4 heterotetramers) was hampered by the endogenous Kir3 subunit expressed in *Xenopus* oocytes [23]. Conversely, many GPCRs couple to Cl_{Ca}, and the change in current flow through this channel can be used as a measure of GPCR activation [24]. Either way, when expressing channels in these cells the potential effects of endogenous currents cannot be automatically discounted.

The pharmacology of channels expressed in oocytes can be different, especially for lipophilic compounds as the yolk sac of the oocytes can sequester compound, thus denying the compound the opportunity to bind with the channel, leading to underestimation of potency. The non-mammalian nature of the post-translational processing can also make certain scientists reluctant to use oocytes. Nevertheless, they have certain advantages in their ease of use – the technical feasibility, true voltage control and the availability of parallel machines to increase throughput are all beneficial. However, continual requirement for cDNA, injection of said cDNA limits the technique to follow up assays, not million-plus compound HTSs.

4.5
Membrane Potential Sensing Dyes

The use of the microtiter plate has become ubiquitous in drug discovery. Though the mechanisms by which the techniques described below report changes in voltage-gated ion channel activity vary widely (e.g., fluorescence, ion flux), they have many advantages and disadvantages in common. The pros and cons are discussed especially with respect to membrane potential-sensitive dyes but many of them (especially that of the lack of voltage control) are common to all plate-based techniques.

4.5.1
Basic Description of Membrane Potential-sensing Dyes

Interest in membrane potential-sensing dyes has undergone a resurgence over the past 5–10 years due to the increase in throughput needed for large scale HTS of companies' compound files. They are a convenient, plate-based functional screening technique that can also be applied to electrogenic transporters. The basic mechanism of action of this technique is that the dye molecules are lipophilic compounds containing a delocalized charge. The dyes respond to a change in membrane potential (driven by the activity of ion channels) with an alteration in intramolecular charge distribution or by transversing the plasma membrane, which causes a change in their fluorescent emission [25].

There are two main types of dyes – slow response and fast response. The latter, typified by the Di-ANEPPS (Amino Naphthyl Ethenyl Pyridinium) series, have differing spectral properties dependent upon voltage due to a change in their intramolecular charge distribution. But despite having time constants (of fluorescent intensity change in response to a change in membrane potential) that are fast enough to follow action potentials, the magnitude of change in their fluorescence is typically too small to register in plate-based assays and so are usually used in conjunction with fluorescent microscopes.

Slow response dyes, typified by the DiBAC [e.g., DiBAC$_4$(3); (bis-(1,3-Dibutyl-barbituric Acid)Trimethine Oxonol)] dyes, have a much slower response time as they need to transverse the entire lipid membrane and bind to intracellular components to elicit an increase in fluorescence emission. A time course of minutes to tens of minutes has the consequence that assays are unresponsive to short-term changes in membrane potential and can lead to long assay times and

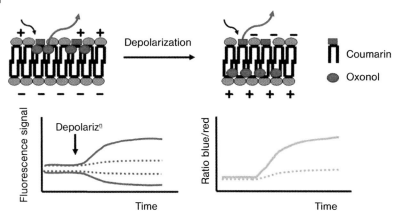

Fig. 4.7 Mechanism of FRET-based membrane potential-sensitive dyes. Top: The red emission DiBAC molecule transverses the plasma membrane in response to depolarization, breaking the FRET and resulting in more blue emission and less red. Bottom left: The fluorescence emissions are captured by photomultiplier tubes and plotted against time. If a blocker is pre-incubated there will be less change in the fluorescence emissions (dashed lines). Bottom right: Time series data is then ratioed. In the presence of a blocker (dashed line) the response will be diminished.

thus are only really useful for studying steady-state changes in membrane potential [26, 27].

Newer dyes, developed by Gonzalez and Tsien [28] respond to changes in membrane voltage in a quicker manner by utilizing Fluorescence Energy Transfer (FRET). In this system, a FRET donor (coumarin-phospholipid; CC2-DMPE) partitions to the extracellular half of the plasma membrane and is excited by 405 nm light. The emission of this dye (460 nm) is in the excitation range of the FRET acceptor [DiBAC$_2$(3)], which emits at 570 nm. Under hyperpolarized conditions the negatively charged DiBAC is close to the CC2-DMPE FRET donor. Upon depolarization the DiBAC molecule transverses to the intracellular side of the lipid bilayer and, as the efficiency of FRET decreases with the sixth power of the distance between the two dyes, the FRET signal is broken. This results in an increase in blue wavelength emission from the CC2-DMPE and a decrease in red emission from the DiBAC (Fig. 4.7). This short translocation results in a responsive time domain (20–200 ms time constants dependent upon the specific DiBAC molecule used) that brings two advantages – quicker assay time than the DiBAC-only assays and better temporal resolution allowing for the measurements of slow (i.e., cardiac) action potentials. Also, the presence of two fluorophores allows for ratiometric measurement that decreases variability between assays.

Washing off of excess dye in assay leads to extra steps and variability and is usually avoided if possible. A no-wash kit containing a fluorescent indicator of membrane potential in conjunction with a quenching agent has recently been released by Molecular Devices, and a side by side comparison of several membrane potential sensitive fluorescent dyes has recently been reported [29]. This dye, which is of unknown chemical composition but is suspected to be DiBAC-like in character, appears to have a relatively quick time course.

An alternative approach that obviates the need for dye loading and wash steps, but introduces requirement of the generation of a stable cell line, is the integration of a fluorescent reporter, such as Green Fluorescent Protein (GFP), into a voltage-gated channel such as a K_V or a Na_V channel [30, 31]. If the GFP is inserted into the correct position, the channel can retain function and fluorescently report physical movements of the channel in response to voltage by changes in fluorescence. Whether there is any effect of changing the structure of the channel on pharmacology is unknown. In future, advantages such as these may lead to targetable (intracellular organelles, distinct cell populations) fluorescent indicators of ion channel activity. However, it is likely that these approaches will also be of too limited sensitivity for the current generation of plate readers as well as suffering the disadvantages of membrane potential sensitive dyes documented below.

4.5.2
Advantages and Disadvantages of Membrane Potential-sensing Dyes

Plate-based assays are of high throughput and can be run in 96-, 384- or higher density well plates. They are amenable to automation and thus HTSs can be run. All plate-based assays can report results in terms of IC_{50}s (or K_is in binding assays) and thus are relatively easy to interpret within a single assay and easy to rank compounds in order of potency. FRET dyes also have the advantage of being ratiometric, leading to less inter-well, -plate and -day variability.

Sensing of membrane potential has several advantages over the measurement of current (as in patch clamp). One is that it there is no pre-requisite to have large expression of channels. In a cell with a high resistance (not many open channels), the opening or closing of a small number of channels will produce a dramatic change in membrane potential (remembering $V = IR$). In fact, FRET-based dye assays for transporters (which have a much lower conductance than ion channels) have been described. Generically, membrane potential-sensitive dyes can respond to both a depolarization and hyperpolarization and so can report activation or block of various voltage-gated channels. Thus, the technique is generic to any ion channel that can effect a change in membrane potential, which enables facile comparison of selectivity screening data if assayed with the same technology.

However, care must be taken when relating changes in membrane potential (voltage) reported by these dyes to other assay formats such as the changes in current measured in voltage clamp experiments. It seems obvious that these two techniques measure different parameters, and thus may give different results, though it is often forgotten and apples are compared with oranges.

Notably, there is no mechanism for the accurate controlling of membrane potential in these plate-based assays. The membrane potential at the start of the assay is set by the activity of the ion channels in the cell in combination with the ionic constituents of the bathing medium. This is in contrast to patch clamp where the voltage (and hence the state of the channel) can be controlled by the experimenter regardless of the composition of the bathing fluid. The activity of the ion channels in the assay (and hence any modification by compound) is measured by eliciting a depolarization by changing the ionic composition of the bathing medium (introduction of Na^+ into Na^+-free buffers or the raising of K^+ concentration; Fig. 4.7). These ionic alterations induce a depolarization towards a thermodynamically defined maximum membrane potential. Any block of the channels by pre-incubation with a compound will induce a depolarization towards that theoretical maximum and thus decrease the measurement made when the ionic alteration is made. Thus the IC_{50} reported will be dominated by the affinity/efficacy of the compound for the state that predominates at the start of the assay. The initial membrane potential is usually in the region of -60 to -30 mV. Therefore, compounds that show voltage-dependent block and have higher affinity for channels at more depolarized potentials are likely to be have underestimated potency in these types of assays.

Any compound-induced depolarization may alter the balance of ion channels in the different conformational states and may lead to a feed-forward loop where further depolarization induces further compound effect. Thus, the response in these assays where membrane potential is not controlled can be nonlinear with respect to compound binding to channels. Increasing the concentrations of Na^+ or K^+ in the bathing solution affects the pharmacology of compounds in patch clamp and thus it might be expected that the same is true in plate-based experiments. However, it is important to remember that the addition of these cations is merely a mechanism by which the assay window is assessed. The compound is pre-incubated with the cells and will have its effect before the ionic conditions are altered. Hence it is not expected that this will be a reason for any unexpected pharmacology for K_V channels but is a possibility for Na_V channels where the initial conditions are usually Na^+-free. An alternative assay format has been proposed where the cells are incubated in physiological Na^+ and the depolarization induced by pharmacological manipulation of the channels to remove inactivation [32].

It is also important to remember that membrane potential sensing dyes measure *changes* in potential rather than an absolute value. Estimations of absolute membrane potential can be made by incubating in the presence of increasing concentrations of K^+ to depolarize towards the thermodynamically defined maximum membrane potential and extrapolation back to the basal conditions.

There are always endogenous ion channels expressed in cell lines, which affect the starting membrane potential. In some circumstances these can be used to the assay developer's advantage, e.g., using endogenous K_V channels in HEK cells to set the starting membrane potential further from the thermodynamically defined maximum membrane potential and thus increase the assay window in Na_V and Ca_V assays [33, 34]. The corollary of this is that any compound that blocks the endogenous K_V channel will depolarize the cells, decrease the assay window, appear as a blocker, and thus will be a false positive. This necessitates some kind of counter screening, which can be difficult to interpret if the efficacy of the compound at the endogenous and target channels is similar.

Voltage-gated channels are often expressed in tissues that fire action potentials and thus undergo regular cycling through the closed, open, inactivated set of states. This is difficult to replicate in a plate-based scenario. Electrical field stimulation of tissues is a well-utilized technique in *in vitro* biology, yet there is a lack of widely available electrical stimulation methods that are applicable to plate-based screening scenarios. Vertex is one company known to have field stimulation capability for their FRET technology and some large pharma companies are believed to have in-house methodology, but there is no commercially available source despite early literature reports [35]. This is of consequence due to the differing affinity of compounds for the different states of the channels and the usefulness of characterizing the use-dependency of compounds. Arguably, this downside is not a weakness in the dyes themselves but due to their incorrect use in screening targets expressed in excitable, action potential firing tissues. The inability to use voltage as a "ligand" or stimulus is negated if the target tissue exhibited graded, slow depolarizations (e.g., vascular smooth muscle, lymphocytes).

One way to bias the assay to report an affinity of a compound on a particular state is to increase the proportion of channels in the desired state. For example, incubation in slightly elevated concentrations of K^+ will depolarize the cells and, therefore, push the channels more towards the inactivated state [34]. Comparison between assays performed in low and elevated K^+ can give an indication of the selectivity of compound for closed and open/inactivated states. The downside to this approach is that the more the cells are depolarized the smaller the assay window will be for measuring any pharmacological effect.

Na_V channels open and inactivate over a time scale of single milliseconds. To study Na_V channel blockers, the channels need to be kept in the open state long enough for a depolarization to occur and for this depolarization to be reliably identified in plate readers that only sample at 1 Hz. Thus, alkaloid compounds such as veratridine or deltamethrin are routinely added to the assay [32]. These compounds act to promote an open state of the channel and prevent inactivation, thus prolonging current flow through the channels to effect a depolarization. However, it is not known how pharmacologically representative of the native open channel this modified state is and certain types of compounds are not identified by these assays [36].

At high (>10 µM) concentrations, interactions can occur between compounds and dye that lead to an apparent inhibition of signal and thus potential false positives. These interactions are chemical series-dependent and thus could be a critical flaw in this approach if they are identified (or not!) in the lead series. For the FRET-based dyes, it is estimated that ~1% of a random chemical file will demonstrate drug–dye interactions at concentrations in excess of 10 µM. It is also possible that the dyes can interact directly with the channel of interest to alter the pharmacology of the channel, e.g., Di-8-ANEPPS block of hERG channels [37]. The dyes can also be somewhat temperature dependent, necessitating the regular use of control compounds.

In summary, the advantages of throughput and ease of use need to be weighed against any potential compromises in the pharmacology induced by the limitations of these assays. Compounds showing strong voltage-dependence of block or strong dependence upon cation concentration are likely to be misrepresented. In a hit identification program these assays can be useful (as there may be limited screening alternatives, false positives can be identified in a subsequent screen and a certain proportion of false negatives can be accepted), but in a safety pharmacology assay that is required to determine the concentrations at which an effect is likely to occur *in vivo* these limitations may critically undermine the use of membrane-potential sensing dyes.

4.6
Binding

One of the most ubiquitous assay formats used in drug discovery is the radiometric binding assay. The basic requirements are for a displaceable ligand of a saturable binding site that has an affinity within the physiological range. Typically, binding assays take the format of the displacement of a radiolabeled ligand and thus there exists a pre-requisite that there is a known ligand available. For a detailed description of the technique the reader is directed to one of the many available texts on binding techniques, e.g., Refs. [38–40].

In many respects, the advantages and disadvantages of screening ion channels in this manner are similar to other proteins such as GPCRs. The assay format gives little or no information on the functional effect of the ligand upon the channel (blocker, potentiator, no effect) and allosteric ligands – those that bind a site remote or distinct from the site utilized by the labeled ligand – are unlikely to be identified. However, binding assays possess several advantages in that they are relatively easy to set up and run, have high throughput and so can support large-scale HTS and are relatively easy to interpret in terms of IC_{50} or binding constant, K_i. Recently, due to environmental, safety and throughput concerns, radioactive techniques are being replaced with fluorescence assays such as fluorescent polarization.

One example of a parallel between GPCRs and ion channels that is demonstrated in binding assays is state-dependent (conformational-dependent) binding. The presence/absence of GTP in a binding assay can place a GPCR in an R/R* state that can lead to different affinities being recorded for agonist/antagonist binding. When configuring ion channel binding screens one needs to validate the assay to ensure that the channel is reporting the desired pharmacology. As the channel binding site is expressed (usually) in a heterologous cell, the membranes of which have been fractionated to decrease non-specific binding and are present in the assay as micelles of unknown membrane potential, little control exists over the membrane potential (if any exists) and hence there is little control over the state of the channel. Adjustments such as changing the level of the K^+ in the assay buffer (and hence membrane potential) have been attempted to aid in biasing the population of channels one way or another [41].

Despite the caveats mentioned above regarding state-dependence of compound affinity, binding assays can work for voltage-gated channels and should not be dismissed out of hand. For example, [^{3}H]dofetilide binding to h-ERG has an extremely high predictivity for QT effects [42, 43], and binding assays for K_V channels can be used successfully, as can those for Ca_V and Na_V channels [44–48]. One common feature of the ligands used in these assays is that they bind in the pore region of the channels. Previously, in the absence of known, potent small molecule ligands, toxins (such as charybdotoxin or dendrotoxin) were radiolabeled and used in displacement assays. These toxins have high affinity for channels (often single digit nanomolar affinities), so lower total amounts of radioactivity were needed and toxins such as these can be exquisitely specific. It has since been shown that most toxins used bind to the large, rather flat, extracellular face of channels, which is neither the binding site for most small molecule ligands identified so far nor an attractive "beautiful" binding site likely to contain binding sites for yet to be identified small molecule ligands [49, 50]. Thus, although radiolabeled toxins can be useful in identification, localization and purification of ion channels, their usefulness for drug discovery for K^+ channels is questionable. Ion channel binding assays are not limited to pore-blocking molecules. As in the example of gabapentin and pregabalin, therapeutically useful and efficacious ligands, binding to ancillary subunits of voltage-gated ion channels can be identified – in this case the $\alpha2\delta$ subunit of Ca_V channels [51].

As with all binding assays, it is important to remember that one is only likely to identify compounds that bind to the same site as your radiolabeled ligand and, although it may be trite, to relate the concentration of the test compound to a clinically relevant one. Antibiotics, such as erythromycin, cause QT prolonging effects by blocking h-ERG channels but are often termed "false negatives" in such assays as binding because they are tested at concentrations below those therapeutically used.

4.7
Ion Flux

4.7.1
Fluorescent Indicators of Ion Flux

In addition to the above indirect sensors of ion movement (Section 4.5), fluorescent dyes are available that directly alter their spectral characteristics in response to changes in the local concentration of specific ions. These encompass both loaded dyes and genetically-encoded fluorophores. When the concentration of the ion in question increases, a binding event occurs between the dye and the ion, leading to an increase in fluorescence.

The most frequently used are the calcium sensitive dyes pioneered by Tsien [52]. Utilized not only for ion channels but for G-protein coupled receptors they have remained a workhorse of drug discovery as they are relatively easily to use and have an appropriate K_d for physiological levels of Ca^{2+}. There are a large range of Ca^{2+}-sensitive dyes available with a range of mechanisms of action and differing K_ds, e.g., fluo-3, fura-2, aequorin etc. [53]. Dyes sharing the spectral characteristics of fluo-3 have the advantage that robust plate reading equipment in the shape of the Fluorometric Imaging Plate Reader (FLIPR) is available in most pharmaceutical companies [54]. However, it must be remembered that these dyes report the change in bulk cytosolic Ca^{2+}, and that Ca_V channels can lead to significant changes in local sub-sarcolemmmal $[Ca^{2+}]$ that, due to the large concentration of Ca^{2+} buffering proteins in the cytoplasm, may not be translate into large changes in cytosolic compounds. This can lead to Ca^{2+} dyes being less sensitive than other techniques such as patch clamp or binding.

Dyes also exist for permeant ions of other voltage-gated channels – Na^+ (e.g., SBFI, CoroNa Green) and K^+ (PBFI), though they have received less attention than the Ca^{2+} dyes. One reason for this is that the K_Ds of these dyes tend to be affected by the concentration of other cations and thus can alter during the experiment.

4.7.2
Direct Measurement of Ion Flux

Much early work ion channel work was performed with radiolabeled ions (^{45}Ca, ^{22}Na or permeant surrogate ions ^{86}Rb, $[^{14}C]$-guanidium). The cells are loaded with the radiotracer and excess washed off. The compound of interest is pre-incubated, before a stimulation of ion flux by an ionic change as described for membrane potential-sensing dyes. The supernatant or cells (or both) can be isolated and the amount of radioactivity in each compartment measured relative to control. Due to the high selectivity of voltage-gated ion channels for their respective ions, the flux is specific for the ion involved, though this approach suffers the disadvantages of only reporting net flux of ion, and with little or no temporal information. Recently, environmental and safety implications of the use of high energy radioisotopes have lead to this technique fading from favor. However, by the use of atomic absorption spectroscopy it is possible to configure flux assays with non-radioactive isotopes [55].

Both fluorescent and direct measurement of ion flux have the same advantages of scale and ease of use as other plate-based assays and, in the case of radioactive ion flux, can be performed in laboratories without specialized electrophysiological equipment. However, they still suffer from the lack of electrical stimulation/voltage control as described above.

4.8
What Technologies Cannot be Used ... Yet?

In addition to the above descriptions of frequently used assays, it can also be instructive to briefly describe some of the other assay techniques that are not (yet) easily applicable to the primary screening of ion channels, mainly because they possess low throughput. Once primary and secondary (selectivity) screening has identified compounds of interest, then these assays may be of use in clarifying the mechanism of action of compounds.

As discussed above, manual patch clamp offers the highest amount of quality data, but with the lowest throughput. Another mechanism by which ion channels can be voltage clamped is by inserting channels into an artificial (black lipid) membrane. Even though the advantage of studying the channel in isolation (even from intracellular components) is gained, this technique is critically limited by the throughput, which is even lower than manual patch clamp.

Integrative tissue assays can also be amenable to electrophysiology and can be useful to study the effect of ligands on the channel of interest in the presence of a full complement of other native ion channels. The presence of other channels, and the fact that the target channel is not overexpressed in a cell line, has the consequence that the experiment is less removed from the physiological situation. The electrophysiological integrative tissue assays tend to be extracellular recording of electrical activity. The cardiac Purkinje fiber assay is recommended as an assay to be considered as part of an integrated risk assessment for the propensity of a novel pharmaceutical agents to cause cardiac arrhythmias due to block of the h-ERG channel [56]. This is by no means the only tissue that can be utilized as an integrative assay, but is discussed to illustrate some points for generic consideration. The presence of other channels in the tissue can complicate interpretation of the experiment. In the example of the Purkinje fiber, h-ERG blockers cause an increase in the length of the action potential. However, concomitant Na_V or Ca_V blocking activity can mask any h-ERG prolonging effects or produce complicated, multiphasic concentration–response curves [57]. When this assay is used as a safety assay to identify h-ERG blockers, it must be remembered that compounds that enhance Na_V or Ca_V currents will have the same functional effect of prolonging the duration of the action potential. Assays such as these are of great use in investigating the efficacy and safety of compounds identified by higher throughput assays. They can be used as a stepping stone between *in vitro* and *in vivo* experiments – the Purkinje fiber assay can be used between h-ERG patch clamp and *in vivo* dog haemodynamic studies. They can also help show whether a compound is likely to be active in the pathophysiological state if tissues from animal models of disease are used.

Crystal structures of soluble proteins are of use in rational drug design. Membrane bound proteins are of much greater technical difficulty to crystallize than enzymes. At the time of writing, the number of solved voltage-gated ion channel structures greatly outnumbers the number of GPCR structures. However, the few known GPCR crystal structures have a ligand bound, albeit the light sensitive pigment retinol as the crystal structures are those of rhodopsin. A relatively safe assumption has been made that the binding site for many endogenous ligands for class A GPCRs is the same as that of rhodopsin. The binding site for the vast majority of ion channel ligands is unknown with respect to the three-dimensional structure as they have not been crystallized in the presence of a pharmaceutically relevant ligand. However, drug binding sites have been implied by elegant biophysical experiments and site-directed mutagenesis. In a voltage-gated ion channel drug discovery environment the conclusions from these types experiments are fraught with difficulty due to the lack of knowledge of the exact relationship between primary sequence and tertiary protein structure of mammalian voltage-gated channels. The channels of known structure are of prokaryotic origin and rely upon a four-fold symmetry to refine the structure. The vast majority of voltage gated channels do not display *exact* fourfold symmetry – in heterotetrameric channels consisting of two differing subunits there is evidence for non-equivalence of the sub-units [58]. The large differences in structure between open, closed and inactivated states of channels, coupled with the present inability to crystallize mammalian channels and the low throughput of obtaining prokaryotic channel structures, means that crystal structures of voltage-gated ion channels are currently of limited use in rational drug design.

4.9
Summary

Voltage-gated ion channels are receiving much press in the drug-discovery literature due to the advances in screening technologies and the perception that they are an under-exploited, but well validated, target class having many marketed pharmaceuticals targeted to them. The cloning of many different subtypes of channels leads to optimism for more specific therapeutics though there remains the challenge to obtain selectivity between different channels. The field of voltage-gated ion channel screening is no different to others in that the data content given by the assays is roughly inversely proportional to the throughput. It is expected that throughput of informative, high quality assays will increase, as has happened with other screening methodologies, through improvements in current methodologies or future technical innovations. Currently, no technology fills all the requirements of being a high throughput, affordable, functional screening methodology in which the conformational state of the channel can be controlled. The throughput exists in techniques such as membrane potential sensing dyes though the electrical control is, critically, missing. New advances in planar patch may fill all the biological requirements but at a cost that is prohibitive to high throughput screening of millions of compounds. Thus, for voltage-gated ion channels, as with every other protein target, a robust, logical screening sequence is required that applies differing technologies at different stages of the discovery process.

References

1. Vilardaga, J.P., et al., Measurement of the millisecond activation switch of G protein-coupled receptors in living cells. *Nat. Biotechnol.*, **2003**. 21(7), 807–812.
2. Abbott, G.W., et al., MiRP1 forms IKr potassium channels with HERG and is associated with cardiac arrhythmia. *Cell*, **1999**. 97(2), 175–187.
3. Jones, E.M., et al., Cardiac IKr channels minimally comprise hERG 1a and 1b subunits. *J. Biol. Chem.*, **2004**. 279(43), 44 690–44 694.
4. Hamill, O.P., et al., Improved patch-clamp techniques for high-resolution current recording from cells and cell-free membrane patches. *Pflugers Archiv – Eur. J. Physiol.*, **1981**. 391(2), 85–100.
5. The Axon Guide: www.axon.com/ mr_Axon_Guide.html.
6. Cahalan, M., E. Neher, Patch clamp techniques: an overview. *Methods Enzymol.*, **1992**. 207, 3–14.
7. Bean, B.P., Two kinds of calcium channels in canine atrial cells. Differences in kinetics, selectivity, and pharmacology. *J. Gen. Physiol.*, **1985**. 86(1), 1–30.
8. Sanguinetti, M.C., N.K. Jurkiewicz, Two components of cardiac delayed rectifier K^+ current. Differential sensitivity to block by class III antiarrhythmic agents. *J. Gen. Physiol.*, **1990**. 96(1), 195–215.
9. Zhou, Z., et al., Properties of HERG channels stably expressed in HEK 293 cells studied at physiological temperature. *Biophys. J.*, **1998**. 74(1), 230–241.
10. Dorian, P., D. Newman, Rate dependence of the effect of antiarrhythmic drugs delaying cardiac repolarization: an overview. *Europace*, **2000**. 2(4), 277–285.
11. Hondeghem, L.M., B.G. Katzung, Anti-arrhythmic agents: the modulated receptor mechanism of action of sodium and calcium channel-blocking drugs. *Annu. Rev. Pharmacol. Toxicol.*, **1984**. 24, 387–423.
12. Bean, B.P., Nitrendipine block of cardiac calcium channels: high-affinity binding to the inactivated state. *Proc. Natl. Acad. Sci. U.S.A.*, **1984**. 81(20), 6388–6392.
13. Bean, B.P., C.J. Cohen, R.W. Tsien, Li-docaine block of cardiac sodium chan-

nels. *J. Gen. Physiol.*, **1983**. 81(5), 613–642.
14. Schmidt, C., M. Mayer, H. Vogel, A chip-based biosensor for the functional analysis of single ion channels. *Angew. Chem. Int. Ed.*, **2000**. 39(17), 3137–3140.
15. Sorota, S., et al., Characterization of a hERG screen using the IonWorks HT: comparison to a hERG rubidium efflux screen. *Assay Drug Dev. Technol.*, **2005**. 3(1), 47–57.
16. Schroeder, K., et al., Ionworks HT: a new high-throughput electrophysiology measurement platform. *J. Biomol. Screen*, **2003**. 8(1), 50–64.
17. Dubin, A.E., et al., Identifying modulators of hERG channel activity using the PatchXpress planar patch clamp. *J. Biomol. Screen*, **2005**. 10(2), 168–181.
18. Stefani, E., F. Bezanilla, Cut-open oocyte voltage-clamp technique. *Methods Enzymol.*, **1998**. 293, 300–318.
19. Hilgemann, D.W., C.C. Lu, Giant membrane patches: improvements and applications. *Methods Enzymol.*, **1998**. 293, 267–280.
20. Pehl, U., et al., Automated higher-throughput compound screening on ion channel targets based on the Xenopus laevis oocyte expression system. *Assay Drug Dev. Technol.*, **2004**. 2(5), 515–524.
21. Shieh, C.C., et al., Automated Parallel Oocyte Electrophysiology Test station (POETs): a screening platform for identification of ligand-gated ion channel modulators. *Assay Drug Dev. Technol.*, **2003**. 1(5), 655–663.
22. Nowak, M.W., et al., In vivo incorporation of unnatural amino acids into ion channels in Xenopus oocyte expression system. *Methods Enzymol.*, **1998**. 293, 504–529.
23. Hedin, K.E., N.F. Lim, D.E. Clapham, Cloning of a Xenopus laevis inwardly rectifying K+ channel subunit that permits GIRK1 expression of IKACh currents in oocytes. *Neuron*, **1996**. 16(2), 423–429.
24. Szekeres, P.G., Functional assays for identifying ligands at orphan G protein-coupled receptors. *Receptors Channels*, **2002**. 8(5–6), 297–308.

25. Haugland, R.P., M.T.Z. Spence, Probes for membrane potential, in *The Handbook: A Guide to Fluorescent Probes and Labeling Technologies*, ed. R.P. Haugland, M.T.Z. Spence. Invitrogen Corp, Carlsbad CA, **2005**.

26. Whiteaker, K.L., et al., Validation of FLIPR membrane potential dye for high throughput screening of potassium channel modulators. *J. Biomol. Screen*, **2001**. 6(5), 305–312.

27. Terstappen, G.C., et al., Pharmacological characterisation of the human small conductance calcium-activated potassium channel hSK3 reveals sensitivity to tricyclic antidepressants and antipsychotic phenothiazines. *Neuropharmacology*, **2001**. 40(6), 772–783.

28. Gonzalez, J.E., R.Y. Tsien, Improved indicators of cell membrane potential that use fluorescence resonance energy transfer. *Chem. Biol.*, **1997**. 4(4), 269–277.

29. Wolff, C., B. Fuks, P. Chatelain, Comparative study of membrane potential-sensitive fluorescent probes and their use in ion channel screening assays. *J. Biomol. Screen*, **2003**. 8(5), 533–543.

30. Guerrero, G., et al., Tuning FlaSh: redesign of the dynamics, voltage range, and color of the genetically encoded optical sensor of membrane potential. *Biophys. J.*, **2002**. 83(6), 3607–3618.

31. Ataka, K., V.A. Pieribone, A genetically targetable fluorescent probe of channel gating with rapid kinetics. *Biophys. J.*, **2002**. 82(1 Pt 1), 509–516.

32. Felix, J.P., et al., Functional assay of voltage-gated sodium channels using membrane potential-sensitive dyes. *Assay Drug Dev. Technol.*, **2004**. 2(3), 260–268.

33. Zhu, G., et al., Identification of endogenous outward currents in the human embryonic kidney (HEK 293) cell line. *J. Neurosci. Methods*, **1998**. 81(1–2), 73–83.

34. Xia, M., et al., *Assay Methods for State-dependent Calcium Channel Agonists/Antagonists*. WO2004/033647 **2004**.

35. Burnett, P., et al., Fluorescence imaging of electrically stimulated cells. *J. Biomol. Screen*, **2003**. 8(6), 660–667.

36. Middleton, R.E., et al., Two tarantula peptides inhibit activation of multiple sodium channels. *Biochemistry*, **2002**. 41(50), 14734–14747.

37. Hardy, M.E.L., et al., Drug-induced action potential prolongation in isolated cardiac myocytes recorded using a voltage-sensitive dye. *49th Biophysical Society Meeting*, Log Beach CA, **2005**.

38. Seethala, R., Receptor screens for small molecule agonist and antagonist discovery, in *Handbook of Drug Screening*, ed. R. Seethala, P.B. Fernandes. Marcel Dekker Inc. pp. 189–264. New York NY, **2001**.

39. Kenakin, T., Pharmacological assays used in screening for therapeutic ligands, in *Quantitative Molecular Pharmacology and Informatics in Drug Discovery*, ed. T. Lutz, T. Kenakin, John Wiley & Sons, Inc. pp. 97–134. Chichester UK, **1999**.

40. McKinney, M., Practical aspects of radioligand binding, in *Current Protocols in Pharmacology*, ed. S.J. Enna, et al., John Wiley & Sons, Inc. pp. 1.3.1–1.3.33. Chichester UK, **2002**.

41. Diaz, G.J., et al., The [3H]dofetilide binding assay is a predictive screening tool for hERG blockade and proarrhythmia: Comparison of intact cell and membrane preparations and effects of altering [K+]o. *J. Pharmacol. Toxicol. Methods*, **2004**. 50(3), 187–199.

42. Fermini, B., A.A. Fossa, The impact of drug-induced QT interval prolongation on drug discovery and development. *Nat. Rev. Drug Discov.*, **2003**. 2(6), 439–447.

43. Greengrass, P.M., M. Stewart, C.M. Wood, *Affinity-assay for the Human Erg Potassium Channel*. WO200302127 **2003**.

44. Lee, H.R., W.R. Roeske, H.I. Yamamura, High affinity specific [3H](+)PN 200-110 binding to dihydropyridine receptors associated with calcium channels in rat cerebral cortex and heart. *Life Sci.*, **1984**. 35(7), 721–732.

45. Schoemaker, H., S.Z. Langer, [3H]diltiazem binding to calcium channel antagonists recognition sites in rat cerebral cortex. *Eur. J. Pharmacol.*, **1985**. 111(2), 273–277.

46. Reynolds, I.J., A.M. Snowman, S.H. Snyder, (-)-[3H]desmethoxyverapamil labels multiple calcium channel modulator receptors in brain and skeletal muscle membranes: differentiation by temperature and dihydropyridines. *J. Pharmacol. Exp. Ther.*, **1986**. 237(3), 731–738.

47. Felix, J.P., et al., Identification and biochemical characterization of a novel nortriterpene inhibitor of the human lymphocyte voltage-gated potassium channel, Kv1.3. *Biochemistry,* **1999**. 38(16), 4922–4930.

48. Brown, G.B., 3H-Batrachotoxinin-A benzoate binding to voltage-sensitive sodium channels: inhibition by the channel blockers tetrodotoxin and saxitoxin. *J. Neurosci.,* **1986**. 6(7), 2064–2070.

49. Hopkins, A.L., C.R. Groom, The druggable genome. *Nat. Rev. Drug Discov.,* **2002**. 1(9), 727–730.

50. Hanner, M., et al., The beta subunit of the high conductance calcium-activated potassium channel. Identification of residues involved in charybdotoxin binding. *J. Biol. Chem.,* **1998**. 273(26), 16 289–16 296.

51. Gee, N.S., et al., The novel anticonvulsant drug, gabapentin (Neurontin), binds to the alpha2delta subunit of a calcium channel. *J. Biol. Chem.,* **1996**. 271(10), 5768–5776.

52. Tsien, R.Y., A non-disruptive technique for loading calcium buffers and indicators into cells. *Nature,* **1981**. 290(5806), 527–528.

53. Haugland, R.P., M.T.Z. Spence, Indicators for Ca^{2+}, Mg^{2+}, Zn^{2+} and other metal ions, in *The Handbook: A Guide to Fluorescent Probes and Labeling Technologies*, ed. R.P. Haugland, M.T.Z. Spence. Invitrogen Corp, Carlsbad CA, **2005**.

54. Schroeder, K., B. Neagle, FLIPR: A new instrument for accurate, high throughput optical screening. *J. Biomol. Screen,* **1996**. 1(2), 75–80.

55. Terstappen, G.C., Functional analysis of native and recombinant ion channels using a high-capacity nonradioactive rubidium efflux assay. *Anal. Biochem.,* **1999**. 272(2), 149–155.

56. *The Non-Clinical Evaluation of the Potential for Delayed Ventricular Repolarization (QT interval) by Human Pharmaceuticals S7B.* **2005**: www.ich.org.

57. Martin, R.L., et al., The utility of hERG and repolarization assays in evaluating delayed cardiac repolarization: influence of multi-channel block. *J. Cardiovasc. Pharmacol.,* **2004**. 43(3), 369–379.

58. Silverman, S.K., H.A. Lester, D.A. Dougherty, Asymmetrical contributions of subunit pore regions to ion selectivity in an inward rectifier K^+ channel. *Biophys. J.,* **1998**. 75(3), 1330–1339.

5
Calcium Channels

5.1
Overview of Voltage-gated Calcium Channels
Clinton Doering and Gerald Zamponi

5.1.1
Introduction

Influx of calcium ions into cells mediates numerous intracellular events. For example, calcium is required for contraction of skeletal muscle and for excitation–contraction coupling in the heart; influx of calcium at nerve terminals in response to action potentials allows vesicle fusion and release of neurotransmitters; calcium signaling is required in non-excitable tissues to allow for release of hormones and immune responses; sensory systems rely on calcium flux to detect stimuli; gene transcription can be mediated via calcium influx; cell death can be induced through calcium dependent cascades. Calcium ions also activate various enzymes such as protein kinases and signaling cascades, further underscoring the role of calcium ions as intracellular messenger molecules. Precise control of calcium influx is therefore critical, and, as such, cytosolic calcium entry pathways are targets for numerous pharmacological agents. Here, we overview voltage-gated calcium channels (VGCCs), a principal route through which calcium ions enter the cytosol.

5.1.2
Native and Cloned Calcium Channels: Nomenclature and Classification

Initial identification of voltage-dependent calcium entry into muscle cells by Fatt and Katz [1] and later Fatt and Ginsborg [2] led to the idea that, like sodium and potassium, calcium entry into cells occurs via proteinaceous pores selective for calcium ions. In the ensuing years the advent of patch-clamp techniques allowed for the identification of voltage-dependent calcium fluxes in numerous tissues, including all excitable cells and non-excitable cells responsible for hormone release and immune responses (reviewed in Ref. [3]).

Initially, currents carried by VGCCs were classified based on their biophysical and pharmacological properties. Low voltage activated (LVA) currents activated

Voltage-Gated Ion Channels as Drug Targets. Edited by D. Triggle
Copyright © 2006 WILEY-VCH Verlag GmbH & Co. KGaA, Weinheim
ISBN 3-527-31258-7

near resting membrane potentials while high voltage activated (HVA) currents required larger membrane depolarization and activate at more positive potentials [4, 5]. Within the HVA class, individual calcium currents could be isolated based on their sensitivities to various pharmacological agents, such as dihydropyridines, as well as toxins isolated from fish-hunting cone snails and spider venoms. Molecular biological and biochemical techniques led to the isolation and subsequent cloning of the first VGCC in the mid 1980s [6]. Today, ten different VGCC genes have been identified and cloned, and the currents recorded from these channels expressed in heterologous expression systems correspond to calcium currents recorded from native tissues.

Currently, VGCCs are classified into three families, Ca_v1, Ca_v2, and Ca_v3, with Ca_v1 and Ca_v2 families corresponding to HVA calcium channels and Ca_v3 to LVA calcium channels (summarized in Table 5.1.1; reviewed in [7]). Four family members comprise the Ca_v1 family, $Ca_v1.1$ (formerly α_{1S}), $Ca_v1.2$ (formerly α_{1C}), $Ca_v1.3$ (formerly α_{1D}) and $Ca_v1.4$ (formerly α_{1F}). The Ca_v1 family has also been defined biophysically as L-type because of the long-lasting and large amplitude single channel opening events (typically, single channel conductances are on the order of >20 pS) and pharmacologically because of their high sensitivities to certain dihydropyridines (reviewed in [8]; also, see Chapter 5.3 in this book). Ca_v1 channels are also sensitive to phenylalkylamines and benzothiazepines. Three family members comprise the Ca_v2 family, $Ca_v2.1$ (formerly α_{1A}; P/Q-type; sensitive to ω-agatoxin IVA), $Ca_v2.2$ (formerly α_{1B}; N-type; sensitive to ω-conotoxin GVIA), and $Ca_v2.3$ (formerly α_{1E}; R-type; sensitive to SNX-482). Ca_v2 family members typically have single channel conductances on the order of 10–20 pS and opening events considerably briefer than L-type channels, with the exception of P-type channels that display very long lasting openings. Finally, three members comprise the Ca_v3 family, $Ca_v3.1$ (formerly α_{1G}), $Ca_v3.2$ (formerly α_{1H}), and $Ca_v3.3$ (formerly α_{1I}). The Ca_v3 family has also been defined biophysically as T-type because of the tiny and transient single channel opening events, with very brief openings and single channel conductances typically < 10 pS. T-type channels are potently inhibited by nickel ions and show some sensitivity to dihydropyridines but, to date, no truly selective Ca_v3 inhibitor is available.

5.1.3
Distribution of VGCCs and their Physiological Roles

$Ca_v1.1$ is the only VGCC identified to date that is not found in the nervous system. This channel is heavily abundant in skeletal muscle [6], although recent evidence suggests this channel may also be found in immune cells [9]. $Ca_v1.1$ channels serve as the voltage sensor for excitation–contraction (EC) coupling in skeletal muscle cells, likely through an association with sarcoplasmic reticulum ryanodine receptors that release calcium from internal stores to initiate contraction. Yet, it is now well established that calcium influx through $Ca_v1.1$ is itself not required for EC coupling. In contrast, calcium influx through $Ca_v1.2$ channels is an essential step during EC coupling in cardiac muscle. This calcium channel is also

Tab. 5.1.1 Summary of VGCC and ancillary subunit nomenclature, and pharmacology.

		Gene name	Protein name	Former name	Family	Pharmacology
VGCC α_1 subunit	HVA	CACNA1S	Ca$_v$1.1	α_{1S}	L-type	Dihydropyridines Phenylalkylamines Benzothiazepines
		CACNA1C	Ca$_v$1.2	α_{1C}		
		CACNA1D	Ca$_v$1.3	α_{1D}		
		CACNA1F	Ca$_v$1.4	α_{1F}		
		CACNA1A	Ca$_v$2.1	α_{1A}	P/Q-type	ω-aga IVA
		CACNA1B	Ca$_v$2.2	α_{1B}	N-type	ω-Conotoxins GVIA and MVIIA
		CACNA1E	Ca$_v$2.3	α_{1E}	R-type	SNX-482
	LVA	CACNA1G	Ca$_v$3.1	α_{1G}	T-type	Kurtoxin[a] Dihydropyridines[a] Ni^{2+}[a]
		CACNA1H	Ca$_v$3.2	α_{1H}		
		CACNA1I	Ca$_v$3.3	α_{1I}		
VGCC α_2 subunit		CACNA2D1	α_2-δ_1			Gabapentin? Pregabalin?
		CACNA2D2	α_2-δ_2			
		CACNA2D3	α_2-δ_3			
		CACNA2D4	α_2-δ_4			
VGCC β subunit		CACNB1	β_1			
		CACNB2	β_2			
		CACNB3	β_3			
		CACNB4	β_4			
VGCC γ subunit		CACNG1	γ_1			
		CACNG2	γ_2			
		CACNG3	γ_3			
		CACNG4	γ_4			
		CACNG5	γ_5			
		CACNG6	γ_6			
		CACNG7	γ_7			
		CACNG8	γ_8			

[a] Not selective.

found in various other tissues and cells, including lung, brain, retina, and immune cells [9, 10, reviewed in 11]. Antibody staining reveals that in nervous tissue $Ca_v1.2$ channels are concentrated near the soma, although they are also found at lower levels in axons and dendrites [12]. Influx of Ca^{2+} through this channel has also been shown to mediate gene transcription via a signal transduction cascade involving the mitogen-activated protein (MAP) kinase pathway in a process termed excitation–transcription coupling [13, 14]. $Ca_v1.3$ channels are found in the retina, where they may be important in cone-mediated visual transduction pathways [15, 16]. They are also essential in mechanosensation pathways, especially in cochlear hair cells that mediate the processing of sound into electrical signals in the ear [12, 17–20]. Finally, $Ca_v1.3$ channel activity is an essential mediator of secretion from islet cells and other endocrine tissues. $Ca_v1.4$ channels were originally thought to be confined to retinal rod cells and important for low light rod-mediated visual transduction [21, 22], although additional evidence points to a wider distribution, including thymus, spleen, spinal cord, skeletal muscle, and non-excitable immune cells [9, 23–25]. We are only now beginning to understand the function of these channels outside the visual transduction pathway.

Ca_v2 VGCCs are distributed throughout the nervous system and are thought to be important in synaptic transmission because they tend to be clustered in presynaptic terminals, and their inhibition via pharmacological means suppresses neurotransmitter release (reviewed in [26, 27]). $Ca_v2.1$ was originally identified in Purkinje cells and cerebellar granule cells [28, 29]. Interestingly, alternative splicing leads to two isoforms with vastly different biophysical and pharmacological properties: P-types, which display slow inactivation kinetics and are highly sensitive to ω-agatoxin IVA, and Q-types which are faster inactivating and less sensitive to ω-agatoxin IVA [30, 31]. It is not clear whether these splice variants have distinct roles in neurotransmitter release. $Ca_v2.2$ is widely distributed throughout the nervous system and is involved in neurotransmission pathways responsible for pain, especially in dorsal root ganglion cells. Classically these channels are defined by their sensitivities to ω-conotoxins GVIA and MVIIA (now approved as a therapeutic agent, PRIALT, for treatment of chronic pain) [32–35], although more recently selective blockers such as ω-conotoxin CVID have been found (reviewed in [8, 36]). $Ca_v2.3$ channels, originally identified as residual current resistant to known HVA calcium channel blockers (dihydropyridines, conotoxins, agatoxins), are distributed in proximal dendrites, presynaptic termini, and cardiac tissue [37–40]. They are sensitive to a tarantula venom fraction known as SNX-482, but their precise physiological function remains enigmatic.

Ca_v3 channels activate near resting membrane potentials [e.g., 4, 41–46]. Interestingly, although a high percentage of channels are inactivated near resting membrane potentials these channels exhibit a window current allowing for calcium influx in response to only minor membrane depolarizations [47, 48]. They are found in numerous tissues, including brain, cardiac and skeletal muscle, and endocrine cells, and they may play a key role in establishing oscillating firing patterns and rebound bursts [e.g., 49–52]. No selective blockers of T-type channels have been reported, but Ca_v3 currents can be blocked by certain dihydropyridines, Ni^{2+}, and kurtoxin, a peptide toxin isolated from a scorpion species [53, 54].

Taken together, it is well established that there are multiple types of voltage-gated calcium channels with unique cellular and subcellular distributions and which mediate specific physiological functions.

5.1.4
Structure of VGCC α_1 Subunits

VGCCs are heteromultimers composed of a pore-forming α_1 subunit that defines the channel subtype, plus ancillary α_2-δ, β, and γ subunits. The α_1 subunits are structurally homologous to voltage-gated sodium and potassium channels (reviewed in [27, 55]). Unlike potassium channels, however, both sodium and calcium channels contain four domains encoded by a single gene. Each domain consists of six transmembrane helices, with intracellular N- and C-termini (Fig. 5.1.1). Within each domain, the fourth membrane-spanning alpha helix (S4 segment) contains positively charged arginine and lysine residues spaced every three to four amino acids. Based on mutagenesis and crystallization studies this region

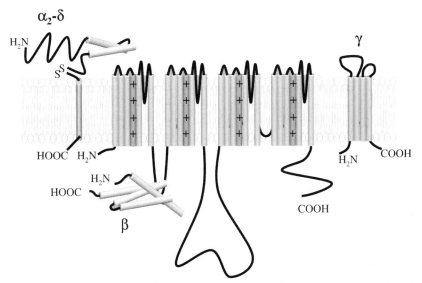

Fig. 5.1.1 Schematic diagram of VGCC α_1 subunits and ancillary subunits. The pore-forming α_1 subunit, encoded by a single gene, consists of four domains, each consisting of six transmembrane helices. The fourth helix in each domain contains positively charged residues spaced every three to four amino acids and is thought to form the voltage sensor. The re-entrant loop between the S5 and S6 trans- membrane helices in each domain is thought to line the channel pore, and each contains a critical glutamic acid residue responsible for selectivity of divalent ions over monovalent ions. Also shown are α_2-δ and γ subunits. The beta interacting domain interacts with 30 residues in the channel I-II intracellular linker alpha interacting domain.

is thought to form the voltage-sensor that translocates within the membrane in response to voltage changes [56, 57]. As a result of this translocation, the reentrant p-loops between S5 and S6 segments, believed to line the pore of the channel, change conformation and allow for opening and closing of the channel. Unlike potassium channels, whose p-loops carry the highly conserved valine-glycine-tyrosine-glycine selectivity filter signature sequence [56], VGCC S5-S6 linkers each contain a critical glutamic acid residue (four in total) that forms a negative ring of charge and allows for selectivity of divalent cations over monovalent ions [58–62]. However, other amino acid residues located within the outer vestibule of the channels also appear to contribute to the permeation characteristics of VGCCs [63, reviewed in 64].

Upon sustained depolarization, both LVA and HVA calcium channels enter an inactivated state, with differing rates for different channel subtypes. Voltage-dependent inactivation occurs through at least two different mechanisms, while calcium-dependent inactivation (CDI) occurs through a conserved mechanism. Fast voltage-dependent inactivation (millisecond time scale) has been proposed to occur via a hinged-lid mechanism involving the I-II linker as a putative gating particle, although other regions of the channel, in particular the S6 regions, appear to be important (reviewed in [65, 66]); a slower voltage-dependent inactivation (minute time scale) may occur via complicated conformational changes but these are poorly understood. Influx of calcium ions through VGCCs can also lead to CDI, a phenomenon that is lost when currents are recorded in the presence of Ba^{2+} ions as the charge carrier. Previously thought to be a selective characteristic of Ca_v1 channels, more recent evidence suggests that CDI is also a property of Ca_v2 family members. This process occurs via a conserved mechanism involving interaction of the C-terminus of the VGCC with Ca^{2+}-sensing protein calmodulin [67]. Interestingly, $Ca_v1.4$ appears to show the least sensitivity to CDI of all Ca_v1 and Ca_v2 family members in heterologous expression systems [24, 68, 69] despite containing many of the structural features associated with CDI in other channel subtypes. $Ca_v1.4$ channels are likely involved in transmitter release at ribbon synapses of the retina and allow for tonic calcium influx under resting (dark) conditions, thus requiring these channels to maintain a persistent level of activity. It therefore appears that CDI is important for channels where calcium influx is of a transient nature, and less important when calcium signaling needs to be sustained for prolonged periods ($Ca_v1.4$ in the retina). Section 5.1.5.4 discusses the mechanism of CDI in more detail.

5.1.4.1 Ancillary VGCC Subunits

While the gene encoding the calcium channel α_1 subunit is sufficient to form a functional pore allowing for calcium influx, ancillary proteins can interact with the channel to alter channel kinetics (both activation and inactivation), pharmacology, voltage-sensitivity, and targeting to the plasma membrane [70]. Biochemical studies have shown that when HVA channels are isolated from native tissues, four separate proteins with distinct molecular weights can be identified: an

α_1 subunit of $\sim$175–225 kDa, an α_2-δ subunit of $\sim$170 kDa, a β-subunit of $\sim$50 kDa, and a γ subunit of $\sim$30 kDa (reviewed in [27, 71–75]). These ancillary subunits are believed to associate primarily with HVA channels, although they may also functionally interact with LVA channels [76, 77]. As may be expected, these subunits can be detected in the same tissues that functionally express VGCC α_1 subunits.

Four genes (CACNA2Dx) have currently been identified that code for different α_2-δ proteins (α_2-δ_1, α_2-δ_2, α_2-δ_3, α_2-δ_4), with several potential splice variants (reviewed in [78]). The proteins encoded by these genes undergo post-translational modifications whereby the peptide is proteolytically cleaved into two fragments joined via disulfide bonds to yield a membrane-spanning δ peptide (27 kDa) and extracellular α_2 peptide (143 kDa) [79]. As a result of its extracellular domains, the protein can be heavily glycosylated. Expression studies have shown that the α_2-δ protein alters current kinetics and current densities when coexpressed with HVA α_1 subunits, although to different degrees with different channels [80; reviewed in 71]. This family of subunits is important pharmacologically because drugs such as gabapentin and pregabalin may exert their action on calcium channels by binding directly to α_2-δ [81; reviewed in 8].

Four genes encoding β-subunits (CACNBx) have been identified (β_1, β_2, β_3, β_4), along with various splice variants (reviewed in [73, 74]). Unlike α_2-δ proteins, β-subunits appear to be exclusively cytosolic in nature, with the exception of β_{2a} whose N-terminus contains a pair of cysteine residues that can be palmitoylated, therefore resulting in membrane insertion [82, 83]. The β subunit core resembles membrane-associated guanylate kinase homologs with conserved interacting SH3 and guanylate kinase (GK) domains [84]. Residues in the GK domain participate in the formation of a deep hydrophobic groove for high-affinity binding of high voltage-activated calcium channels [85, 86]. The recently published crystal structure of the β subunit–VGCC I-II linker shows that binding of the β subunit to the channel is critically dependent on a functional association of the SH3 and GK regions [86, 87]. This association requires a short sequence stretch (termed the beta interaction domain, or BID, approximately 30 amino acids long) that aids the stability of the SH3-GK complex [86]. The β-subunit associates with the α interaction domain (AID, $\sim$18 highly conserved amino acids located on the I-II linker of the α_1 subunit) to possibly mask an endoplasmic reticulum retention sequence in the I-II linker, thereby promoting targeting of the channel to the plasma membrane [88–90]. Additionally, coexpression of the β subunit dramatically alters channel kinetics, especially inactivation properties (e.g., β_2 dramatically slows inactivation, while β_1 and β_3 speed inactivation) with different channels affected to varying degrees by the various β isoforms [80, 83 , reviewed in 72].

Eight genes encoding skeletal muscle and neuronal γ subunits (CACNGx) have been identified (γ_1, γ_2, γ_3, γ_4, γ_5, γ_6, γ_7, γ_8) (reviewed in [72]). While it has been observed that γ subunits alter channel kinetics for both HVA and LVA channels (but not channel densities) (reviewed in [75]), systematic studies are lacking. The γ subunits consist of four transmembrane domains with intracellular N- and C-termini. γ Subunits are also known to interact with other synaptic proteins, such as AMPA receptors [91], and therefore they may exert effects beyond VGCCs and serve to coordinate signaling between VGCCs and other proteins.

5.1.5
VGCC Modulation

In addition to modulation by ancillary subunits, numerous other proteins alter VGCC currents both *in vivo* and *in vitro*. The literature on calcium channel modulation is vast, and here we briefly discuss only some of the major regulatory pathways. For a more detailed overview of calcium channel modulation, refer to Bannister et al. [92], Nirdosh and Zamponi [93], and Lee and Catterall [94]. Figure 5.1.2 illustrates some of the key interaction sites where modulatory proteins interact with VGCCs.

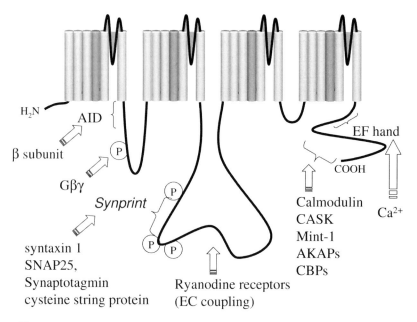

P Phosphorylation sites

Fig. 5.1.2 Schematic of VGCC, showing important protein interaction sites involved in channel modulation. Intracellular linkers contain numerous sites for phosphorylation events, some of which are shown. Phosphorylation of ancillary subunits also affects channel activity (not shown for simplicity, see text for more detail). The I-II linker contains the AID domain that binds the ancillary β subunits. Gβγ binds to the I-II linker to inhibit current in some channel subtypes, but phosphorylation by PKC can antagonize this interaction. The II-III linker interacts with numerous proteins. The *synprint* motif of the linker interacts with various proteins involved in vesicle fusion such as syntaxin-1. The C-terminal portion of the channel also interacts with numerous proteins, such as calmodulin, calcium binding proteins, A kinase anchoring proteins, Mint-1, and CASK1.

5.1.5.1 Kinases and Phosphatases

Phosphorylation by protein kinases (PKs) of serine and threonine residues can alter the biophysical characteristics of VGCCs in a calcium channel subtype dependent manner. PKs can be activated via numerous signaling cascades that are beyond the scope of the present discussion. Needless to say, most experiments are based on using pharmacological agents that activate certain cascades, but, notably, often these drugs are nonspecific and activate a plethora of other enzymes, thus clouding the interpretation of experimental results. Additionally, many of the pathways described below are highly integrated with one another, allowing for precise control of calcium influx through all VGCC subtypes even within the same cell.

Most information regarding the role of cyclic AMP dependent protein kinase (PKA) in regulating VGCC function is based upon experiments involving skeletal muscle ($Ca_v1.1$) and cardiac ($Ca_v1.2$) VGCCs. Numerous experiments have shown that PKA phosphorylation of these channels can alter channel gating kinetics [e.g., see 95], targeting [96], and current size [97, 98]. In muscle tissue, activation of PKA signaling cascades dramatically upregulates Ca_v1 channel function due to phosphorylation of either the α_1 [99, 100, reviewed in 101] or ancillary β [102] and γ [103] subunits. Therefore, precise control of PKA activity is essential for normal physiology of cardiac and skeletal muscle. As outlined below, PKA kinase anchoring proteins (AKAPs) are required for appropriate PKA action on L-type channels, which is why robust PKA modulation of transiently expressed channels can only be observed upon coexpression of AKAP18 [104]. While the effects of PKA on Ca_v1 function have been widely studied, less is known about its effect on other channel subtypes. Ca_v2 channels are phosphorylated by PKA in response to endogenous hippocampal electrical activity [105], which may alter the sensitivity of these channels to phosphatidylinositol 4,5-bisphosphate, a compound that helps to stabilize channel activity *in vivo* [106].

Activation of protein kinase C (PKC) can result in an increase in current through $Ca_v2.2$ channels [107, 108]. These channels are strongly susceptible to inhibition by G proteins (discussed below), and phosphorylation of residues contained within the $G\beta\gamma$ binding motif in the I-II linker region of the channel antagonizes this inhibition [109]. PKC also phosphorylates the domain II-III linker region of $Ca_v2.2$ channels, which counteracts the inhibitory effects of syntaxin 1A (discussed below) on channel activity, thus resulting in overall enhancement of channel activity [110]. Hence, $Ca_v2.2$ channels are generally enhanced in their activities upon PKC dependent phosphorylation. $Ca_v2.3$ channels are also upregulated by PKC following muscarinic or phorbol ester stimulation [108, 111, 112]. Interestingly, Ca_v1 and some Ca_v3 channels can be inhibited by PKC activation [113, 114] while some $Ca_v3.2$ channels are enhanced [115], possibly due to phosphorylation at different sites. These results suggest that PKC can selectively upregulate certain channel subtypes over others, even within the same cell.

Calmodulin-dependent protein kinases (CaM kinases) enhance Ca_v3 currents in endocrine cells by causing a hyperpolarizing shift in the activation profile of these channels [116–118]. Importantly, not all Ca_v3 family members are equally affected by CaM kinases; $Ca_v3.2$ is highly sensitive while $Ca_v3.1$ is not [119]. CaM kinase also phosphorylates the syntaxin 1A binding site on the $Ca_v2.2$ channel, leading to a loss of syntaxin binding in biochemical experiments [120]. The functional significance of these effects has yet to be resolved.

Tyrosine kinases affect multiple VGCC subtypes (reviewed in [121]). For example, there is a net downregulation of $Ca_v1.2$ channel activity in the heart when tyrosine kinases are activated that lead to α-adrenergic inhibition of $Ca_v1.2$-enhancing β-adrenergic responses [122]. $Ca_v1.3$ channel activity is enhanced when tyrosine kinases are activated in brain and retinal pigment epithelial cells [123]. $Ca_v2.1$ channels are inhibited by tyrosine kinases in adrenal chromaffin cells [124], and $Ca_v2.2$ channels in dorsal root ganglion neurons are inhibited via $GABA_B$ receptors through a tyrosine kinase pathway [125].

As may be expected, given that VGCC activity is sensitive to phosphorylation by various kinases, channel activity is also sensitive to dephosphorylation by protein phosphatases [reviewed by 126]. For example, calcineurin (protein phosphatase 2B) has been shown to inhibit $Ca_v1.2$ currents [127] and modulates $Ca_v2.2$ currents by selectively regulating pertussis toxin-sensitive G-protein-coupled receptor-mediated inhibition of the channel [128]. Thus, phosphatases help refine calcium signaling *in vivo* by helping to balance phosphorylation levels and contribute to the integration of various signaling pathways with one another. For example, activation of $Gα_s$ through β-adrenergic receptors leads to increases in cAMP, which then activates PKA, leading to phosphorylation of HVA calcium channels and subsequently facilitating current through these channels [97]. In contrast, activation of dopamine receptors results in activation of protein phosphatase 1 (again, via a PKA cascade), which leads to dephosphorylation of HVA channels and inhibits current [98]. Therefore, it is difficult to make generalizations about how individual signaling cascades affect VGCCs as they should never be viewed in isolation. Most importantly, however, given that some channel subtypes can be upregulated while others are downregulated, this allows for specific regulation of multiple calcium signals within the same cell.

5.1.5.2 Gβγ Proteins

Besides triggering second messenger cascades such as kinase activity, the activation of G protein coupled receptors results in direct regulation of VGCC activity via G-protein subunits, a key mechanism for regulating synaptic calcium entry and thus neurotransmitter release [129, reviewed in 130, 93]. Gβγ subunits directly interact with the I-II linker of $Ca_v2.1$ and $Ca_v2.2$ calcium channels, and cause voltage-dependent, membrane-delimited inhibition of current through these channels [109, reviewed in 27, 130]. The Gβγ binding site overlaps with the AID region of the channel I-II linker, and therefore Gβγ effects can also be modulated by the presence of different calcium channel β subunits [74, 83]. The N- and C-termini of the channel have been shown to affect the voltage-dependence of G-protein inhibition. Additional complexity arises from the fact that the five known Gβ proteins inhibit calcium currents to different degrees [83], and

PKC-dependent phosphorylation of the N-type channel I-II linker selectively an-tagonizes $G\beta_1$-mediated responses [108]. As a result, integration of multiple mod-ulatory pathways (e.g., kinases, phosphatases, G-proteins) is highly complex and not fully understood.

5.1.5.3 Synaptic Proteins

Presynaptic proteins thought to be important in membrane fusion have also been shown to modulate current through VGCCs (reviewed in [131]). For example, syn-taxin 1 interacts with the II-III linker of $Ca_v2.1$ and $Ca_v2.2$ channels at a consen-sus synaptic protein interaction (*synprint*) motif and causes current inhibition by triggering hyperpolarizing shifts in half-inactivation potential [132–135, reviewed in 27]. Additionally, SNAP-25, synaptotagmin 1, and cysteine string protein inter-act with *synprint* and alter calcium influx [136–138]. These interactions also affect $G\beta\gamma$ regulation of the channel and are sensitive to PKC-dependent phosphoryla-tion [110, 120], providing yet another level of complexity and a feedback mechan-ism that allows for fine tuning and precise control of calcium influx. Additionally, adaptor proteins such as CASK and Mint-1 interact with the C-terminus of Ca_v2 channels to affect targeting of the channels to synaptic terminals and colocalize them with the synaptic release machinery [139, 140].

5.1.5.4 Calmodulin, AKAPs, CBPs, RyRs

As previously mentioned, the calcium binding protein calmodulin interacts with the C-terminus of HVA calcium channels and allows for CDI (although, to differ-ing degrees between various HVA members). Calmodulin appears to bind up to four calcium ions, with two high affinity and two low affinity sites [67]. When the high affinity sites are occupied, Ca_v1 channels display CDI, whereas low affinity sites need to be occupied for Ca_v2 channels to show CDI. As a result, Ca_v1 chan-nels respond to small localized calcium rises induced by calcium entry through each individual channel, whereas Ca_v2 family members respond only to larger, global rises in intracellular calcium concentration [141]. It is unknown if calmo-dulin can interact with Ca_v3 channels in a similar mechanism. The VGCC C-ter-minus also contains an EH hand motif, which may allow for direct calcium bind-ing to the calcium channel and subsequent modulation [67, 142].

AKAPs are important modulators of VGCCs affected by PKA activity because they act as scaffolds to colocalize PKA with the channel (reviewed in [143]). For example, AKAP18 enhances PKA phosphorylation of $Ca_v1.2$ channels when coex-pressed, thereby increasing L-type currents in these cells [104] as previously dis-cussed. Interestingly, however, AKAPs can by themselves affect channel activity by enhancing channel expression levels in the plasma membrane. For example, AKAP79 interacts with $Ca_v1.2$ and targets the channel to the membrane, thereby promoting surface expression, but in a PKA-independent manner [96]. Thus, AKAPs add another level of complexity to modulation of VGCCs by PKA.

Finally, several other proteins are known to interact with VGCCs and modulate current activity. For example, calcium binding proteins (CBPs) interact with the C-terminus of calcium channels, modulating current activity through the channels. CBP1 interacts with $Ca_v2.1$ channels and causes a depolarizing shift in activation profile and speeds channel inactivation [144], while CBP4 has been shown to bind to $Ca_v1.4$ and to hyperpolarize the activation profile of these channels [145]. Lastly, the Ca_v1 II-III linker is critical for the interaction between the channel at the membrane and the ryanodine receptors in sarcoplasmic reticulum, allowing for EC coupling in skeletal and cardiac muscle [e.g., see 146, 147]. Recall that influx of ions through Ca_v1 channels is not required for skeletal muscle EC coupling, but it is required for cardiac EC coupling.

5.1.6
VGCCs: Channelopathies and Pathologies

Precise control of calcium influx through VGCCs is critical for proper cellular responses and normal physiology. Alterations of calcium signaling, either by altering the calcium itself or by altering interacting proteins that modulate channel activity can perturb this delicate balance. To date, numerous calcium channel related pathophysiological conditions have been identified and correlated to alterations in VGCC function, some of which are discussed below. Mutations in calcium channels can either lead to reduced function, complete loss of function (no calcium influx), or gain of function (increased calcium influx), and, hence, the phenotypes can also show varying degrees of severity depending on the mutation.

Ca_v1 family members are distributed in both excitable and nonexcitable tissues, and therefore, not surprisingly, a wide array of conditions have been correlated with these channels [reviewed in 11]. For example, mutations in $Ca_v1.1$, the major calcium channel subtype in skeletal muscle, can lead to hypokalemic periodic paralysis [148, 149], and knockout mice [150] die from asphyxiation because the diaphragm fails to contract. Knockout mice for $Ca_v1.2$ die at day E14.5 [151], as this channel is critical for EC coupling in the heart. Given the importance of $Ca_v1.3$ in mechanosensation in hair cells, knockout mice for this gene are deaf and also show cardiac arrhythmias [20]. In humans, mutations in $Ca_v1.4$ are implicated in incomplete X-linked congenital stationary night blindness (CSNB2) [21, 22, 23] (as previously mentioned, this channel is also found in nonexcitable cells such as plasma and mast cells [24], but it is not known what effect mutations in $Ca_v1.4$ have with respect to these cells in CSNB2 patients).

Ca_v2 family members are predominantly presynaptic and mutations in these channels lead to altered synaptic physiology. For example, mutations in $Ca_v2.1$ have been associated with episodic ataxia type 2, spinocerebellar ataxia type 6, and familial hemiplegic migraine [152–154], while knockout mice die a few weeks after birth [155]. Knockout of $Ca_v2.2$ in mice leads to decreased pain transmission, indicating that this channel is a suitable therapeutic target for analgesics [156–158]. Knockout mice for $Ca_v2.3$ also show altered sensitivities to pain [159,

160]. Interestingly, polycystic kidney disease protein 2, which when mutated leads to polycystic kidney disease, shows high sequence homology in its six transmembrane and C-terminal domains to the corresponding $Ca_v2.3$ regions, although any link between the two proteins remains unknown [161].

Ca_v3 VGCCs are activated near resting membrane potentials. Mutations in these channels may affect the amount of calcium influx during minor membrane potential fluctuations, and can have profound consequences. Knockout of $Ca_v3.1$ in mice leads to reduced spike and wave discharges and hence resistance to epileptic seizures; therefore, this channel may be a promising therapeutic target in the treatment of epilepsy [162]. Mutations in $Ca_v3.2$ have been correlated with childhood absence and idiopathic generalized epilepsies [163–165], while knockout mice also show impaired vascular relaxation [166] and knockdown studies show mice with reduced sensitivities to pain [167]. To date, no published information is available for $Ca_v3.3$ knockouts or mutations.

Finally, several mutations in ancillary calcium channel subunits have been identified in mice that lead to significant phenotypic alterations. For example, mutations in α_2-δ_2 give rise to the *ducky* mouse, a phenotype with severe ataxia and paroxysmal dyskinesia [168, 169]. Loss of β_1 leads to perinatal death because of a loss of skeletal EC coupling [170, 171], while β_2 loss is embryonic lethal because this subunit is required for cardiac EC coupling [172]. Insertion of four nucleotides in the β_4 gene near a splice donor site produces a truncated protein that cannot bind to the α_1 subunit in mice (termed the *lethargic* mouse phenotype) and results in lethargy, ataxia, and seizures [173]. Mutations in γ subunits result in the *stargazer* and *waggler* mouse phenotypes, which are characterized by ataxia, spike wave seizures, and abnormal gait [174–176]. These mutations may lead to upregulation of presynaptic calcium channel currents ($Ca_v2.1$), thus contributing to seizure activity [161].

5.1.7
Summary

VGCCs are members of a diverse family with diverse distribution, biophysical, and pharmacological properties. Molecular biological and biochemical techniques have given us insight into how these channels function *in vivo*, and how alterations in channel function lead to various diseases and conditions. Experiments involving knockout and mutant mice have confirmed the potential of VGCCs as useful therapeutic targets, especially in the treatment of pain, hypertension, and epilepsy. Presently, several pharmaceutical agents, either on the market or in various stages of clinical trials, target VGCCs in the treatment of various conditions. The three ensuing chapters in this book detail our knowledge and current understanding of several VGCCs.

References

1. Fatt, P., Katz, B. *J. Physiol.* **1953**, 120, 171–204.
2. Fatt, P., Ginsborg, B. L. *J. Physiol.* **1958**, 142, 516–543.
3. Tsien, R. W., Ellinor, P. T., Horne, W. A. *Trends Pharmacol. Sci.* **1991**, 12, 349–354.
4. Hagiwara, S., Ozawa, S., Sand, O. *J. Gen. Physiol.* **1975**, 65, 617–644.
5. Fox, A. P., Krasne, S. *J. Physiol.* **1984**, 356, 491–505.
6. Tanabe, T., Takeshima, H., Mikami, A., Flockerzi, V., Takahashi, H., Kangawa, K., Kojima, M., Matsuo, H., Hirose, T., Numa, S. *Nature* **1987**, 328, 313–318.
7. Hille, B. Voltage-gated calcium channels. *Ion Channels of Excitable Membranes*, Sinauer Associates, Inc., Sunderland, MA, **2001**.
8. Doering, C. J., Zamponi, G. W. *J. Bioenerg. Biomembr.* **2003**, 35, 491–505.
9. Badou, A., Basavappa, S., Desai, R., Peng, Y. Q., Matza, D., Mehal, W. Z., Kaczmarek, L. K., Boulpaep, E. L., Flavell, R. A. *Science* **2005**, 307, 117–121.
10. Stokes, L., Gordon, J., Grafton, G. *J. Biol. Chem.* **2004**, 279, 19566–19573.
11. Striessnig, J., Hoda, J. C., Koschak, A., Zaghetto, F., Mullner, C., Sinnegger-Brauns, M. J., Wild, C., Watschinger, K., Trockenbacher, A., Pelster, G. *Biochem. Biophys. Res. Commun.* **2004**, 322, 1341–1346.
12. Hell, J. W., Westenbroek, R. E., Warner, C., Ahlijanian, M. K., Prystay, W., Gilbert, M. M., Snutch, T. P., Catterall, W. A. *J. Cell Biol.* **1993**, 123, 949–962.
13. Dolmetsch, R. E., Pajvani, U., Fife, K., Spotts, J. M., Greenberg, M. E. *Science* **2001**, 294, 333–339.
14. Dolmetsch, R. *Sci. STKE* **2003**, 2003, E4.
15. Firth, S. I., Morgan, I. G., Boelen, M. K., Morgans, C. W. *Clin. Exp. Ophthalmol.* **2001**, 29, 183–187.
16. Morgans, C. W. *Eur. J. Neurosci.* **1999**, 11, 2989–2993.
17. Bell, D. C., Butcher, A. J., Berrow, N. S., Page, K. M., Brust, P. F., Nesterova, A., Stauderman, K. A., Seabrook, G. R., Nurnberg, B., Dolphin, A. C. *J. Neurophysiol.* **2001**, 85, 816–827.
18. Koschak, A., Reimer, D., Huber, I., Grabner, M., Glossmann, H., Engel, J., Striessnig, J. *J. Biol. Chem.* **2001**, 276, 22100–22106.
19. Sinnegger-Brauns, M. J., Hetzenauer, A., Huber, I. G., Renstrom, E., Wietzorrek, G., Berjukov, S., Cavalli, M., Walter, D., Koschak, A., Waldschutz, R., Hering, S., Bova, S., Rorsman, P., Pongs, O., Singewald, N., Striessnig, J. J. *J. Clin. Invest.* **2004**, 113, 1430–1439.
20. Platzer, J., Engel, J., Schrott-Fischer, A., Stephan, K., Bova, S., Chen, H., Zheng, H., Striessnig, J. *Cell* **2000**, 102, 89–97.
21. Strom, T. M., Nyakatura, G., Apfelstedt-Sylla, E., Hellebrand, H., Lorenz, B., Weber, B. H., Wutz, K., Gutwillinger, N., Ruther, K., Drescher, B., Sauer, C., Zrenner, E., Meitinger, T., Rosenthal, A., Meindl, A. *Nat. Genet.* **1998**, 19, 260–263.
22. Bech-Hansen, N. T., Naylor, M. J., Maybaum, T. A., Pearce, W. G., Koop, B., Fishman, G. A., Mets, M., Musarella, M. A., Boycott, K. M. *Nat. Genet.* **1998**, 19, 264–267.
23. Fisher, S. E., Ciccodicola, A., Tanaka, K., Curci, A., Desicato, S., D'urso, M., Craig, I. W. *Genomics* **1997**, 45, 340–347.
24. McRory, J. E., Hamid, J., Doering, C. J., Garcia, E., Parker, R., Hamming, K., Chen, L., Hildebrand, M., Beedle, A. M., Feldcamp, L., Zamponi, G. W., Snutch, T. P. *J. Neurosci.* **2004**, 24, 1707–1718.
25. Kotturi, M. F., Carlow, D. A., Lee, J. C., Ziltener, H. J., Jefferies, W. A. *J. Biol. Chem.* **2003**, 278, 46949–46960.
26. Dunlap, K., Luebke, J. I., Turner, T. J. *Trends Neurosci.* **1995**, 18, 89–98.
27. Catterall, W. A. *Annu. Rev. Cell Dev. Biol.* **2000**, 16, 521–555.
28. Llinas, R., Sugimori, M., Lin, J. W., Cherksey, B. *Proc. Natl. Acad. Sci. U.S.A.* **1989**, 86, 1689–1693.
29. Mintz, I. M., Bean, B. P. *Neuropharmacology* **1993**, 32, 1161–1169.
30. Starr, T. V., Prystay, W., Snutch, T. P. *Proc. Natl. Acad. Sci. U.S.A.* **1991**, 88, 5621–5625.
31. Bourinet, E., Soong, T. W., Sutton, K., Slaymaker, S., Mathews, E., Monteil, A.,

Zamponi, G. W., Nargeot, J., Snutch, T. P. *Nat. Neurosci.* **1999**, 2, 407–415.

32. Olivera, B. M., McIntosh, J. M., Cruz, L. J., Luque, F. A., Gray, W. R. *Biochemistry* **1984**, 23, 5087–5090.

33. Cruz, L. J., Johnson, D. S., Olivera, B. M. *Biochemistry* **1987**, 26, 820–824.

34. Adams, M. E., Myers, R. A., Imperial, J. S., Olivera, B. M. *Biochemistry* **1993**, 32, 12 566–12 570.

35. Feng, Z. P., Doering, C. J., Winkfein, R. J., Beedle, A. M., Spafford, J. D., Zamponi, G. W. *J. Biol. Chem.* **2003**, 278, 20 171–20 178.

36. Miljanich, G. P. *Curr. Med. Chem.* **2004**, 11, 3029–3040.

37. Latour, I., Hamid, J., Beedle, A. M., Zamponi, G. W., Macvicar, B. A. *Glia* **2003**, 41, 347–353.

38. Zhang, J. F., Randall, A. D., Ellinor, P. T., Horne, W. A., Sather, W. A., Tanabe, T., Schwarz, T. L., Tsien, R. W. *Neuropharmacology* **1993**, 32, 1075–1088.

39. Bourinet, E., Stotz, S. C., Spaetgens, R. L., Dayanithi, G., Lemos, J., Nargeot, J., Zamponi, G. W. *Biophys. J.* **2001**, 81, 79–88.

40. Vajna, R., Klockner, U., Pereverzev, A., Weiergraber, M., Chen, X., Miljanich, G., Klugbauer, N., Hescheler, J., Perez-Reyes, E., Schneider, T. *Eur. J. Biochem.* **2001**, 268, 1066–1075.

41. Lacerda, A. E., Perez-Reyes, E., Wei, X., Castellano, A., Brown, A. M. *Biophys. J.* **1994**, 66, 1833–1843.

42. Lambert, R. C., McKenna, F., Maulet, Y., Talley, E. M., Bayliss, D. A., Cribbs, L. L., Lee, J. H., Perez-Reyes, E., Feltz, A. *J. Neurosci.* **1998**, 18, 8605–8613.

43. Perez-Reyes, E. *J. Bioenerg. Biomembr.* **1998**, 30, 313–318.

44. Cribbs, L. L., Lee, J. H., Yang, J., Satin, J., Zhang, Y., Daud, A., Barclay, J., Williamson, M. P., Fox, M., Rees, M., Perez-Reyes, E. *Circ. Res.* **1998**, 83, 103–109.

45. Perez-Reyes, E., Cribbs, L. L., Daud, A., Lacerda, A. E., Barclay, J., Williamson, M. P., Fox, M., Rees, M., Lee, J. H. *Nature* **1998**, 391, 896–900.

46. Perez-Reyes, E. *Cell Mol. Life Sci.* **1999**, 56, 660–669.

47. McRory, J. E., Santi, C. M., Hamming, K. S., Mezeyova, J., Sutton, K. G., Baillie, D.

L., Stea, A., Snutch, T. P. *J. Biol. Chem.* **2001**, 276, 3999–4011.

48. Beedle, A. M., Hamid, J., Zamponi, G. W. *J. Membr. Biol.* **2002**, 187, 225–238.

49. Craig, P. J., Beattie, R. E., Folly, E. A., Banerjee, M. D., Reeves, M. B., Priestley, J. V., Carney, S. L., Sher, E., Perez-Reyes, E., Volsen, S. G. *Eur. J. Neurosci.* **1999**, 11, 2949–2964.

50. Perez-Reyes, E. *Physiol. Rev.* **2003**, 83, 117–161.

51. Murbartian, J., Arias, J. M., Perez-Reyes, E. *J. Neurophysiol.* **2004**, 92, 3399–3407.

52. Yunker, A. M., McEnery, M. W. *J. Bioenerg. Biomembr.* **2003**, 35, 533–575.

53. Chuang, R. S., Jaffe, H., Cribbs, L., Perez-Reyes, E., Swartz, K. J. *Nat. Neurosci.* **1998**, 1, 668–674.

54. Sidach, S. S., Mintz, I. M. *J. Neurosci.* **2002**, 22, 2023–2034.

55. Catterall, W. A. *Ann. New York Acad. Sci.* **1993**, 707, 1–19.

56. Doyle, D. A., Morais, C. J., Pfuetzner, R. A., Kuo, A., Gulbis, J. M., Cohen, S. L., Chait, B. T., Mackinnon, R. *Science* **1998**, 280, 69–77.

57. Jiang, Y., Lee, A., Chen, J., Ruta, V., Cadene, M., Chait, B. T., Mackinnon, R. *Nature* **2003**, 423, 33–41.

58. Kim, M. S., Morii, T., Sun, L. X., Imoto, K., Mori, Y. *FEBS Lett.* **1993**, 318, 145–148.

59. Mikala, G., Bahinski, A., Yatani, A., Tang, S., Schwartz, A. *FEBS Lett.* **1993**, 335, 265–269.

60. Tang, S., Mikala, G., Bahinski, A., Yatani, A., Varadi, G., Schwartz, A. *J. Biol. Chem.* **1993**, 268, 13 026–13 029.

61. Yang, J., Ellinor, P. T., Sather, W. A., Zhang, J. F., Tsien, R. W. *Nature* **1993**, 366, 158–161.

62. Wu, X. S., Edwards, H. D., Sather, W. A. *J. Biol. Chem.* **2000**, 275, 31 778–31 785.

63. Feng, Z. P., Hamid, J., Doering, C., Jarvis, S. E., Bosey, G. M., Bourinet, E., Snutch, T. P., Zamponi, G. W. *J. Biol. Chem.* **2001**, 276, 5726–5730.

64. Sather, W. A., McCleskey, E. W. *Annu. Rev. Physiol.* **2003**, 65, 133–159.

65. Stotz, S. C., Jarvis, S. E., Zamponi, G. W. *J. Physiol.* **2004**, 554, 263–273.

66. An, M. T., Zamponi, G. W. Voltage-dependent inactivation of voltage gated calcium channels. *Voltage-Gated Calcium*

Channels (ed Zamponi G.W.), Kluwer Academic/Plenum Publishers, Georgetown, Texas, **2005**.

67. Liang, H., DeMaria, C. D., Erickson, M. G., Mori, M. X., Alseikhan, B. A., Yue, D. T. *Neuron* **2003**, 39, 951–960.

68. Koschak, A., Reimer, D., Walter, D., Hoda, J. C., Heinzle, T., Grabner, M., Striessnig, J. *J. Neurosci.* **2003**, 23, 6041–6049.

69. Baumann, L., Gerstner, A., Zong, X., Biel, M., Wahl-Schott, C. *Invest. Ophthalmol. Vis. Sci.* **2004**, 45, 708–713.

70. Walker, D., De Waard, M. *Trends Neurosci.* **1998**, 21, 148–154.

71. Klugbauer, N., Marais, E., Hofmann, F. *J. Bioenerg. Biomembr.* **2003**, 35, 639–647.

72. Arikkath, J., Campbell, K. P. *Curr. Opin. Neurobiol.* **2003**, 13, 298–307.

73. Dolphin, A. C. *J. Bioenerg. Biomembr.* **2003**, 35, 599–620.

74. Richards, M. W., Butcher, A. J., Dolphin, A. C. *Trends Pharmacol. Sci.* **2004**, 25, 626–632.

75. Black, J. L., III *J. Bioenerg. Biomembr.* **2003**, 35, 649–660.

76. Lacinova, L., Klugbauer, N. *Arch. Biochem. Biophys.* **2004**, 425, 207–213.

77. Dubel, S. J., Altier, C., Chaumont, S., Lory, P., Bourinet, E., Nargeot, J. *J. Biol. Chem.* **2004**, 279, 29 263–29 269.

78. Klugbauer, N., Lacinova, L., Marais, E., Hobom, M., Hofmann, F. *J. Neurosci.* **1999**, 19, 684–691.

79. De Jongh, K. S., Warner, C., Catterall, W. A. *J. Biol. Chem.* **1990**, 265, 14 738–14 741.

80. Yasuda, T., Chen, L., Barr, W., McRory, J. E., Lewis, R. J., Adams, D. J., Zamponi, G. W. *Eur. J. Neurosci.* **2004**, 20, 1–13.

81. Wang, M., Offord, J., Oxender, D. L., Su, T. Z. *Biochem. J.* **1999**, 342 (Pt 2), 313–320.

82. Qin, N., Platano, D., Olcese, R., Costantin, J. L., Stefani, E., Birnbaumer, L. *Proc. Natl. Acad. Sci. U.S.A.* **1998**, 95, 4690–4695.

83. Feng, Z. P., Arnot, M. I., Doering, C. J., Zamponi, G. W. *J. Biol. Chem.* **2001**, 276, 45 051–45 058.

84. Takahashi, S. X., Miriyala, J., Colecraft, H. M. *Proc. Natl. Acad. Sci. U.S.A.* **2004**, 101, 7193–7198.

85. Chen, Y. H., Li, M. H., Zhang, Y., He, L. L., Yamada, Y., Fitzmaurice, A., Shen, Y., Zhang, H., Tong, L., Yang, J. *Nature* **2004**, 429, 675–680.

86. Van Petegem, F., Clark, K. A., Chatelain, F. C., Minor, D. L., Jr. *Nature* **2004**, 429, 671–675.

87. Opatowsky, Y., Chomsky-Hecht, O., Kang, M. G., Campbell, K. P., Hirsch, J. A. *J. Biol. Chem.* **2003**, 278, 52 323–52 332.

88. Pragnell, M., De Waard, M., Mori, Y., Tanabe, T., Snutch, T. P., Campbell, K. P. *Nature* **1994**, 368, 67–70.

89. Chien, A. J., Zhao, X., Shirokov, R. E., Puri, T. S., Chang, C. F., Sun, D., Rios, E., Hosey, M. M. *J. Biol. Chem.* **1995**, 270, 30 036–30 044.

90. Bichet, D., Cornet, V., Geib, S., Carlier, E., Volsen, S., Hoshi, T., Mori, Y., De Waard, M. *Neuron* **2000**, 25, 177–190.

91. Chen, L., El Husseini, A., Tomita, S., Bredt, D. S., Nicoll, R. A. *Mol. Pharmacol.* **2003**, 64, 703–706.

92. Bannister, R. A., Ulises, M., Adams, B. A. Phosphorylation-dependent regulation of voltage-gated Ca^{2+} channels. *Voltage-Gated Calcium Channels* (ed. Zamponi, G.W.), Kluwer Academic/Plenum Publishers, New York, **2005**.

93. Nirdosh, A., Zamponi, G. W. Determinants of G protein inhibition of presynaptic calcium channels. *Voltage-Gated Calcium Channels* (ed. Zamponi, G.W.), Kluwer Academic/Plenum Publishers, New York, **2005**.

94. Lee, A., Catterall, W. A. (2005) Ca^{2+}-dependent modulation of voltage-gated Ca^{2+} channels. *Voltage-Gated Calcium Channels* (ed. Zamponi, G.W.), Kluwer Academic/Plenum Publishers, New York, **2005**.

95. Carbone, E., Carabelli, V., Cesetti, T., Baldelli, P., Hernandez-Guijo, J. M., Giusta, L. *Pflugers Arch.* **2001**, 442, 801–813.

96. Altier, C., Dubel, S. J., Barrere, C., Jarvis, S. E., Stotz, S. C., Spaetgens, R. L., Scott, J. D., Cornet, V., De Waard, M., Zamponi, G. W., Nargeot, J., Bourinet, E. *J. Biol. Chem.* **2002**, 277, 33 598–33 603.

97. Hartzell, H. C., Mery, P. F., Fischmeister, R., Szabo, G. *Nature* **1991**, 351, 573–576.

98. Surmeier, D. J., Bargas, J., Hemmings, H. C., Jr., Nairn, A. C., Greengard, P. *Neuron* **1995**, 14, 385–397.

99. Perets, T., Blumenstein, Y., Shistik, E., Lotan, I., Dascal, N. *FEBS Lett.* **1996**, 384, 189–192.

100. Naguro, I., Nagao, T., Adachi-Akahane, S. *FEBS Lett.* **2001**, 489, 87–91.

101. Kamp, T. J., Hell, J. W. *Circ. Res.* **2000**, 87, 1095–1102.

102. Bunemann, M., Gerhardstein, B. L., Gao, T., Hosey, M. M. *J. Biol. Chem.* **1999**, 274, 33 851–33 854.

103. Held, B., Freise, D., Freichel, M., Hoth, M., Flockerzi, V. *J. Physiol.* **2002**, 539, 459–468.

104. Fraser, I. D., Tavalin, S. J., Lester, L. B., Langeberg, L. K., Westphal, A. M., Dean, R. A., Marrion, N. V., Scott, J. D. *EMBO J.* **1998**, 17, 2261–2272.

105. Hell, J. W., Yokoyama, C. T., Breeze, L. J., Chavkin, C., Catterall, W. A. *EMBO J.* **1995**, 14, 3036–3044.

106. Wu, L., Bauer, C. S., Zhen, X. G., Xie, C., Yang, J. *Nature* **2002**, 419, 947–952.

107. Swartz, K. J., Merritt, A., Bean, B. P., Lovinger, D. M. *Nature* **1993**, 361, 165–168.

108. Stea, A., Soong, T. W., Snutch, T. P. *Neuron* **1995**, 15, 929–940.

109. Zamponi, G. W., Bourinet, E., Nelson, D., Nargeot, J., Snutch, T. P. *Nature* **1997**, 385, 442–446.

110. Jarvis, S. E., Zamponi, G. W. *J. Neurosci.* **2001**, 21, 2939–2948.

111. Bannister, R. A., Melliti, K., Adams, B. A. *Mol. Pharmacol.* **2004**, 65, 381–388.

112. Blumenstein, Y., Kanevsky, N., Sahar, G., Barzilai, R., Ivanina, T., Dascal, N. *J. Biol. Chem.* **2002**, 277, 3419–3423.

113. Marchetti, C., Brown, A. M. *Am. J. Physiol.* **1988**, 254, C206–C210.

114. McCullough, L. A., Egan, T. M., Westfall, T. C. *Am. J. Physiol.* **1998**, 274, C1290–C1297.

115. Park, J. Y., Jeong, S. W., Perez-Reyes, E., Lee, J. H. *FEBS Lett.* **2003**, 547, 37–42.

116. Chen, X. L., Bayliss, D. A., Fern, R. J., Barrett, P. Q. *Am. J. Physiol.* **1999**, 276, F674–F683.

117. Barrett, P. Q., Lu, H. K., Colbran, R., Czernik, A., Pancrazio, J. J. *Am. J. Physiol. Cell Physiol.* **2000**, 279, C1694–C1703.

118. Lu, L. H., Barrett, A. M., Cibula, J. E., Gilmore, R. L., Fennell, E. B., Heilman, K. M. *J. Neurol. Neurosurg. Psychiatry* **2000**, 69, 820–823.

119. Wolfe, J. T., Wang, H., Perez-Reyes, E., Barrett, P. Q. *J. Physiol.* **2002**, 538, 343–355.

120. Yokoyama, C. T., Sheng, Z. H., Catterall, W. A. *J. Neurosci.* **1997**, 17, 6929–6938.

121. Davis, M. J., Wu, X., Nurkiewicz, T. R., Kawasaki, J., Gui, P., Hill, M. A., Wilson, E. *Am. J. Physiol. Heart Circ.Physiol* **2001**, 281, H1835–H1862.

122. Belevych, A. E., Nulton-Persson, A., Sims, C., Harvey, R. D. *J. Physiol.* **2001**, 537, 779–792.

123. Rosenthal, R., Thieme, H., Strauss, O. *FASEB J.* **2001**, 15, 970–977.

124. Weiss, J. L., Burgoyne, R. D. *J. Biol. Chem.* **2001**, 276, 44 804–44 811.

125. Diverse-Pierluissi, M., Remmers, A. E., Neubig, R. R., Dunlap, K. *Proc. Natl. Acad. Sci. U.S.A.* **1997**, 94, 5417–5421.

126. Yakel, J. L. *Trends Pharmacol. Sci.* **1997**, 18, 124–134.

127. Day, M., Olson, P. A., Platzer, J., Striessnig, J., Surmeier, D. J. *J. Neurophysiol.* **2002**, 87, 2490–2504.

128. Zhu, Y., Yakel, J. L. *J. Neurophysiol.* **1997**, 78, 1161–1165.

129. Dunlap, K., Fischbach, G. D. *J. Physiol.* **1981**, 317, 519–535.

130. Dolphin, A. C. *Pharmacol. Rev.* **2003**, 55, 607–627.

131. Zamponi, G. W. *J. Pharmacol. Sci.* **2003**, 92, 79–83.

132. Sheng, Z. H., Rettig, J., Takahashi, M., Catterall, W. A. *Neuron* **1994**, 13, 1303–1313.

133. Sutton, K. G., McRory, J. E., Guthrie, H., Murphy, T. H., Snutch, T. P. *Nature* **1999**, 401, 800–804.

134. Lu, Q., Atkisson, M. S., Jarvis, S. E., Feng, Z. P., Zamponi, G. W., Dunlap, K. *J. Neurosci.* **2001**, 21, 2949–2957.

135. Jarvis, S. E., Barr, W., Feng, Z. P., Hamid, J., Zamponi, G. W. *J. Biol. Chem.* **2002**, 277, 44 399–44 407.

136. Zhong, H., Yokoyama, C. T., Scheuer, T., Catterall, W. A. *Nat. Neurosci.* **1999**, 2, 939–941.

137. Magga, J. M., Jarvis, S. E., Arnot, M. I., Zamponi, G. W., Braun, J. E. *Neuron* **2000**, 28, 195–204.

138. Jarvis, S. E., Magga, J. M., Beedle, A. M., Braun, J. E., Zamponi, G. W. *J. Biol. Chem.* **2000**, 275, 6388–6394.

139. Spafford, J. D., Munno, D. W., Van Nierop, P., Feng, Z. P., Jarvis, S. E., Gallin, W. J., Smit, A. B., Zamponi, G. W., Syed, N. I. *J. Biol. Chem.* **2003**, 278, 4258–4267.

140. Maximov, A., Bezprozvanny, I. *J. Neurosci.* **2002**, 22, 6939–6952.

141. Zamponi, G. W. *Neuron* **2003**, 39, 879–881.

142. de Leon, M., Wang, Y., Jones, L., Perez-Reyes, E., Wei, X., Soong, T. W., Snutch, T. P., Yue, D. T. *Science* **1995**, 270, 1502–1506.

143. Michel, J. J., Scott, J. D. *Annu. Rev. Pharmacol. Toxicol.* **2002**, 42, 235–257.

144. Lee, A., Westenbroek, R. E., Haeseleer, F., Palczewski, K., Scheuer, T., Catterall, W. A. *Nat. Neurosci.* **2002**, 5, 210–217.

145. Haeseleer, F., Imanishi, Y., Maeda, T., Possin, D. E., Maeda, A., Lee, A., Rieke, F., Palczewski, K. *Nat. Neurosci.* **2004**, 7, 1079–1087.

146. Schuhmeier, R. P., Gouadon, E., Ursu, D., Kasielke, N., Flucher, B. E., Grabner, M., Melzer, W. *Biophys. J.* **2005**, 88, 1765–1777.

147. Ouardouz, M., Nikolaeva, M. A., Coderre, E., Zamponi, G. W., McRory, J. E., Trapp, B. D., Yin, X., Wang, W., Woulfe, J., Stys, P. K. *Neuron* **2003**, 40, 53–63.

148. Ptacek, L. J., Tawil, R., Griggs, R. C., Engel, A. G., Layzer, R. B., Kwiecinski, H., McManis, P. G., Santiago, L., Moore, M., Fouad, G., *Cell* **1994**, 77, 863–868.

149. Monnier, N., Procaccio, V., Stieglitz, P., Lunardi, J. *Am. J. Hum. Genet.* **1997**, 60, 1316–1325.

150. Adams, B. A., Tanabe, T., Mikami, A., Numa, S., Beam, K. G. *Nature* **1990**, 346, 569–572.

151. Seisenberger, C., Specht, V., Welling, A., Platzer, J., Pfeifer, A., Kuhbandner, S., Striessnig, J., Klugbauer, N., Feil, R., Hofmann, F. *J. Biol. Chem.* **2000**, 275, 39 193–39 199.

152. Ophoff, R. A., Terwindt, G. M., Vergouwe, M. N., van Eijk, R., Oefner, P. J., Hoffman, S. M., Lamerdin, J. E., Mohrenweiser, H. W., Bulman, D. E., Ferrari, M., Haan, J., Lindhout, D., van Ommen, G. J., Hofker, M. H., Ferrari, M. D., Frants, R. R. *Cell* **1996**, 87, 543–552.

153. Kraus, R. L., Sinnegger, M. J., Koschak, A., Glossmann, H., Stenirri, S., Carrera, P., Striessnig, J. *J. Biol. Chem.* **2000**, 275, 9239–9243.

154. Wappl, E., Koschak, A., Poteser, M., Sinnegger, M. J., Walter, D., Eberhart, A., Groschner, K., Glossmann, H., Kraus, R. L., Grabner, M., Striessnig, J. *J. Biol. Chem.* **2002**, 277, 6960–6966.

155. Jun, K., Piedras-Renteria, E. S., Smith, S. M., Wheeler, D. B., Lee, S. B., Lee, T. G., Chin, H., Adams, M. E., Scheller, R. H., Tsien, R. W., Shin, H. S. *Proc. Natl. Acad. Sci. U.S.A.* **1999**, 96, 15 245–15 250.

156. Kim, C., Jun, K., Lee, T., Kim, S. S., McEnery, M. W., Chin, H., Kim, H. L., Park, J. M., Kim, D. K., Jung, S. J., Kim, J., Shin, H. S. *Mol. Cell Neurosci.* **2001**, 18, 235–245.

157. Hatakeyama, S., Wakamori, M., Ino, M., Miyamoto, N., Takahashi, E., Yoshinaga, T., Sawada, K., Imoto, K., Tanaka, I., Yoshizawa, T., Nishizawa, Y., Mori, Y., Niidome, T., Shoji, S. *Neuroreport* **2001**, 12, 2423–2427.

158. Saegusa, H., Kurihara, T., Zong, S., Kazuno, A., Matsuda, Y., Nonaka, T., Han, W., Toriyama, H., Tanabe, T. *EMBO J.* **2001**, 20, 2349–2356.

159. Saegusa, H., Kurihara, T., Zong, S., Minowa, O., Kazuno, A., Han, W., Matsuda, Y., Yamanaka, H., Osanai, M., Noda, T., Tanabe, T. *Proc. Natl. Acad. Sci. U.S.A.* **2000**, 97, 6132–6137.

160. Saegusa, H., Matsuda, Y., Tanabe, T. *Neurosci. Res.* **2002**, 43, 1–7.

161. Ashcroft, F. M. Voltage-gated Ca^{2+} channels. *Ion Channels and Disease*, Academic Press, San Diego **2000**.

162. Kim, D., Song, I., Keum, S., Lee, T., Jeong, M. J., Kim, S. S., McEnery, M. W., Shin, H. S. *Neuron* **2001**, 31, 35–45.

163. Chen, Y., Lu, J., Pan, H., Zhang, Y., Wu, H., Xu, K., Liu, X., Jiang, Y., Bao, X., Yao, Z., Ding, K., Lo, W. H., Qiang, B., Chan, P., Shen, Y., Wu, X. *Ann. Neurol.* **2003**, 54, 239–243.

164. Khosravani, H., Altier, C., Simms, B., Hamming, K. S., Snutch, T. P., Mezeyova, J., McRory, J. E., Zamponi, G. W. *J. Biol. Chem.* **2004**, 279, 9681–9684.

165. Heron, S. E., Phillips, H. A., Mulley, J. C., Mazarib, A., Neufeld, M. Y., Berkovic, S.

F., Scheffer, I. E. *Ann. Neurol.* **2004**, 55, 595–596.

166. Chen, C. C., Lamping, K. G., Nuno, D. W., Barresi, R., Prouty, S. J., Lavoie, J. L., Cribbs, L. L., England, S. K., Sigmund, C. D., Weiss, R. M., Williamson, R. A., Hill, J. A., Campbell, K. P. *Science* **2003**, 302, 1416–1418.

167. Bourinet, E., Alloui, A., Monteil, A., Barrere, C., Couette, B., Poirot, O., Pages, A., McRory, J., Snutch, T. P., Eschalier, A., Nargeot, J. *EMBO J.* **2005**, 24, 315–324.

168. Barclay, J., Balaguero, N., Mione, M., Ackerman, S. L., Letts, V. A., Brodbeck, J., Canti, C., Meir, A., Page, K. M., Kusumi, K., Perez-Reyes, E., Lander, E. S., Frankel, W. N., Gardiner, R. M., Dolphin, A. C., Rees, M. *J. Neurosci.* **2001**, 21, 6095–6104.

169. Brodbeck, J., Davies, A., Courtney, J. M., Meir, A., Balaguero, N., Canti, C., Moss, F. J., Page, K. M., Pratt, W. S., Hunt, S. P., Barclay, J., Rees, M., Dolphin, A. C. *J. Biol. Chem.* **2002**, 277, 7684–7693.

170. Gregg, R. G., Messing, A., Strube, C., Beurg, M., Moss, R., Behan, M., Sukhareva, M., Haynes, S., Powell, J. A., Coronado, R., Powers, P. A. *Proc. Natl. Acad. Sci. U.S.A.* **1996**, 93, 13 961–13 966.

171. Strube, C., Beurg, M., Powers, P. A., Gregg, R. G., Coronado, R. *Biophys. J.* **1996**, 71, 2531–2543.

172. Ball, S. L., Powers, P. A., Shin, H. S., Morgans, C. W., Peachey, N. S., Gregg, R. G. *Invest. Ophthalmol. Vis. Sci.* **2002**, 43, 1595–1603.

173. Burgess, D. L., Jones, J. M., Meisler, M. H., Noebels, J. L. *Cell* **1997**, 88, 385–392.

174. Letts, V. A., Mahaffey, C. L., Beyer, B., Frankel, W. N. *Proc. Natl. Acad. Sci. U.S.A.* **2005**, 102, 2123–2128.

175. Chen, L., Chetkovich, D. M., Petralia, R. S., Sweeney, N. T., Kawasaki, Y., Wenthold, R. J., Bredt, D. S., Nicoll, R. A. *Nature* **2000**, 408, 936–943.

176. Chen, L., Bao, S., Qiao, X., Thompson, R. F. *Proc. Natl. Acad. Sci. U.S.A.* **1999**, 96, 12 132–12 137.

5.2
Drugs Active at T-type Ca^{2+} Channels

Thomas M. Connolly and James C. Barrow

5.2.1
Introduction

A typical cell lives under a large calcium gradient. They are bathed in an extracellular milieu that contains approximately 2 mM calcium, while their resting intracellular calcium concentration is only about 100 nM. Upon cellular activation, the internal calcium concentration increases up to micromolar levels, attaining even higher levels in some cellular micro regions. The increase in intracellular calcium can come from release of calcium from intracellular stores, or from outside the cell through ligand-operated channels (typically permeable to sodium and/or calcium ions) or highly selective voltage-gated calcium channels. The control of Ca^{2+} movement across the membrane by voltage-gated calcium channels is highly efficient and tightly regulated. These channels are activated by a change in membrane potential that results in a conformational change in the structure of the channel, allowing the Ca^{2+} influx down the electrochemical gradient. In excitable cells, voltage-gated calcium channels contribute to cellular signaling. They regulate calcium entry and mediate the linkage of an electrical activity change to a biochemical signal that results from the increased intracellular calcium. This signaling can lead to such functions as contraction, neurotransmitter or hormone secretion, cellular excitability, including activation of other potassium or chloride channels, and regulation of rhythmicity and gene expression [1].

Voltage-gated calcium channels belong to the transmembrane ion channel superfamily that includes the K$^+$ and Na$^+$ channels. Members of the calcium channel family are characterized based on their electrophysiological [2–4] and molecular properties [4, 5]. They are found in the membranes of excitable cells, including central and peripheral neurons, skeletal and smooth muscle and some secretory cells such as pancreatic cells. The calcium channel family is divided into two major groups, the low voltage activated (LVA) and the high voltage activated (HVA), based on the amount of cellular depolarization required for activation. HVA calcium channels open at more depolarized potentials than low voltage activated, or T-type, channels, which are characterized by activation at more polarized potentials nearer to the cells resting potential. This chapter will focus on the pharmacology of the T-type calcium channel.

T-type calcium channels include three subtypes, Cav 3.1 (or α1G), Cav3.2 (α1H) and Cav 3.3 (α1I), which are the products of three distinct genes, CACNA1G, CACNA1H, and CACNA1I, respectively. They show ~58% overall amino acid identity between subtypes [6], which increases to ~90% identity when using the CLUSTAL program to compare only the transmembrane regions [4]. α1G is expressed most significantly in the brain, from which it was originally cloned [7], and is in particularly high abundance in the thalamus [8, 9]. It is also expressed to a lesser extent in peripheral tissues [7, 10]. α1H was cloned from a human heart library [11] and a medullary thyroid carcinoma line [12]. It is expressed peripherally in the heart, liver, kidneys, pancreas [11] and adrenal cortex [13]. It is also expressed in the brain in the thalamic reticular nucleus, in peripheral sensory neurons in the dorsal root ganglia, and in mechanosensitive neurons [8]. The α1I subtype was cloned from rat [14] and human [15] brain. It is found in high levels in the thalamic reticular neurons, with lower expression levels in the spinal cord [8]. As with all voltage-gated calcium channels, T-type calcium channels exist as splice variants, with considerable variability identified for the G and I subtypes. This alternative splicing can alter the biophysical, physiological and pharmacological properties [15–19], the exact implication of which is not known.

The existence of the LVA, or T-type calcium channels, was first suggested in electrophysiological studies by the identification of distinct calcium currents in starfish eggs [20] and later in sensory neurons [21] and cardiac tissues [22, 23]. They have several characteristics that distinguish them from other voltage-gated calcium channels [24]. The conductance of their currents is small compared to other calcium channels (transient or T-type). Second, they can conduct either calcium or barium equally as well, compared to high-voltage activated channels, which conduct greater amounts of barium. Third, the cellular potentials at which they activate and inactivate (–70 to –50 mV) are more negative than that for other channels (thus, low voltage activated channels). Fourth, T-type currents inactivate rapidly (tiny and transient) relative to other channels. This rate is itself voltage dependent. They deactivate (channel returning to the resting condition) more slowly than for most calcium channels. Finally, recombinant T-type calcium channels composed of only the α subunit, which contains the pore and voltage sensing region, display characteristics of the native T-type channel [14, 25]. The active form of most other types of calcium channels consists of multiple subunits including α plus β, α2δ, and γ [4, 5]. Taken together, the properties of T-type calcium channels are consistent with a role in bursting and rhythmicity, in areas such as the thalamus and SA node in the heart, and not in functions such as transmitter release or contraction, which require considerable and long lasting calcium elevation.

5.2.2
Methodology

Various assays are typically used to determine the pharmacology of compounds on T-type calcium channels. Patch-clamping, in which the current that is carried across the membrane through the channels is directly measured, is the most frequently used. Studies are typically done on native channels. More recently, recombinant expression of the channels in oocytes or mammalian cell lines has been described [6, 14, 18, 19, 25]. The expression of the channels has allowed the use of other methods that utilize fluorescence of calcium binding dyes [25, 26] to monitor channel function, a method particularly useful for high-throughput screening. A few recent abstracts suggest that these channels are also amenable to high-throughput voltage clamping on new 384-well automated patch clamp devices [27, 28].

5.2.2.1 Pharmacology of T-type Calcium Channels

Historically, various pharmacological tools, including various toxins and organic compounds, were used to classify calcium channels. While many of these diverse compounds were reported as T-type channel inhibitors, their selectivity for these channels versus other types is quite variable. This divergence in potencies is largely because many of these studies were carried out on native channels. These preparations could contain heterogeneous subtype expression within the same tissue, or have differing subunit compositions in the different tissues, each of which could have a profound effect on the pharmacology of the inhibitors. Likewise, the use of different test conditions, such as holding potential, charge carrier, protocol timing, etc., could also markedly affect the inhibitory profile of a compound. Investigation in recombinant systems simplifies this process, but also has the liability of not always having all the cellular component signaling pathways present that might be associated with the native channels. This chapter will try to point out each of these caveats when possible.

5.2.2.1.1 Inorganic Ions

T-type calcium channels are inhibited to various extents by micromolar amounts of various inorganic cations. Ni^{2+} potently blocks $\alpha1H$ currents with potencies in the $10-40\,\mu M$ range [29], but is a much weaker inhibitor (IC_{50}s of $200-1000\,\mu M$) of the $\alpha1G$ and $\alpha1I$ subtypes [18, 29, 30]. The block of $\alpha1H$ by Ni^{2+} is less effective at more positive testing potentials, suggesting that gating is effected. The effect of Ni^{2+} on $\alpha1H$ is consistent with its ability to block T-type currents in rabbit sino-atrial nodes, cardiac myocytes and sensory neurons of the dorsal root ganglia, suggesting a prominent role for $\alpha1H$ in these tissues. The block by Ni^{2+} of some neuronal T-type currents was less potent and quite variable [31], suggesting that other T-type subtypes may be present in the neuronal tissues. Consistent with this result Ni^{2+} weakly inhibited other neuronal calcium channels, $\alpha1A$, $\alpha1B$, $\alpha1C$, and $\alpha1E$ [32]. Inhibition of these types of calcium channels was highly dependent on assay conditions, such as the charge carrier used, holding potential and co-expression of β subunit. Overall, the results suggest that only $\alpha1H$ is selectively blocked by Ni^{2+}.

Various other metal ions, including $2+$",$4 > Cu^{2+}$, Pb^{2+} and Zn^{2+}, inhibit-T-type calcium channels, with $\alpha1H$ being the most sensitive [33]. Cadmium is a more effective blocker than Ni^{2+} of the $\alpha1G$ channel [30]. Finally, trivalent cations such as yttrium, scandium, holmium amongst others are nanomolar inhibitors of $\alpha1G$ channels when Ba^{2+} is used as the charge carrier [34].

5.2.2.1.2 Toxins

Various spider and snail toxins, ω-conotoxin-GVIVA, ω-conotoxin MVIIA, ω-conotoxin CVID, ω-agatoxin IVA, are effective tools to selectively block P- and N-type high voltage activated calcium channels. They are useful research tools and were used in the classification of these channels. T-type currents are insensitive to these toxins [12, 4, 24] and to TTX [29]. Kurtoxin, a 63 amino acid protein from the *Parabuthus transvaalicus* scorpion, was reported to bind to the α1G channel with high affinity and to inhibit it and the α1H channel with high potencies, 15–60 nM [35]. Over the same concentration range it did not inhibit other calcium channel currents, including α1B, α1A, α1E or α1C, while showing some slowing of rat sodium channel activation and inactivation kinetics. Overall, it is the only toxin that demonstrates selectivity for T-type calcium channels and as such is a research tool.

5.2.2.1.3 Antiepileptics

Epilepsy is thought to involve neuronal hypersensitivity and in some specific types is manifested as spike wave discharges. Some of the drugs used to treat epilepsy are thought to work through either inhibition of Na^+ or Ca^{2+} channels. Ethosuximide, a drug used for absence seizures, was initially shown to partially block T-type currents in thalamic neurons (IC_{50} = 0.2 mM) at therapeutically relevant ~0.5 mM concentrations [36]. Methsuximide is also efficacious for treatment of absence seizures, and its active metabolite, methyl-phenyl-succinimide (MPS), also blocks T-type currents [37]. Several structurally related compounds that are ineffective *in vivo* are inactive on the channels. These findings support the role of the blockade of T-type channels in the treatment of absence epilepsy. More recent studies on recombinantly expressed T-type channels indicate that ethosuximide and MPS block all three subtypes (0.3–1.2 mM IC_{50}s) and in a state dependent manner, with higher affinity for inactivated channels [38].

Hydantoin epileptic drugs, phenytoin and phenobarbital, block Na^+ channels. They also were reported to have variable effects on T-type currents. Phenytoin partially inhibited dorsal root ganglion (DRG) T-currents (IC_{50} of 8 μM), but fully blocked recombinant α1G (IC_{50} of 140 μM), while it only partially (45% maximum) blocked recombinant α1H (IC_{50} of 8 μM) [39, 40]. Furthermore, other epileptics such as valproic acid only partially inhibited recombinant α1G and α1H at a therapeutically relevant concentration [39] and zonisamide only partially blocked T-type currents [40, 41]. Their role as T-type antagonists in epilepsy treatment is less certain.

5.2.2.1.4 Anesthetics

Various anesthetics are reported to block T-type calcium currents. Unlike many of the other inhibitors discussed, these compounds demonstrate the same potency across tissues and between recombinant channels and native currents. The anesthetics propofol, octanol and isoflurane blocked T-type Ca^{2+} currents in DRG and recombinant α1G over concentration ranges of ~12–300 μM [39]. For isoflurane this is a therapeutically relevant concentration. In some other systems the blockade of T-type Ca^{2+} currents was only partial by these agents [40]. As blocking T-type channels would reduce neuronal excitability, it is possible that this contributes to their mechanism of action.

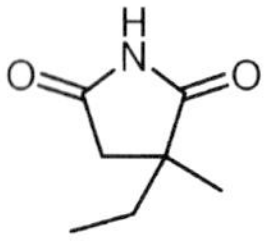

ethosuximide

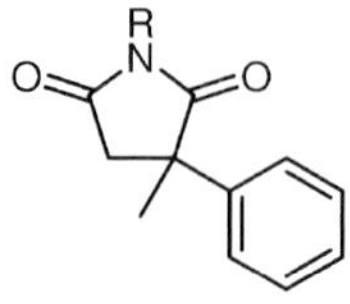

R=Me methsuximide
R=H methyl phenyl succinimide (MPS)

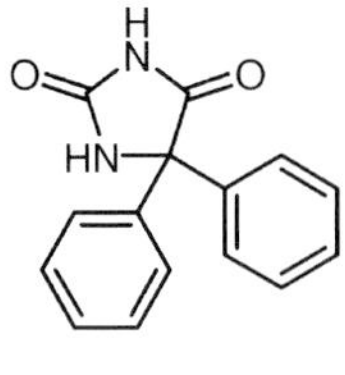

phenytoin

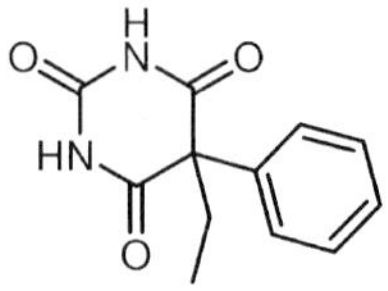

phenobarbital

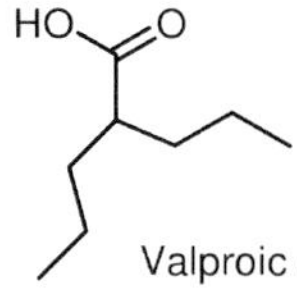

Valproic Acid

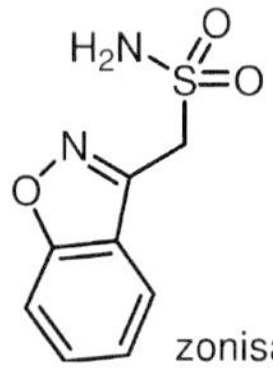

zonisamide

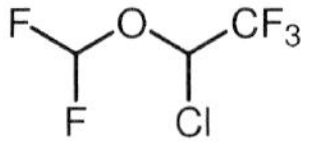

Isoflurane

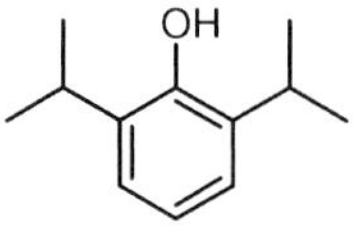

propofol

Nitrous oxide (N_2O) is a widely used anesthetic and analgesic. It blocks T-type currents in sensory neuron DRGs with an IC_{50} of ~45%, and with a maximal inhibition of ~40% [42]. High voltage activated currents were not inhibited under these subanesthetic conditions. In the same studies N_2O blocked recombinant α1H currents but had no effect on α1G currents. In each case the inhibition had no use or voltage dependent component. Taken together, blockade of T-type currents may be a component of the anesthetic effects of N_2O.

5.2.2.1.5 Neuroleptics

Antipsychotics are thought to work primarily through their inhibitory action on D2 dopaminergic receptors. Many of these drugs were shown to inhibit various neuronally expressed calcium channels, although the potencies and effects are variable in different studies. In studies on recombinantly expressed channels, pimozide and penfluridol potently inhibit all three T-type channels, α1G, α1H, and α1I, with similar potencies (IC_{50}s of 30–100 nM) [43]. This potency range is slightly greater than that for these compounds on native currents, is more potent than that for their effects on L-type channels [44], and is at a clinically relevant concentration range. Overall, the diphenylmethylpiperazine flunarizine, is a weaker blocker of the T-type Ca^{2+} currents and it blocks α1G and α1I more potently than it does α1H, with IC_{50}s of 530, 837, and 3550 nM, respectively [43]. Lastly, haloperidol, a potent antipsychotic used to treat schizophrenia, blocks all three subtypes, but it is the weakest of the antipsychotics (IC_{50} of ~1200 nM).

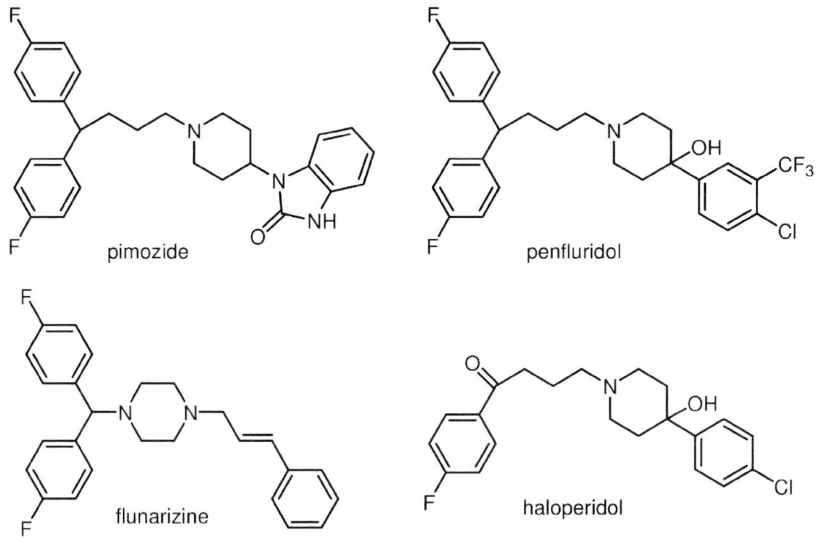

pimozide

penfluridol

flunarizine

haloperidol

For all compounds the inhibition is voltage dependent and has a greater block on the inactivated state of the channels.

5.2.2.1.6 Dihydropyridines (DHPs) and Phenylalkylamines

DHPs are L-type calcium channel blockers that are widely used in the treatment of hypertension and angina [45]. While some native T-type channels have shown limited sensitivity to some DHPs [46], usually micromolar concentrations are required while only nanomolar concentrations are needed to inhibit the L-type channel [45]. For this reason the typical neutral DHPs such as isridipine, nifedipine and imodipine are thought to produce their vasodilatory effects solely though L-type channels. Likewise the charged DHP amlodipine is effective in patients at 15 nM and yet it blocks T-type currents with an IC$_{50}$ of 31 μM [47]. A report of a series of nifedipine analogs with long hydrocarbon chains appended to the aromatic ring, exemplified by PPK-12, demonstrated that one could obtain similar potencies on transiently expressed L and T-type channels, with ~0.5 and 1.6 μM IC$_{50}$s, respectively [48]. In recent studies carried out on recombinantly expressed channels in *Xenopus* oocytes, four DHPs, amlodipine, benidipine, barnidipine, and efonidipine blocked both L and α1G T-type currents, while nifedipine, nitrendipine, nimodipine and felodipine showed no inhibition of the T-type currents under the same conditions [49]. In these studies the potencies on the L-type channel currents were weaker than those reported in studies using mammalian systems and the comparable potencies between T- and L-currents were only observed at the resting, hyperpolarized state of the channel. Tests at more depolarized conditions revealed significant increases in potencies on the L-type channel only. In another study on recombinant channels expressed in both *Xenopus* oocytes and mammalian cells, efonidipine and its S-(+) isomer inhibited both

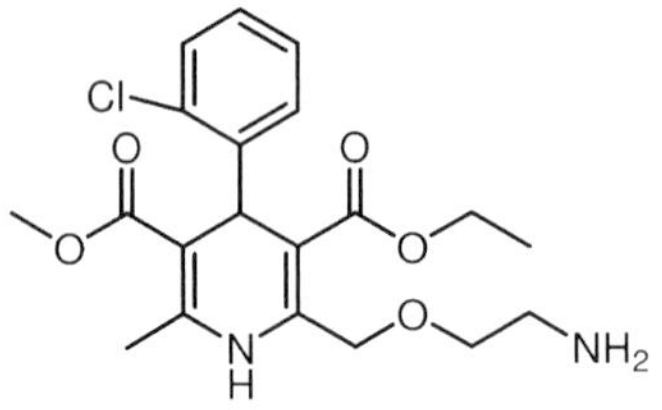

amlodipine

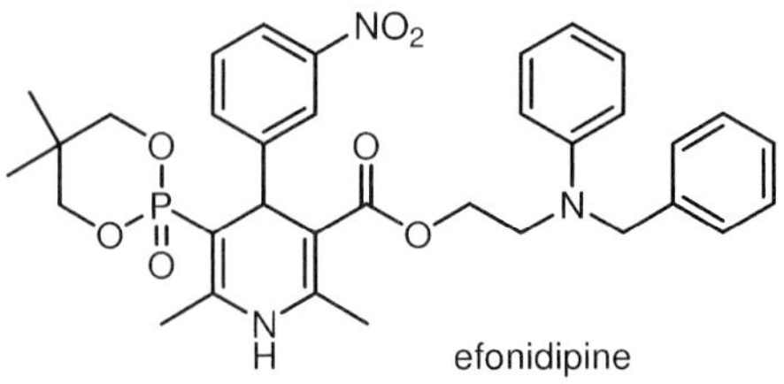

efonidipine

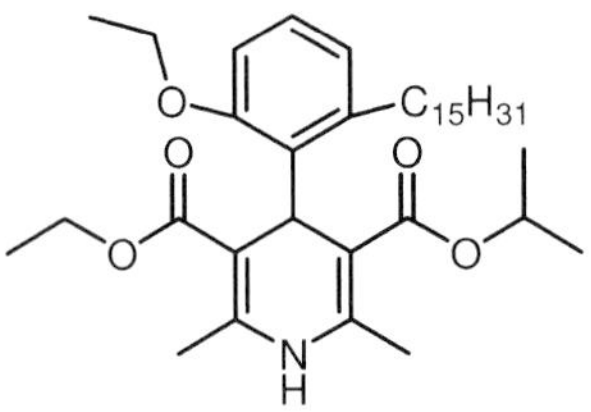

PPK-12

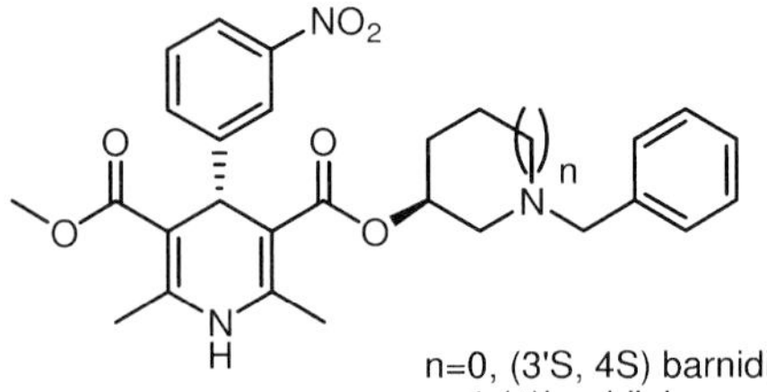

n=0, (3'S, 4S) barnidipine
n=1 (±)benidipine

L- and T-type calcium channels. In contrast, the R-(–) isomer inhibited T-type channels with similar potency as its S-(+) isomer, but was a much less potent antagonist of the L-type channel [50]. This study defines the R-(–) isomer of efonidipine as a selective T-type calcium channel antagonist.

The phenylalkylamine verapamil is only a weak inhibitor of α1G current, 17% at 1 μM [12] and 6 to 29 μM on α1H [25], depending on the conditions of the assay. These weak potencies suggest that it is unlikely that verapamil (effective concentration of < 1 μM) is working through blockade of the T-type channel.

5.2.2.1.7 Mibefradil

Mibefradil is a tetralol derivative for which there is a wide literature suggesting that it is a selective T-type calcium channel antagonist [51, 52]. It was developed by Hoffman-LaRoche as Ro 40-5967 and briefly marketed as a treatment for angina and hypertension under the trade name Posicor™ [53, 54]. Mibefradil was reported to be ~10–30-fold selective for T-type over L-type currents in vascular [55] and cardiac [56] myocytes, and selective for T-type over P-type currents in Purkinje neurons [57]. This selectivity, however, was not observed in all neurons [58]. The potency of mibefradil to block T-type calcium currents has been reported to be over a wide range, from ~ 100 nM in vascular smooth muscle [55] and neuronal Purkinje cells [57] to ~ 5 μM in mouse spermatogenic cells [59]. In a systematic study on recombinant forms of the T-type calcium channels, mibefradil fully blocked all three subtypes with potencies of ~ 1 μM [60]. Blockade of L-type channels occurred at approximately 13-fold lower potency. These studies used 10 mM Ba^{2+} as the charge carrier. When Ca^{2+} was used as charge carrier the potencies of inhibition shifted significantly to 0.14–0.27 μM, a range similar to that observed for inhibition of T-type current in many native tissues. In addition to the significant effect charge carrier has on mibefradil's potency, holding potential also influ-

mibefradil

ences its inhibitory potency. A shift in holding potential from slightly depolarized to hyperpolarized caused a 10-fold decrease in potency in Purkinje neurons [57]. Different pulse protocols can also alter the potency of blockade by mibefradil [61], suggesting a use dependent component to its inhibitory mechanism. Taken together, mibefradil is selective for T-type channels in some tissues and under some experimental conditions, while it may inhibit other types of calcium channels or other types of channels as well under other conditions [62–68]. In lieu of other compounds, it remains the prototype T-type calcium channel blocker.

5.2.2.1.8 Other Antagonists
To better understand the role of T-type calcium channels in biological processes, several groups have attempted to prepare new pharmacological tools with better potency and selectivity than the aforementioned agents. One well-studied example is the flunarizine analog U-92032 [69] which is three-orders of magnitude more potent at T-Type channels than methylphenylsuccinimide (MPS) [70].

Further modifications of flunarizine have resulted in the preparation of TH-1177, which was used for exploration of the role of T-type calcium channels in cancer cell proliferation [71]. TH-1177 blocked T-type calcium current in PC3 prostate cancer cells (IC$_{50}$ = 2.3 µM) and reduced proliferation of the same cells (IC$_{50}$ = 14 µM). The linkage between calcium current and proliferation was further supported by studies of other known calcium channel blockers (e.g., flunarizine, nicardipine, and mibefradil) and structural analogs of TH-1177.

Using a combinatorial library approach inspired by flunarizine, U-92032 and mibefradil, an isoxazole–piperazine library was constructed, resulting in a series of novel T-type calcium channel antagonists exemplified by compound 1. These compounds had potency similar to mibefradil (1 µM) for inhibition of recombinantly expressed α1G calcium channels [72]. Also investigated as T-type calcium channel antagonists were a series of 3,4-dihydroquinazolinones, of which KYS05044 was the most potent (IC$_{50}$ = 0.5 µM) and with the best selectivity versus N-type calcium channels [73]. These studies led to a 5-feature, three-dimensional pharmacophore model for the T-type calcium channel, which may prove useful in designing more potent derivatives [74].

The endogenous cannabinoid anandamide is a potent inhibitor of T-type calcium channel currents [75]. In voltage clamp experiments carried out at a −80 mV holding potential, anandamide blocked α1G, α1H, and α1I currents

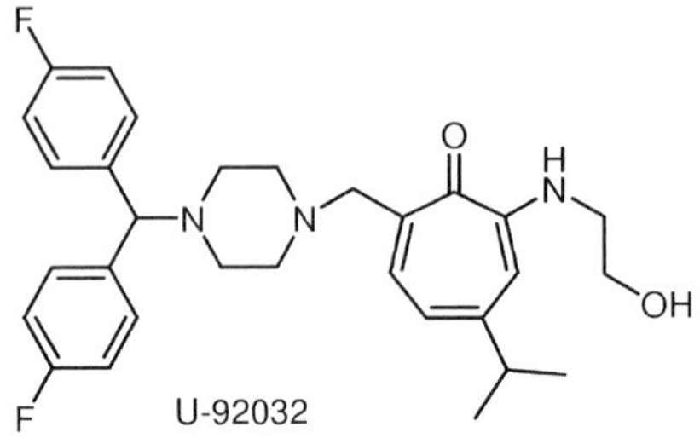

U-92032

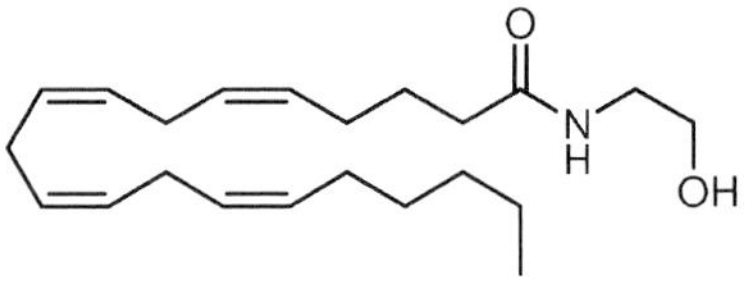

anandamide

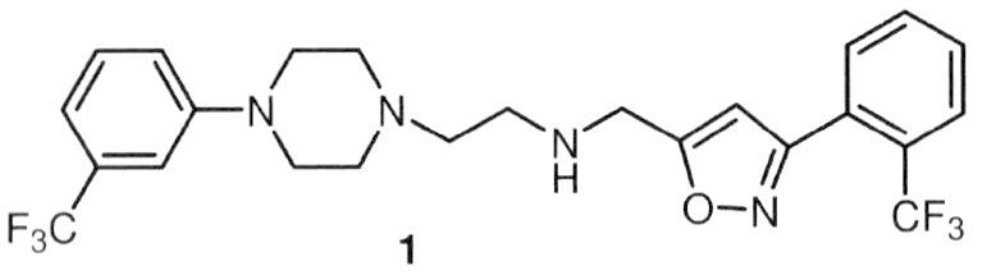

1

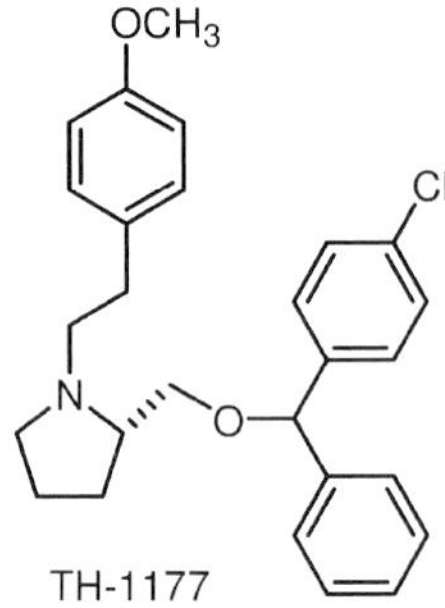

TH-1177

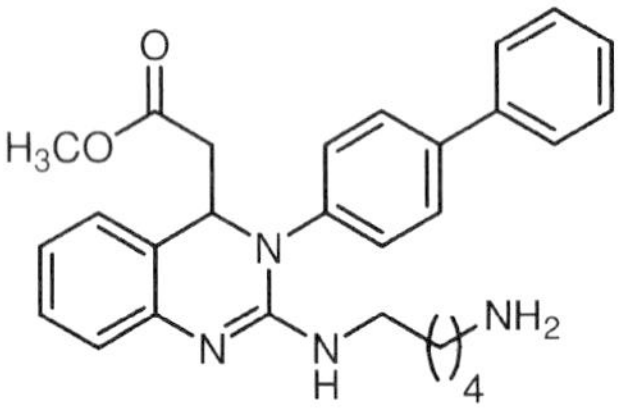

KYS05044

with IC$_{50}$s of 0.33, 1.1, and 4.15 µM, respectively. Anandamide preferentially binds to the inactivated state of the channel, as evidenced by its lower potency at more negative holding potentials. It also accelerated the activation and inactivation kinetics of T-type calcium channels, with the most profound effects on the α1I subtype. This is the first endogenous substance to be reported as a direct calcium channel blocker, and the consequence of this activity remains to be determined.

5.2.3
Indications

Based on the localization of the expression of T-type calcium channels, their functional properties, and the pharmacological effects of the available T-type channel antagonists, several roles for these channels and future uses for antagonists are suggested. These include the treatment of hypertension, cellular proliferation and cases of abnormal neuronal firing.

For cardiovascular indications, mibefradil (Posicor™) was shown to effectively lower blood pressure and heart rate in patients without producing the reflex tachycardia or loss in inotropy often associated with other calcium channel blockers

[53, 54]. This favorable profile was attributed to selective inhibition of T-type calcium channels. Because of these experiences with mibefradil it was suggested that T-type calcium channel antagonists are effective anti-hypertensive agents. Unfortunately, this working hypothesis was based largely on results with mibefradil, and not with any other T-type antagonists. As discussed above, mibefradil is selective for T-type channels under some conditions; however, it blocks other calcium channels, including L and N types as well as Na^+, K^+ and non-selective cation currents [62–68]. The contribution of L-type blockade to its overall efficacy is suggested to be largely underestimated [76]. Indeed, it may be the combination of inhibition of multiple channels by mibefradil that contributes to its favorable *in vivo* cardiovascular profile.

Consistent with a more limited role for T-type calcium channels in the regulation of normal cardiovascular function, α1H knockout mice had no change in EKG or heart rate [77]. Indeed, these mice actually had an increased contraction of their coronary smooth muscle, suggesting a role for these T-type channels in their relaxation process. In the only other reported T-type channel knockout animal, α1G knockout mice, the cardiovascular phenotype in these animals was not reported [78]. T-type calcium channels possibly play a more subtle role in blood pressure regulation through regulation of normal kidney function. T-type calcium channels were identified in expression studies and functionally in both afferent and efferent renal arterioles [78]. In contrast, L-type calcium channels are reported to only be expressed in renal afferent arterioles [79]. In functional studies, low doses of pimozide and mibefradil dilated both afferent and efferent renal arterioles, while the L-type antagonist diltiazem only dilated the afferent arterioles [80]. Overall these studies are significant as L-type antagonists have not been beneficial in models of renal failure, while mibefradil demonstrated significant benefits [81, 82]. In support of a beneficial effect of T-type channel blockade in the kidney, the dihydropyridine dual L- and T-type calcium channel antagonist efonidipine also demonstrated renal protective properties similar to those of angiotensin converting enzyme (ACE) inhibitors in studies in people [83].

Several lines of evidence suggest that T-type calcium channels are important in regulating growth and differentiation. Channel expression levels are often present at higher densities in embryonic and neonatal systems, where there is significant growth, and reduced in mature cells, including neuronal and cardiac atrial and ventricular myocytes [24]. In models of cardiac hypertrophy and cardiomyopathy mibefradil reduced remodeling [84], suggesting a role for T-type channels in these growth processes. Mibefradil also inhibited electrical remodeling in atria in dogs, while diltiazem [85] or verapamil [86] were without effect, suggesting a role for T-type calcium channels in this process. Studies with TH-1177 also suggest a role for T-type calcium channels in some types of cancer [71], where the role of a voltage-gated channel in an electrically non-excitable cell is less clear.

A general area where the role for T-type calcium channels is suggested and where a T-type antagonist would be useful is in the regulation of spontaneously active cells in the nervous system. The initial finding of significant expression of T-type currents in dorsal root ganglia neurons suggested that these channels are likely important for sensory processing. Ethosuximide administered intraperitoneally, but not intrathecally, was efficacious in thermal and tactile neuropathic pain models in nerve injured rats [87, 88]. Systemic or local administration of mibefradil also inhibited basal, mechanical and thermal nociception and thermal hyperalgesia in the spinal nerve ligation model [87, 89]. In further support of a role of T-type channels in pain processing, intrathecal administration of antisense oligonucleotides for the α1H calcium channel resulted in antinociceptive, anti-hyperalgesic and anti-allodynic effects [90]. Lastly, the response of α1G knockout mice to thermal or mechanical stimuli was the same as that of wild-type mice, while they demonstrated a markedly enhanced response to visceral pain [78]. This observation, in contrast to the studies with antagonists, indicates an anti-nociceptive function for this α1G subtype in visceral pain processing. Overall, the studies with channel inhibitors and gene expression modification suggest a role for T-type calcium channels in pain processing pathways and that this role may differ depending on the pathway and subtype of channel involved.

Sensory transmission in states of pain is only one area in which alterations of neuronal firing involving T-type channels may occur. Thalamic dysrhythmias have been implicated in various clinical states, including tinnitus, Parkinson's, depression and epilepsy [91]. Examination of the α1G knockout mice indicated that thalamo-cortical relay neurons from these animals were not capable of burst firing, while tonic firing was normal [92]. They were also unable to generate spike-wave discharges in response to GABA receptor activation. Burst firing is a major characteristic of generalized absence epilepsy. These results and the clinical experience with ethosuximide support a role for T-type channels in this type of epilepsy. Cortical rhythmogenesis has a strong effect on routine brain functions such as attention and vigilance, with the thalamus playing a key role in the transition from desynchronized to synchronized activity [93, 94]. T-type calcium channels are thought to be involved in the generation of the low threshold spikes that depolarize the cell membrane and incite burst firing. In support of a role for these channels in this process, the general [95] and thalamic [96] α1G knockout mice demonstrated an altered vigilance pattern compared to wild-type mice. Further examination of this process and the role of T-type channels is suggested.

5.2.4
Conclusions

Recent years have seen a marked increase in knowledge on the localization, function, and properties of the T-type calcium channels. They are typically expressed in spontaneously active cells such as cardiac and neuronal pacemaker cells, thalamic neurons and some hormone secreting cells. Defining the true functional role for these calcium channels has suffered from the lack of potent and selective antagonists. As more information becomes available on their function they become a more attractive target for further drug discovery efforts. Defining their true therapeutic potential will require identification of more selective compounds than the ones highlighted in this chapter, including selectivity against other ion channels, HVA calcium channels and the different T-type subtypes. When progress is made in this area we can better evaluate their full therapeutic potential.

References

1. Tsien, R.W., Wheeler, D.B. Voltage-gated calcium channels. In *Calcium as a Cellular Regulator*, Carafoli, E., Klee, C. Eds., Oxford University Press, New York, pp. 171–199, **1999**.

2. Randall, A., Tsien, R.W. Pharmacological dissection of multiple types of Ca^{2+} channel currents in rat cerebellar granule neurons. *J. Neurosci.* **1995**, 15, 2995–3012.

3. Hofmann, F, Biel, M., Flockerzi, V. Molecular basis for Ca^{2+} channel diversity. *Annu. Rev. Neurosci.* **1994**, 17, 399–418.

4. Ertel, E.A., Campbell, K.P., Harpold, M.M., Hofmann, F., Mori, Y., Perez-Reyes, E., Schwartz, A., Snutch, T.P., Tanabe, T., Birnbaumer, L., Tsien, R.W., Catteral, W.A. Nomenclature of voltage-gated calcium channels. *Neuron* **2000**, 25, 533–535.

5. Catterall, W.A. Functional subunit structure of voltage-gated calcium channels. *Science* **1991**, 253, 1499–1553.

6. Cribbs, L.L., Gomora, J.C., Daud, A.N., Lee, J.H., Perez-Reyes, E. Molecular cloning and functional expression of $Ca_v3.1c$, a T-type calcium channel from human brain. *FEBS Lett.* **2000**, 466, 54–58.

7. Perez-Reyes, E., Cribbs, L.L., Daud, A., Lacerda, A.E., Barclay, J., Williamson, M.P., Fox, M., Rees, M., Jung-Ha, Lee. Molecular characterization of a neuronal low-voltage-activated T-type calcium channel, *Nature* **1998**, 391, 896–900.

8. Tally, E.M., Cribbs, L.L., Lee, J.H., Daud, A., Perez-Reyes, E., Baylisss, D.A. Differential distribution of three members of a gene family encoding low voltage-activated (T-type) calcium channels. *J. Neurosci.* **1999**, 19, 1895–1911.

9. Yunker, A.M.R., Sharp, A.H., Sundarraj, V., Ranganathan, V., Copeland, T.D., Mcenery, M.W. Immunological characterization of T-type voltage-dependent calcium channel Cav3.1 (alpha1G) and Cav3.3 (alpah1I) isoforms reveal differences in their localization, expression, and neural development, *Neuroscience* **2003**, 117, 321–335.

10. Monteil, A., Chemin, J., Bourinet, E., Mennessier, G., Lory, P., Nargeot, J., Molecular and functional properties of the human alpha (1G) subunit that forms T-type calcium channels. *J. Biol. Chem.* **2000**, 275, 6090–6100.

11. Cribbs, L.L., Lee, J.H., Yang, J., Satin, J., Zhang, Y., Daud, A., Barclay, J., Williamson, M.P., Fox, M., Rees, M., Perez-Reyes, E. Cloning and characterization of $α1H$ from human heart, a member of the T-type calcium channel gene family. *Circ. Res.* **1998**, 83, 103–109.

12. Williams, M.E., Washburn, M.S., Hans, M., Urrutia, A., Brust, P.F., Prodanovich, P., Harpold, M.M., Stauderman, K.A. Structure and functional characterization of a novel human low-voltage activated calcium channel. *J. Neurochem.* **1999**, 72, 791–799.

13. Schrier, A.D., Wang, H., Talley, E.M., Perez-Reyes, E., Barrett, P.Q. $α1H$ T-type Ca^{2+} channel is the predominant subtype expressed in bovine and rat zona glomerulosa. *Am. J. Physiol. Cell Physiol.* **2001**, 280, C265–C272.

14. Lee, J.H., Daud, A.N., Cribbs, L.L., Lacerda, A.E., Pereverzev, A., Klockner, U., Schneider, T., Perez-Reyes, E. Cloning and expression of a novel member of the low voltage-activated T-type calcium channel family. *J. Neurosci.* **1999**, 19, 1912–1921.

15. Mittman, S., Guo, J., Emerick, M.C., Agnew, W.S. Structure and alternative splicing of the gene encoding $α1I$, a human brain T calcium channel $α$ 1 subunit. *Neurosci. Lett.*, **1999**, 269, 121–124.

16. Chemin. J., Monteil, A., Dubel, S., Nargeot, J., Lory, P. The $α1I$ T-type calcium channel exhibits faster gating properties when overexpressed in neuroblastoma/glioma NG 108-15 cells. *Eur. J. Neurosci.* **2001**, 14, 1678–1686.

17. Mittman, S., Guo, J., Agnew, W.S. Structure and alternative splicing of the gene encoding α1G, a human brain T calcium channel α1 subunit. *Neurosci. Lett.* **1999**, 274, 143–146.

18. Monteil, A., Chemin, J., Leuranguer, V., Altier, C., Mennessier, G., Bourinet, E., Lory, P., Nargeot, J. Specific properties of T-type calcium channels generated by the human α1I subunit. *J. Biol. Chem.* **2000**, 275, 16 530–16 535.

19. Murbartian, J., Arias, J.M., Lee, J.-H., Gomora, J.C., Perez-Reyes, E. Alternative splicing of the rat Cav3.3 T-type calcium channel gene produces variants with distinct functional properties. *FEBS Lett.* **2002**, 528, 272–278.

20. Hagiwara, S. Ozawa, S. Voltage clamp analysis of two inward current mechanisms in the egg cell membranes of a starfish. *J. Gen. Physiol.* **1975**, 65, 617–644.

21. Carbone, E., Lux, H.D. A low voltage-activated calcium conductance in embryonic chick sensory neurons. *Biophys. J.* **1984**, 46, 413–418.

22. Bean, B.P. Two kinds of calcium channels in canine atrial cells. Differences in kinetics, selectivity and pharmacology. *J. Gen. Physiol.* **1985**, 86, 1–30.

23. Benham, C.D., Hess, P., Tsien, R.W. Two types of calcium channels in single smooth muscle cells from rabbit ear artery studied with whole cell and single-channel recordings. *Circ. Res.* **1987**, 61S, I-10–I16.

24. Ertel, E.A. Pharmacology of Cav3 (T-type) channels, in *Calcium Channel Pharmacology* McDonough (ed.), Kluwer Academic/Plenum **2004**, pp. 183–236.

25. Xia, M., Imredy, J.P., Santarelli, V.P., Liang, H.A. Condra, C.L., Bennett, P. Koblan, K.S., Connolly, T.M. Generation and characterization of a cell line with inducible expression of $Ca_v3.2$ (T-type) channels. *Assay Drug Develop. Technol.* **2003**, 1, 637–645.

26. Zheng, W., Spencer, R.H., Kiss, L. High throughput assay technologies for ion channel drug discovery. *Assay Drug Develop. Technol.* **2004**, 2, 543–552.

27. Santarelli, V. Kiss, L., Connolly, T.M., Renger, J.J. Development of a high-throughput planar patch assay for T-type antagonists. *Biophysical Society Meeting, Abstract*, **2005**, #2549.

28. Qian, Y-X., Callamaras, N., McGivern, J.G. A high-throughput electrophysiological assay for T-type Ca^{2+} calcium channels on Ionworks HT. *Molecular Devices User Meeting*, **2004**.

29. Lee, J-H, Gomora, J.C., Cribbs, L.L, Perez-Reyes, E. Nickel block of three cloned T-type calcium channels: low concentrations selectively block α1H. *Biophys. J.*, **1999**, 77, 3034–3042.

30. Lacinova, L., Klugbauer, N., Hofman, F. Regulation of calcium channel α1G subunit by divalent cations and organic blockers. *Neuropharmacolgy* **2000**, 39, 1254–1266.

31. Huguenard, J.R. Low-threshold calcium currents in central nervous system neurons. *Annu. Rev. Physiol.* **1996**, 58, 329–348.

32. Zamponi, G.W., Bourinet, E., Snutch, T.P. Nickel block of a family of neuronal calcium channels: subtype- and subunit-dependent action at multiple sites. *J. Membr. Biol.* **1996**, 151, 77–90.

33. Joeng, S-W., Park, B-G., Park, J-Y., Lee, J-W., Lee, J-H. Divalent metals differentially block cloned T-type calcium channels. *Mol. Neurosci.* **2003**, 14, 1537–1540.

34. Beedle, A.M., Hamid, J., Zamponi, G.W. Inhibition of transiently expressed low- and high-voltage-activated calcium channels by trivalent metal cations. *J. Membr. Biol.* **2002**, 187, 225–238.

35. Chuang, R.S-I., Jaffe, H., Cribbs, L., Perez-Reyes, E., Swartz, K.J. Inhibition of T-type voltage-gated calcium channels by a new scorpion toxin. *Nat. Neurosci.* **1998**, 1, 668–674.

36. Coulter, D.A., Huguenard, J.R., Prince, D.A. Characterization of ethosuximide reduction of low-threshold calcium current in thalamic neurons. *Ann. Neurol.* **1989**, 25, 582–593.

37. Coulter, D.A., Huguenard, J.R., Prince, D.A. Differential effects of petit mal anticonvulsants and convulsants on thalamic neurons: calcium current reduction. *Br. J. Pharmacol.* **1990**, 100, 800–806.

38. Gomora, J.C.V., Daud, A.N., Weiergraber, M., Perez-Reyes, E. Block of cloned human T-type calcium channels by succinimide antiepileptic drugs. *Mol. Pharm.* **2001**, 60, 1121–1132.

39. Todorovic, S.M., Perez-Reyes, E., Lingle, C.J. Anticonvulsants but not general anesthetics have differential blocking effects on different T-type current variants. *Mol. Pharm.* **2000**, 58, 98–108.

40. Heady, T.N., Gomora, J.C., Macdonald, T.L., Perez-Reyes, E. Molecular pharmacology of T-type channels. *Jpn. J. Pharmacol.* **2001**, 85, 339–350.

41. Kito, K.M., Maehara, M., Watanabe, K. Mechanisms of T-type calcium channel blockade by zonisamide. *Seizure* **1996**, 5, 115–119.

42. Todorovic, S.M., Jevtovic-Todorovic, V., Mennerick, S., Perez-Reyes, E., Zorumski, C.F. Cav3.2 is a molecular substrate for inhibition of T-type calcium currents in rat sensory neurons by nitrous oxide. *Mol. Pharm.* **2001**, 60, 603–610.

43. Santi, C.M., Cayabyab, F.S., Sutton, K.G., McRory, J.E., Mezeyova, J., Hamming, K.S., Parker, D., Stea, A., Snutch, T.P. Differential inhibition of T-type calcium channels by neuroleptics. *J. Neurosci.* **2002**, 22, 396–403.

44. Enyeart, J.J., Biafi, B.A., Mlinar, B. Preferential block of T-type calcium channels by neuroleptics in neural crest-derived rat and human C cell lines. *Mol. Pharmacol.* **1992**, 42, 364–372.

45. Triggle, D.J. 1,4-Dihydropyridine calcium channel ligands: selectivity of action. The roles of pharmacokinetics, state-dependent interactions, channel isoforms, and other factors. *Drug Develop. Res.* **2003**, 58, 5–17.

46. Akaike, N., Kostyuk, P.G., Osipchuk, Y.V. Dihydropyridine-sensitive low-threshold calcium channels in isolated rat hypothalamic neurons. *J. Physiol.* **1989**, 412, 181–195.

47. Perchenet, L., Benardeau, A., Ertel, E.A. Pharmacological properties of Cav3.2, a low voltage-activated Ca2+ channel cloned from human heart. *Naunyn Schmiedebergs Arch. Pharmacol.* **2000**, 361, 590–599.

48. Kumar, P.P., Stotz, S.C., Paramashivappa, R., Beedle, A.M., Zamponi, G.W., Rao, A.S. Synthesis and evaluation of a new class of nifedipine analogs with T-type calcium channel blocking activity. *Mol. Pharmacol.* **2002**, 61, 649–658.

49. Furukawa, T., Nukada, T., Miura, R., Ooga, K., Honda, M., Watanabe, S., Koganewawa, S., Isshiki, T. Differential blocking action of dihydropyridines Ca^{2+} antagonists on a T-type Ca^{2+} channel 1G expressed in *Xenopus* oocytes. *J. Cardiovsasc. Pharmacol.* **2005**, 45, 241–246.

50. Furukawa, T., Miura, R., Honda, M., Kamiya, N., Mori, Y., Takeshita, S., Isshiki, T., Nukada, T. Identification of R(-)-isomer of efonidipine as a selective blocker of T-type Ca^{2+} channels *Br. J. Pharmacol.* **2004**, 143, 1050–1057.

51. Clozel, J.P., Ertel, S.I., Ertel, E.A. Discovery and main pharmacological properties of the novel calcium antagonist mibefradil (Ro 40-5967), the first selective T-type calcium channel blocker. *J. Hypertens.* **1997**, 15, S17–25.

52. Hefti, F., Clozel, J-P., Osterrieder, W. Antihypertensive properties of the novel calcium channel antagonist (1S,2S)-2-[2-[[3-(2-benzimidazolyl)propyl]methylamino]ethyl]-6-fluoro-1,2,3,4-tetrhydro-1-isopropyl-2-naphthyl methoxyacetate dihdrochloride in rat models of hypertension. *Arzneim. Forsch./Drug Res.* **1990**, 40, 417–421.

53. Oparil, S. Mibefradil, a T-channel-selective calcium channel antagonist. Clinical trials in hypertension. *Am. J. Hypertens.* **1998**, 11, 88S–94S.

54. Massie, B.M. Mibefradil: a selective T-type calcium antagonist. *Am. J. Cardiol.* **1997**, 80, 23I–32I.

55. Mishra, S.K., Hermsmeyer, K. Selective inhibition of T-type Ca^{2+} channels by Ro 40-5967. *Circ. Res.* **1994**, 75, 144–148.

56. Bernardeau, A., Ertel, E.A. Selective block of myocardial T-type calcium channels by mibefradil: a comparison with the 1, 4-dihydopyridine amlodipine, in *Low-voltage Activated T-Type Calcium Channels*, Tsien, R.W, Clozel, J-P, Nargeot, J, eds, Adis International Limited, Chester, England, pp. 386–394, **1998**.

57. McDonough, S.I., Bean, B.P. Mibefradil inhibition of T-type calcium channels in cerebellar purkinje neurons. *Mol. Pharm.*, **1998**, 54, 1080–1087.

58. Randall, A.D., Tsien, R.W., Contrasting biophysical and pharmacological properties of T-type and R-type calcium channels. *Neurpopharmacology* **1997**, 36, 879–893.

59. Arnoult, C., Villaz, M. et al. Pharmacological properties of the T-type Ca^{2+} current of mouse spermatogenic cells. *Mol. Pharmacol.* **1998**, 53, 1104–1111.

60. Martin, R.L., Lee, J-H., Cribbs, L.L., Perez-Reyes, E., Hanck, D.A. Mibefradil block of cloned T-type calcium channels. *J. Pharmacol. Exp. Ther.*, **2000**, 295, 302–308.

61. Leuranguer, V., Mangoni, M.E., Nargeot, J., Richard, S. Inhibition of T-type and L-type calcium channels by mibefradil: physiologic and pharmacologic bases of cardiovascular effects. *J. Cardiovasc. Pharmacol.* **2001**, 37, 649–661.

62. Perez-Reyes, E. Paradoxical role of T-type calcium channels in coronary smooth muscle. *Mol. Intervent.*, **2004**, 4, 16–18.

63. Viana, F., Van den Bosch, L., Missiaen, L., Vandenberghe, W., Droogmands, G., Nilius, B., Robbrecht, W. Mibefradil (Ro 40-5967) blocks multiple types of voltage-gated calcium channels in cultured rat spinal motor neurones. *Cell Calcium* **1997**, 22, 299–311.

64. Eller, P., Berjukov, S., Wanner, S., Huber, I., Hering, S., Knaus, H.G., Toth, G., Kimball, S.D., Striessing, J. High affinity interaction of mibefradil with voltage-gated calcium and sodium channels. *Br. J. Pharmacol.* **2000**, 130, 669–677.

65. Gomora, J.C., Enyeart, J.A., Enyeaart, J.J. Mibefradil potently blocks ATP-activated K^+ channels in adrenal cells. *Mol. Pharmacol.* **1999**, 56, 1192–1197.

66. Perchenet, L., Clement-Chomienne, O. Characterization of mibefradil block of the human heart delayed rectifier hKv1.5. *J. Pharmacol. Exp. Ther.* **2000**, 295, 771–778.

67. Liu, J.H., Bijienga, P., Occhiodoro, T., Fischer-Lougheed, J., Bader, C.R., Bernheim, L. Mibefradil (Ro 40-5967) inhibits several Ca^{2+} and K^+ currents in human fusion-competent myoblasts. *Br. J. Pharmacol.* **1999**, 126, 245–250.

68. Koh., S.D., Monaghan, K., Ro, S., Mason, H.S., Kenyon, J.L., Sanders, K.M. Novel voltage-dependent non-selective cation conductance in murine colonic myocytes. *J. Physiol. (London)* **2001**, 533, 341–355.

69. Ito, C., Im, W.B., Takagi, H., Takahashi, M., Tsuzuki, K., Liou, S., Kunihara, M. U-92032, a T-type Ca^{2+} channel blocker and antioxidant reduces neuronal ischemic injuries. *Eur. J. Pharmacol.* **1994**, 257, 203–210.

70. Porcello, D.M., Smith, S.D., Huguenard, J.R. Actions of U-92032, a T-type Ca^{2+} channel antagonist, supports a functional linkage between I_T and slow intrathalamic rhythms. *J. Neurophysiol.* **2003**, 89, 177–185.

71. Gray, L.S., Perez-Reyes, E., Gamorra, J.C., Haverstick, D.M., Shattock, M., McLatchie, L., Harper, J., Brooks, G., Heady, T., Macdonald, T.L., The role of voltage gated T-type Ca^{2+} channel isoforms in mediating "capacitative" Ca^{2+} entry in cancer cells. *Cell Calcium* **2004**, 36, 489–497.

72. Jung, H.K., Doddareddy, M.R., Cha, J.H., Rhim, H., Cho, Y.S., Koh, H.Y., Jung, B.Y., Pae, A.N. Synthesis and biological evaluation of novel T-type Ca^{2+} channel blockers. *Bioorg. Med. Chem. Lett.* **2004**, 14, 3965–3970.

73. Rhim, H., Lee, Y.S., Park, S.J., Chung, B.Y., Lee, J.Y. Synthesis and biological activity of 3,4-dihydroquinazolines for selective T-type Ca^{2+} channel blockers. *Bioorg. Med. Chem. Lett.* **2005**, 15, 283–286.

74. Doddareddy, M.R., Jung, H.K., Lee, J.Y., Lee, Y.S., Cho, Y.S., Koh, H.Y., Pae, A.N. First pharmacophoric hypothesis for T-type calcium channel blockers. *Bioorg. Med. Chem.* **2004**, 12, 1605–1611.

75. Chemin, J., Monteil, A., Perez-Reyes, E., Nargeot, J., Lory, P. Direct inhibition of T-type calcium channels by the endogenous cannabinoid anandamide. *EMBO J.* **2001**, 20, 7033–7040.

76. Leuranger, V., Mangoni, M.E., Nargeot, J., Richard, S. Inhibiton of T-type and L-type calcium channels by mibefradil: physiologic and pharmacologic bases of cardiovascular effects. *J. Cardiovasc. Pharmacol.* **2001**, 37, 649–661.

77. Chen, C-C., Lamping, K.G., Nuno, D.W., Basresi, R., Prouty, S.J., Lavoie, J.L., Cribbs, L.L., England, S.K., Sigmund, C.D., Weiss, R.M., Williamson, R.A., Hill, J.A., Campbell, K.P. Abnormal coronary function in mice deficient in α1H T-type

Ca^{2+} channels. *Science,* **2003**, 302, 1416–1418.

78. Kim, D., Park, D., Choi, S., Lee, S., Sun, M., Kim, C., Shin, H-S. Thalamic control of visceral nociception mediated by T-type Ca^{+2} channels. *Science* **2003**, 302, 117–119.

79. Hansen, P.B., Jensen, B.L., Andreasen, D., Skott, O. Differential expression of T- and L-type voltage dependent calcium channels in renal resistance vessels. *Circ. Res.* **2001**, 89, 630–638.

80. Feng, M-G., Li, M., Navar, L.G. T-type calcium channels in the regulation of afferent and efferent arterioles in rats. *Am. J. Physiol. Renal Physiol.* **2004**, 286, F331–F337.

81. Karam, H., Clozel, J-P., Bruneval, P., Gonzalez, M-F., Menard, J. Contrasting effects of selective T- and L-type calcium channel blockade on glomerular damage in DOCA hypertensive rats. *Hypertension* **1999**, 34, 673–678.

82. Menard, J., Karam, H., Veniant, M., Heudes, D., Bruneval, P., Clozel, J-P. Effects of calcium blockade on end-organ damage in experimental hypertension. *J. Hypertension* **1997**, 15S, S-19–S30.

83. Hayashi, K., Kumagai, H., Saruta, T. Effect of Efonidipine and ACE inhibitors on proteinuria in human hypertension with renal impairment, *Am. J. Hypertension* **2003**, 16, 116–122.

84. Schmitt, R., Clozel, J.P. Iberg, N., Buhler, F.R. Mibefradil prevents neointima formation after vascular injury in rats. Possible role of the blockade of the T-type voltage-operated calcium channel. *Arterioscler. Thromb. Vasc. Biol.* **1995**, 15, 1161–1165.

85. Fareh, S., Benardeau, A., Nattel, S. Differential efficacy of L- and T-type calcium channel blockers in preventing tachycardia-induced atrial remodeling in dogs. *Cardiovasc. Res.* **2001**, 49, 762–770.

86. Ohashi, N., Mitamura, H., Tanimoto, K., Fukuda, Y., Kinebuchi, O., Kurita, Y., Shiroshita-Takeshita, A., Miyoshi, S., Hara, M., Takatsuki, S., Ogawa, S. A comparison between calcium channel blocking drugs with different potencies for T- and L-type channels in preventing atrial electrical remodeling. *J. Cardiovasc. Pharmacol.* **2004**, 44, 386–392.

87. Dogrul, A., Gardell, L.R., Ossipov, M.H., Tulunay, C.F., Lai, J., Porreca, F. Reversal of experimental neuropathic pain by T-type calcium channel blockers. *Pain,* **2003**, 105, 159–168.

88. Flatters, S.J., Bennett, G.J. Ethosuximide reverses paclitaxel- and vincristine-induced painful peripheral neuropathy. *Pain,* **2004**, 109, 150–161.

89. Todoric, S.M., Meyenburg, A., Jevtovic-Todorovic, V. Mechanical and thermal antinociception in rats following systemic administration of mibefradil, a T-type calcium channel blocker. *Brain Res.* **2002**, 951, 336–340.

90. Bourinet, E., Alloui, A., Monteil, A., Barrere, C., Couette, B., Poirot, O., Pages, A., McRory, J. Snutch, T.P., Eschalier, A., Nargeot, J. Silencing of the Cav 3.2 T-type calcium channel gene in sensory neurons demonstrates its major role in nociception. *EMBO J.* **2005**, 24, 315–324.

91. Llinas, R.R., Ribary, U., Jeanmonod, D., Kronberg, E., Mitra, P.P. Thalamocortical dysrythmia: a neurological and neuropsychiatric syndrome characterized by magnetoencephalography. *Proc. Natl. Acad. Sci. U.S.A.,* **1999**, 96, 15 222–15 227.

92. Kim, D., Song, I., Keum, S., Lee, T., Jeong, M-J., Kim, S-S., McEnery, M.W., Shin, H-S. Lack of the burst firing of thalamocortical relay neurons and resistance to absence seizures in mice lacking α1G T-type Ca^{2+} channels. *Neuron,* **2001**, 31, 35–45.

93. Steriade, M., McCormick, D.A., Sejnowski, T.J. Thalamocortical oscillations in the sleeping and aroused brain. *Science,* **1993**, 262, 679–685.

94. McCormick, D.A., Bal, T. Sleep and arousal: thalamocortical mechanisms. *Annu. Rev. Neurosci.* **1997**, 20, 185–215.

95. Lee, J., Kim, D., Shin, H-S. Lack of delta waves and sleep disturbances during non-rapid eye movement sleep in mice lacking α1G-subunit of T-type calcium channels. *Proc. Natl. Acad. Sci. U.S.A.,* **2004**, 101, 18 195–18 199.

96. Anderson, M.P., Mochizuki, T., Xie, J., Fischler, W., Manger, J.P., Talley, E.M., Scammell, T.E., Tonegawa, S. Thalamic Cav3.1 T-type Ca^{2+} channel plays a crucial role in stabilizing sleep. *Proc. Natl. Acad. Sci. U.S.A.,* **2005**, 102, 1743–1748.

5.3
L-type Calcium Channels

David J. Triggle

5.3.1
Introduction

The discovery of the groups of drugs collectively referred to as *"Calcium antago-nists"*, *"Calcium channel blockers"*, or *"Calcium entry blockers"* is intimately linked to an understanding of the roles of cellular calcium. Calcium plays dual roles in cellular function. It serves as an inward current-carrying species mediating cellular depolarization, and thus contributing to the control of cellular events through membrane potential control as in the plateau current of the cardiac ac-tion potential. Calcium serves also as a key cellular intermediate mediating stim-ulus–response coupling in its many forms, including excitation–contraction and stimulus–secretion coupling. Thus, such drugs will depress both cellular excitabil-ity *and* stimulus–response coupling. Voltage-gated calcium channels constitute a large family with at least five major classes, namely L-type ($Ca_V1.1–1.4$), P/Q-type ($Ca_V2.1$), N-type ($Ca_V2.2$.), R-type ($Ca_V2.3$) and T-type ($Ca_V3.1–3.3$); there will also be a corresponding series of calcium antagonists (one series for each channel class) and likely a parallel series of calcium activators – drugs that maintain the channel in the open state [1; Chapter 3]. Although molecules, both small synthetic drugs and large peptides or toxins, exist for each channel class the term *"calcium antagonist"* or its equivalent has been historically employed for drugs that interact with the L-type channel since these represent the most extensively investigated *and* clinically available agents [2]. Pharmacological study of the L-type channel is now a mature field, and fundamental research has largely shifted to other members of the calcium channel family.

The current extensive investigations on other calcium channel classes will likely produce drugs that are both pharmacologically selective and therapeutically useful in several areas, including both neuronal and cardiovascular diseases (Chapters 8.1 and 8.2). Some reasons for the slower development of drugs active at other channel classes have been discussed [3]. Of related interest is the fact that the L-type channel has been best characterized pharmacologically by small synthetic molecule ligands that exhibit potency and selectivity, whereas the other members of the voltage-gated calcium channel family are best characterized by potent natu-rally occurring peptide toxins, or by non-selective small molecules [3].

5.3.2
Drugs that Interact with L-type Channels

The discovery and classification of this class of drugs owes much to the work of Albrecht Fleckenstein and his colleagues who showed that prenylamine and ver-apamil produced electromechanical uncoupling in the heart, that these effects mi-micked those following the withdrawal of extracellular calcium, and that the inhi-

bitory effects of these drugs could be blocked by procedures, including the elevation of extracellular calcium levels, that increase calcium mobilization (reviewed
in [4]). Similarly, Theophile Godfraind and his colleagues showed that diphenyl-
piperazine-containing molecules, including cinnarizine and flunarizine, blocked
excitation–contraction coupling in smooth muscle (reviewed in [5]). Subsequently,
these findings were extended to structurally diverse agents, including diltiazem
and nifedipine, suggesting the remarkable diversity of molecules that interact
with the L-type channel, a suggestion amply borne out by many subsequent investigations [6–8]. These agents act dominantly in the cardiovascular system and are
widely employed in the treatment of hypertension, angina and selectively for
some arrhythmias and peripheral vascular disorders. Although these drugs
have a common target, the L-type calcium channel, they exhibit considerable selectivity of action between cardiac and vascular actions (Table 5.3.1) that underlies
their differential therapeutic utility. This selectivity of action is considerably enhanced in second-generation agents of the 1,4-dihydropyridine class. The differential selectivity between the three principal classes of calcium antagonists has
several origins, including interaction at distinct binding sites associated with
topographically distinct regions of the channel α_1 subunit [9–11; Fig. 5.3.1].

Tab. 5.3.1 Calcium channel antagonists. cardiovascular effects and therapeutic uses.

Parameter	Nifedipine	Verapamil	Diltiazem
Coronary tone	+++ decrease	++ decrease	++ decrease
Coronary flow	– – – increase	– – increase	– – increase
Peripheral vasodilatation	– – – increase	– – increase	– increase
Cardiac rate	++ increase[a]	– decrease	– decrease
Cardiac contractility	No effect[b]	– decrease	– decrease
AV node conduction	No effect	– decrease	– decrease
AV node ERP	No effect	– decrease	– decrease
Therapeutic use			
Angina, exertional	+++	++	++
Angina, variant	+++	++	++
Angina, unstable	+	++	++
Arrhythmia, PSVT	NU	+++	++
Atrial fibrillation/flutter	NU #	++	++
Hypertension	+++	++	++
Raynaud's disease	++	++	++
Cerebral vasospasm	++[c]	NU	NU

[a], [b] May be increased following rapid administration due to reflex sympathetic activation.
[c] Nimodipine (Nimotop).
+ = frequency of use or strength of effect;   NU = not used.

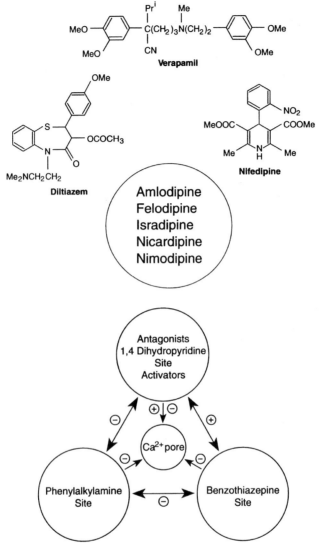

Fig. 5.3.1 Structural formulae of calcium antagonists and a schematic arrangement of the three allosterically linked drug binding sites.

The structure–activity relationships of molecules around the L-type channel have been the subject of many reviews [6–8, 11–13]. This chapter will therefore not be exhaustive, but rather will review highlights and more recent findings and for the older literature will refer to the large number of available and comprehensive reviews.

5.3.2.1 Major Structural Classes of Drugs

The principal (and the most selective) classes of drugs are represented by the pro-totypical verapamil (phenylalkylamine), diltiazem (benzothiazepine) and nifedi-pine (1,4-dihydropyridine) (Fig. 5.3.1). The nifedipine class is also represented by several clinically available second-generation agents. Of some interest is the ability of the 1,4-dihydropyridine nucleus to serve as a *"privileged structure"*, cap-able of interacting when appropriately substituted with various cellular targets [11, 14]. Although several analogs of both verapamil and diltiazem have been studied, none have achieved clinical utility [15, 16].

5.3.2.2 Other Structural Classes of Drugs

Diverse molecular structures, several of which are included in Fig. 5.3.2, also pos-sess blocking activity at L-type channels, although often without the selectivity of the molecules depicted in Fig. 5.3.1. These structures have been reviewed else-where [6–8, 11–13].

5.3.3
Specific Drug Classes

The 1,4-dihydropyridine class of molecule has been particularly well investigated, but the verapamil and diltiazem structures have been progressively less well stud-ied. There are likely three underlying reasons – ease and versatility of synthesis of the 1,4-dihydropyridine structure, the potency of many 1,4-dihydropyridines, and the existence of *both* antagonist and activator properties in this series of com-pounds. At least two principal factors complicate current interpretations of struc-ture–activity relationships of these calcium channel antagonists (and other classes of drugs active at ion channels) – state-dependent binding properties whereby the potency of the molecule depends on the state of the channel as determined by membrane potential (for further discussion see Chapter 3), and the tissue-depen-dent presence of isoforms of the oligomeric complex whereby comparisons between tissues and even within a single tissue are complicated because of the presence of two or more isoforms with different binding properties. No comprehensive structure–activity relationships are available for a molecularly defined channel complex under voltage-controlled conditions.

5.3.3.1 Benzothiazepine Class

Structure–activity studies around the basic skeleton of diltiazem have revealed a few molecules with greater potency [6–8]. Critical features of the diltiazem struc-ture (Fig. 5.3.1) include the stereochemistry with the levorotatory *cis*-enantiomer being the most potent, the presence of the dimethylaminoethyl group (primary and tertiary amines have significantly less activity), and the presence of the sub-stituted phenyl group at C2. Substantial change can be affected at C3 and substi-tuents can be introduced at C7 and C8 – the 8-chloro analog is more potent than

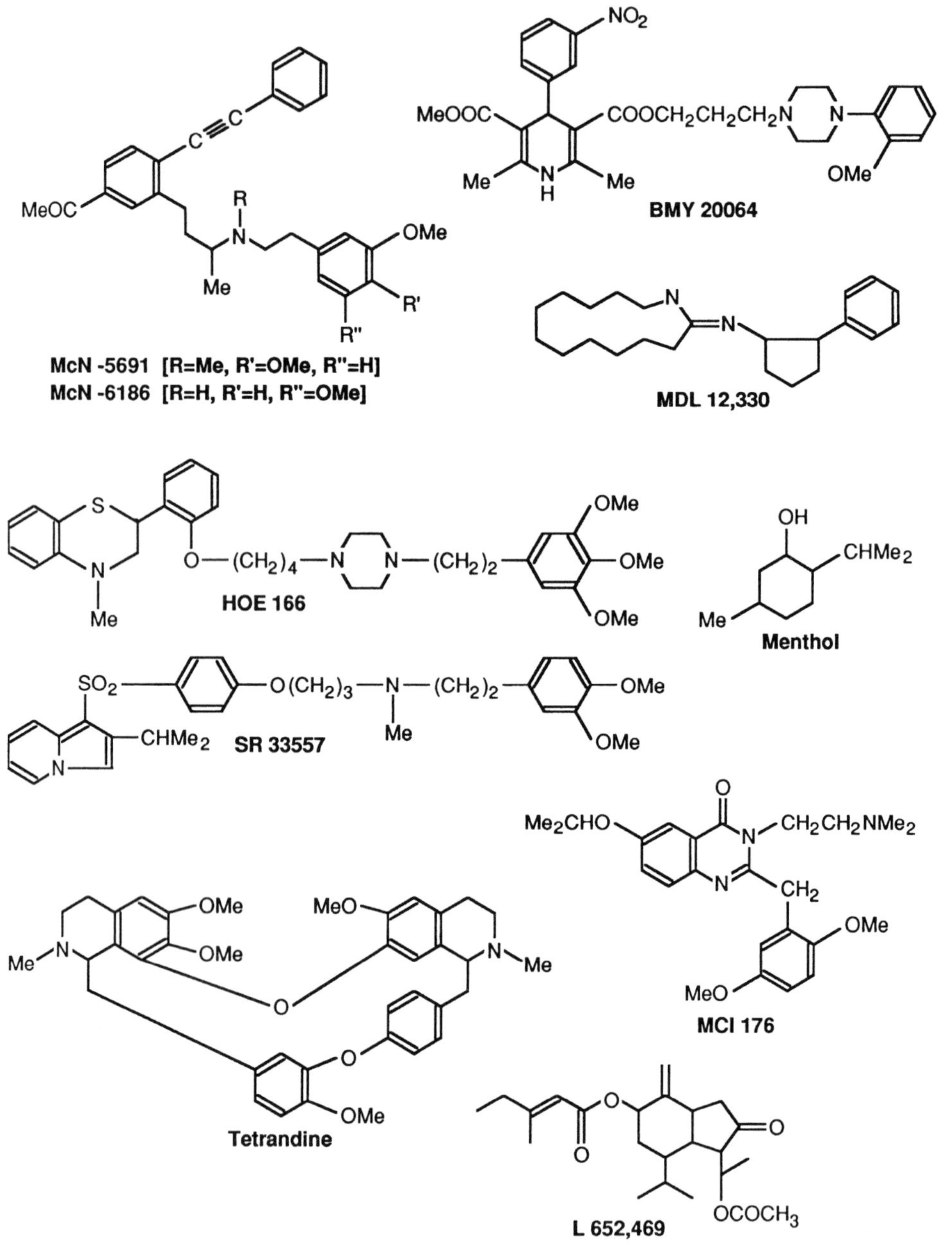

McN -5691 [R=Me, R'=OMe, R"=H]
McN -6186 [R=H, R'=H, R"=OMe]

BMY 20064

MDL 12,330

HOE 166

SR 33557

Menthol

MCI 176

Tetrandine

L 652,469

Fig. 5.3.2 Molecular structures of agents active as antagonists at L-type channels.

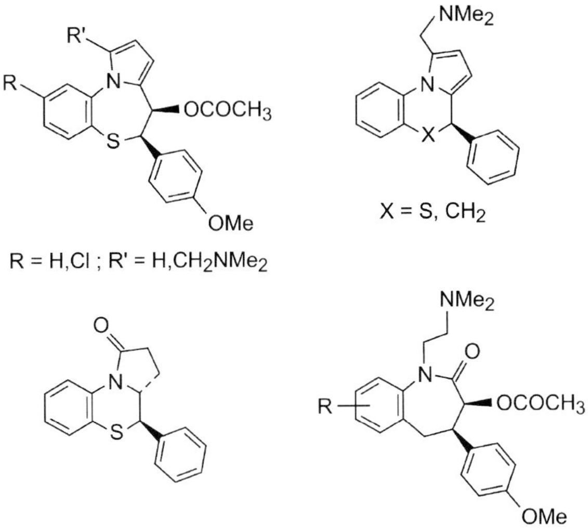

Fig. 5.3.3 Analogs of diltiazem.

R = H,Cl ; R' = H,CH$_2$NMe$_2$

X = S, CH$_2$

diltiazem. Various ring analogs of diltiazem have been prepared (Fig. 5.3.3), but these studies have not led to major advances in potency or efficacy over diltiazem itself, which remains the only clinically available member of this class [17–20].

5.3.3.2 Phenylalkylamine Class

Like diltiazem, verapamil is the only clinically available member of its class, but unlike diltiazem is available only in the racemic form. There are, however, significant differences in the pharmacodynamic and pharmacokinetic behavior of the enantiomeric forms (reviewed in [21, 22]). The (–)-S enantiomer of verapamil is some 10 times more potent than the (+)-R enantiomer in depressing atrio-ventricular conduction and depressing smooth muscle contractility. However, there are both pharmacokinetic and pharmacodynamic aspects to this stereoselectivity. Because of stereoselective first-pass metabolism that is greater for the S-enantiomer verapamil appears to be less potent following oral administration than after intravenous administration. Additionally, the stereoselectivity of verapamil action is different in vascular and cardiac muscle, being greater in cardiac than smooth muscle, primarily because of the lower cardiodepressant activity of the R- relative to the S-enantiomer [23] Reviews of structure–activity relationships in this series are available and some analogs of verapamil are depicted in Fig. 5.3.4 [6–8, 24]. The activity of verapamil is enhanced by chorine and trifluoromethyl substitution and desmethoxyverapamil, anipamil and ronipamil also have enhanced activity relative to verapamil.

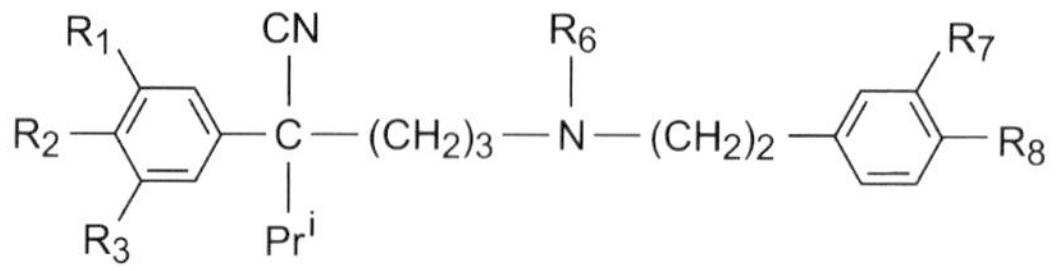

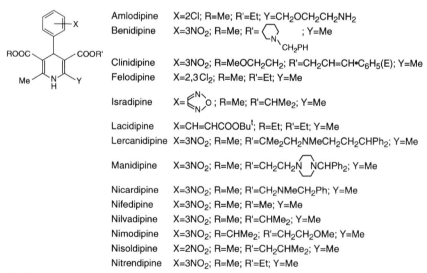

Fig. 5.3.4 Analogs of verapamil (if unspecified, R = H).

Verapamil:	R₂=OMe; R₃=OMe;	R₆=Me;	R₇=R₈=OMe
D600:	R₁=R₂=R₃=OMe;	R₆=Me;	R₇=R₈=OMe
DesmethoxyVP:	R₂=R₃=OMe;	R₆=Me;	R₇=OMe
Anipamil:	R₃=OMe;	R₆=Me;	R₇=OMe
Ronipamil:	R₆=Me		
#18:	R₁=R₂=Cl;	R₆=Me;	R₇=R₈=OMe
#19:	R₁=CF₃;	R₆=Me;	R₇=R₈=OMe

5.3.3.3 1,4-Dihydropyridine Class

1,4-Dihydropyridines are the best explored of the calcium channel antagonists and they also represent the largest clinically available class (Fig. 5.3.5). As a class they exhibit several interesting characteristics, including the highest potency, the ability to exhibit significantly varying degrees of vascular:cardiac selectivity, including the availability of cardiac-selective agents, the ability to exhibit regional vascular bed selectivity, the ability in appropriately substituted compounds to exhibit potent channel activator properties, stereoselectivity where the enantiomers may exhibit both qualitatively and quantitatively different biological activities, the ability of the 1,4-dihydropyridine nucleus to serve as a privileged structure to gen-

Amlodipine	X=2Cl; R=Me; R'=Et; Y=CH₂OCH₂CH₂NH₂
Benidipine	X=3NO₂; R=Me; R'=(ring) ; Y=Me
Clinidipine	X=3NO₂; R=MeOCH₂CH₂; R'=CH₂CH=CH•C₆H₅(E); Y=Me
Felodipine	X=2,3Cl₂; R=Me; R'=Et; Y=Me
Isradipine	X=(ring) ; R=Me; R'=CHMe₂; Y=Me
Lacidipine	X=CH=CHCOOBuᵗ; R=Et; R'=Et; Y=Me
Lercanidipine	X=3NO₂; R=Me; R'=CMe₂CH₂NMeCH₂CH₂CHPh₂; Y=Me
Manidipine	X=3NO₂; R=Me; R'=CH₂CH₂N NCHPh₂; Y=Me
Nicardipine	X=3NO₂; R=Me; R'=CH₂NMeCH₂Ph; Y=Me
Nifedipine	X=3NO₂; R=Me; R'=Me; Y=Me
Nilvadipine	X=3NO₂; R=Me; R'=CHMe₂; Y=Me
Nimodipine	X=3NO₂; R=CHMe₂; R'=CH₂CH₂OMe; Y=Me
Nisoldipine	X=2NO₂; R=Me; R'=CH₂CHMe₂; Y=Me
Nitrendipine	X=3NO₂; R=Me; R'=Et; Y=Me

Fig. 5.3.5 1,4-Dihydropyridine antagonists.

erate non-channel activities, and the actions of some 1,4-dihydropyridines to serve as NO generators. Several reviews are available [6, 11, 12, 25–27].

The principal features defining the structure–activity relationships in the 1,4-dihydropyridine series are depicted in Fig. 5.3.6. Ester substituents at C3 and C5 confer optimum antagonist activity, although considerable variation in ester group size is permissible and can generate both maintenance and enhancement of activity relative to nifedipine, the simplest member of the series. The ester groups when dissimilar confer chirality at C4 with stereoselectivity of interaction. Quite generally, stereoselectivity increases with increasing difference in ester group bulk, which is consistent with significantly different ester binding sites in the port and starboard areas of the receptor [6, 12, 28]. Such differentiation also applies when chirality is conferred by asymmetric substitution at C2 and C6. Thus, nitrendipine (Fig. 5.3.5) has a modest stereoselectivity index of between 5 and 10 in both binding and functional studies, whereas amlodipine (Fig. 5.3.5) has a stereoselectivity (–)/(+) of approximately 1000 [12, 28, 29]. A similar degree of stereoselectivity exists for tiamdipine analogs of amlodipine [30]. A series of nifedipine analogs bearing charged quaternary ammonium function (Fig. 5.3.7) served as molecular rulers to probe the location of the 1,4-dihydropyridine binding site within the membrane-spanning α_1-subunit [31].

The presence of the aromatic ring and its substituents at C4 are of particular importance to biological activity. The absence of a phenyl ring or its substitution by aliphatic or alicyclic groups results in loss of activity. Similarly, replacement of the phenyl ring with heterocyclic nuclei, furan, pyridine, thiophene etc, also confers loss of activity [32]. Optimum activity is associated with a substituted phenyl

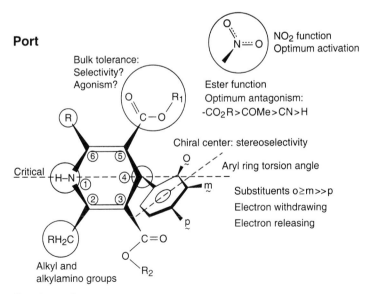

Fig. 5.3.6 Structural requirements for 1,4-dihydropyridine activation and antagonism.

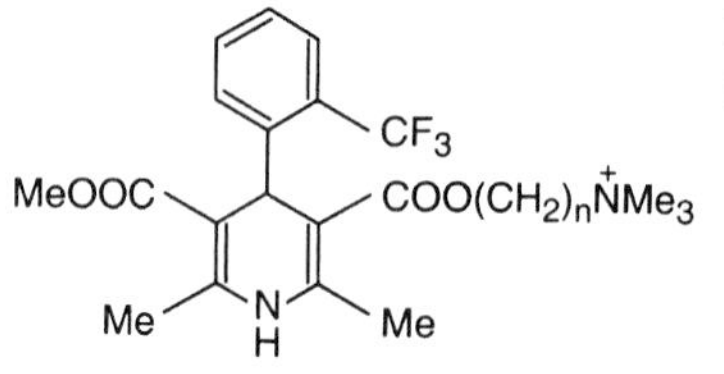

Fig. 5.3.7 A 1,4-dihydropyridine with a permanent positive charge on the side-chain.

ring at C4 with small electron-withdrawing substituents in the *o*- and *m*-positions. Substitution into the *p*-position is detrimental to activity.

Appropriately substituted 1,4-dihydropyridines show potent channel activator properties with relatively small molecular change. Optimum activator properties are associated with a nitro group at C3 and weaker activator properties are associated with lactone groups at this position. Such substitution confers chirality at C4 whereby one enantiomer possesses activator properties and the other enantiomer antagonist properties (Fig. 5.3.8) [11, 26, 33]. Such *qualitative* differences in stereoselectivity of interaction contrast with the more commonly observed quantitative differences [23] and are consistent with the enantiomers interacting with different conformations of the binding site associated with the open (activator) and inactivated (antagonist) state of the channel [23, 26, 33]. Other activator structures include a benzoylpyrrole series typified by FPL 64176 [34–36], the benzacocine CGP 48506 [37], and the phytohormone abscisic acid [38] (Fig. 5.3.9).

The C2 methyl group of nifedipine is replaceable with various substituents. The aminoethylmethoxy group in amlodipine (Fig. 5.3.5) both maintains activity and confers the desirable pharmacokinetic properties of long duration and slow onset and offset of action [28, 39]. The NH function of the 1,4-dihydropyridine ring is essential to activity. Replacement by methyl or other functions abolishes activity and oxidation of the ring to pyridine (the major metabolic route) is also associated with a complete loss of activity. This is consistent with a critical H-bonding inter-

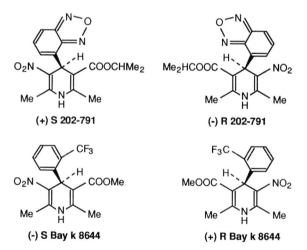

(+) S 202-791

(-) R 202-791

(-) S Bay k 8644

(+) R Bay k 8644

Fig. 5.3.8 Stereoisomeric 1,4-dihydropyridine activator:antagonist pairs.

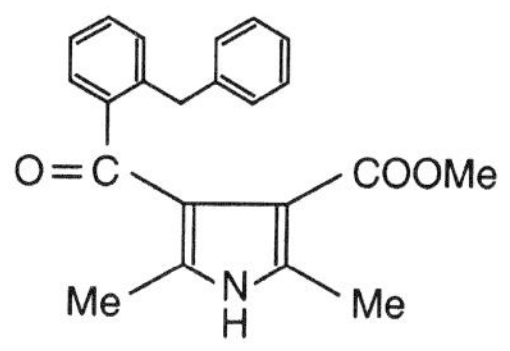

Fig. 5.3.9 L-Type calcium channel activators.

FPL 64176

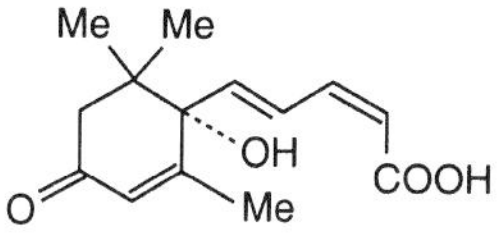

Abscisic Acid

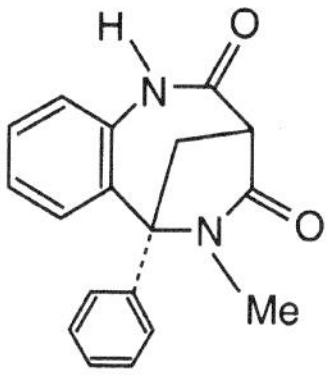

CGP 48506

action of this NH function [27, 40, 41]. The 1,4-dihydropyrimidine ring can substitute effectively for the 1,4-dihydropyridine ring as an isosteric replacement and such dihydropyrimidines are approximately equipotent to the corresponding 1,4-dihydropyridines and have a similar solid and solution state conformation [42].

Several studies have attempted to delineate the conformational requirements for 1,4-dihydropyridine action and an extensive collection of solid state X-ray and solution-state NMR data is available [6, 11, 12, 26, 43–45]. Several general conclusions can be drawn. The 1,4-dihydropyridine moiety has a boat conformation with the substituted 4-phenyl ring in the pseudoaxial orientation: *"port"* and *"starboard"* space can be defined (Fig. 5.3.6) where substituent interactions at the C3 and C5 positions define antagonist and activator properties. In principle, substituents may have antiperiplanar (ap) or synperiplanar (sp) orientations (Fig. 5.3.10): the study of rigid analogs indicates that optimum activity is associated with the sp orientation [12, 26, 46]. Similarly, the C3 and C5 ester substituents can be oriented in the cis- or trans-orientations (Fig. 5.3.10): here also the study of rigid analogs suggests that the cis-orientation is preferred [12, 26]. Of particular interest, a linear relationship exists between antagonist activity and the degree of planarity of the 1,4-dihydropyridine ring – activity increasing with increasing ring planarity (Fig. 5.3.11) [12, 43–45, 47]. This correlation is consistent with a stacking interaction between the 1,4-dihydropyridine ring and an aromatic residue of the binding site.

QSAR studies have been performed for series of 1,4-dihydropyridines [48–52]. The most extensive analysis is that by Coburn et al. [48] whose extensive compound set of 46 phenyl-substituted 1,4-dihydropyridines has been used by several groups. These workers found the correlation given by Eq. (1), indicating that steric interactions at the meta′ position are more unfavorable than at the meta position and that steric interactions are still less favorable at the para position of the phenyl ring.

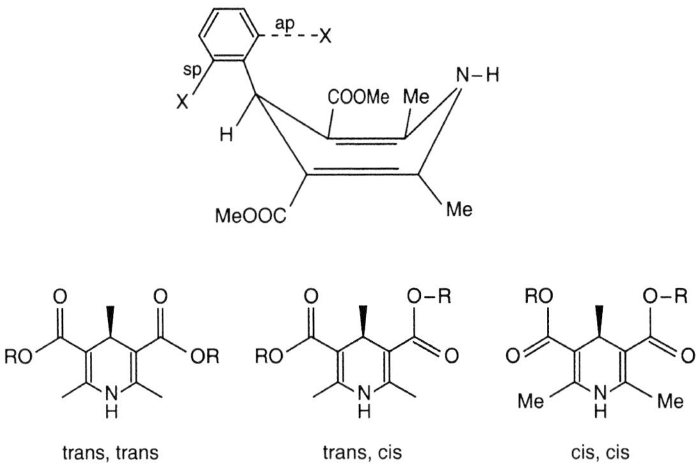

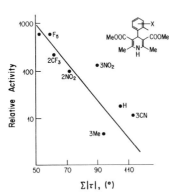

trans, trans   trans, cis   cis, cis

Fig. 5.3.10 Conformational arrangements in 1,4-dihydropyridines (see text for further details).

Fig. 5.3.11 Relationship between calcium channel antagonism and the planarity of the 1,4-dihydropyridine ring (sum of torsion angles around the ring).

$$\log 1/IC_{50} = 0.69\pi + 1.96\sigma_m - 0.44L_{meta} - 3.26B_{1para} - 1.51L_{meta'}$$
$$+ 14.23; \; n = 46, \; r = 0.90 \tag{1}$$

From these structure–activity relationships a number of speculations have been made concerning the mode of interaction of 1,4-dihydropyridines at the receptor site. These speculations remain just that for several reasons; first, the biological data are drawn almost exclusively from tissue experiments and likely refer to interaction at two or more isoforms of the channel; second, the 1,4-dihydropyridines exhibit prominent state-dependence of interaction and the biological data does not, unlike electrophysiological data, necessarily define adequately the equilibria between the several channel states.

The clinically available 1,4-dihydropyridines are all vascular selective, albeit to varying degrees (Section 5.3.3.3.1). However, cardioselective 1,4-dihydropyridines are known, including derivatives in the amlodipine series [53] (Table 5.3.2). A

Tab. 5.3.2 Vascular:cardiac selectivity of amlodipine analogs [53].

C2 substituent	Vascular:cardiac ratio
$CH_2OCH_2CH_2NH_2$ (Amlodipine)	10:1
$CH_2OCH_2CH_2NHCH_2$ $H_2NCH_2CH_2NHCO$	1:1
$CH_2OCH_2CH_2NHCONH$-3-HO- pyrimidine	1:10

Tab. 5.3.3 Activator and antagonist properties of 4-(2-pyridyl) substituted 1,4-dihydropyridines [32].

Pyridine substituent		1,4-DHP substituent		$10^6 \times EC_{50}$ (M) Ant. Smooth	$10^6 \times EC_{50}$ (M) m. Act., heart
C2	C5	C3	C5		
Me	H	NO_2	COOPr^i	25	18
Me	H	NO_2	COOEt	17	12
H	Me	NO_2	COOPr^i	2.8	9.9
Cl	H	NO_2	COOPr^i	17	5.9

series of 1,4-dihydropyridines bearing a C4 2-pyridyl group have both activator and antagonist properties, being cardiostimulant and vasorelaxant, respectively, in their actions, a potential benefit for cardiac failure (Table 5.3.3) [32].

Regrettably few data are available for structure–activity studies on expressed single isoforms of calcium channels: such data will be helpful both for further clarifying existing data and for the possible generation of new 1,4-dihydropyridines with enhanced tissue selectivity. Welling et al. [54, 55] have examined the cardiac and smooth muscle splice variants of the $Ca_V1.2$ a1 subunit. Their data show that the smooth muscle splice variant was more sensitive to low concentrations of nisoldipine, showed greater block at negative membrane potentials and developed faster block at depolarized membrane potentials. Table 5.3.4 compares the activities of several 1,4-dihydropyridines at native $Ca_V1.2a$ and $Ca_V1.2b$ isoforms [56]: these data reveal greater differences in activity at polarized levels of membrane potential (resting channel block).

5.3.3.3.1 Vascular:Cardiac Selectivity and State-dependent Interactions of 1,4-Dihydropyridines

In principle, the tissue selectivity (vascular:cardiac activity ratio) of action of the calcium channel antagonists depends upon several factors, including the mode of calcium mobilization in the tissue(s) in question, pharmacokinetic factors,

Tab. 5.3.4 Interaction of 1,4-dihydropyridines with splice variants of L-type channel [56].

	K_D −100 mV (nM)	K_D −50 mV (nM)
	α_{1Ca}	α_{1Cb}
Nifedipine	47	10
Nisoldipine	2.1	0.56
(+)PN200110	15	2.1
SDZ 207-180	91	100

class and subclass of calcium channel(s) activated, the pathological state of the tissue and the state-dependence of antagonist interaction at the drug binding site(s) (for reviews see [1, 3, 11, 57]). In practice, all of these aspects are important and it is likely that they act in combination rather than in isolation. Nonetheless, the state-dependent interactions are of particular interest, whereby the affinity of the molecular and/or its access to the drug binding site is governed by the state of the channel (Chapter 3).

Voltage-dependent interactions of 1,4-dihydropyridines have been well studied, with data available from electrophysiological, radioligand binding, and pharmacological studies (reviewed in [3, 11, 57, 58]). The 1,4-dihydropyridines exhibit voltage-dependent binding consistent with preferential interaction at an open or inactivated channel state. This interaction is highly dependent upon the substitution pattern of the dihydropyridine, as indicated in Fig. 5.3.12 for three series

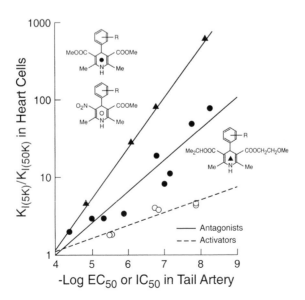

Fig. 5.3.12 Correlations between voltage-dependent binding (ratio of dissociation constants for binding in polarized and depolarized cardiomyocytes) and activity as activator or antagonist in rat tail artery strips.

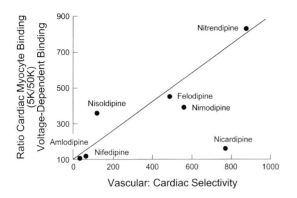

Fig. 5.3.13 Relationship between voltage-dependent binding and vascular:cardiac selectivity for clinically available 1,4-dihydropyridines. (Data from Sun and Triggle, Ref. [62].)

of molecules. Quite generally, 1,4-dihydropyridine activators show little voltage-dependence and the antagonists show greater and structure-dependent voltage-dependence (Fig. 5.3.12) [59, 60]. The relative lack of voltage-dependence for 1,4-dihydropyridine activators is quite consistent with the observation that a switch in properties from activator to antagonist occurs with increasing membrane depolarization [61]. For a series of clinically available 1,4-dihydropyridines an approximately linear correlation exists between the extent of voltage-dependent binding and vascular:cardiac selectivity (Fig. 5.3.13) [62], suggesting that voltage-dependent interactions are a principal factor in determining vascular:cardiac selectivity.

5.3.3.3.2 1,4-Dihydropyridines and NO-releasing Activities
There is increasing evidence that 1,4-dihydropyridines decrease vascular tension via mechanisms other than interaction at the L-type calcium channel. One of these mechanisms may involve nitric oxide, the diffusible relaxing factor [63]. Clinical hypertension may be due, in part, to nitric oxide deficiency [64]. There appear to be at least two classes of 1,4-dihydropyridines that act through nitric oxide. The first contain nitrooxyalkylester moieties (or other NO-releasing species) at the 3 or 5 positions of the 1,4-dihydropyridine ring: these agents release nitric oxide from these substituents and may be regarded as "NO donors" [*inter alia*, 65–67] (Fig. 5.3.14). However, several other 1,4-dihydropyridines that lack specific NO or nitrite donor functions also appear to have a component of their vasorelaxant activity mediated through nitric oxide, and it has been proposed that this may be a "class effect" [68–70]. Of particular interest is the stereoselectivity of this process: the enantiomers of amlodipine differ in their nitric oxide releasing ability. The less active R-amlodipine releases NO in an L-NAME sensitive manner and the pharmacologically active (on L-type channels) S-enantiomer is without effect [70, 71]. The mechanism of this effect remains uncertain, but several speculations have been advanced: these include activation of NO synthase through calcium entry via stretch-activated channels, the intermediacy of kinins and the role of angiotensin receptors.

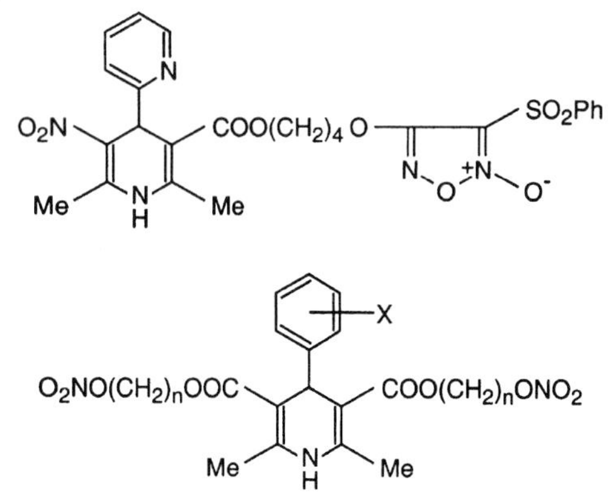

Fig. 5.3.14 NO-releasing
1,4-dihydropyridines.

5.3.3.3.3 1,4-Dihydropyridine Actions at Non-calcium Channel Sites

The 1,4-dihydropyridine nucleus constitutes a "privileged structure" – capable, according to the substitution pattern, of interacting selectively with various receptors and ion channels (reviewed in [11, 14]). These actions include interaction with the NO system (Section 5.3.3.3.2.), anti-oxidant actions that may be related to their antiatherogenic properties, interaction with other classes of calcium channels as well as with potassium and sodium channels and at various pharmacological receptors. Many of these actions occur at supra-pharmacological concentrations and are not likely to contribute to the therapeutic profile, save for the anti-oxidant effects during prolonged administration and the NO-releasing actions.

5.3.4
Other Drug Classes Active at Ca$_V$1 Channels

A remarkably diverse set of molecules interacts with the L-type channel (Fig. 5.3.2). Reviews of this field are available and few new directions have been charted [1, 6, 7, 72, 73].

5.3.4.1 Simple Aliphatic Molecules

Some long-chain aliphatic molecules have demonstrated activity at the L-type channel. These include farnesol, a 15-C isoprenoid that is a component of farnesyl pyrophosphate, a critical intermediate in the mevalonic acid pathway, as well as simple aliphatic amines and alcohols, including dodecanol [$CH_3(CH_2)_{11}OH$] and dodecylamine [$CH_3(CH_2)_{11}NH_2$] [74–77]. Of some interest also is the plasticizer Tinuvin 770 [bis(2,2,6,6-tetramethyl-4-piperidyl)sebacate] that exhibits nanomolar potency [78]. Menthol, a component of peppermint oil, is active at both L- and T-type channels and this activity may underlie its gastrointestinal relaxant properties (reviewed in [79]). Other small molecules introduced since the last reviews include the antifungal agent clotrimazole, 1-[(2-chlorophenyl)diphenyl-

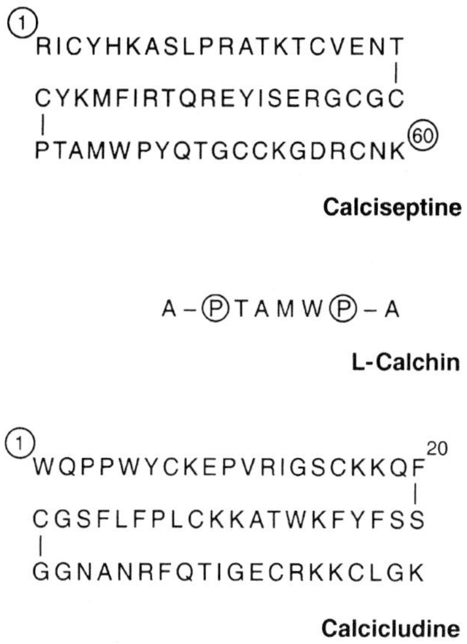

①
RICYHKASLPRATKTCVENT
                                        |
CYKMFIRTQREYISERGCGC
|
PTAMWPYQTGCCKGDRCNK⑥⁰

Calciseptine

Fig. 5.3.15 Peptide toxins active at L-type channels.

A – ⓅTAMWⓅ – A

L-Calchin

①   20
WQPPWYCKEPVRIGSCKKQF
                                    |
CGSFLFPLCKKATWKFYFSS
|
GGNANRFQTIGECRKKCLGK

Calcicludine

methyl]imidazole [80], and the benzimidazole mibefradil (Posicor), introduced and subsequently withdrawn as an antihypertensive agent active at both L- and T-channels [81–84].

5.3.4.2 Peptide Toxins

Although the L-type channel has been best characterized by small synthetic molecules, in contrast to other members of the high voltage-gated class, several peptide toxins are active at the L-type channel (Fig. 5.3.15). Calciseptine, a 60 amino acid residue peptide with four disulfide bridges, from the black mamba snake is active in both cardiac and vascular smooth muscle and appears to interact competitively at the 1,4-dihydropyridine site [85, 86]. Calcicludine, also a 60 residue peptide, from the green Mamba snake is less specific and blocks various calcium channels in the 10–100 nM range [87]. Most conotoxins interact at the N-, and P/Q-type channels, but ω-conotoxin-TxVII is specific for L-type channels and in the 10^{-7}–10^{-5} M range [88], and ω-agatoxin IIIA blocks both L- and N-type channels with nanomolar potency [89, 90]. Additionally KP4, which is a 104 residue peptide encoded by the UMV4 virus, blocks L-type channels in a voltage-independent manner and has a similar potency to that of calciseptine [91].

5.3.4.3 Vitamin D

Much evidence indicates that the steroid hormones exhibit rapid non-genomic actions in addition to their better-known roles as transcription factor activators [92]. Several lines of evidence indicates that vitamin 1α,25-dihydroxyvitamin D3 and its metabolites can activate an L-type 1,4-dihydropyridine-sensitive calcium current

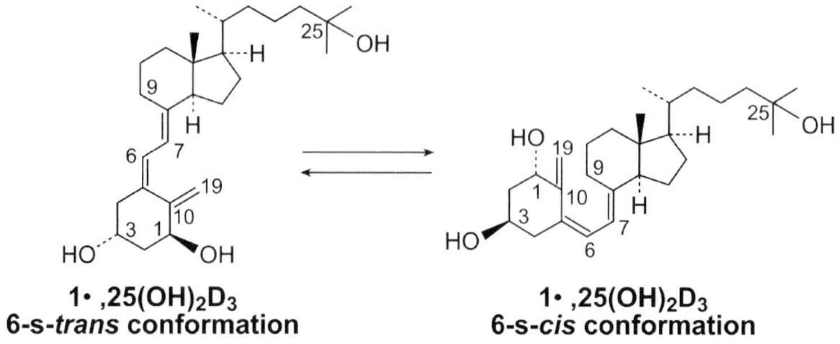

Fig. 5.3.16 The cis- and trans-arrangements of $1\alpha,25(OH)_2$-vitamin D [95].

in several cell lines, including osteosarcoma and preosteoblasts [93, 94], and in smooth muscle [95]. It has been suggested that different conformations of the molecule may mediate the rapid and the genomic response [93, 94]. Thus, the 6-s-cis analog (Fig. 5.3.16) was a potent activator of rapid responses in ROS 17/.8 cells, whereas the 6-s-trans analog was inactive in mediating both genomic and non-genomic responses [95]. Of related interest is the observation that $1\alpha,25$-dihydroxyvitamin D_3 offers both neuroprotection and down-regulation of L-type calcium channels in hippocampal neurons [96].

5.3.5
Drug Interactions at Non-α-subunit Sites

Although the α_1-subunit clearly represents the dominant site of interaction for drugs that modulate the activity of voltage-gated calcium channels, recent observations indicate that other subunits of the heteromeric complex may also play an important role. Since the functions and expression of these channels depend on the specific combinations of subunits, it is likely that drugs that interact at non-α_1-subunits may offer potential for selectivity of action additional to that from the α_1-subunit alone [97–99]. Thus, Zamponi et al. observed that the blockade by local anesthetics of the $Ca_V2.1$. $Ca_V2.2$ and $Ca_V1.2$ channels was dependent on the β-subunit that was co-expressed and on amino acid substitutions in the I-II cytoplasmic linker, which is consistent with both α_1- and β-subunit sites being involved in drug binding [100].

5.3.5.1 $\alpha_2\delta$ Subunit
Gabapentin (Neurontin), an agent used clinically in the treatment of epilepsy and neurogenic pain, and its analog pregabalin appear to interact, despite their structural similarity to GABA, with the $\alpha_2\delta$ subunit of the N-type channel [101–105]. A structure–activity relationship is observed, including stereoselectivity, and since the $\alpha_2\delta$ subunit is up-regulated in pain models this interaction likely accommodates the clinical use of these agents in pain [Chapter 5.4].

5.3.5.2 α₁-β Subunit Site of Interaction

Interaction between the α₁- and the β-subunit is reversible and is characterized by an 18-residue sequence in the intracellular loop separating domains I and II of the α₁ subunit and a 30-residue sequence in the β subunit [106, 107]. Peptides of 21- and 68-residues in length and containing the α₁ interaction sequence promote dissociation of the α₁β complex [108]. Such peptides, although of mechanistic utility, are not likely to be useful as drug molecules.

However, recent observations with the anti-schistosomal agent praziquantel (Fig. 5.3.17) are very consistent with small molecule modulation of α₁-β subunit interaction. Praziquantel is efficacious against flatworms of the genus *Schistosoma* where it promotes calcium influx and subsequent muscle paralysis [109, 110]. β-Subunits from *Schistosoma mansoni* and *S. japonicum*, SmCa$_V$βA and SjCa$_V$β, appear to be inhibitory when co-expressed with mammalian Ca$_V$2.3 or jellyfish Ca$_V$1 subunits and this inhibition was reversed by praziquantel, suggesting a mechanism for the drug that involves promotion of α₁-β subunit dissociation [111]. The unique sensitivity of these schistosmal β subunits has been attributed to the absence of conserved serine residues that confer protein kinase C sensitivity [112]. Elimination of these residues in the rat β2a subunit confers praziquantel sensitivity when the mutated β subunit is co-expressed with Ca$_V$2.3 [113].

Of closely related interest is WAY 141520 (Fig. 5.3.17), discovered in a two-hybrid assay as capable of blocking the interaction between human Ca$_V$2.2 and β₃ subunits [114]. Although of low potency, K_I = 95 μM, this molecule did exhibit selectivity and was able to inhibit channel function. Both praziquantel and WAY 141520 indicate that small molecules are capable of inhibiting protein–protein association and suggest the feasibility of further extension of this method of blocking channel protein function [115].

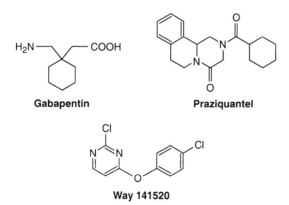

Gabapentin   **Praziquantel**

Way 141520

Fig. 5.3.17 Molecules that interact at non-α₁ subunits (see text for further discussion).

5.3.6
Calcium Antagonism through Gene Delivery

The β-subunits of the oligomeric voltage-gated calcium channel family are members of the membrane-associated guanylate kinase family (MAGUK) of proteins. These are scaffolding proteins that interact with other proteins, including RGK family of Ras-like GTPases. These proteins, as the GTP-bound forms, interact with the $Ca_v b$ subunits to inhibit calcium current (reviewed in [116]). The delivery of Gem, a member of the RGK family, by adenovirus-mediated gene delivery has now been accomplished in ventricular cardiac cells to produce a significant inhibition of the L-type current and a reduction in the plateau component of the action potential [117]. Such focal delivery offers an interesting potential for cell-specific blockade of calcium currents.

References

1. D. J. Triggle, *Eur. J. Pharmacol.* **1999**, 375, 311–325.
2. D. J. Triggle, *Stroke*, **1990**, 21(Supppl. IV), IV-49–IV-58.
3. D. J. Triggle, *ASSAY Drug Develop. Technol.*, **2003**, 1, 719–733.
4. A. Fleckenstein, *Calcium Antagonism in Smooth and Cardiac Muscle. Experimental Facts and Therapeutic Observations.* John Wiley and Sons, Inc., New York, **1983**.
5. T. Godfraind, R. Miller, M. Wibo, *Pharmacol. Rev.* **1968**, 33, 1209–1217.
6. R. A. Janis, P. Silver, D. J. Triggle, *Adv. Drug Res.*, **1987**, 16, 309–591.
7. D. Rampe, D. J. Triggle, *Prog. Drug Res.* **1993**, 40, 191–238.
8. M. Romero, I. Sander, M. D. Pujol, *Curr. Med. Chem. – Cardiovas. Hematolog. Agents* **2003**, 1, 113–141.
9. J. Striessnig, M. Grabner, J. Mitterdorfer, S. Hering, M. J. Sinnager, H. Glossman, *Trends Pharmacol. Sci.* **1998**, 19, 108–115.
10. W. A. Catterall, *Annu. Rev. Cell Dev. Biol.* **2000**, 16, 521–555.
11. D. J. Triggle, in *Pharmacology of Calcium Channels*, ed. S. McDonough Kluwer Academic Press, New York, **2004**, pp. 21–72.
12. D. J. Triggle, D. A. Langs, R. A. Janis, *Med. Res. Rev.* **1989**, 9, 123–186.
13. C. J. Doering, G. W. Zamponi, *J. Bioenerget. Biomembr.*, **2003**, 35, 491–505.
14. D. J. Triggle, *Mini-Rev. Med. Chem.* **2003**, 3, 137–147.
15. F. Corelli, F. Manetti, A. Tafi, G. Campiani, V. Nacci, M. Hotta, *J. Med. Chem.*, **1997**, 40, 125–131.
16. K. J. Schleifer, E. Tob, *Pharmaceut. Res.*, **1999**, 16, 1506–1513.
17. G. Campiani, I. Fiorini, M. P. De Filipis, S. M.Ciani, A. Garafoli, V. Nacci, G. Giogi, A. Sega, M. Botta, A. Chiarini, R. Budriesi, G. Bruni, M. R. Romero, C. Manzoni, T. Menini, *J. Med. Chem.*, **1996**, 39, 2922–2938.
18. K. Yamamto, M. Fujita, K. Tabashi, K. Kawashima, E. Kato, M. Oya, J. Iwao, *J. Med. Chem.*, **1988**, 31, 919–930.
19. S. D. Kimball, D. M. Floyd, J. Das, J. T. Hunt, J. Krapcho, G. Rovnyak, K. J. Duff, R. V. Lee, C. F. Tuck, S. A. Hedberg, S. Morlenad, R. J. Brittain, D. M. McMullen, D. E. Normandin, G. G. Cucinotta, *J. Med. Chem.*, **1992**, 35, 780–793.
20. G. Campiani, A. Garofalo, I. Firoini, M. Botta, V. Nacci, A. Tafi, A. Chiarini, R. Budriesi, G. Bruni, M. R. Romeo, *J. Med. Chem.*, **1995**, 38, 4393–4410.
21. Y. W. Kwon, D. J. Triggle, *Chirality* **1991**, 3, 393–404.
22. D. J. Triggle, in *Chirality in Natural and Applied Science*, eds., W. J. Lough, I. Wainer, pp. 109–138. Blackwell, Oxford and London, **2002**.
23. F. T. van Amsterdam, J. Zaagsma, *Naunyn-Schmied. Arch. Pharmacol.* **1988**, 337, 213–219.

24. R. Mannhold, R. Rodenkirchen, R. Bayer, *Prog. Pharmacol.*, **1982**, 5, 25–42.

25. F. Bossert, W. Vater, *Med. Res. Rev.*, **1989**, 9, 291–324.

26. S. Goldmann, J. Stoltefuss, *Angew. Chem., Int. Ed.*, **1991**, 30, 1539–1578.

27. B. S. Zhorov, E. V. Folkmann, V. S. Ananthanaryanna, *Arch. Biochem. Biophys.* **2001**, 393, 22–41.

28. J. E. S. Arrowsmith, S. F. Campbell, P. E. Cross, J. K. Stubbs, R. A. Burges, D. G. Gardiner, K. J. Blackburn, *J. Med. Chem.*, **1986**, 29, 1696–1702.

29. G. T. Bolger, P. Gengo, R. Klockowski, E. Luchowski, H. A. Siegel, R. A. Janis, D. J. Triggle, *J. Pharmacol. Exp. Therap.* **1983**, 225, 291–310.

30. Y.-W. Kwon, Q. Zhong, X.-Y. Wei, D. J. Triggle, *Naunyn-Schmied. Arch. Pharmacol.* **1990**, 341, 128–136.

31. R. Perikumar, S. Padmanabhan, A. Rutledge, S. Singh, D. J. Triggle, *J. Med. Chem.* **2000**, 43, 2906–2914.

32. N. Iqbal, M. R. Akula, D. Vu, W. C. Matowe, C. A. McEwen, M. W. Wolowyck, E. E. Knaus, *J. Med. Chem..* **1998**, 41, 1827–1837.

33. W. Zheng, J. Stoltefuss, S. Goldman, D. J. Triggle, *Mol. Pharmacol.* **1992**, 41, 535–541.

34. K. McKechnie, P. G. Killinback, I. Naga, S. E. O'Connor, G. W. Smith, D. G. Wattam, E. Wells, Y.-M. Whitehead, G. E. Williams, *Br. J. Pharmacol.* **1989**, 98, 673P.

35. D. Rampe, B. Anderson, V. Raoier-Pryor, T. Li, R. C. Dage, *J. Pharmacol. Exp. Therap.* **1993**, 265, 1125–1130.

36. W. Zheng, D. Rampe, D. J. Triggle, *Mol. Pharmacol.* **1991**, 40, 734–741.

37. S. Herzog, *Eur. J. Pharmacol.*, **1986**, 295, 113–117.

38. P. J. White, *Biochim. Biophys. Acta* **2000**, 1465, 171–189.

39. P. A. Meredith, H. L. Elliott, *Clin. Pharmacokinet.* **1992**, 22, 22–31.

40. D. A. Langs P. D. Strong, D. J. Triggle, *J. Comput. Aid. Mol. Design* **1990**, 4, 215–230.

41. D. A. Langs, Y.-W. Kwon, P. D. Strong, D. J. Triggle, *J. Comput. Aid. Mol. Design* **1991**, 5, 95–106.

42. K. S. Atwal, G. C. Rovnyak, J. Schwartz, S. Moreland, A. Hedberg, J. Gougolas, M. Malley, D. Floyd, *J. Med. Chem.* **1990**, 33, 1510–1515.

43. D. A. Langs, D. J. Triggle, *Mol. Pharmacol.*, **1985**, 27, 544–548.

44. R. Fossheim, K. Svarteng, A. Mostad, C. Romming, E. Shefter, D. J. Triggle, *J. Med. Chem.*, **1981**, 25, 126–131.

45. A. M. Triggle, E. Shefter, D. J. Triggle, *J. Med. Chem.*, **1980**, 23, 1442–1445.

46. G. C. Rovnyak, N. Anderson, J. Gougolas, A. Hedberg, S. D. Kimball, S. Moreland, M. Porubean, A. Pudzianowski, *J. Med. Chem.*, **1988**, 31, 936–944.

47. R. Fossheim, A. Joslyn, A. J. Solo, E. Luchowski, A. Rutledge, D. J. Triggle, *J. Med. Chem.*, **1988**, 31, 300–305.

48. R. A. Coburn, M. Wierzba, M. J. Suto, A. J. Solo, A. M. Triggle, D. J. Triggle, *J. Med. Chem.*, **1988**, 31, 2103–2107.

49. R. Mannhold, R. Rodenkirchen, *Prog. Pharmacol.*, **1982**, 5, 25–42.

50. M. Mahmoudian, W. G. Richards, *J. Pharm. Pharmacol.*, **1986**, 38, 272–281.

51. B. Hemmateenejad, R. Miri, M. Akhond, M. Shamsipur, *Arch. Pharm. Pharm. Med. Chem.* **2002**, 10, 472–480.

52. K.-J. Schleifer, E. Tot, *Quant. Struct-Act. Relat.* **2002**, 21, 239–248.

53. D. Alker, S. F. Campbell, P. E. Cross, R. A. Burges, A.J. Carter, D. G. Gardiner, *J. Med. Chem.*, **1980**, 33, 585–591.

54. A. Welling, A. Ludwig, S. Zimmer, N. Klugbauer, V. Flockerzi, F. Hofmann, *Circ. Res.* **1997**, 81, 526–532.

55. A. Welling, Y.-W. Kwon, E. Bosse, V. Flockerzi, F. Hofmann, R. S. Kass, *Circ. Res.*, **1993**, 73, 974–980.

56. N. Morel, V. Burgi, O. Feren, J.-P. Gomez, M.-O. Christen, T. Godfraind, *Br. J. Pharmacol.*, **1998**, 125, 1005–1012.

57. D. J. Triggle, in *Calcium Antagonists in Clinical Medicine*, ed. M. Epstein, pp. 1–32. Hanley, Belfus Inc., Philadelphia, PA, **2002**.

58. D. J. Triggle, in Molecular and Cellular Mechanisms of Antiarrhythmic Agents, ed. L. M. Hondeghem, pp. 269–291. Futura Publishing, Mt. Kiscoe, New York, **1989**.

59. X.-Y. Wei, E. M. Luchowski, A. Rutledge, C. M. Su, D. J. Triggle, *J. Pharmacol. Exp. Therap.*, **1986**, 239, 144–153.

60. X.-Y. Wei, A. Rutledge, D. J. Triggle, *Mol. Pharmacol.* **1989**, 35, 541–552.

61. R. S. Kass, *Circ. Res.* **1987**, 61(Suppl. I), 1–5.

62. J. P. Sun, D. J. Triggle, *J. Pharmacol. Exp. Therap.* **1995**, 274, 419–428.

63. S. Dhein, A. Salameh, R. Berkels, W. Klaus, *Drugs* **1999**, 58, 397–404.

64. G. D. Thomas, W. Zhang, R. G. Victor, *J. Am. Med. Assoc.* **2001**, 285, 2055–2057.

65. T. Ogawa, A. Nakazoto, K. Tsuchida, K. Hatayama, *Chem. Pharm. Bull.* **1993**, 41, 1049–1054.

66. D. Vo, J.-T. Nguyen, C.-A. McEwen, R. Shan, E. E. Knaus, *Drug Dev. Res.* **2002**, 56, 1–16.

67. S. Visentin, B. Rolando, A. Di. Stilo, R. Fruttero, M. Novaro, E. Carbone, C. Roussel, N. Vanthuyne, A. Gasco, *J. Med. Chem.*, **2004**, 47, 2688–2693.

68. A. Salameh, G. Schomecker, K. Breitkopf, *Br. J. Pharmacol.* **1996**, 118, 1899–1904.

69. F. Crespi, E. Vecchiato, C. Lazzarini, M. Andreoli, G. Gaviraghi, *J. Cardiovasc. Pharmacol.* **2002**, 39, 471–477.

70. X.-P. Zhang, K. E. Loke, S. Mital, S. Chahwala, T. H. Hintze, *J. Cardiovasc. Pharmacol.* **2002**, 39, 208–214.

71. R. P. Mason, P. Marche, T. H. Hintze, *Arterioscle. Thromb. Vas. Biol.*, **2003**, 23, 2155–2163.

72. M. Spedding, *Trends Pharmacol. Sci.* **1995**, 16, 139–142.

73. J. Striessnig, *Cell Physiol. Biochem.* **1999**, 9, 242–269.

74. A. M. Beedle, G. W. Zamponi, *Biophys. J.* **2000**, 79, 260–270.

75. J.-B. Roulett, H. Xue, J. Chapman, P. McDougal, C. M. Roulett, D. A. McCarron, *J. Clin. Invest.* **1996**, 97, 2384–2390.

76. J.-B. Roulett, U. C. Luft, H. Xue, J. Chapman, R. Bychkov, C. M. Roulett, F. C. Luft, H. Haller, D. A. McCarron, *J. Biol. Chem.*, **1997**, 32 240–32 446.

77. J.-B. Roulett, R. L. Spaetgens, T. Burlingame, Z.-P. Feng, G. W. Zamponi, *J. Biol. Chem.*, **1999**, 274, 25 439–25 446.

78. H. S. Glossman, A. Hering, A. Savchenko, W. Berger, K. Friedrich, M. L. Garcia, M. A. Goetz, J. M. Liesch, D. L. Zink, G. J. Kaczorowski, *Proc. Natl. Acad. Sci. U.S.A.* **1993**, 90, 9253–9257.

79. M. Hawthorn, J. Ferrante, E. Luchowski, A. Rutledge, X.-Y. Wei, D. J. Triggle, *J. Alimen. Pharmacol.* **1988**, 2, 101–118.

80. L. R. Benzaquan, C. Brugnarai, H. R. Byers, S. Gattani-Delli, J. A. Halperin, *Nat. Med.* **1995**, 1, 534–540.

81. T. F. Luscher, J.-P. Clozel, G. Noll, *J. Hypertension* **1997**, 15(Suppl. 3), S11–S18.

82. G. Mehrke, X. G. Zong, V. Flockerzi, F. Hofmann, *J. Pharmacol. Exp. Therap.*, **1994**, 271, 1483–1488.

83. C. Jiminez, E. Bourinet, V. Leuranguer, S. Richard, T. P. Snutch, J. Nargeot, *Neuropharmacology* **2000**, 39, 1–10.

84. G. Bernatchez, R. Sauve, L. Parent, *J. Membr. Biol.* **2001**, 184, 143–159.

85. J. R. De Weille, H. Schweitz, P. Moes, A. Tartor, M. Lazdunski, *Proc. Natl. Acad. Sci. U.S.A.*, **1991**, 88, 2437–2440.

86. O. Yasuda, S. Morimoteo, Y. Chen, B. Jiang, T. Kimura, S. Satickibera, E. Koh, K. Fukuo, S. Kitano, T. Ogihara, *Biochem. Biophys. Res. Commun.*, **1993**, 194, 587–594.

87. H. C. Schweitz, C. Herteaux, P. Bois, D. Moinier, G. Romey, M. Lazdunski, *Proc. Natl. Acad. Sci. U.S.A.*, **1994**, 91, 878–882.

88. M. Fainzilber, J. C. Lodder, R. C. Van den Schors, K. W. Li, Z. Yu, A. L. Burlingame, W. M. P. Geraerts, K. S. Kits, *Biochemistry*, **1996**, 35, 8748–8752.

89. E. A. Ertel, M. M. Smith, M. D. Leibowitz, C. J. Cohen, *J. Gen. Physiol.*, **1994**, 103, 731–753.

90. E. A. Ertel, V. A. Warren, M. E. Adams, P. K. Griffin, C. J. Cohen, M. M. Smith, *Biochemistry*, **1994**, 33, 5098–5108.

91. M. J. Gage, G. Rane, G. H. Hockerman, T. J. Smith, *Mol. Pharmacol.* **2002**, 61, 936–944.

92. A. W. Norman, M. T. Mizwicki, D. P. G. Norma, *Nat. Rev. Drug Discovery* **2004**, 3, 27–41.

93. J. M. Caffrey, M. C. Farach-Carson, *J. Biol. Chem.*, **1989**, 264, 20265–20274.

94. W. Li, R. L. Duncan, N. J. Kain, M. C. Farach-Carson, *Am. J. Physiol.*, **1997**, 273, E599–E605.

95. A. W. Norman, W. H. Okamura, M. W. Hammond, J. E. Bishop, M. C. Dormanen, R. Bouillon, H. van Balen, A. L. Ridall, E. Daane, R. Khoury, M. C. Farach-Carson, *Mol. Endocrinol.* **1997**, 11, 1518–1531.

96. L. D. Brewer, V. Thinault, K.-C. Chen, M. C. Langrub, P. W. Landfield, N. M. Porter, *J. Neurosci.* **2001**, 21, 98–108.

97. S. Hering, *Trends Pharmacol. Sci.* **2002**, 23, 509–513.

98. J. Arikkath, K. P. Campbell, *Curr. Opin. Neurobiol.* **2003**, 13, 298–307.

99. S. X. Takahashi, S. Mittman, H. M. Colecraft, *Biophys. J.* **2003**, 84, 3007–3021.

100. G. W. Zamponi, T. W. Soong, E. Bourinet, T. P. Snutch, *J. Neurosci.* **1996**, 16, 2430–2443.

101. J. S. Bryans, N. Davies, N. S. Gee, G. S. Ratcliff, D. C. Hornwell, C. O. Keen, A. J. Morrell, R. J. Oles, J. C. O'Toole, G. M. Perkins, L. M. Singh, N. Suman-chauban, J. O'Neill, *J. Med. Chem.*, **1998**, 41, 1838–1845.

102. N. S. Gee, J. P. Brown, V. U. K. Dissanyake, J. Offord, R. Thurlow, G. N. Woodruff, *J. Biol. Chem.*, **1996**, 271, 5768–5776.

103. E. N. Maraus, N. Klugbauer, F. Hofmann, *Mol. Pharmacol.*, **2002**, 59, 1243–1248.

104. K. G. Sutton, T. P. Snutch, *Drug Dev. Res.* **2002**, 54, 167–172.

105. Z. D. Luo, S. R. Chaplan, E. S. Higuera, L. S. Sorkin, K. A. Stauderman, M. E. Williams, T. L. Yaksh, *J. Neurosci.* **2001**, 21, 1868–1873.

106. M. De Waard, M. Pragnell, K. P. Campbell, *Neuron* **1994**, 13, 495–503.

107. M. M. Pragnell, M. Denbard, Y. Mori, T. Tanabe, T. P. Snutch, K. P. Campbell, *Nature* **1994**, 368, 67–70.

108. D. Bichet, C. Lecompte, J.-M. Sabatier, R. Felix, M. De Waard, *Biochem. Biophys. Res. Commun.* **2000**, 277, 729–735.

109. C. A. Redman, A. Robertson, P. G. Fallon, J. Modha, J. R. Kusel, M. J. Doenhoff, R. J. Martin, *Parasitol. Today* **1996**, 12, 14–22.

110. D. G. McNeil, Jr., *New York Times*, **2004**, November 2.

111. A. B. Kohn, P. A. V. Anderson, J. M. Roberts-Misterly, R. M. Greenberg, *J. Biol. Chem.*, **2001**, 276, 36873–36876.

112. B. Kohn, J. M. Roberts-Misterly, P. A. V. Anderson, N. Kahn, R. M. Greenberg, *Parasitology* **2003**, 127, 349–356.

113. A. B. Kohn, J. M. Roberts-Misterly, P. A. V. Anderson, R. M. Greenberg, *Int. J. Parasitol.*, **2003**, 33, 1303–1308.

114. K. Young, S. Lin, L. Sun, E. Lee, M. Modi, S. Hellings, M. Husbands, B. Ozenberger, R. Franco, *Nat. Biotechnol.* **1988**, 16, 948–950.

115. T. Berg, *Angew. Chem. Int. Ed.* **2003**, 42, 2462–2481.

116. R. C. Balijepalli, J. D. Foell, T. J. Kamp, *Circ. Res.*, **2004**, 95, 337–339.

117. M. Murata, E. Cingolani, A. D. McDonald, J. K. Donahue, E. Marban, *Circ. Res.* **2004**, 95, 398–405.

5.4
N-type Calcium Channels

Terrance P. Snutch

5.4.1
Introduction

N-type calcium channels represent a target of wide interest from both the scientific and clinical communities. Compared to that for the equally clinically relevant high threshold L-type (four genes) and low-threshold T-type calcium channels (three genes), the N-type channel is encoded by a single gene and likely represents a comparatively more homogeneous pharmacological target (reviewed in [1]). The major pore-forming α_{1B} (Cav2.2) subunit of the N-type calcium channel ($\sim$2300 amino acid; $\sim$260 kDa) was first described from rat brain (rbB-I, rbB-II) and has since been cloned from various species, including rabbit (BIII) and human ($\alpha_{1B-1,2}$) [2–5]. Biochemical studies have reported the association of Cav2.2 with the $\alpha_2\delta$-1 subunit and any of the β_1, β_3 and β_4 subunits. Co-expression of Cav2.2 with different β subunits variously affects several channel properties, including the levels of surface expression, kinetic and voltage-dependent properties, and modulation by G-proteins and second-messengers [6–8]. It remains unclear whether the native N-type channel complex includes an ancillary γ subunit similar to that for the skeletal muscle L-type channel complex, and exogenous expression does not require this particular protein to fully reconstitute N-type channel biophysical and pharmacological properties. Several regions of alternative splicing of the Cav2.2 subunit have been identified and include the domain I-II linker, domain II-III linker, IIIS3-IIIS4, IVS3-IVS4 and the carboxyl terminus. Splicing alters channel current–voltage relations and kinetics and there is accumulating evidence for cell-specific expression of spice variants [9, 10]. In one instance, expression of the Cav2.2 e37a splice isoform in dorsal root ganglia correlates with a subset of nociceptive neurons and may prove especially relevant as a potential therapeutic pain target [11]. Alternative splicing also appears to affect N-type channel interactions with intracellular synaptic proteins such as Mint1, CASK, syntaxin and SNAP-25 [1].

RNA expression studies show that Cav2.2 N-type channel transcripts are exclusively expressed in neurons and neurally-derived cells such as neuroendocrine cells [2]. In the central nervous system, N-type channels are widely expressed in most regions, including the cerebral cortex, hippocampus, forebrain, midbrain, cerebellum, brainstem and spinal cord. Immunohistochemical staining shows that at the subcellular level N-type channels are mainly clustered along dendritic regions and on some cell bodies, and are also found at a subset of presynaptic terminals in both the central and peripheral nervous systems [12]. The case for involvement of N-type channels in the pathophysiology of neuropathic pain arises largely from the fact that they are found highly concentrated in the dorsal root ganglia (DRG) cell bodies and the synaptic terminals that they make into the dorsal horn of the spinal cord [13, 14]. These primary afferents are implicated in various painful stimuli, including mechanical, thermal and inflammatory, and block of N-type channels in the DRGs potently attenuates chronic and neuropathic pain conditions.

N-type channel activity can be modulated by activation of a number of G-protein coupled receptors, with most appearing to mediate a decrease in channel activity (e.g., somatostatin, opioid, $GABA_B$, cannabinoid, neuropeptide Y, dopamine, adenosine) [8, 15–18]. This receptor-mediated inhibitory regulation has been examined in detail and involves the release of $G_{\beta\gamma}$ from the trimeric $G_{-\alpha\beta\gamma}$ complex, subsequent physical binding of $G_{\beta\gamma}$ to the domain I-II linker of the Cav2.2 subunit (in 1:1 stoichiometry) and a stabilizing of the closed state [19–21]. At presynaptic terminals the net effect is to decrease neurotransmitter release in response to action potentials, which also has a major modulatory affect on synaptic processes such as pain transmission. The regulation of synaptic activity mediated by N-type channels is highly dynamic as N-type channel activity can also be upregulated via the protein-kinase C dependent phosphorylation of Cav2.2 while G-protein dependent inhibition can itself be antagonized by PKC [7, 19, 22–24].

Genetic deletion of the Cav2.2 gene shows that mice lacking functional N-type channels gene possess a normal life span and surprisingly few detectable behavioral modifications compared to wild-type animals (there was partial lethality in one of three derived strains but no deaths in the other two strains) [25–28]. In one study, an increase in basal mean atrial pressure and heart rate were apparent although in another strain examined there was no evidence for cardiovascular alterations. None of the three Cav2.2 gene deletion lines exhibited any of the ataxia or other behavioral alterations associated with intrathecal Prialt administration (see below). Importantly, regarding proof-of-concept for the N-type channel as a therapeutic target, all three Cav2.2 knock-out strains exhibit attenuated abilities to respond to inflammatory and neuropathic pain insults. To date, and somewhat surprisingly given the preponderance of naturally occurring mutations in other voltage-gated ion channel genes, there have been no natural mutations reported in any mammalian $Ca_V2.2$ gene.

5.4.2
N-type Calcium Channel Pharmacology

The identification of high affinity, selective N-type channel blockers is a critical step towards the dissection of channel distribution and developmental expression patterns as well as helping to define biophysical properties such as gating and permeation. Critically, these agents also contribute to defining physiological functions in both the normal and diseased states and a major goal has been to develop N-type calcium channel blocks that can be utilized clinically. Examination of the scientific and patent literature shows that targeting N-type channels is of interest in a wide variety of human pathophysiological conditions, including chronic pain states such as diabetic neuropathy, post-herpetic and trigeminal neuralgias, cancer pain and osteoarthritis, as well as for stroke and traumatic brain injury.

There is also considerable interest towards developing N-type blockers to treat gastrointestinal and genitourinary disorders (e.g., inflammatory bowel disease, overactive bladder), mood disorders related to anxiety and obsessive compulsiveness, and addictions related to alcohol, nicotine and amphetamines.

The development of N-type blockers for clinical usage needs to consider multiple parameters, including physiochemical properties related to the target site of action (e.g., crossing the blood–brain barrier since most N-type channels appear localized in the CNS), high specificity and selectivity over other calcium channels and other non-specific targets, a requirement for oral bioavailability in most disease indications, and the need for a wide therapeutic index (ratio of relative toxicity to relative efficacy). Crucially related to later point, successfully developed clinical ion channel blocking agents such as the L-type calcium channel antagonists for cardiovascular disease and migraine and the anti-arrhythmic sodium channel blockers all have the mechanistic property of acting in a state-dependent manner ([29, 30] and see also Chapter 3). It is highly likely that clinically efficacious N-type blockers will require a similarly state-dependent action to minimize potential adverse effects related to N-type channels localized to the peripheral nervous system and that possibly affect sympathetic regulation.

5.4.3
Inorganic Cations

Many types of divalent and trivalent metal cations block calcium currents, with the most widely utilized being cadmium (Cd^{2+}) and nickel (Ni^{2+}). Cd^{2+} primarily acts to physically occlude the pore region of both high threshold and low threshold calcium channels and likely involves interactions with the ring of negatively charged residues (glutamates and aspartates) that control calcium permeation [31, 32]. Cd^{2+} completely blocks N-type channels at low micromolar concentrations ($IC_{50} \sim 5\,\mu M$) but also similarly blocks all other types of high- and low-threshold calcium channels. Block of native and cloned N-type currents by Ni^{2+} roughly follows 1:1 kinetics and occurs in the mid- to high-micromolar range ($IC_{50}s \sim 200–750\,\mu M$). Ni^{2+} appears to act through both pore blockade and also shifting the voltage-dependence of activation to more depolarized potentials [33]. While widely used to discriminate between high- and low-voltage-activated calcium channel subtypes, Ni^{2+} ions are actually quite a poor pharmacological tool. There are reports of N-type current blockade by other cations, including, mercury (Hg^{2+}), zinc (Zn^{2+}), lead (Pb^{2+}) and aluminum (Al^{3+}) with $IC_{50}s$ generally ranging from the low- to mid-micromolar, although most of these studies have been performed on native cells with mixed types of calcium currents, thus complicating determination of the exact affinities to N-type channels [34–38]. Yttrium (Y^{3+}) has been reported to block exogenously expressed N-type channels with an $IC_{50} \sim 600\,nM$ although block is non-specific as this trivalent metal cation also blocks T-type ($IC_{50} \sim 700\,nM$), P/Q-type ($IC_{50} \sim 770\,nM$), L-type ($IC_{50} \sim 390\,nM$) and R-type channels ($IC_{50} \sim 275\,nM$); all measured in 20 mM Ba saline [39].

Notably, in many instances channel block by inorganic ions is strongly voltage-dependent – thus the reported IC_{50}s can change depending upon the holding and test potentials. Along these lines, factors that affect the gating of high-threshold calcium channels such as alternative splicing and β subunit co-expression can also alter apparent blocking affinities. Furthermore, blockade is often also dependent upon the both concentration of permeant ion and the extracellular surface charge; thus reported IC_{50}s can also vary significantly, dependent upon the specific measurement conditions.

5.4.4
Peptide Blockers

Many peptide toxins that block voltage-gated calcium channels have been described from numerous species of hunting cone snails and venomous spiders. The calcium channel blocking cone snail peptides (ω-conotoxins) tend to be relatively smaller, in the range of 22 to 30 amino acids, while the spider toxins range from ~36 to 80 residues. The ω-conotoxins are synthesized as precursor molecules containing signal and propeptide sequences and undergo disulfide coupled folding in the lumen of the endoplasmic reticulum. In general, the various venomous peptides are positively charged (+5 to +7), are folded into a 3-dimensinal configuration by highly conserved disulfide linkages and block with 1:1 kinetics. There appears to be a similar three-dimensional (3D) motif structure common across many of the peptide toxins and consists of an anti-parallel, triple-stranded β-sheet that is stabilized by a "cysteine knot" motif [40].

5.4.4.1 **Specific N-type Blockers**
Much of the definitive work describing the distribution and physiological functions of N-type channels results directly from the pioneering work of Olivera and colleagues concerning their description of the ω-conotoxin family of peptide toxins from marine hunting cone snails [41, 42]. Although there is considerable sequence variability between individual conopeptides, the ω-conotoxins possess a conserved pattern of cysteines (C-C-CC-C-C) and also usually contain several post-translationally modified amino acids (e.g., hydroxyproline, γ-carboxyglutamate). Examination of the 3D structure and folding patterns of several ω-conotoxins shows that formation of the correct disulfide bonds is largely determined by noncovalent interactions rather than steric effects [43].

The most widely utilized of all cone snail toxins is the 27 amino acid ω-conotoxin-GVIA, isolated from *Conus geographus* and which is a potent, selective and irreversible inhibitor of N-type channels (Fig. 5.4.1) [44]. ω-Conotoxin-GVIA binds to N-type channels in 1:1 stoichiometry with reported K_ds in the range of 10 to 100 pM. At concentrations of between 300 nM and 1 μM ω-conotoxin-GVIA completely blocks exogenously expressed cloned N-type channels from rat, human, rabbit and mouse. Complete and irreversible block of native N-type currents in various rodent and human preparations occurs at between 100 nM

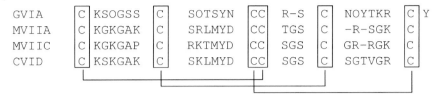

GVIA	C	KSOGSS	C	SOTSYN	CC	R-S	C	NOYTKR	C	Y
MVIIA	C	KGKGAK	C	SRLMYD	CC	TGS	C	-R-SGK	C	
MVIIC	C	KGKGAP	C	RKTMYD	CC	SGS	C	GR-RGK	C	
CVID	C	KSKGAK	C	SKLMYD	CC	SGS	C	SGTVGR	C	

Fig. 5.4.1

to 3 µM although generally block is complete at 300 to 500 nM and there is little need to go to higher concentrations. ω-Conotoxin-GVIA appears to act via direct binding to the outer mouth of the channel pore region and to physically occlude permeation [45, 46]. On the Cav2.2 subunit, the extracellular region defined by the domain III S5-S6 linker and, in particular, glycine 1326 are critical to ω-cono-toxin-GVIA interaction [45, 47]. In ω-conotoxin-GVIA itself, structure–function analyses show that the hydroxyl moiety of tyrosine-13 is essential for ω-conotox-in-GVIA activity on the N-type channel [48–50]. Interestingly, with the exceptions of tyrosine-13 and lysine-2, replacement any of the remaining residues with ala-nine did not significantly affect toxin binding, suggesting that toxin-channel inter-action largely occurs through these two residues. Alanine substitutions at the other positions were, however, shown to reduce the potency of N-type channel blockade, supporting the notion that while considerable flexibility exists within the toxin binding domain there are separable sites involved in the functional occlusion of the channel pore [27, 28].

Utilizing ω-conotoxin-GVIA as a definitive pharmacological tool, N-type chan-nels have been implicated in mediating neurotransmitter release in various cen-tral and peripheral neurons, the sympathetic regulation of the circulatory system, contributing to higher behaviors such as activity and vigilance state control, as well as an involvement in various inflammatory and neuropathic pain states [51, 52]. Notably, at low micromolar concentrations ω-conotoxin-GVIA also rever-sibly inhibits both L-type calcium channels and nicotinic acetylcholine receptors [53, 54].

The *Conus magus* peptidic toxin ω-conotoxin-MVIIA has also been widely used to define the contributions of N-type channels in various animal model and human pain conditions. ω-Conotoxin-MVIIA, also known as SNX-111, Ziconotide and Prialt, is a 25 residue cationic peptide that blocks N-type channels in a rever-sible manner [55]. Structurally, ω-conotoxin-MVIIA does not contain any post-translationally modified residues (other than an amidated C-terminal cysteine). Examination of native and cloned N-type channels shows that ω-conotoxin-MVIIA blocks completely at concentrations between 1 to 100 nM. In rat models, intrathecal administration of either ω-conotoxin GVIA or ω-conotoxin MVIIA shows strong analgesic effects on inflammatory pain, post-surgical pain, thermal hyperalgesia and mechanical allodynia [30, 56, 57]. In humans, the intrathecal administration of Prialt (ω-conotoxin MVIIA) to patients unresponsive to in-trathecal opiates significantly reduces pain scores in several chronic conditions, including cancer and HIV-related pain [58, 59]. In several case studies, Prialt has been shown to be effective for the management of intractable spasticity following spinal cord injury [60].

Side effects of intrathecal administration of Prialt in humans include nystagmus, sedation, confusion, auditory and visual hallucinations and severe agitation. Peripherally, intravenous administration of Prialt to humans results in significant orthostatic hypotension [61]. It is unknown whether these adverse affects are a direct result of N-type channel blockade or because at higher concentrations ($> 1 \mu M$) Prialt can also non-selectively inhibit other calcium channels and receptors [47, 62, 63].

More recently, Lewis and coworkers have reported that ω-conotoxin-CVID (AM336), a 27 amino acid peptide from *Conus catus*, is a potent blocker of N-type currents and displaces radiolabeled ω-conotoxin-GVIA binding at potencies similar to that for both ω-conotoxin-GVIA and ω-conotoxin-MVIIA [63]. Structurally, ω-conotoxin-CVID displays a distinct 3D orientation in two of its loops, which distinguishes it from ω-conotoxin-GVIA and ω-conotoxin-MVIIA. Interestingly, in animal models AM336 results in potent dose-dependent anti-nociception although compared to ω-conotoxin MVIIA (Prialt) it appears to possess a markedly better therapeutic index (~5-fold) in that rodents display a decreased incidence of the serpentine tail movements and whole body shaking [64].

The venom from the Chinese bird spider *Ornithoctonus huwena* contains a 33 amino acid peptide called huwentoxin-I that blocks N-type channels with high affinity ($EC_{50} \sim 100$ nM) and selectivity [65]. NMR analysis shows that huwentoxin-I exhibits a similar anti-parallel, triple-stranded β-sheet and cysteine knot structure to that of the ω-conotoxins. In a rat model of inflammatory pain huwentoxin-I appears to be as effective as ω-conotoxin-MVIIA in both the early and late phases of the formalin response but exhibits a significantly lower degree of motor dysfunction and ataxia than ω-conotoxin MVIIA (Prialt).

The rank order of toxin potency of both N-type channel binding and blockade between various ω-conotoxins differs amongst different cellular and tissue preparations and is suggestive of pharmacologically distinct N-type channels in the central and peripheral nervous systems [66–69]. Other factors that may affect discrepancies between functional and binding assays include conditions that may alter the conformational state of the channel such as membrane potential (e.g., the membrane potential in isolated membrane preparations is zero) and interaction with cytoskeletal anchoring proteins. In addition, reliable toxin affinities can only be obtained under equilibrium conditions and thus some of the variability reported in the literature concerning binding and blocking constants may reflect the fact that different toxins likely require different incubation periods and also that individual cellular, slice and tissue preparations reach equilibrium at distinct rates. In many instances toxin binding is also highly sensitive to the concentration of external ions such as calcium, barium, sodium and magnesium, thus any comparison of potencies needs to also consider the composition of the external saline.

5.4.4.2 **Non-specific N-type Channel Blockers**

The *Conus magus* peptide toxin ω-conotoxin MVIIC (Fig. 5.4.1) is a potent blocker of N-type currents and likely competes at the same binding site as ω-conotoxin GVIA ($IC_{50} \sim 3{-}20$ nM). The onset of ω-conotoxin MVIIC block is rapid ($\tau \sim 1$ s) and, similar to that for other ω-conotoxins, can be affected by the concentration of external divalent cations [70]. ω-Conotoxin MVIIC has been widely used to block N-type channels although this peptide blocks P/Q-type channels with a similar binding affinity and thus cannot be used as a selective pharmacological agent in many native preparations.

At saturating concentrations (200 nM) the *C. magus* peptide ω-agatoxin-IIIA produces partial block of both N-type and P/Q-type channels. Although acting as a pore blocker, it appears to act a distinct binding site on N-type channels compared to that for ω-conotoxin GVIA and ω-conotoxin MVIIC [71]. ω-Agatoxin-IIIA also completely blocks neuronal and cardiac L-type channels ($K_d \sim 1$ nM) but not neuronal T-type channels at concentrations up to 300 nM [72].

Several toxins from the venom of the spider *Phoneutria nigriventer* have been characterized, including, ω-PnTx-II, ω-Ptx-IIA, and PnTx3-6. PnTx3-6 is a 55 residue peptide shown to block both K^+-evoked increases in intracellular calcium and calcium-dependent glutamate release from synaptosomes [73]. Examination of cloned and native mammalian calcium channels shows that PnTx3-6 reversibly and completely inhibits N-type ($IC_{50} = 120$ nM), incompletely inhibits R-type ($IC_{50} = 135$ nM), P/Q-type ($IC_{50} = 260$ nM) and L-type ($IC_{50} = 610$ nM) channels, and has no affect on Cav3.1 T-type currents (up to 1 μM). Block of the high-threshold channels by PnTx3-6 is consistent with a 1:1 channel–toxin interaction and is without affect on either kinetic or voltage-dependent properties.

The *Segestria florentina* spider toxin SNX-325 is a 49 residue peptide, structurally similar to ω-agatoxin IVA, that blocks high-threshold calcium channels at micromolar concentrations but not sodium or potassium channels [74]. In the nanomolar range SNX-325 appears most selective for N-type channels ($IC_{50} \sim 30$ nM) and less so for L-type (IC_{50} s $\approx$ low micromolar), P/Q-type (IC_{50} s $\approx$ low micromolar) and R-type (IC_{50} s $\approx$ low micromolar).

The venom of the hunting spider *Filistata hibernalis* contains a 74 amino acid peptide (DW13.3) most closely structurally related to curtotoxin and that potently and reversibly blocks all high-threshold calcium channels, including the N-type ($IC_{50} = 14$ nM) [75]. Block occurs with 1:1 kinetics, is only partial even at high concentrations, and the binding site appears to overlap with that of ω-conotoxin-MVIIC but not the site responsible for ω-agatoxin IVA binding. Blocking affinity varies by nearly an order of magnitude between cloned N-type channels expressed in HEK cells versus *Xenopus* oocytes, suggesting that cell-specific expression factors such as glycosylation patterns may affect toxin binding. At a 230 nM test concentration DW13.3 does not affect the native T-type currents found in GH3 cells.

The venom of the tarantula *Grammostola spatulata* contains a 36 residue peptide (ω-grammotoxin SIA) that blocks both native N-type and P-type calcium channels with high affinity (complete block > 50 nM). Similar to that for ω-agatoxin IVA blockade of P-type channels, ω-grammotoxin SIA acts as a gating modifier, shifting the voltage-dependence of activation more depolarized although the two peptides do not appear to compete for the same binding site.

Kurtoxin is a 63 residue peptide toxin from the venom of the scorpion *Parabuthus transvaalicus* and is most closely related to the α-scorpion toxins that slow inactivation of voltage-gated sodium channels. Kurtoxin was originally identified as a reversible and selective T-type channel blocker (K_d ~15 nM), acting as a gating modifier and exhibiting greater than 600-fold selectivity over exogenously expressed high-threshold calcium channels [76]. Subsequently, in native central and peripheral neurons, kurtoxin was shown to partially inhibit high-threshold N-type (K_d ~ 450 nM), L-type (K_d ~70 nM) and P/Q-type currents (K_d ~ 14 nM) [77]. The gating effects of kurtoxin on the various high-threshold calcium channels are complex, differentially altering activation kinetics and voltage-dependence and even facilitating steady-state P-type currents.

The funnel web spider toxin ω-agatoxin IVA is a gating modifier that is widely utilized as a potent and selective blocker of native P-type channels (K_d ~ 1 nM) and also at higher bath concentrations (~ 0.5–1 μM) to block Q-type channels. In fact, ω-agatoxin IVA also appears to non-selectively block N-type channels at this higher concentration – thus pharmacological manipulations aimed at blocking Q-type currents in native cells expressing multiple high threshold calcium channel subtypes must be performed with some caution [78].

Table 5.4.1 describes some of the identified peptidic toxins known to inhibit N-type calcium channels. The large number of peptide toxins found in both hunting spiders and cone snails that non-specifically block N-type and other high-threshold calcium channels likely reflects the fact that the channels are derived from a common ancestor and share similar structural features involved in overall channel architecture and functions such as permeation and gating. Pairwise comparisons amongst several peptide toxins on cerebellar Purkinje cells shows that N-type channels possess two distinct toxin-binding sites (in comparison to three separate sites on P-type channels) [71]. That a number of the toxins themselves do not exhibit significant sequence homology and yet similarly block the various calcium channels suggests that conserved tertiary structure and charge distribution are most critical to blockade. Overall, while of significant interest in and of themselves as tools for probing both channel structure–function and the biophysical mechanisms of toxin blockade, many of these non-specific blockers are limited with regard to their usefulness as pharmacological tools concerning the study of native cells expressing multiple subtypes of calcium channels.

Tab. 5.4.1 N-type channel peptide antagonists.

Compound	Structure	*In vitro*	*In vivo*	Ref.
ω-Conotoxin GVIA *Conus geographus*	27 Amino acids; 3 disulfide bonds; +5 net positive charge	Irreversible block of N-type currents at ∼ 100 pM to 1 µM; pore blocker; interacts with Cav2.2 domain III S5-S6; tyrosine-13 is critical to binding to N-type channel	Intrathecal administration attenuates tactile allodynia in Chung model; no effect in acute tail-flick model	41, 46, 47, 55, 123, 124
ω-Conotoxin MVIIA (a.k.a.: SNX-111; ziconotide; Prialt) *Conus magus*	25 Amino acids; 3 disulfide bonds; +6 net positive charge	Reversible block of N-type (IC$_{50}$ ∼ 1–40 nM); reversibly block L-, P/Q- and R-types at low [µM]; pore blocker; interacts with Cav2.2 domain III S5-S6; tyrosine-13 is critical to binding to N-type channel	Intrathecal administration is anti-nociceptive in rat neuropathic and inflammatory models; no effect in acute tail-flick model; Prialt is approved in humans for severe intractable pain	41, 47, 55, 125–127
ω-Conotoxin MVIIC (SNX-230) *Conus magus*	26 Amino acids; 3 disulfide bonds; +7 net positive charge	Blocks both P/Q- and N-types at low [nM]; pore blocker; tyrosine-13 is critical to binding to the channel		46, 70, 71, 128, 129
ω-Conotoxins CVIA, CVIB, CVIC, CVID (AM336) *Conus catus*	25–27 Amino acids; 3 disulfide bonds; +5 to +7 net positive charge	Pore blockers; inhibit electrically stimulated vas deferens contractions (IC$_{50}$s ∼ 20–600 nM); CVID reversibly blocks α$_{1B}$ + β$_3$ N-type currents in *Xenopus* oocytes (IC$_{50}$ ∼ 2 nM) and inhibits substance P release in rat spinal cord slices (EC$_{50}$ ∼ 20 nM) and EPSPs from ganglionic neurons (IC$_{50}$ ∼ 30 nM); CVID exhibits ∼ 100× greater selectivity for N-type over P/Q-type compared to ω-conotoxin MVIIA (Prialt)	Intrathecal CVID (AM336) is anti-nociceptive in rat Freund's complete adjuvant inflammatory model; attenuates tactile allodynia in Chung model; no effect in acute tail-flick model; shows improved therapeutic window compared to ω-conotoxin MVIIA (Prialt)	63, 64
ω-Conotoxins SVIA and SVIB *Conus striatus*	24 and 26 Amino acids; 3 disulfide bonds; +5 (SVIA) and +6 net positive (SVIB) charge	Pore blockers; SVIA blocks N-type at low [nM]; SVIB blocks both P/Q- and N-types at low [nM]	SVIB is lethal in mice by intracerebral injection	129, 130
ω-Grammotoxin SIA *Grammostola spatulata*	36 Amino acids	Reversible block of N- and P/Q-types at > 50 nM; gating modifier		131, 132

Tab. 5.4.1 N-type channel peptide antagonists (continued).

Compound	Structure	*In vitro*	*In vivo*	Ref.
ω-Agatoxin IIIA *Agenelopsis aperta*	76 Amino acids; 6 disulfide bonds	Blocks N-type, P-type, and L-type channels ($K_d \sim 1$ nM)		72, 133–135
ω-Agatoxin IVA *Agenelopsis aperta*	48 Amino acids; 4 disulfide bonds	Generally used as a selec- tive P/Q-type blocker (IC$_{50}$ $\sim$ 1–10 nM) but will block N-type at low [μM]; gating modifier	Minimal affects in ani- mals models of pain	136
ω-Phonetoxin IIA *Phoneutria nigri- venter*	76 Amino acids; 7 disulfide bonds; structural homol- ogy to ω-agatoxins IIIA and IIIB	Irreversible block of N- and P/Q-types at low- mid [nM]; reversible block of R-type		137, 138
SNX-325 *Segestria florentina*	49 Amino acids	High-affinity to N-type (IC$_{50}$ $\sim$ 30 nM); L-, P/Q- and R-types at low [μM]; inhibits norepinephrine release (IC$_{50}$ $\sim$ 0.5 nM)		74
Huwentoxin-I *Ornithoctonus huwena*	33 Amino acids	Blocks N-type currents in NG108-15 cells (EC$_{50}$ $\sim$ 100 nM)	Anti-nociceptive in rat formalin inflammatory model by intrathecal administration	65
Kurtoxin *Parabuthus trans- vaalicus*	63 Amino acids	Partially blocks native N-type ($K_d \sim 450$ nM); blocks L-, P/Q- and T-types K_ds $\sim$ 14 to 70 nM; com- plex gating modifier		76, 77
DW13.3 *Filistata hibernalis*	74 Amino acids; 6 disulfide bonds	Non-selective blocker of high threshold channels; in *Xenopus* oocytes: P/Q (IC$_{50}$ = 4 nM); N-type (IC$_{50}$ = 14 nM); L-type IC$_{50}$ = 27 nM); R- type (IC$_{50}$ = 96 nM); in HEK cells N-type IC$_{50}$ = 2 nM); no effect on T-type currents (up to 230 nM); pore blocker		75
PnTx3-6 *Phoneutria nigriventer*	55 Amino acids	Non-selective blocker of L-, P/Q-, R- and N-types at IC$_{50}$s $\sim$ 120 to 610 nM; inhibits Ca-dependent glutamate release from synaptosomes		73, 139

5.4.5
Small Organic Molecule N-type Blockers

While several of the peptide toxins bind with very high affinity and selectivity to block N-type channels, the limitation of delivery via intravenous and intrathecal routes makes it unlikely that they will become widely used therapeutic agents for the treatment of disorders such as neuropathic pain. In this regard, there has been a concerted effort to identify and develop small organic molecules that selectively target N-type channels, that are orally available and that do no exhibit adverse affects such as the ataxia and orthostatic hypotension observed with Prialt (ω-conotoxin MVIIA) administration.

5.4.5.1 Dialkyl-dipeptidylamines/Substituted l-Amino Acids

Hu and colleagues initially identified the N,N-dialkyl-dipeptidylamines as N-type channel blockers by high-throughput screening and then explored a series of specific derivatives with varying substitutions (Table 5.4.2) [79–82]. Examining native N-type currents in human neuroblastoma IMR32 cells, potency for N-type channel blockade was optimized with the best compounds having IC_{50}s between ~40 and 400 nM. Generally, the potency of N-type blockade correlates most strongly with both increasing lipophilicity and, in the case of alkyl substituents at the N-terminus, the size of the group is critical (4-tert-butylbenzyl is more potent than methyl). Interestingly, while a few compounds exhibit a broad potency range (over 300-fold), most derivatives fall within a relatively narrow range, suggesting that the high-affinity drug-N-type channel interaction site is tolerable of many types of substitutions. Upon intravenous administration several compounds show efficacy in a mouse audiogenic seizure model although there does not seem to be a strong correlation between N-type current blockade in IMR32 cells and *in vivo* efficacy. Notably, (1) IMR32 cells express a complex set of calcium currents, making it difficult to select out the pure N-type component for the purposes of SAR analyses, and (2) many compounds of this class also block native sodium currents and thus the N,N-dialkyl-dipeptidylamines may not represent a highly promising class of pure N-type blockers.

Through screening of compound libraries against calcium influx into IMR32 cells, Seko and colleagues identified N-(t-butoxycarbonyl)-L-aspartic acid as a novel structural motif with potential N-type channel blocking activity (Table 5.4.2) [83]. Subsequent SAR around several substituted L-amino acids revealed L-cysteine-based derivatives showing the highest N-type channel blocking affinities and that also exhibit efficacy in the rat formalin model of inflammatory pain [83–85]. Most substituted L-cysteines exhibit a relatively narrow potency range measured against calcium influx through N-type channels in IMR32 cells (IC_{50}s ~ 0.4 to 5 µM) and also block L-type calcium currents within a similarly narrow range (N-type:L-type ratios generally range from ~1:1 to 1:6). While it is difficult to define a predictive pattern of substituents versus N-type blocking activity in IMR32 cells, a phenoxybenzyl moiety made a crucial contribution towards

N-type blockade and the best compound of this class possessed an $IC_{50} \sim 140$ nM and a 12-fold selectivity over the L-type channel. Subsequent structural modifications to improve physicochemical properties (decrease $\log P$) by introducing a basic nitrogen resulted in enhanced water solubility, reasonable selectivity over the L-type channel (N:L $\sim$ 12- to 15-fold), and oral availability as measured by analgesic activity in both the rat formalin model and the rat chronic constriction model of neuropathic pain. One of these compounds, the substituted L-cysteine derivative ONO-2921 (Table 5.4.2), attenuates C-fiber-evoked responses induced by carrageenan inflammation, alters Aβ-fiber responses under neuropathic conditions, and is generally efficacious in several rat models for inflammatory and chronic pain ($A_{50}s \sim 10$–20 mg kg^{-1} p.o.) [86, 141]. Similar to that for the peptidic N-type channel blockers, there appears to be little effect of ONO-2921 administration in rat acute pain models (e.g., hot plate, paw pressure) [86]. Of particular re-

Tab. 5.4.2 Small organic molecule N-type channel blockers.

Compound class	Structure	In vitro	In vivo	Ref.
N,N-Dialkyl-dipeptidyl-amines		Inhibit N-type currents in IMR-32 cells ($IC_{50}s \sim$ 0.04–2 µM); block both N-type calcium and sodium currents in rat SCGs	Anti-seizure activity in audiogenic DBA/2 mouse model (80–100% activity at 10–30 mg kg^{-1} i.v.)	79–82
Substituted L-amino acids		Inhibit N-type channels in IMR32 cells ($IC_{50}s \sim$ 0.12 to 6 µM); up to 12-fold selectivity over L-type	Oral availability; analgesic in the rat formalin and chronic constriction models	83–85, 140
	ONO-2921	State-dependent blocker; inhibits N-type ($IC_{50} \sim$ 325 nM) and R-type ($IC_{50} \sim$ 500 nM); selective over L- and P/Q-types ($IC_{50}s$ > 5 µM)	Orally available; efficacious in rat pain models mediated via formalin and nerve injury; reduces C-fiber evoked responses induced by carrageenan; no affect on acute pain; no adverse cardiovascular affects	86, 141

Tab. 5.4.2 Small organic molecule N-type channel blockers (continued).

Compound class	Structure	*In vitro*	*In vivo*	Ref.
Substituted piperidines, piperazines, morpholines	 Flunarizine	Blocks sympathetic N-type channel (IC$_{50}$ ~ 3 μM) and cloned N-type channel (IC$_{50}$ ~ 2–6 μM)	Widely used clinical antipsychotic	107
	 Fluspiriline	Reversible block of N-type channels in PC12 cells (IC$_{50}$ ~ 30 nM); blocks N-type and P-type channels in cerebellar Purkinje cells (IC$_{50}$s ~ 2–6 μM)	Widely used clinical antipsychotic	87–89
	 Substituted piperidines and piperazines	Block N-type (IC$_{50}$s ~ 0.03 to > 20 μM); selective over P/Q- and L-types (10 > 1000-fold)	Orally available; efficacious in rat inflammatory (formalin, carrageenan) and neuropathic (Chung, Bennett) pain models; no adverse cardiovascular affects	90–96
	 4-Aminopiperidines	Inhibit N-type currents in IMR-32 cells (IC$_{50}$s ~ 0.2–120 μM); block both N-type calcium and sodium currents in rat SCGs	Anti-seizure activity in DBA/2 mouse model (80–100% activity at 30 mg kg^{-1} i.v.); anti-nociceptive in mouse acetic acid writhing assay (ED$_{50}$ ~ 6 mg kg^{-1} i.v);	97–99
	 SB-201823-A	Reversibly inhibits N-type channel (IC$_{50}$ ~ 11 μM) and also sodium channels (IC$_{50}$ ~ 5 μM)	Neuroprotective against global and focal ischemia in rodents	100–102

Tab. 5.4.2 Small organic molecule N-type channel blockers (continued).

Compound class	Structure	*In vitro*	*In vivo*	Ref.
Long-chain unsaturated hydrocarbons	Farnesol	Blocks all cloned channels in HEK cells at low [µM]: L-type >N-type >R-type >P/Q-type; open channel blocker of N-type; shifts N-type steady-state inactivation more hyperpolarized	Naturally occurring metabolite in brain; no animal data regarding N-type affects; alters cardiovascular tone and blood pressure via L-type channel	104
Aliphatic monoamines	Dodecylamine	Blocks cloned N-type channel in HEK cells at high [nM]: also blocks L-, P/Q-, and R-types at low [µM]; open channel blocker; shifts N-type steady-state inactivation more negative	No animal data	105
Imidazolines	Antazoline	Reversible and voltage-dependent block of N-type channels in primary neurons (IC$_{50}$ ~ 110 µM); blocks P/Q- (IC$_{50}$ ~ 10 µM) and R-types (IC$_{50}$ ~ 210 µM)	Block NMDA receptors at low concentrations; neuroprotective effects	106
Volatile anesthetics	Isofluorane / Halothane	At 0.6 to 0.8 mM reversibly and partially inhibit cloned N-, P/Q-, L- and R-type channels	Widely used clinically as general anesthetics	114–117, 119

Tab. 5.4.2 Small organic molecule N-type channel blockers (continued).

Compound class	Structure	*In vitro*	*In vivo*	Ref.
Dihydropyri-dines	Cilnidipine (FRC-8653)	Mixed N- and L-type blocker (IC$_{50}$ ~ 0.5–2 µM); inhibits K$^+$-induced release of catecholamines and glutamate	Reduces infarct size in rat focal ischemia model; used clinically to treat hypertension	107, 108, 110, 142
Ethanol	Ethanol	Block N-type, P/Q-type, L-type and T-type (50–200 mM)	Wide range of physiological and behavioral affects; N-type target implicated in reinforcing and rewarding proper-ties of ethanol	120–122

levance there are no adverse affects on blood pressure, locomotor activity or other general behaviors at doses up to 300 mg kg^{-1} orally. *In vitro*, ONO-2921 blocks both N-type (IC$_{50}$ ~ 325 nM) and R-type (IC$_{50}$ ~ 500 nM) channels but appears to be selective over the L- and P/Q-types (IC$_{50}$s > 5 µM).

5.4.5.2 Substituted Piperidines/Piperazines

Several groups have independently identified what appears to be a promising high-affinity binding site on the N-type channel for piperidine, piperazine and morpholine-based compounds. Several classes of clinically utilized agents interact non-specifically and at relatively low affinity with both low- and high-voltage-activated classes of calcium channels. These include the tertiary amine-based local anesthetics (e.g., flecainide, fomocaine) and a number of antipsychotics tradition-ally thought to exert their clinical effects through dopamine D2 receptors (e.g., pimozide, penfluridol, haloperidol, fluspiriline) [87–89]. Investigating the effects of local anesthetics on the then recently cloned high voltage-activated calcium channel family, Zamponi and coworkers noticed that archetypal tertiary amine-based anesthetics such as procaine and diethylcarbamazine reacted poorly (IC$_{50}$s ~ 2–5 mM) with calcium channels but that atypical anesthetics such as fo-mocaine and flecainide exhibited higher affinities (IC$_{50}$s ~ 100–800 µM) [90]. Keying in on the fact that the atypical anesthetics possess a tertiary amine as part of a piperidine, piperazine or morpholine ring (rather than an alkyl chain), the authors found evidence for subtype specificity amongst various agents, with R-type channels showing the highest degree of block and N-type channels generally the least. SAR analyses with commercially available compounds such as pen-fluridol and flunarizine indicated that high-affinity piperidine/piperazine-based binding requires the protonated species, follows 1:1 kinetics, is partially reversible, and likely occurs on the cytoplasmic side of the channel complex [90, 91].

While the piperidine/piperazine/morpholine class of molecules exhibit subtype specificity amongst the cloned high voltage-activated calcium channel isoforms (with the R-type being the most sensitive), the design of a biphenyl-substituted pharmacophore and subsequent SAR analyses show that it is possible to obtain high affinity and selective N-type blockers (Table 5.4.2) [91–94]. Certain embodiments of this pharmacophore appear to be preferred N-type blockers and include derivatives with the Z to Y distance preferentially being from ~3 to 20 Å, with Y containing at least one aromatic moiety, and if X is (CH_2) then n preferentially between 1 and 4 (see Table 5.4.2) [91, 93, 94]. There is also considerable structural flexibility allowed and the linker between Z and Y may contain amines, carbonyls or amides, and also R1 and R2 may be alkyl, aryl or alkylaryl, contain heteroatoms and also be independently substituted. The best compounds of this class exhibit high N-type affinity ($IC_{50}s$ ~ 30 to 150 nM) and selectivity over both P/Q-type (10- to 15-fold) and L-type calcium channels (30- to >1000-fold). Of particular relevance, the lead compounds block N-type channels in a state- and frequency-dependent manner and are efficacious by oral administration in various animal models for inflammatory and neuropathic pain [95, 96]. Similar to that for the substituted L-amino acid class of N-type blockers, the highest affinity and selective substituted piperidine/piperazines appear to have few deleterious affects on motor activity, cardiovascular function or general CNS-related behaviors.

Independent studies by Hu, Ryder and colleagues show that a series of substituted 4-aminopiperidine derivatives result in block of native human N-type currents in IMR32 cells over a wide concentration range ($IC_{50}s$ ~ 0.2–120 μM; Table 5.4.2). At the R1 position, the most potent compounds possessed either an O-benzyl substitution, an N-benzyl containing non-polar substituents at the para position, a cyclohexylmethyl or a 3,3-dimethylbutyl ($IC_{50}s$ ~ 0.2–0.4 μM) [97]. The 3,3-dimethylbutyl-substituted analogue showed activity in the mouse DBA/2 audiogenic seizure model upon intravenous administration (80% response at 30 mg kg^{-1}), suggesting that it crosses the blood–brain barrier. Further 4-piperidine derivatives of this series exhibit high N-type channel blockade in IMR32 cells with lipophilic substitutions of a certain size generally being the most advantageous [97–99]. There is a relatively poor correlation between *in vitro* blockade of N-type currents and activity in audiogenic seizure animals, suggesting the possibility of less than ideal physicochemical characteristics (e.g., high log$P$, low aqueous solubility and poor pharmacokinetic properties). Similar to that for the N,N-dialkyl-dipeptidylamines, there is evidence that many of the molecules in this class also block voltage-gated sodium channels.

SB-201823-A, 4-[2-(3,4-dichlorophenoxy)ethyl]-1-pentyl piperidine, reversibly blocks both native calcium currents in C2D7 cells and exogenously expressed cloned human N-type channels in a concentration-dependent manner (IC_{50} ~ 11 μM; Table 5.4.2). *In vivo* SB-201823-A protects against global ischemia in a gerbil model when administered 30 min post-occlusion [100] and also decreases in-farct volume in both rat and mouse models of focal ischemia [101]. Overall, SB-201823-A is a moderately potent N-type channel blocker that also inhibits sodium currents with a somewhat higher affinity (IC$_{50}$ ~ 5 µM) [102].

Taken together, several groups have independently shown that certain substi-tuted piperidines and piperazines bind with high affinity to N-type calcium chan-nels and that with specific structural modifications selectivity over other high threshold calcium channels can be obtained. Of particular note, some members of this class of compounds are both efficacious in animal models of inflammatory and neuropathic pain and compared to the ω-conotoxins have the benefits of being both orally available and generally exhibit fewer adverse affects at therapeu-tic doses. One note of caution concerning comparisons between the reported IC$_{50}$s of these compounds is that the intracellular environment appears critical. The presence of internal chloride ions in the patch pipette can alter the apparent affinity to N-type channels (by as much 10- to 40-fold), as can tonic inhibition by G-proteins (by ~3-fold) [103].

5.4.5.3 Unsaturated Hydrocarbons and Aliphatic Monoamines

Farnesol is an isoprenoid intermediate in the mevalonate pathway produced by dephosphorylation of farnesyl pyrophosphate. Farnesol plays roles in cell growth and differentiation, apoptosis, the synthesis of ubiquinone and cholesterol, and as a regulator of vascular tone and blood pressure. Farnesol blocks the cloned rat brain N-type calcium channel expressed in HEK293 cells with IC$_{50}$s as low as 1 to 3 µM although it also blocks other high threshold calcium channels, with the L-type being the most potently inhibited [104]. Interestingly, the mechanism of block appears different between L-type and N-type channels, with L-type block-ade largely being to the resting state of the channel, while N-type block is due to a combination of open channel block at higher concentrations and a large shift of the steady-state inactivation profile to more hyperpolarized potentials at lower concentrations. Brain levels of farnesol *in vivo* are sufficient to mediate N-type channel blockade although it remains to be demonstrated that activity of the me-valonate pathway can affect N-type channels *in vivo*.

Examination of several linear 12-carbon molecules structurally related to farne-sol shows that the functional group attached to the dodecyl backbone is a critical determinant of N-type blocking affinity. The addition of an amine head group re-sults in an ~30-fold increase in blockade and led Beedle and Zamponi to identify dodecylamine as a novel high-affinity N-type channel blocker [105]. The length of the carbon chain backbone is also critical and the SAR profile exhibits a U-shaped curve with chain linear lengths of 12 and 13 carbons being the highest affinity N-type blockers and progressively less blockade with carbon chain lengths C9–C11 and C14–C18. Similar to that for farnesol, dodecylamine inhibition is not specific as L-, P/Q- and R-type channels are also blocked. Mechanistically, N-type channel inhibition occurs predominantly through the open channel and inactivated states while that for the L-type channel occurs via resting state blockade.

5.4.5.4 Imidazolines

Imidazolines display a wide variety of physiological effects in the central and peripheral nervous systems, and although their molecular targets certainly include NMDA receptors they also appear to target other types of ion channels. Of particular note, the imidazoline antazoline exhibits rapid, fully reversible and voltage-dependent blockade of N-type channels in primary striatal neurons (IC_{50} ~ 110 μM). Interestingly, the NMDA receptor antagonist MK-801 appears to interact at a common competitive binding site on the N-type channel, which is suggestive of a conserved albeit lower affinity site between these distinct classes of ion channels. The effect of antazoline is not particularly specific to N-type channels as P/Q-type (IC_{50} ~ 10 μM) and R-type (IC_{50} ~ 210 μM) channels are also blocked [106].

5.4.5.5 Dihydropyridines

Largely used to selectively target L-type calcium channels (Chapter 5.3), several dihydropyridines (DHPs) also interact with other high- and low-threshold channels [107]. Cilnidipine is a DHP with mixed N-type and L-type blocking activity that is used clinically in Japan as an anti-hypertensive agent [108, 109]. In animals, cilnidipine protects against focal brain ischemia at relatively low doses that are independent of any effects on cerebral blood flow (100 μg kg^{-1} i.p.). Cilnidipine also attenuates glutamate and catecholamine release and has been proposed to exert its neuroprotective effects selectively through its action on N-type channels [110, 111].

5.4.5.6 Volatile Anaesthetics

Volatile anesthetics are used clinically to produce general anesthesia and also exhibit significant hypnotic and amnestic effects. They likely have several molecular targets, including $GABA_A$ receptors, and there is good evidence that they affect excitatory synaptic transmission, possible through altering calcium channel activity [112–114]. Some studies have suggested that N-type channels are relatively more sensitive to volatile anesthetics than other calcium channel subtypes [115, 116] although other data suggests that this class of molecules does not discriminate between any particular high-voltage activated calcium channel or amongst the three classes of T-type channels [114]. Examining cloned neuronal calcium channels in *Xenopus* oocytes, Kamatchi and coworkers showed that both halothane and isoflurane reversibly inhibit peak whole cell currents and increase inactivation kinetics amongst all high threshold channels in a concentration-dependent manner (L-, P/Q-, N- and R-types) [117]. Nikonorov et al. [118] have suggested that the differences between data from native cells compared to that for exogenously expressed cloned high-voltage activated calcium channels may be because the G-protein dependent activation of N-type channels acts to decrease the ability of isoflurane to inhibit channel activity.

Relevant to a proposed action of volatile anesthetics on neurotransmitter release, Takei and coworkers examined the effects of halothane in mice lacking the N-type gene [119]. Results show an increased sensitivity in the knock-out animals compared to wild type as assessed both behaviorally and by an increased halothane-induced depression of hippocampal Schaffer collaterall-CA1 field EPSPs, suggesting that halothane targets N-type channels involved in glutamate release.

5.4.5.7 Ethanol

Interaction with voltage-gated calcium channels is a well described effector for the physiological affects of ethanol ingestion [120]. It is more generally known that ethanol both directly inhibits L-type currents (in the range of 10 to 50 mM) and also indirectly through altering circulating levels of vasopressin. Less well known, acute ethanol exposure partially blocks N-type channels (maximal ~45%) at physiologically relevant concentrations (50 to 100 mM) and also partially attenuates ω-conotoxin GVIA-sensitive dopamine release (200 mM) and K^+ evoked increases in intracellular calcium. P/Q-type currents are also blocked by ethanol, and for both N-type and P/Q-type channels the inhibitory effects of ethanol are antagonized by activation of protein kinase A [121]. N-type channel blockade is slow onset (~ 8 to 10 min) and only partially reversible and it remains to be determined whether acute exposure to ethanol acts directly on N-type channels or indirectly via metabolic pathways involving channel phosphorylation. Interestingly, chronic ethanol exposure has recently been shown to induce a specific Cav2.2 spice variant (lacking exon 31a) and that exhibits faster activation kinetics and a shift in voltage-dependent activation to more negative potentials [122].

5.4.6
Conclusions

The highest affinity and selective N-type calcium channel blockers remain the peptide toxins designed by Mother Nature. While the cone snail and hunting spider toxins have proven extremely useful in defining N-type channel biochemical composition, spatial and temporal expression patterns, as well as helping elucidate physiological functions, the peptidic nature of these molecules remains an obstacle to the development of therapeutic agents that can be widely administered. To date, the most promising small organic molecule N-type blockers that are likely be orally available appear to be the substituted piperidines/piperazines and L-amino acid based classes of molecules. The pyrrolidine-based anticonvulsant levetiracetam and GABA analogues gabapentin and pregabalin also appear to be interesting candidates as possible N-type channel blockers, although they likely act indirectly through associated proteins and therefore make difficult targets for SAR analyses on N-type channel activity.

Biophysically distinct types of N-type currents can be generated by several means, including alternative splicing of the Cav2.2 subunit, co-expression with distinct β subunits and differential association with various adaptor, modulatory and synaptic vesicle proteins. Utilizing different ω-conotoxins as probes, there is accumulating evidence that there also exist pharmacologically distinct subtypes of N-type channels in several native cellular preparations. Of particular relevance, these biophysical and pharmacological differences may form the basis for the development of state-dependent blockers that distinguish between central and peripheral N-type channels.

Acknowledgments

I thank Ms. Cynthia Chow and Dr. Hassan Pajouhesh for help with the figure and tables. T. P. Snutch is Professor in the Michael Smith Laboratories, University of British Columbia. Work in the Michael Smith Laboratories is supported by an operating grant from the Canadian Institutes for Health Research and by a Canadian Research Tier 1 Chair in Biotechnology-Genomics.

References

1. Snutch, T. P., Peloquin, J., Mathews, E., McRory, J. Molecular properties of voltage-gated calcium channels. *Voltage-Gated Calcium Channels*, Kluwer Academic/Plenum Publishers: New York, **2005**, pp. 113–151.

2. Dubel, S. J., Starr, T. V., Hell, J., Ahlijanian, M. K., Enyeart, J. J., Catterall, W. A., Snutch, T. P. Molecular cloning of the alpha-1 subunit of an omega-conotoxin-sensitive calcium channel. *Proc. Natl. Acad. Sci. U.S.A.* **1992**, *89*, 5058–5062.

3. Stea, A., Dubel, S. J., Snutch, T. P. alpha 1B N-type calcium channel isoforms with distinct biophysical properties. *Ann. New York Acad. Sci.* **1999**, *868*, 118–130.

4. Williams, M. E., Brust, P. F., Feldman, D. H., Patthi, S., Simerson, S., Maroufi, A., McCue, A. F., Velicelebi, G., Ellis, S. B., Harpold, M. M. Structure and functional expression of an omega-conotoxin-sensitive human N-type calcium channel. *Science* **1992**, *257*, 389–395.

5. Fujita, Y., Mynlieff, M., Dirksen, R. T., Kim, M. S., Niidome, T., Nakai, J., Friedrich, T., Iwabe, N., Miyata, T., Furuichi, T., et al. Primary structure and functional expression of the omega-conotoxin-sensitive N-type calcium channel from rabbit brain. *Neuron* **1993**, *10*, 585–598.

6. Cahill, A. L., Hurley, J. H., Fox, A. P. Coexpression of cloned alpha(1B), beta(2a), and alpha(2)/delta subunits produces non-inactivating calcium currents similar to those found in bovine chromaffin cells. *J. Neurosci.* **2000**, *20*, 1685–1693.

7. Stea, A., Soong, T. W., Snutch, T. P. Determinants of PKC-dependent modulation of a family of neuronal calcium channels. *Neuron* **1995**, *15*, 929–940.

8. Bourinet, E., Soong, T. W., Stea, A., Snutch, T. P. Determinants of the G protein-dependent opioid modulation of neuronal calcium channels. *Proc. Natl. Acad. Sci. U.S.A.* **1996**, *93*, 1486–1491.

9. Lin, Z., Haus, S., Edgerton, J., Lipscombe, D. Identification of functionally distinct isoforms of the N-type Ca2+ channel in rat sympathetic ganglia and brain. *Neuron* **1997**, *18*, 153–166.

10. Lin, Z., Lin, Y., Schorge, S., Pan, J. Q., Beierlein, M., Lipscombe, D. Alternative splicing of a short cassette exon in alpha1B generates functionally distinct N-type calcium channels in central and peripheral neurons. *J. Neurosci.* **1999**, *19*, 5322–5331.

11. Bell, T. J., Thaler, C., Castiglioni, A. J., Helton, T. D., Lipscombe, D. Cell-specific alternative splicing increases calcium

channel current density in the pain pathway. *Neuron* **2004**, *41*, 127–138.

12. Westenbroek, R. E., Hell, J. W., Warner, C., Dubel, S. J., Snutch, T. P., Catterall, W. A. Biochemical properties and subcellular distribution of an N-type calcium channel alpha 1 subunit. *Neuron* **1992**, *9*, 1099–1115.

13. Kerr, L. M., Filloux, F., Olivera, B. M., Jackson, H., Wamsley, J. K. Autoradiographic localization of calcium channels with [125I]omega-conotoxin in rat brain. *Eur. J. Pharmacol.* **1988**, *146*, 181–183.

14. Gohil, K., Bell, J. R., Ramachandran, J., Miljanich, G. P. Neuroanatomical distribution of receptors for a novel voltage-sensitive calcium-channel antagonist, SNX-230 (omega-conopeptide MVIIC). *Brain Res.* **1994**, *653*, 258–266.

15. Soldo, B. L., Moises, H. C. mu-Opioid receptor activation inhibits N- and P-type Ca2+ channel currents in magnocellular neurones of the rat supraoptic nucleus. *J. Physiol.* **1998**, *513 (Pt 3)*, 787–804.

16. Pan, X., Ikeda, S. R., Lewis, D. L. Rat brain cannabinoid receptor modulates N-type Ca2+ channels in a neuronal expression system. *Mol. Pharmacol.* **1996**, *49*, 707–714.

17. Sun, L., Miller, R. J. Multiple neuropeptide Y receptors regulate K+ and Ca2+ channels in acutely isolated neurons from the rat arcuate nucleus. *J. Neurophysiol.* **1999**, *81*, 1391–1403.

18. Shapiro, M. S., Hille, B. Substance P and somatostatin inhibit calcium channels in rat sympathetic neurons via different G protein pathways. *Neuron* **1993**, *10*, 11–20.

19. Zamponi, G. W., Bourinet, E., Nelson, D., Nargeot, J., Snutch, T. P. Crosstalk between G proteins and protein kinase C mediated by the calcium channel alpha1 subunit. *Nature* **1997**, *385*, 442–446.

20. Zamponi, G. W., Snutch, T. P. Decay of prepulse facilitation of N type calcium channels during G protein inhibition is consistent with binding of a single Gbeta subunit. *Proc. Natl. Acad. Sci. U.S.A.* **1998**, *95*, 4035–4039.

21. Patil, P. G., de Leon, M., Reed, R. R., Dubel, S., Snutch, T. P., Yue, D. T. Elementary events underlying voltage-dependent G-protein inhibition of N-type

calcium channels. *Biophys. J.* **1996**, *71*, 2509–2521.

22. Hamid, J., Nelson, D., Spaetgens, R., Dubel, S. J., Snutch, T. P., Zamponi, G. W. Identification of an integration center for cross-talk between protein kinase C and G protein modulation of N-type calcium channels. *J. Biol. Chem.* **1999**, *274*, 6195–6202.

23. Swartz, K. J. Modulation of Ca2+ channels by protein kinase C in rat central and peripheral neurons: disruption of G protein-mediated inhibition. *Neuron* **1993**, *11*, 305–320.

24. Swartz, K. J., Merritt, A., Bean, B. P., Lovinger, D. M. Protein kinase C modulates glutamate receptor inhibition of Ca2+ channels and synaptic transmission. *Nature* **1993**, *361*, 165–168.

25. Ino, M., Yoshinaga, T., Wakamori, M., Miyamoto, N., Takahashi, E., Sonoda, J., Kagaya, T., Oki, T., Nagasu, T., Nishizawa, Y., Tanaka, I., Imoto, K., Aizawa, S., Koch, S., Schwartz, A., Niidome, T., Sawada, K., Mori, Y. Functional disorders of the sympathetic nervous system in mice lacking the alpha 1B subunit (Cav 2.2) of N-type calcium channels. *Proc. Natl. Acad. Sci. U.S.A.* **2001**, *98*, 5323–5328.

26. Kim, C., Jun, K., Lee, T., Kim, S. S., McEnery, M. W., Chin, H., Kim, H. L., Park, J. M., Kim, D. K., Jung, S. J., Kim, J., Shin, H. S. Altered nociceptive response in mice deficient in the alpha(1B) subunit of the voltage-dependent calcium channel. *Mol. Cell Neurosci.* **2001**, *18*, 235–245.

27. Saegusa, H., Kurihara, T., Zong, S., Kazuno, A., Matsuda, Y., Nonaka, T., Han, W., Toriyama, H., Tanabe, T. Suppression of inflammatory and neuropathic pain symptoms in mice lacking the N-type Ca2+ channel. *EMBO J.* **2001**, *20*, 2349–2356.

28. Saegusa, H., Matsuda, Y., Tanabe, T. Effects of ablation of N- and R-type Ca(2+) channels on pain transmission. *Neurosci. Res.* **2002**, *43*, 1–7.

29. Bean, B. P. Nitrendipine block of cardiac calcium channels: high-affinity binding to the inactivated state. *Proc. Natl. Acad. Sci. U.S.A.* **1984**, *81*, 6388–6392.

30. Hille, B. *Ion Channels of Excitable Membranes*, Sinauer Associates, Inc., Sunderland, MA, **2001**.

31. Lansman, J. B. Blockade of current through single calcium channels by trivalent lanthanide cations. Effect of ionic radius on the rates of ion entry and exit. *J. Gen. Physiol.* **1990**, *95*, 679–696.

32. Lansman, J. B., Hess, P., Tsien, R. W. Blockade of current through single calcium channels by Cd2+, Mg2+, and Ca2+. Voltage and concentration dependence of calcium entry into the pore. *J. Gen. Physiol.* **1986**, *88*, 321–347.

33. Zamponi, G. W., Bourinet, E., Snutch, T. P. Nickel block of a family of neuronal calcium channels: subtype- and subunit-dependent action at multiple sites. *J. Membr. Biol.* **1996**, *151*, 77–90.

34. Busselberg, D., Michael, D., Evans, M. L., Carpenter, D. O., Haas, H. L. Zinc (Zn2+) blocks voltage gated calcium channels in cultured rat dorsal root ganglion cells. *Brain Res.* **1992**, *593*, 77–81.

35. Busselberg, D., Michael, D., Platt, B. Pb2+ reduces voltage- and N-methyl-D-aspartate (NMDA)-activated calcium channel currents. *Cell Mol. Neurobiol.* **1994**, *14*, 711–722.

36. Busselberg, D., Pekel, M., Michael, D., Platt, B. Mercury (Hg2+) and zinc (Zn2+): two divalent cations with different actions on voltage-activated calcium channel currents. *Cell Mol. Neurobiol.* **1994**, *14*, 675–687.

37. Busselberg, D., Platt, B., Haas, H. L., Carpenter, D. O. Voltage gated calcium channel currents of rat dorsal root ganglion (DRG) cells are blocked by Al3+. *Brain Res.* **1993**, *622*, 163–168.

38. Busselberg, D., Platt, B., Michael, D., Carpenter, D. O., Haas, H. L. Mammalian voltage-activated calcium channel currents are blocked by Pb2+, Zn2+, and Al3+. *J. Neurophysiol.* **1994**, *71*, 1491–1497.

39. Beedle, A. M., Hamid, J., Zamponi, G. W. Inhibition of transiently expressed low- and high-voltage-activated calcium channels by trivalent metal cations. *J. Membr. Biol.* **2002**, *187*, 225–238.

40. Norton, R. S., Pallaghy, P. K. The cystine knot structure of ion channel toxins and related polypeptides. *Toxicon* **1998**, *36*, 1573–1583.

41. Olivera, B. M., Gray, W. R., Zeikus, R., McIntosh, J. M., Varga, J., Rivier, J., de Santos, V., Cruz, L. J. Peptide neurotoxins from fish-hunting cone snails. *Science* **1985**, *230*, 1338–1343.

42. Terlau, H., Olivera, B. M. Conus venoms: a rich source of novel ion channel-targeted peptides. *Physiol. Rev.* **2004**, *84*, 41–68.

43. Price-Carter, M., Gray, W. R., Goldenberg, D. P. Folding of omega-conotoxins. 1. Efficient disulfide-coupled folding of mature sequences in vitro. *Biochemistry* **1996**, *35*, 15 537–15 546.

44. Cruz, L. J., Olivera, B. M. Calcium channel antagonists. Omega-conotoxin defines a new high affinity site. *J. Biol. Chem.* **1986**, *261*, 6230–6233.

45. Ellinor, P. T., Zhang, J. F., Horne, W. A., Tsien, R. W. Structural determinants of the blockade of N-type calcium channels by a peptide neurotoxin. *Nature* **1994**, *372*, 272–275.

46. Boland, L. M., Morrill, J. A., Bean, B. P. omega-Conotoxin block of N-type calcium channels in frog and rat sympathetic neurons. *J. Neurosci.* **1994**, *14*, 5011–5027.

47. Feng, Z. P., Doering, C. J., Winkfein, R. J., Beedle, A. M., Spafford, J. D., Zamponi, G. W. Determinants of inhibition of transiently expressed voltage-gated calcium channels by omega-conotoxins GVIA and MVIIA. *J. Biol. Chem.* **2003**, *278*, 20 171–20 178.

48. Kim, J. I., Takahashi, M., Ogura, A., Kohno, T., Kudo, Y., Sato, K. Hydroxyl group of Tyr13 is essential for the activity of omega-conotoxin GVIA, a peptide toxin for N-type calcium channel. *J. Biol. Chem.* **1994**, *269*, 23 876–23 878.

49. Lew, M. J., Flinn, J. P., Pallaghy, P. K., Murphy, R., Whorlow, S. L., Wright, C. E., Norton, R. S., Angus, J. A. Structure-function relationships of omega-conotoxin GVIA. Synthesis, structure, calcium channel binding, and functional assay of alanine-substituted analogues. *J. Biol. Chem.* **1997**, *272*, 12 014–12 023.

50. Nielsen, K. J., Adams, D. A., Alewood, P. F., Lewis, R. J., Thomas, L., Schroeder, T., Craik, D. J. Effects of chirality at Tyr13 on the structure-activity relationships of omega-conotoxins from Conus magus. *Biochemistry* **1999**, *38*, 6741–6751.

51. Vanegas, H., Schaible, H. Effects of antagonists to high-threshold calcium channels upon spinal mechanisms of pain, hyperalgesia and allodynia. *Pain* **2000**, *85*, 9–18.

52. Dunlap, K., Luebke, J. I., Turner, T. J. Exocytotic Ca2+ channels in mammalian central neurons. *Trends Neurosci.* **1995**, *18*, 89–98.

53. Mermelstein, P. G., Surmeier, D. J. A calcium channel reversibly blocked by omega-conotoxin GVIA lacking the class D alpha 1 subunit. *Neuroreport* **1997**, *8*, 485–489.

54. Fernandez, J. M., Granja, R., Izaguirre, V., Gonzalez-Garcia, C., Cena, V. omega-Conotoxin GVIA blocks nicotine-induced catecholamine secretion by blocking the nicotinic receptor-activated inward currents in bovine chromaffin cells. *Neurosci. Lett.* **1995**, *191*, 59–62.

55. Olivera, B. M., Cruz, L. J., de Santos, V., LeCheminant, G. W., Griffin, D., Zeikus, R., McIntosh, J. M., Galyean, R., Varga, J., Gray, W. R., et al. Neuronal calcium channel antagonists. Discrimination between calcium channel subtypes using omega-conotoxin from Conus magus venom. *Biochemistry* **1987**, *26*, 2086–2090.

56. Wang, Y. X., Gao, D., Pettus, M., Phillips, C., Bowersox, S. S. Interactions of intrathecally administered ziconotide, a selective blocker of neuronal N-type voltage-sensitive calcium channels, with morphine on nociception in rats. *Pain* **2000**, *84*, 271–281.

57. Wang, Y. X., Pettus, M., Gao, D., Phillips, C., Scott Bowersox, S. Effects of intrathecal administration of ziconotide, a selective neuronal N-type calcium channel blocker, on mechanical allodynia and heat hyperalgesia in a rat model of postoperative pain. *Pain* **2000**, *84*, 151–158.

58. Mathur, V. S. A new pharmacological class of drug for the management of pain. *Semin. Anesthesia, Perioperative Med. Pain* **2000**, *19*, 67–75.

59. Staats, P. S., Yearwood, T., Charapata, S. G., Presley, R. W., Wallace, M. S., Byas-Smith, M., Fisher, R., Bryce, D. A., Mangieri, E. A., Luther, R. R., Mayo, M., McGuire, D., Ellis, D. Intrathecal ziconotide in the treatment of refractory pain in patients with cancer or AIDS: a randomized controlled trial. *J. Am. Med. Assoc.* **2004**, *291*, 63–70.

60. Ridgeway, B., Wallace, M., Gerayli, A. Ziconotide for the treatment of severe spasticity after spinal cord injury. *Pain* **2000**, *85*, 287–289.

61. McGuire, D., Bowersox, S., Fellmann, J. D., Luther, R. R. Sympatholysis after neuron-specific, N-type, voltage-sensitive calcium channel blockade: first demonstration of N-channel function in humans. *J. Cardiovasc. Pharmacol.* **1997**, *30*, 400–403.

62. Herrero, C. J., Garcia-Palomero, E., Pintado, A. J., Garcia, A. G., Montiel, C. Differential blockade of rat alpha3beta4 and alpha7 neuronal nicotinic receptors by omega-conotoxin MVIIC, omega-conotoxin GVIA and diltiazem. *Br. J. Pharmacol.* **1999**, *127*, 1375–1387.

63. Lewis, R. J., Nielsen, K. J., Craik, D. J., Loughnan, M. L., Adams, D. A., Sharpe, I. A., Luchian, T., Adams, D. J., Bond, T., Thomas, L., Jones, A., Matheson, J. L., Drinkwater, R., Andrews, P. R., Alewood, P. F. Novel omega-conotoxins from Conus catus discriminate among neuronal calcium channel subtypes. *J. Biol. Chem.* **2000**, *275*, 35 335–35 344.

64. Scott, D. A., Wright, C. E., Angus, J. A. Actions of intrathecal omega-conotoxins CVID, GVIA, MVIIA, and morphine in acute and neuropathic pain in the rat. *Eur. J. Pharmacol.* **2002**, *451*, 279–286.

65. Chen, J. Q., Zhang, Y. Q., Dai, J., Luo, Z. M., Liang, S. P. Antinociceptive effects of intrathecally administered huwentoxin-I, a selective N-type calcium channel blocker, in the formalin test in conscious rats. *Toxicon* **2005**, *45*, 15–20.

66. Sanger, G. J., Ellis, E. S., Harries, M. H., Tilford, N. S., Wardle, K. A., Benham, C. D. Rank-order inhibition by omega-conotoxins in human and animal autonomic nerve preparations. *Eur. J. Pharmacol.* **2000**, *388*, 89–95.

67. Boot, J. R. Differential effects of omega-conotoxin GVIA and MVIIC on nerve stimulation induced contractions of guinea-pig ileum and rat vas deferens. *Eur. J. Pharmacol.* **1994**, *258*, 155–158.

68. Hong, S. J., Chang, C. C. Calcium channel subtypes for the sympathetic and parasympathetic nerves of guinea-pig atria. *Br. J. Pharmacol.* **1995**, *116*, 1577–1582.

69. Waterman, S. A. Role of N-, P- and Q-type voltage-gated calcium channels in transmitter release from sympathetic neurones in the mouse isolated vas deferens. *Br. J. Pharmacol.* **1997**, *120*, 393–398.

70. McDonough, S. I., Swartz, K. J., Mintz, I. M., Boland, L. M., Bean, B. P. Inhibition of calcium channels in rat central and peripheral neurons by omega-conotoxin MVIIC. *J. Neurosci.* **1996**, *16*, 2612–2623.

71. McDonough, S. I., Boland, L. M., Mintz, I. M., Bean, B. P. Interactions among toxins that inhibit N-type and P-type calcium channels. *J. Gen. Physiol.* **2002**, *119*, 313–328.

72. Mintz, I. M., Venema, V. J., Adams, M. E., Bean, B. P. Inhibition of N- and L-type Ca2+ channels by the spider venom toxin omega-Aga-IIIA. *Proc. Natl. Acad. Sci. U.S.A.* **1991**, *88*, 6628–6631.

73. Vieira, L. B., Kushmerick, C., Hildebrand, M. E., Garcia, E., Stea, A., Cordeiro, M. N., Richardson, M., Gomez, M. V., Snutch, T. P. Inhibition of high voltage-activated calcium channels by spider toxin PnTx3-6. *J. Pharmacol. Exp. Ther.* **2005**, *314*, 1370–1377.

74. Newcomb, R., Palma, A., Fox, J., Gaur, S., Lau, K., Chung, D., Cong, R., Bell, J. R., Horne, B., Nadasdi, L., et al. SNX-325, a novel calcium antagonist from the spider Segestria florentina. *Biochemistry* **1995**, *34*, 8341–8347.

75. Sutton, K. G., Siok, C., Stea, A., Zamponi, G. W., Heck, S. D., Volkmann, R. A., Ahlijanian, M. K., Snutch, T. P. Inhibition of neuronal calcium channels by a novel peptide spider toxin, DW13.3. *Mol. Pharmacol.* **1998**, *54*, 407–418.

76. Chuang, R. S., Jaffe, H., Cribbs, L., Perez-Reyes, E., Swartz, K. J. Inhibition of T-type voltage-gated calcium channels by a new scorpion toxin. *Nat. Neurosci.* **1998**, *1*, 668–674.

77. Sidach, S. S., Mintz, I. M. Kurtoxin, a gating modifier of neuronal high- and low-threshold Ca channels. *J. Neurosci.* **2002**, *22*, 2023–2034.

78. Sidach, S. S., Mintz, I. M. Low-affinity blockade of neuronal N-type Ca channels by the spider toxin omega-agatoxin-IVA. *J. Neurosci.* **2000**, *20*, 7174–7182.

79. Hu, L. Y., Ryder, T. R., Rafferty, M. F., Cody, W. L., Lotarski, S. M., Miljanich, G. P., Millerman, E., Rock, D. M., Song, Y., Stoehr, S. J., Taylor, C. P., Weber, M. L., Szoke, B. G., Vartanian, M. G. N,N-dialkyl-dipeptidylamines as novel N-type calcium channel blockers. *Bioorg. Med. Chem. Lett.* **1999**, *9*, 907–912.

80. Hu, L. Y., Ryder, T. R., Rafferty, M. F., Dooley, D. J., Geer, J. J., Lotarski, S. M., Miljanich, G. P., Millerman, E., Rock, D. M., Stoehr, S. J., Szoke, B. G., Taylor, C. P., Vartanian, M. G. Structure-activity relationship of N-methyl-N-aralkyl-peptidylamines as novel N-type calcium channel blockers. *Bioorg. Med. Chem. Lett.* **1999**, *9*, 2151–2156.

81. Hu, L. Y., Ryder, T. R., Nikam, S. S., Millerman, E., Szoke, B. G., Rafferty, M. F. Synthesis and biological evaluation of substituted 4-(OBz)phenylalanine derivatives as novel N-type calcium channel blockers. *Bioorg. Med. Chem. Lett.* **1999**, *9*, 1121–1126.

82. Ryder, T. R., Hu, L. Y., Rafferty, M. F., Millerman, E., Szoke, B. G., Tarczy-Hornoch, K. Multiple parallel synthesis of N,N-dialkyldipeptidylamines as N-type calcium channel blockers. *Bioorg. Med. Chem. Lett.* **1999**, *9*, 1813–1818.

83. Seko, T., Kato, M., Kohno, H., Ono, S., Hashimura, K., Takimizu, H., Nakai, K., Maegawa, H., Katsube, N., Toda, M. Structure-activity study and analgesic efficacy of amino acid derivatives as N-type calcium channel blockers. *Bioorg. Med. Chem. Lett.* **2001**, *11*, 2067–2070.

84. Seko, T., Kato, M., Kohno, H., Ono, S., Hashimura, K., Takimizu, H., Nakai, K., Maegawa, H., Katsube, N., Toda, M. Structure-activity study of L-cysteine-based N-type calcium channel blockers: optimization of N- and C-terminal substituents. *Bioorg. Med. Chem. Lett.* **2002**, *12*, 915–918.

85. Seko, T., Kato, M., Kohno, H., Ono, S., Hashimura, K., Takimizu, H., Nakai, K., Maegawa, H., Katsube, N., Toda, M. Structure-activity study of L-amino acid-based N-type calcium channel blockers. *Bioorg. Med. Chem.* **2003**, *11*, 1901–1913.

86. Takimizu, H., Nakai, K., Sasamura, T., Ito, Y., Seko, T., Kagamiishi, Y., Maegawa, H., Tateishi, N., Katsube, N., Kamanaka, Y. A novel N-type calcium channel blocker, ONO-2921, attenuates persistent pain in rats. In *11th World Congress on Pain*: Sydney, Australia, **2005**, Program No. 623-P229.

87. Enyeart, J. J., Biagi, B. A., Mlinar, B. Preferential block of T-type calcium channels by neuroleptics in neural crest-derived rat and human C cell lines. *Mol. Pharmacol.* **1992**, *42*, 364–372.

88. Sah, D. W., Bean, B. P. Inhibition of P-type and N-type calcium channels by dopamine receptor antagonists. *Mol. Pharmacol.* **1994**, *45*, 84–92.

89. Grantham, C. J., Main, M. J., Cannell, M. B. Fluspirilene block of N-type calcium current in NGF-differentiated PC12 cells. *Br. J. Pharmacol.* **1994**, *111*, 483–488.

90. Zamponi, G. W., Soong, T. W., Bourinet, E., Snutch, T. P. Beta subunit coexpression and the alpha1 subunit domain I-II linker affect piperidine block of neuronal calcium channels. *J. Neurosci.* **1996**, *16*, 2430–2443.

91. Snutch, T., Zamponi, G. Calcium channel blockers: *US Pat. 6,011,035*, **2000**.

92. Santi, C. M., Cayabyab, F. S., Sutton, K. G., McRory, J. E., Mezeyova, J., Hamming, K. S., Parker, D., Stea, A., Snutch, T. P. Differential inhibition of T-type calcium channels by neuroleptics. *J. Neurosci.* **2002**, *22*, 396–403.

93. Snutch, T. Partially saturated calcium channel blockers: *US Pat. 6,492,375B2*, **2002**.

94. Snutch, T. Preferentially substituted calcium channel blockers: *US Pat. 6,387,897B1*, **2002**.

95. Snutch, T. P., Feng, Z. P, Doering, C., Cayabyab, F., Janke, D., Parker, D. B., Belardetti, F., Morimoto, B., Vanderah, T., Zamponi, G. W. and Porreca, F. A new class of N-type calcium channel blocker efficacious in aninmal models of chronic pain, *Soc. Neurosci. Abstr.*, **2001**, Vol. 27, Program No. 465-1.

96. Snutch, T. P. Calcium Channel Therapeutic Targets: Beyond the L-type. *Society for Biomolecular Screening 11th Annual Conference and Exhibition*, Geneva, Switzerland **2005**, p. 127.

97. Ryder, T. R., Hu, L. Y., Rafferty, M. F., Lotarski, S. M., Rock, D. M., Stoehr, S. J., Taylor, C. P., Weber, M. L., Miljanich, G. P., Millerman, E., Szoke, B. G. Structure-activity relationship at the proximal phenyl group in a series of non-peptidyl N-type calcium channel antagonists. *Bioorg. Med. Chem. Lett.* **1999**, *9*, 2453–2458.

98. Hu, L. Y., Ryder, T. R., Rafferty, M. F., Feng, M. R., Lotarski, S. M., Rock, D. M., Sinz, M., Stoehr, S. J., Taylor, C. P., Weber, M. L., Bowersox, S. S., Miljanich, G. P., Millerman, E., Wang, Y. X., Szoke, B. G. Synthesis of a series of 4-benzyloxyaniline analogues as neuronal N-type calcium channel blockers with improved anticonvulsant and analgesic properties. *J. Med. Chem.* **1999**, *42*, 4239–4249.

99. Hu, L. Y., Ryder, T. R., Rafferty, M. F., Taylor, C. P., Feng, M. R., Kuo, B. S., Lotarski, S. M., Miljanich, G. P., Millerman, E., Siebers, K. M., Szoke, B. G. The discovery of [1-(4-dimethylamino-benzyl)-piperidin-4-yl]-[4-(3,3-dimethylbutyl)-phenyl]-(3-methyl-but-2-enyl)-amine, an N-type Ca+2 channel blocker with oral activity for analgesia. *Bioorg. Med. Chem.* **2000**, *8*, 1203–1212.

100. Benham, C. D., Brown, T. H., Cooper, D. G., Evans, M. L., Harries, M. H., Herdon, H. J., Meakin, J. E., Murkitt, K. L., Patel, S. R., Roberts, J. C., et al. SB 201823-A, a neuronal Ca2+ antagonist is neuroprotective in two models of cerebral ischaemia. *Neuropharmacology* **1993**, *32*, 1249–1257.

101. Barone, F. C., Lysko, P. G., Price, W. J., Feuerstein, G., al-Baracanji, K. A., Benham, C. D., Harrison, D. C., Harries, M. H., Bailey, S. J., Hunter, A. J. SB 201823-A antagonizes calcium currents in central neurons and reduces the effects of focal ischemia in rats and mice. *Stroke* **1995**, *26*, 1683–1689, discussion 1689–1690.

102. O'Neill, M. J., Bath, C. P., Dell, C. P., Hicks, C. A., Gilmore, J., Ambler, S. J., Ward, M. A., Bleakman, D. Effects of Ca2+ and Na+ channel inhibitors in vitro and in global cerebral ischaemia in vivo. *Eur. J. Pharmacol.* **1997**, *332*, 121–131.

103. Zamponi, G. W. Cytoplasmic determinants of piperidine blocking affinity for N-type calcium channels. *J. Membr. Biol.* **1999**, *167*, 183–192.

104. Roullet, J. B., Spaetgens, R. L., Burlingame, T., Feng, Z. P., Zamponi, G. W. Modulation of neuronal voltage-gated calcium channels by farnesol. *J. Biol. Chem.* **1999**, *274*, 25 439–25 446.

105. Beedle, A. M., Zamponi, G. W. Block of voltage-dependent calcium channels by aliphatic monoamines. *Biophys. J.* **2000**, *79*, 260–270.

106. Milhaud, D., Fagni, L., Bockaert, J., Lafon-Cazal, M. Inhibition of voltage-gated Ca2+ channels by antazoline. *Neuroreport* **2002**, *13*, 1711–1714.

107. Uneyama, H., Takahara, A., Dohmoto, H., Yoshimoto, R., Inoue, K., Akaike, N. Blockade of N-type Ca2+ current by cilnidipine (FRC-8653) in acutely dissociated rat sympathetic neurones. *Br. J. Pharmacol.* **1997**, *122*, 37–42.

108. Oike, M., Inoue, Y., Kitamura, K., Kuriyama, H. Dual action of FRC8653, a novel dihydropyridine derivative, on the Ba2+ current recorded from the rabbit basilar artery. *Circ. Res.* **1990**, *67*, 993–1006.

109. Takahara, A., Fujita, S., Moki, K., Ono, Y., Koganei, H., Iwayama, S., Yamamoto, H. Neuronal Ca2+ channel blocking action of an antihypertensive drug, cilnidipine, in IMR-32 human neuroblastoma cells. *Hypertens. Res.* **2003**, *26*, 743–747.

110. Takahara, A., Konda, T., Enomoto, A., Kondo, N. Neuroprotective effects of a dual L/N-type Ca(2+) channel blocker cilnidipine in the rat focal brain ischemia model. *Biol. Pharm. Bull.* **2004**, *27*, 1388–1391.

111. Uneyama, H., Uchida, H., Yoshimoto, R., Ueno, S., Inoue, K., Akaike, N. Effects of a novel antihypertensive drug, cilnidipine, on catecholamine secretion from differentiated PC12 cells. *Hypertension* **1998**, *31*, 1195–1199.

112. Takenoshita, M., Steinbach, J. H. Halothane blocks low-voltage-activated calcium current in rat sensory neurons. *J. Neurosci.* **1991**, *11*, 1404–1412.

113. Study, R. E. Isoflurane inhibits multiple voltage-gated calcium currents in hippocampal pyramidal neurons. *Anesthesiology* **1994**, *81*, 104–116.

114. McDowell, T. S., Pancrazio, J. J., Lynch, C., 3rd Volatile anesthetics reduce low-voltage-activated calcium currents in a thyroid C-cell line. *Anesthesiology* **1996**, *85*, 1167–1175.

115. Hirota, K., Fujimura, J., Wakasugi, M., Ito, Y. Isoflurane and sevoflurane modulate inactivation kinetics of Ca2+ currents in single bullfrog atrial myocytes. *Anesthesiology* **1996**, *84*, 377–383.

116. Kameyama, K., Aono, K., Kitamura, K. Isoflurane inhibits neuronal Ca2+ channels through enhancement of current inactivation. *Br. J. Anaesth.* **1999**, *82*, 402–411.

117. Kamatchi, G. L., Chan, C. K., Snutch, T., Durieux, M. E., Lynch, C., 3rd Volatile anesthetic inhibition of neuronal Ca channel currents expressed in Xenopus oocytes. *Brain Res.* **1999**, *831*, 85–96.

118. Nikonorov, I. M., Blanck, T. J., Recio-Pinto, E. G-protein activation decreases isoflurane inhibition of N-type Ba2+ currents. *Anesthesiology* **2003**, *99*, 392–399.

119. Takei, T., Saegusa, H., Zong, S., Murakoshi, T., Makita, K., Tanabe, T. Increased sensitivity to halothane but decreased sensitivity to propofol in mice lacking the N-type Ca2+ channel. *Neurosci. Lett.* **2003**, *350*, 41–45.

120. Walter, H. J., Messing, R. O. Regulation of neuronal voltage-gated calcium channels by ethanol. *Neurochem. Int.* **1999**, *35*, 95–101.

121. Solem, M., McMahon, T., Messing, R. O. Protein kinase A regulates inhibition of N- and P/Q-type calcium channels by ethanol in PC12 cells. *J. Pharmacol. Exp. Ther.* **1997**, *282*, 1487–1495.

122. Newton, P. M., Tully, K., McMahon, T., Connolly, J., Dadgar, J., Treistman, S. N., Messing, R. O. Chronic ethanol exposure induces an N-type calcium channel splice variant with altered channel kinetics. *FEBS Lett.* **2005**, *579*, 671–676.

123. Olivera, B. M., McIntosh, J. M., Cruz, L. J., Luque, F. A., Gray, W. R. Purification and sequence of a presynaptic peptide toxin from Conus geographus venom. *Biochemistry* **1984**, *23*, 5087–5090.

124. Wagner, J. A., Snowman, A. M., Biswas, A., Olivera, B. M., Snyder, S. H. Omega-conotoxin GVIA binding to a high-affinity receptor in brain: characterization,

calcium sensitivity, and solubilization. *J. Neurosci.* **1988**, *8*, 3354–3359.

125. Malmberg, A. B., Yaksh, T. L. Voltage-sensitive calcium channels in spinal nociceptive processing: blockade of N- and P-type channels inhibits formalin-induced nociception. *J. Neurosci.* **1994**, *14*, 4882–4890.

126. Wang, Y. X., Bezprozvannaya, S., Bowersox, S. S., Nadasdi, L., Miljanich, G., Mezo, G., Silva, D., Tarczy-Hornoch, K., Luther, R. R. Peripheral versus central potencies of N-type voltage-sensitive calcium channel blockers. *Naunyn Schmiedebergs Arch. Pharmacol.* **1998**, *357*, 159–168.

127. Bowersox, S. S., Gadbois, T., Singh, T., Pettus, M., Wang, Y. X., Luther, R. R. Selective N-type neuronal voltage-sensitive calcium channel blocker, SNX-111, produces spinal antinociception in rat models of acute, persistent and neuropathic pain. *J. Pharmacol. Exp. Ther.* **1996**, *279*, 1243–1249.

128. Hillyard, D. R., Monje, V. D., Mintz, I. M., Bean, B. P., Nadasdi, L., Ramachandran, J., Miljanich, G., Azimi-Zoonooz, A., McIntosh, J. M., Cruz, L. J., et al. A new Conus peptide ligand for mammalian presynaptic Ca2+ channels. *Neuron* **1992**, *9*, 69–77.

129. Woppmann, A., Ramachandran, J., Miljanich, G. P. Calcium channel subtypes in rat brain: biochemical characterization of the high-affinity receptors for omega-conopeptides SNX-230 (synthetic MVIIC), SNX-183 (SVIB), and SNX-111 (MVIIA). *Mol. Cell Neurosci.* **1994**, *5*, 350–357.

130. Ramilo, C. A., Zafaralla, G. C., Nadasdi, L., Hammerland, L. G., Yoshikami, D., Gray, W. R., Kristipati, R., Ramachandran, J., Miljanich, G., Olivera, B. M., et al. Novel alpha- and omega-conotoxins from Conus striatus venom. *Biochemistry* **1992**, *31*, 9919–9926.

131. McDonough, S. I., Lampe, R. A., Keith, R. A., Bean, B. P. Voltage-dependent inhibition of N- and P-type calcium channels by the peptide toxin omega-grammotoxin-SIA. *Mol. Pharmacol.* **1997**, *52*, 1095–1104.

132. Lampe, R. A., Defeo, P. A., Davison, M. D., Young, J., Herman, J. L., Spreen, R. C., Horn, M. B., Mangano, T. J., Keith, R. A. Isolation and pharmacological characteri-

zation of omega-grammotoxin SIA, a novel peptide inhibitor of neuronal voltage-sensitive calcium channel responses. *Mol. Pharmacol.* **1993**, *44*, 451–460.

133. Ertel, E. A., Smith, M. M., Leibowitz, M. D., Cohen, C. J. Isolation of myocardial L-type calcium channel gating currents with the spider toxin omega-Aga-IIIA. *J. Gen. Physiol.* **1994**, *103*, 731–753.

134. Mintz, I. M. Block of Ca channels in rat central neurons by the spider toxin omega-Aga-IIIA. *J. Neurosci.* **1994**, *14*, 2844–2853.

135. Yan, L., Adams, M. E. The spider toxin omega-Aga IIIA defines a high affinity site on neuronal high voltage-activated calcium channels. *J. Biol. Chem.* **2000**, *275*, 21 309–21 316.

136. Mintz, I. M., Venema, V. J., Swiderek, K. M., Lee, T. D., Bean, B. P., Adams, M. E. P-type calcium channels blocked by the spider toxin omega-Aga-IVA. *Nature* **1992**, *355*, 827–829.

137. Cassola, A. C., Jaffe, H., Fales, H. M., Castro Afeche, S., Magnoli, F., Cipolla-Neto, J. omega-Phonetoxin-IIA: a calcium channel blocker from the spider phoneutria nigriventer. *Pflugers Arch.* **1998**, *436*, 545–552.

138. Dos Santos, R. G., Van Renterghem, C., Martin-Moutot, N., Mansuelle, P., Cordeiro, M. N., Diniz, C. R., Mori, Y., De Lima, M. E., Seagar, M. Phoneutria nigriventer omega-phonetoxin IIA blocks the Cav2 family of calcium channels and interacts with omega-conotoxin-binding sites. *J. Biol. Chem.* **2002**, *277*, 13 856–13 862.

139. Vieira, L. B., Kushmerick, C., Reis, H. J., Diniz, C. R., Cordeiro, M. N., Prado, M. A., Kalapothakis, E., Romano-Silva, M. A., Gomez, M. V. PnTx3-6 a spider neurotoxin inhibits K+-evoked increase in [Ca2+](i) and Ca2+-dependent glutamate release in synaptosomes. *Neurochem. Int.* **2003**, *42*, 277–282.

140. Seko, T., Kato, M., Kohno, H., Ono, S., Hashimura, K., Takenobu, Y., Takimizu, H., Nakai, K., Maegawa, H., Katsube, N., Toda, M. L-Cysteine based N-type calcium channel blockers: structure-activity relationships of the C-terminal lipophilic moiety, and oral analgesic efficacy in rat pain models. *Bioorg. Med. Chem. Lett.* **2002**, *12*, 2267–2269.

141. Dickenson, A., Maegawa, H., Suzuki, R. Effects of ONO-2921, a non-peptide N-type calcium channel blocker on dorsal horn neuronal responses in rat models of formalin, inflammation and neuropathy. In *11th World Congress on Pain*: Sydney, Australia, **2005**, Program No. 625-P231.

142. Takahara, A., Fujita, S., Moki, K., Ono, Y., Koganei, H., Iwayama, S., Yamamoto, H. Neuronal Ca2+ channel blocking action of an antihypertensive drug, cilnidipine, in IMR-32 human neuroblastoma cells. *Hypertens Res.* **2003**, *26*, 743–747.

6
Sodium Channels

6.1
Molecular, Biophysical and Functional Properties of Voltage-gated Sodium Channels
Douglas S. Krafte, Mark Chapman, and Ken McCormack

6.1.1
Introduction

In 1952 Hodgkin and Huxley published a series of papers describing, for the first time, the properties of voltage-gated Na channels. Utilizing a breakthrough technology known as the voltage-clamp technique (Cole 1949), Alan Hodgkin and Andrew Huxley "dissected" the membrane currents responsible for generation of the squid axon action potential [Huxley A.L. and Hodgkin A.F. (1952 a–d)]. In our current period of science, where the trend is to move from the reductionist to the systems approach in biology, it remains refreshing to read the Hodgkin and Huxley papers and marvel at their insight and inductive conclusions based on the experimental data obtained. These studies remain one of the foundations of our modern understanding of membrane biophysics and voltage-gated sodium channels.

Voltage-gated sodium channels are key membrane proteins responsible for action potential generation and impulse propagation in excitable cells. All cells establish asymmetry in ionic gradients through energy-driven pumps in the plasma membrane and then use these energy sources to drive cellular function and processes. Of the major ions in the cellular milieu, Na and Ca are maintained at higher concentrations outside the cell and K is maintained at a higher concentration within the cell. Since the resting membrane permeability of most cells is driven by the potassium gradients, the membrane potential at rest is negative (i.e., intracellular potential relative to extracellular potential). In excitable cells such as neurons, the membrane potential can approach the equilibrium potential for K, which is in the −90 to −100 mV range. When a selectively permeable channel opens, the membrane potential will be driven towards the equilibrium potential for the permeable ion or ions. Since sodium is at higher concentrations outside the cell, the Na equilibrium potential (E_{Na}) is approximately +60 mV. Thus one can think of closed sodium channels as a switch coupled to a vast potential energy store. Opening of the sodium channels in the membrane will rapidly de-

Voltage-Gated Ion Channels as Drug Targets. Edited by D. Triggle
Copyright © 2006 WILEY-VCH Verlag GmbH & Co. KGaA, Weinheim
ISBN 3-527-31258-7

flect the membrane potential to E_{Na} and the change in membrane potential is only limited by the approach to equilibrium, the gating of the channel and the contribution from other ion channels and conductances in the plasma membrane. Physiologically, voltage-gated sodium channels are responsible for initiating excitatory events such as neuronal firing in most neurons of the central and peripheral nervous system, excitation of both cardiac and skeletal muscle, as well as endocrine cell excitability.

6.1.2
Primary and Tertiary Structure

6.1.2.1 General Membrane Topology
The first voltage-gated sodium channel was cloned in 1984 by Noda et al. (1984) and since that time ten family members have been cloned from various species (Table 6.1.1). Sodium channels are ~260 kDa membrane proteins that consist of α subunits and typically at least one β subunit. The molecular machinery responsible for gating and permeability resides in the α-subunit, which is shown schematically in Fig. 6.1.1 along with various sites where pharmacological modulation of channels can occur. Noda et al. (1984) originally proposed a topological structure for the α-subunit, which includes four domains, each containing six transmembrane-spanning regions with both the amino and carboxy termini residing on the intracellular side of the membrane. This topological model has been refined over the years, but remains a useful representation of the channel. The pore region is localized to a P-loop between membrane-spanning domains S5 and S6 and encodes residues responsible for defining channel permeation properties, often denoted the selectivity filter. With the possible exception of one family member (hSCN6A), all Na channels are opened or gated by depolarization of the membrane potential followed by closure or inactivation of the channel at sustained depolarization. Once the membrane potential returns to resting values, the channels revert to a closed but primed state for another activation/inactivation cycle. Activation of the channel is regulated by a voltage sensor encoded in the 4th transmembrane-spanning region of each domain where charged amino acids sense changes in the transmembrane field. The inactivation process is primarily governed by the linker region between domains III and IV, which is identifiable

Fig. 6.1.1 General membrane topology of voltage-dependent sodium channels. Upper panel: linear sequence from amino to carboxy termini. Each of the four pseudo-subunit domains are shown with six putative transmembrane segments; the S1-S3 (light blue), the S4 voltage-sensor (green), and S5 and S6 (pink) regions. The P-loop constituting the selectivity filter is in yellow while the red ball denotes the IFM region of the inactivation domain. Regions of the protein demonstrated to influence the activity of several Na$^+$ channel toxins are indicated in the upper panel: site 1, tetrodotoxin (TTX), saxitoxin (STX) and μ-conotoxins; site 2, veratridine, batrachotoxin, grayanotoxin; site 3, α-scorpion, sea anenome and spider toxins; site 4, α-scorpion toxins; site 5, brevetoxin and ciguatoxin. Local anesthetics have been shown to interact with amino acid residues within the S5-S6 region. The lower two panels illustrate the arrangement of the pseudo-subunits within and across domains.

Table 6.1.1 Nomenclature and gene locus for voltage-dependent Na channel genes; high TTX sensitivity indicates block at low nM concentrations while low sensitivity indicates little or no block at $\geq 1\,\mu M$ concentrations.

Gene	Human gene locus	Alternative names	Ionic current	TTX sensitivity
SCN1A	2q24	Type I	Nav1.1	High
SCN2A	2q23-q24.3	Type II	Nav1.2	High
SCN3A	2q24	Type III	Nav1.3	High
SCN4A	17q23.1-q25.3	SkM1, μ1	Nav1.4	High
SCN5A	3p21	SkM2, H1	Nav1.5	Low
SCN6A	2q21-23	Na-G	–	–
SCN8A	12q13	PN4	Nav1.6	High
SCN9A	2q24	PN1	Nav1.7	High
SCN10A	3p22-24	SNS, PN3	Nav1.8	Very low
SCN12A	3p21-24	SNS2, NaN	Nav1.9	Very low

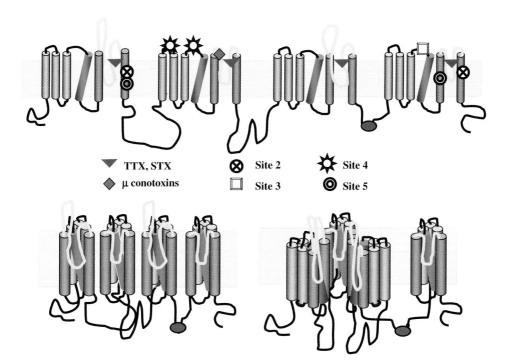

by a signature tri-peptide IFM motif. Enzymatic cleavage of this region of the channel or amino acid substitution of the IFM motif leads to channels, which open but do not undergo fast inactivation. Several neurotoxic compounds that alter the activity of these Na^+ channels have been identified and localized to specific interaction sites with the channel proteins (Cestele, S. and Catterall W.A., 2000).

Notably, the schematic in Fig. 6.1.1 is useful for didactic purposes and has also served the scientific community well in designing experiments to probe structure and function. While our knowledge of the actual three-dimensional structure of sodium channels remains limited, it may be that other structural motifs more accurately represent the actual gating mechanisms of the channel (Jiang et al., 2003). However, these topological maps, such as illustrated in Fig. 6.1.1, remain useful even today.

6.1.2.2 Phylogenetic Analysis

The evolution of vertebrate voltage-gated Na^+ channels may be viewed in the context of a larger four-domain family that also includes voltage-Ca^{2+} channels (Spafford et al., 1999). Figure 6.1.2 illustrates the phyologenetic relationships, utilizing amino acid sequence alignments from all available members of these gene families across a limited number of species, including humans, rat and the pufferfish (*Tetraodon nigroviridis*). In addition, the analysis was anchored with sequences available from jellyfish species. The resulting dendrogram is consistent with the idea that the LVA (low-voltage-activated) or T-type Ca channels were the first to diverge from an ancestral precursor (Perez-Reyes, 2003). The presence of both HVA (high-voltage-activated, L-type) and Na channel genes in jellyfish species indicates that the divergence from an ancestral gene occurred before the origin of these species roughly 800 million years ago. While no T-type channel gene sequences have been isolated from jellyfish to date there are reports of functional T-type currents in jellyfish (e.g., Lin Y.C. and Spencer A.N., 2001).

In humans and rat there are ten Na and ten Ca channel genes while in the pufferfish there are at least ten Na and 18 Ca channels. Of further note, there appears

Fig. 6.1.2 Phylogenetic organization of the Na^+/Ca^{2+} channel gene family. All available human, rat and *Tetraodon nigroviridis* gene products along with two Na^+ and a single Ca^{2+} channel protein sequence from jellyfish species were aligned with the Clustal W sequence alignments using the Megalign software from DNAstar. *T. nigroviridis* Na^+ and Ca^{2+} channel sequences are designated by their Genbank accession numbers and highlighted with a light blue background while jellyfish sequences are offset with the light green background. Genbank accession numbers for the mammalian sequences are: **Human SCN1A-12A**: X65362, X65361, AJ251507, M81758, M77235, M91556, AF225988, X82835, Af117907, AF188679. **Rat SCN1A-12A**: X03638, X03639, Y00766, M26643, M27902, Y09164, AF049239, AF000368, X92184, AJ237852. **Human alpha1a-s**: U79666, M94172, L04569, M76558, L27745, AJ224874, AF126966, AF051946, NM_021096, L33798. **Rat alpha1a-I**: AAC24516, AAD24516, AF394940, AAK72959, AAD31924, NM_053701, AF027984, AF290213, NM_020084. **Rabbit alpha1s**: P07293. ▶

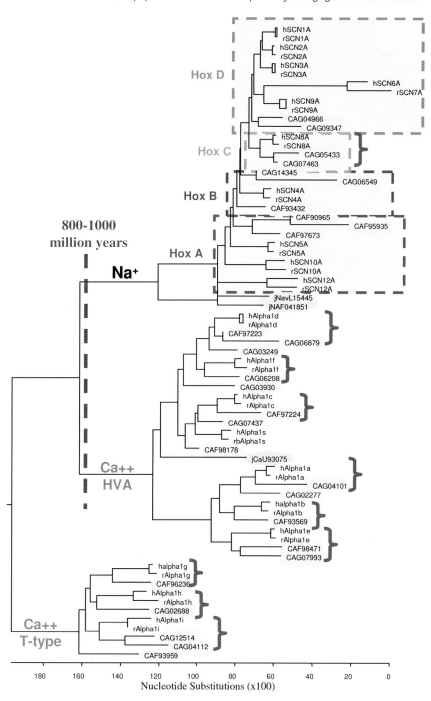

to be at least one and often two pufferfish orthologs for each mammalian Ca channel gene (Fig. 6.1.2). This suggests that each of the Ca channel subtypes had evolved prior to the divergence of a common ancestor between the pufferfish and mammals, roughly 450 million years ago, along with a subsequent whole genome duplication in the ray-finned fish lineage after its separation from tetrapods (Jaillon, O. et al, 2004).

In contrast, with the exception of SCN8A, there appear to be no direct orthologs between the fish and mammalian Na channel genes, which indicates that most Na$^+$ channel genes in mammals evolved after the divergence from ray-finned fishes. Because pufferfish, including *Tetraodon nigroviridis*, produce tetrodotoxin, a potent blocker of many vertebrate Na$^+$ channels, it is possible that the evolution of Na channels in these species has been under unique and significant evolutionary pressures, and that the evolution of Na channels in these organisms is therefore not representative of other fish species. However, phylogenetic analyses of mammalian and six partial Na$^+$ channel sequences from the electric fish *Sternopygus macrurus* also indicated that most Na channel gene duplications in teleost fish and tetrapod lineages took place independently (Lopreato, 2001).

The lineage within the Na$^+$ channel family described here is consistent with recent reports that the channels group along distinct branches that map to their chromosomal location and within specific Hox clusters (Lopreato et al., 2001; Goldin 2002). Duplications within the Hox A and D clusters (Plummer and Meisler, 1999) likely gave rise to the majority of the diversity of mammalian Na channel genes (e.g., SCN1A, SCN2A, SCN3A, SCN6A and SCN9A or SCN5A, SCN10A, SCN12A). While the lineage shown in the figure is consistent with Goldin (2002), it is not consistent with the idea that the cluster of three TTX-resistant genes on chromosome 3, including the DRG specific channels SCN10A and SCN12A, evolved after the rise of the tetrapod lineage as described by Lopreato et al. (2001). Either interpretation may actually be correct, and more definitive characterization of Na channel evolution will be determined with the help of sequences from species arising early within the tetrapod lineage. In addition, future phylogenetic analyses of recently isolated bacterial Na channels composed of a single domain and likely to form functional tetramers may assist in resolving the early events in voltage-gated channel family evolution (Koishi et al., 2004).

Phylogenetic analysis in the context of the super family of voltage-gated Na and Ca channels is useful from an evolutionary perspective. It is also useful to examine the Na channel gene family itself, particularly with respect to functional characteristics. At the primary sequence level the sodium channels are sufficiently similar such that there are not distinct subfamilies (Goldin 2002). However, the genes do cluster into distinct groups, which is also consistent with biophysical and pharmacological properties of the family. The clustering is conserved across species, suggesting an evolutionary divergence of subtype specific properties in voltage-gated sodium channels. Table 6.1.1 lists basic gating kinetics and sensitivity to the puffer fish toxin, tetrodotoxin (TTX), which is a useful tool in characterizing the subtypes of channels within the family. Comparing the properties with the sequence homology, one can categorize the channels into distinct branches.

One branch of the family includes the human genes SCN1A, SCN2A, SCN3A, SCN9A and SCN8A, which are all, primarily, neuronally expressed genes. The channels encoded by these genes gate rapidly and are sensitive to block by low (~10 nM) concentrations of TTX. Depending on the definition of the branch point, some phylogenetic analyses consider SCN8A a separate group since the chromosomal location is different from SCN1A, 2A, 3A and 9A, which are all expressed on chromosome 2. Another distinct branch in the family includes SCN5A, SCN10A and SCN12A. These channels show a more diverse expression pattern within the group, with SCN5A being the primary voltage-dependent Na channel gene expressed in cardiac tissue and SCN10A and SCN12A expressed in peripheral nerves. These three family members are much less sensitive to block by tetrodotoxin and the sensitivity has been mapped to a cysteine residue in the pore region. Substitution of the cysteine by aromatic amino acids expressed in TTX-sensitive channels increases the TTX sensitivity [Heinemann et al. (1992); Satin et al. (1992)]. Of the remaining two gene family members SCN4A encodes a TTX-sensitive channel primarily expressed in skeletal muscle. The sequence alignments indicate that SCN4A constitutes a unique branch in the Na channel gene family consistent with a distinct human chromosomal location on chromosome 17. SCN4A biophysical properties and TTX-sensitivity, however, are most similar to the neuronally expressed genes SCN1A, 2A, 3A and 9A. The final member of the family, SCN6A, is an outlier with respect to biophysical properties. SCN6A encodes a distinct class of channels that are not directly gated by voltage and may play a role in salt intake (Hiyama et al. 2004). Interestingly, SCN6A resides within the same chromosomal cluster as the tetrodotoxin-sensitive channel genes SCN1A, 2A, 3A and 9A.

6.1.3
Sodium Channel Expression

Table 6.1.2 lists the general expression patterns observed among the various sodium channel gene family subtypes. SCN1A, 2A, 3A and 8A are the predominant voltage-gated sodium channels expressed in the central nervous system (CNS) while SCN9A, 10A and 12A are dominant in peripheral nervous system (PNS) under normal conditions. SCN5A encodes the major voltage-gated sodium channel expressed in the heart while SCN4A encodes the major channel expressed in skeletal muscle. Expression of the atypical sodium channel gene, SCN6A, is broader than for other members of the family. Subcellular localization is also listed in Table 6.1.2.

The trends illustrated in Table 6.1.2 are accurate and useful when considering subtype selective expression with respect to physiological and/or pharmacological analysis. The expression patterns become more complex, however, as the granularity of the analysis increases. For example, as noted by Goldin (2003), the SCN1A gene product was originally identified in the rat CNS (Noda et al. 1986), but high levels of expression in the PNS have also been reported (Beckh 1989). SCN8A is primarily a CNS channel, but expression is observed in dorsal

Table 6.1.2 Tissue and cellular distribution of voltage-gated sodium channel gene family members. Tissue distribution listed is for the primary sites of expression although localized expression may exist in other tissues (see text); subcellular localization is based primarily upon reported immunohisto-chemical data.

Gene	Primary site(s) of expression	General subcellular localization
SCN1A	CNS (cortex, hippocampus, cerebellum) DRG, motor neurons	Dendrites, soma
SCN2A	CNS (cortex, hippocampus, cerebellum)	Axons, nerve terminals
SCN3A	Adult CNS (human), embryonic CNS (rat)	Dendrites, soma
SCN4A	Skeletal muscle	Axons
SCN5A	Heart	Intercalated disks
SCN6A	Heart, uterus, muscle, lung, CNS (circumventricular organs)	–
SCN8A	CNS	Nodes of Ranvier (myelinated axons)
SCN9A	PNS (sensory neurons)	Axon terminals
SCN10A	PNS (primarily small diameter neurons)	Cell body, dendrites
SCN12A	PNS (exclusively small diameter neurons)	Broad cellular distribution

root ganglion cells as well (Tzoumaka E. et al., 2000). In addition to the neuronal expression, SCN1A, 2A, 3A, and 8A are expressed in glial cells (Black et al. 1994; Oh et al. 1994). SCN1A, 3A and 8A are also expressed in the heart (Maier et al. 2004) with distinct subcellular localization. Conversely, SCN5A, while predominantly expressed in the heart, is expressed in the limbic system of the brain (Hartmann et al., 1999; Wu et al., 2002). The bottom line is that generalities are only useful to a point and, as specific scientific questions are addressed, detailed expression analysis is essential to understanding the role of voltage-dependent sodium channels in various tissues. This point is also illustrated by the insights gained in knockout mouse studies discussed in Section 6.1.5.2.

Perhaps one of the most interesting observations regarding sodium channel expression patterns is the differential subcellular expression observed for the various genes. This is true even when multiple subtypes are expressed in the same cell. SCN1A gene expression in the CNS leads to protein expression primarily in the soma and dendrites of neurons. Conversely, SCN2A gene expression in the same areas of the CNS leads to channel expression along the axons and at the nerve terminals (Westenbroek et al., 1989). Trimmer and Rhodes (2004) noted that the distinct localization is suggestive of different roles for these two channels in the same cell despite the similarity in biophysical properties. SCN1A-derived channels are likely to be involved in signal integration and action potential initiation in the cell body while SCN2A-derived channels will drive axonal conduction and play a role in transmitter release. Differential subcellular distribution may also be physiologically important in the heart. Maier et al. (2004) studied immunolocalization of both α- and β-subunits in the heart and found the SCN5A-derived channels were predominantly localized to the intercalated disks of ventricular myocytes while SCN1A, 3A and 8A derived channels were localized to T-tubules. Targeting to different subcellular locations may be directed by coupling to different types of β subunits. The authors speculate that SCN5A-derived channels are likely to play a key role in impulse initiation and propagation across the surface of the cell and throughout the heart. The SCN1A, 3A, and 8A-derived channels may be more important for coordinated depolarization and excitation within the cell.

One final aspect of expression that is important in the sodium channel family is differential expression following injury or in disease. Several very good reviews on the role of sodium channels in pain present data on changes in channel expression (e.g., Lai et al., 2004; Baker and Wood, 2001). One of the most widely studied examples is for SCN10A-derived sodium channels in peripheral nerves. Following induced axonal injury, the levels of gene transcription actually decrease in the cell bodies of the injured neurons. However, redistribution in both uninjured (Gold et al., 2003) and injured neurons (Novakovic et al., 1998) may contribute to aberrant electrical activity associated with pain. Another example of channel redistribution playing a role in a pathophysiological process occurs in multiple sclerosis (MS) (Waxman et al. 2004). As noted in Table 6.1.2, SCN8A-derived sodium channels are expressed predominantly at nodes of Ranvier and play a role in the saltutory conduction that occurs along myelinated axons. During the neuro-inflammatory processes that occur in MS, the myelin along the length of the axon is degraded, leading to a redistribution of SCN8A-derived channels along the axon. The biophysical properties of these channels support the presence of a persistent sodium current, which allows more Na to enter the axon. One hypothesis related to neuronal degeneration in MS is that this increased intracellular Na leads to increased Na/Ca exchanger activity, raising intracellular Ca to toxic levels, precipitating degeneration. Other sodium channels, which do not support such a persistent current, do not contribute to the increased intracellular calcium levels. Therefore, simple redistribution of the expression of an existing Na channel may contribute significantly to the neuronal impairment following demyelination in MS.

6.1.4
Biophysical Properties of Voltage-dependent Sodium Channels

As noted in the introduction, sodium channels are gated by changes in membrane potential. The quantitative relationship for this gating phenomenon is illustrated in Fig. 6.1.3, which plots the voltage-dependence of inactivation and activation and also illustrates the kinetics of channel gating during a single depolarizing step in membrane potential. Cells that express functional voltage-gated sodium channels typically rest at negative membrane potentials such that the

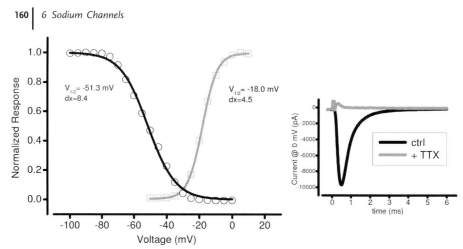

Fig. 6.1.3 Voltage-dependent properties of sodium channels. Activation (□) and inactivation (○) curves are illustrated for TTX-sensitive sodium currents recorded from rat dorsal root ganglion neurons. Data were fit to a two-state Boltzmann distribution to derive the $V_{1/2}$ and *dx* values shown in the figure. Inset: kinetics of a typical sodium current recorded during a step depolarization from −120 to −20 mV in the same neuron. Application of 500 nM tetrodotoxin completely eliminates the current.

steady-state inactivation of the channel is minimal and a significant fraction of channels are available to open. Activation occurs quickly following a depolarization, but at voltages more positive than those necessary to induce inactivation. The result of these properties is that sodium channels tend to open for only brief periods, as illustrated in the Fig. 6.1.3 inset, and close rapidly over the course of several milliseconds. At certain voltages, however, where there is overlap of the activation and inactivation curves, a finite number of channels may be open without inactivating and produce a sustained or window current. The process of recovery from inactivation is also strongly voltage-dependent and can significantly influence the excitability of the cell. This has been very nicely illustrated by Leao et al. (2005), who recorded from presynaptic nerve terminals in the auditory brainstem, which can sustain firing rates up to 1 kHz. These authors show how biophysical properties, as well as a unique subcellular distribution, allow neurons to function at firing rates necessary for appropriate signaling.

One reason voltage-dependent properties are of interest is that pharmacological modulation of sodium channel function has led to successful treatments in various therapeutic applications. Anti-arrhythmics, anti-convulsants and local anesthetics have all been developed that target voltage-dependent sodium channels as a primary mechanism of action. The therapeutic indices of these compounds have often been achieved by capitalizing on state-dependent and/or voltage-dependent block of the channel. Increased excitability, as occurs during arrhythmia or in epileptic seizures, can be associated with increased frequency of action potential firing driven by sodium channel activity. Pharmacological agents that block more effectively at higher firing rates can block the aberrant firing while

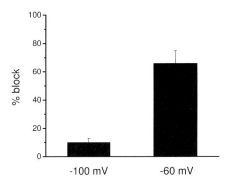

Fig. 6.1.4 Block of TTX-sensitive sodium currents in DRG neurons by mexilitine. Currents were recorded following step depolarizations to −20 mV preceded by 8 s prepulses to either −100 or −60 mV. Control records were obtained and 100 μM mexilitine applied until steady-state block was reached. Percent block was determined compared to control levels; $n = 4$ (mean ± sem).

leaving normal electrical activity intact. This has been particularly important for most sodium channel blocking drugs since they are typically poorly selective among channel subtypes. In principle, an agent that selectively targets a particular sodium channel subtype that shows restricted distribution should be at least as effective and safe, but such compounds have yet to be reported. Gonzalez et al. gives a detailed description of the pharmacological modulation of sodium channels in Chapter 6.2; however, Fig. 6.1.4 serves to introduce the general concepts by illustrating the effect of 100 μM mexilitine on TTX-sensitive sodium currents recorded from a neuron. A much greater channel block results from holding the membrane potential at −60 mV than at −100 mV. Referring to the voltage-dependent properties shown in Fig. 6.1.3, one can see that the channels are primarily in a resting but non-inactivated state at −100 mV while there is significant steady-state inactivation at −60 mV. Since mexilitine is a state-dependent sodium channel blocker and binds to the inactivated state of the channel, greater block occurs as the channel spends more time in depolarized states. Another manifestation of this same property is illustrated in Fig. 6.1.5, which shows the effect of a

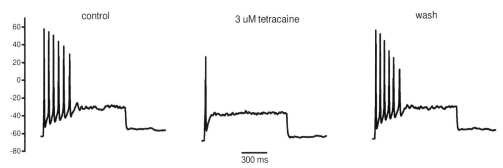

Fig. 6.1.5 Effects of tetracaine on action potential firing in a DRG neuron. Membrane potential changes were recorded following injection of current into single DRG neurons in the presence of 500 nM TTX. Control records were obtained for neurons that showed repetitive firing, and firing rates were assessed after application of 3 μM tetracaine and again following washout. Membrane potential (mV) is shown by the scale on the left. Similar results were obtained in two additional neurons.

state-dependent blocker on neuronal action potential firing. Application of $3\,\mu M$ tetracaine results in decreased firing rates (note the number of spikes) compared to control. The first action potential is usually not affected, but subsequent action potentials are blocked. Use-dependent block is a bit more complicated than simple voltage-dependent block in that on and off rates of the particular compound become important in determining the level of block obtained at different frequencies. Nevertheless, the principles are the same and state-dependent blockers can effectively reduce increased firing rates associated with various pathological states.

6.1.5
Disease Association

6.1.5.1 Channelopathies
Various genetic disorders in humans and other mammals have been linked to specific sodium channel genes (Table 6.1.3). Interestingly, the majority of these mutations are in the two primary muscle genes SCN4A for skeletal muscle and SCN5A for cardiac muscle. This may reflect requisite developmental roles for genes expressed in the nervous system that appear to occur in some non-human species. For example, constitutive knockout of SCN9A in mice is lethal while tissue specific knockouts lead to interesting phenotypes (see below). This, however, is speculative and there may be other reasons that explain the observations.

The myotonias represent the largest group of defined sodium channel mutations in man and occur in the SCN4A gene (Cannon 1997; Lehmann-Horn and Jurkat-Rott 1999). These mutations lead to three different types of myotonia. Hy-

Table 6.1.3 Examples of human diseases or disorders associated with mutations in sodium channel genes; nomenclature for sites of mutations is domain followed by transmembrane spanning region (e.g., IIS4 denotes a mutation in domain II of the 4th transmembrane spanning region).

Gene	Disease/disorder	Site of mutations
SCN1A	General epilepsy with febrile seizures plus (GEFS+)	IIS4, IVS4
SCN4A	Hyperkalemic periodic paralysis	IIS5, IIIS4-S5, IVS1, IVS6
	Paramyotinia congenita	IIS4-S5, IIIS6, III-IV, IVS3, IVS4, IVS4-S5
	Potassium-aggravated myotonias	IS6, IIS6, III-IV, IVS6
SCN5A	Idiopathic ventricular fibrillation long QT syndrome 3 (LQT3)	III/S1-S2, IV/S3-S4 III/S4-S5, III-IV, IVS4
SCN9A	Primary erythermalgia	IIS4-S5

perkalemic periodic paralysis is a condition exacerbated by high serum potassium levels and the disease is characterized by muscle weakness with hyperkalemia. Paramyotonia congenita presents with different clinical signs and such as stiffening of the muscles during exercise or exposure to cold. Both diseases are associated with defects in sodium channel inactivation (Lehmann-Horn et al., 1987; Lehmann-Horn et al., 1995). The third disorder, potassium aggravated myotonia, is characterized by myotonia or muscle stiffness, often following exercise. Many of these mutations have been shown to affect gating although there are also mutations leading to truncated protein production and non-functional channels. These data illustrate the key role sodium channel proteins play in regulating skeletal muscle contraction.

The other prominent area of sodium channel mutations linked to human disease is for the cardiac channel, SCN5A. While the myotonias result in defects in muscle function, the SCN5A mutations lead to defects in cardiac electrical activity that can be life threatening. Long Q-T syndrome type 3 (LQT3) is a rare form of LQT compared to LQT1 and LQT2, which arise from mutations in potassium channels. Several of the mutations characterized to date result in alterations of channel inactivation and lead to more sustained sodium currents, resulting in delayed cardiac repolarization and arrhythmia (Felix 2000). Idiopathic ventricular fibrillation is another genetic disorder linked to SCN5A; however, patients present with normal QT intervals. Two mutations likely lead to loss of function while others affect gating and lead to decreased thresholds of excitability (Chen et al., 1998), making the affected individuals more susceptible to arrhythmic stimuli.

One disease reported to be linked to a peripheral neuronal sodium channel is primary erythermalgia (Cummins et al., 2004). This is a disease where patients experience burning sensations in their extremities associated with redness. The mutations identified by sequence analysis of the SCN9A gene were examined in recombinant expression systems and these mutations were found to result in hyperpolarizing shifts in the activation curve for SCN9A-derived channels and larger currents. Given the expression of SCN9A in nociceptive neurons, such changes in the biophysical properties of the channel would be expected to increase excitability of the nociceptor consistent with the phenotype observed clinically.

Much less has been reported on mutations of CNS channels compared to muscle channels, but there are a few linkages to epilepsy for CNS sodium channels, either directly or indirectly. Mutations in SCN1A have been linked to general epilepsy with febrile seizures plus (GEFS+) type 2 (Escayg et al., 2000) and appear to affect gating based on functional studies and their location within the putative channel structure (Spampanato et al., 2003). In addition, GEFS+ type I results from mutations in a β subunit which normally associates with the α subunit to increase inactivation rates. The mutation prevents the normal modulation of the α subunit when expressed in heterologous expression systems (Wallace et al., 1998). These changes in gating would increase the chances of hyper-excitability in the CNS.

6.1.5.2 **Mouse Studies Provide Additional Insight into Disease Association**

Clearly, from the linkage to human diseases, sodium channels play a key role in regulating skeletal muscle contraction, cardiac rhythm and CNS excitability. For those channels that have not yet been linked to specific genetic diseases, mouse knockout and/or anti-sense studies have provided great insight into physiological and potential pathophysiological roles, particularly for peripheral nervous system channels. A few examples are given in the following paragraphs.

One of the first sodium channels to be characterized in a knockout mouse was SCN10A (Akopian et al., 1999). SCN10A gene expression is largely restricted to small diameter nociceptive neurons although there is also expression in medium and large diameter peripheral neurons. Based on the expression pattern, the knockout would be anticipated to have significant deficits in nociceptive responses. While there was a reduced hyperalgesia in inflammatory pain and an absence of sensitivity to noxious pressure, the responses to other pain stimuli, particularly the development of neuropathic pain, were normal. The authors speculate that this is likely due to compensatory upregulation of other sodium channels and present data indicating upregulation of SCN9A-derived channels. This speculation appears to be correct in that later studies using anti-sense methods showed dramatic effects of SCN10A knockdown in neuropathic pain models (Gold et al., 2003; Lai et al., 2002).

SCN3A is another channel where anti-sense data have been helpful in suggesting a role in regulating pain states. Hains et al. (2003) studied SCN3A expression in the spinal cord and peripheral neurons following spinal contusion injury and found that SCN3A expression increased dramatically in second-order neurons of the dorsal horn. Anti-sense administration prevented upregulation of SCN3A-derived channels and also prevented the development of allodynia in the spinal contusion injury model. SCN3A channels have been reported to undergo upregulation in peripheral neurons in different models of pain, suggesting a possible role in regulation of multiple pain states.

A final example of the utilization of experimental models to define the physiological role of sodium channels subtypes comes from the elegant studies of Nassar et al. (2004) on SCN9A expression. Since constitutive knockout of SCN9A is lethal, this group developed a system using the SCN10A promoter where SCN9A expression was selectively knocked out in nociceptive neurons. The knockout animals developed normally, but had higher thresholds for both mechanical and thermal pain. The most dramatic effects, however, were in responses to inflammatory pain stimuli, which were significantly reduced or eliminated altogether. These results, along with the genetic analysis noted above, strongly suggest SCN9A-derived channels play a role in pain sensation.

6.1.6
Conclusions

Since the initial cloning of sodium channels by Noda et al. (1984) the complete complement of ten family members has been cloned and characterized. Voltage-gated sodium channels play important roles in regulating excitability and muscle contraction in many tissues of the body. Moreover, subtype specific roles in the regulation of different pain states are currently being elucidated. There is a rich history of pharmacologically modulating sodium channel function for the treatment of epilepsy, arrhythmias and acute pain and this is likely to continue to expand. The molecular, biophysical and pharmacological data for specific subtypes of sodium channels also continues to generate interest in the identification and development of newer, more selective modulators for the treatment of human diseases.

References

Akopian A.N., Souslova V., England S., Okuse K., Ogata N., Ure J., Smitt A., Kerr B.J., McMahon S.B., Boyce S., Hill R., Stanfa L.C., Dickenson A.H., Wood J.N. (**1999**) The tetrodotoxin-resistant sodium channel SNS has a specialized function in pain pathways. *Nat. Neurosci.* 2, 541–548.

Beckh S., Noda M., Lubbert H., Numa S. (**1989**) Differential regulation of three sodium channel messenger RNAs in the rat central nervous system during development. *EMBO J.* 8, 3611–3636.

Baker M.D., Wood J.N. (**2001**) Involvement of Na$^+$ channels in pain pathways. *Trends Pharmacol. Sci.* 22(1), 27–31.

Black J.A., Yokoyama S.G., Waxman S.G., Oh Y., Zur K.B., (**1994**) Sodium channel mRNAs in cultured spinal cord astrocytes: in situ hybridization in identified cell types. *Mol. Brain Res.* 23, 235–265.

Cannon S.C. (1997) From mutations to myotonia in sodium channel disorders. *Neuromusc. Disorders* 7, 241–249.

Cestele, S., Catterall W.A. (**2000**) Molecular mechanisms of neurotoxin action on voltage-gated sodium channels. *Biochimie* 82(9–10), 883–892.

Chen Q., Kirsch G.E., Zhang D., Brugada R., Brugada J., Brugada P., Potenza D., Moya A., Borggrefe M., Breithardt G., Ortiz-Lopez R., Wang Z., Antzelevitch C., O'Brien R.E., Schulze-bahr E., Keating M.T., Towbin J.A., Wang G. (**1998**) Genetic basis and molecular mechanism for idiopathic ventricular fibrillation. *Nature* 392, 293–296.

Cole K.S. (1949) Dynamic electrical characteristics of squid axon membrane. *Arch. Sci. Physiol.* 22, 649–670.

Cummins T.R., Dib-Hajj S.D., Waxman S.G. (**2004**) Electrophysiological properties of mutant Na$_v$ 1.7 sodium channels in a painful inherited neuropathy. *J. Neurosci.* 24(38), 8232–8236.

Escayg A., MacDonald B.T., Meisler M.H., Baulac S., Huberfeld G., An-Gourfinkel I., Brice A., LeGuern F., Moulard B., Chaigne D., Buresi C., Malafosse A. (**2000**) Mutations of SCN1A, encoding a neuronal sodium channel, in two families with GEFS+2. *Nat. Genet.* 24, 343–345.

Felix, R. (**2000**) Channelopathies: ion channel defects linked to heritable clinical disorders. *J. Med. Genet.* 37, 729–740.

Gold M.S., Weinreich D., Kim C.S., Wang R., Treanor J., Porreca F., Lai J. (**2003**) Redistribution of Na(V) 1.8 in uninjured axons enables neuropathic pain. *J. Neurosci.* 23, 158–166.

Goldin A.L. (**2002**) Evolution of voltage-gated Na$^+$ channels. *J. Exp. Biol.* 205, 575–584.

Goldin A.L. (**2003**) Resurgence of sodium channel research. *Annu. Rev. Physiol.* 63, 871–894.

Hains B.C., Klein J.P., Saab C.V., Craner M.J., Black J.A., Waxman S.G. (**2003**) Upregulation of sodium channel Nav 1.3 and functional involvement in neuronal hyperexcitability associated with central neuropathic pain after spinal cord injury. *J. Neurosci.* 23(26), 8881–8892.

Hartmann H.A., Colom L.V., Sutherland M.L., Noebels J.L. (**1999**) Selective localization of cardiac SCN5A sodium channels in limbic regions of rat brain. *Nat. Neurosci.* 2, 593–595.

Heinemann S.H., Terlau H., Imoto K. (**1992**) Molecular basis for pharmacological differences between brain and cardiac sodium channels. *Pflugers Arch.* 422, 90–92.

Huxley A.L., Hodgkin A.F. (**1952a**) Measurement of current-voltage relations in the membrane of the giant axon of Loligo. *J. Physiol.* 1, 424–448.

Huxley A.L., Hodgkin A.F. (**1952b**) Currents carried by sodium and potassium ions through the membrane of the giant axon of Loligo. *J. Physiol.* 1, 449–472.

Huxley A.L., Hodgkin A.F. (**1952c**) The components of membrane conductance in the giant axon of Loligo. *J. Physiol.* 1, 473–496.

Huxley A.L., Hodgkin A.F. (**1952d**) The duel effect of membrane conductance in the Giant Axon of Loligo. *J. Physiol.* 1, 497–506.

Hiyama T.Y., Watanabe E., Okado H., Noda M. (**2004**) The subfornical organ is the primary locus of sodium-level sensing by Na_x sodium channels for control of salt-intake behavior. *J. Neurosci.* 24(42), 9276–9281.

Jaillon, O., et al., (**2004**) Genome duplication in the teleost fish Tetraodon nigroviridis reveals the early vertebrate proto-karyotype. *Nature* 431, 946–957.

Jiang Y.X., Ruta V., Chen J.Y., Lee A., MacKinnon R. (**2003**) The principle of gating charge movement in a voltage-dependent K^+ channel. *Nature* 423(6935), 42–48.

Koishi R., Xu H., Ren D., Navarro B., Spiller B.W., Shi Q., Clapham D.E. (**2004**) A superfamily of voltage-gated sodium channels in bacteria. *J. Biol. Chem.* (**2004**) 279(10), 9532–9538.

Lai J., Gold M.S., Kim C.S., Bian D., Ossipov M.H., Hunter J.C., Porreca F. (**2002**) Inhibition of neuropathic pain by decreased expression of the tetrodotoxin-resistant sodium channel, Na_v 1.8. *Pain* 95, 143–152.

Lai J., Porreca F., Hunter J.C., Gold M.S. (**2004**) Voltage-gated sodium channels and hyperalgesia. *Annu. Rev. Pharmacol. Toxicol.* 44, 371–397.

Leao R.M., Kushmerick C., Pinaud R., Renden R., Li G-L., Taschenberger H., Spirou G., Levinson S.R., von Gersdorff, H. (**2005**) Presynaptic Na^+ channels: locus, development, and recovery from inactivation at a high-fidelity synapse. *J. Neurosci.* 25(14), 3724–3738.

Lehmann-Horn F., Jurkat-Rott K. (**1999**) Voltage-gated ion channel and hereditary disease. *Physiolog. Rev.* 79(4), 1317–1372.

Lehmann-Horn F., Kuther G., Ricker K., Grafe P., Ballanyi K., Rudel R. (**1987**) Adynamic episodica hereditaria with myotonia: a non-inactivating sodium current and the effect of extracellular pH. *Muscle Nerve* 10, 363–374.

Lehmann-Horn F., Rudel R. (**1995**) Hereditary nondystrophic myotonias and periodic paralyses. *Curr. Opin. Neurol.* 8, 402–410.

Lopreato G.F., Lu Y., Southwell A., Atkinson N.S., Hillis D.M., Wilcox T.P., Zakon H.H. (**2001**) Evolution and divergence of sodium channel genes in vertebrates. *Proc. Natl. Acad. Sci. U.S.A.* 98(13), 7588–7592.

Lin Y.C., Spencer A.N. (**2001**) Calcium currents from jellyfish striated muscle cells: preservation of phenotype, characterization of currents and channel localization. *J. Exp. Biol.* 204, 3717–3726.

Maier S.K.G., Westenbroek R.E., McCormick K.A., Curtis R., Scheuer T., Catterall W.A. (**2004**) Distinct subcellular localization of different sodium channel α and α subunits in single ventricular myocytes from mouse heart. *Circulation* 109, 1421–1427.

Nassar M.A., Stirling L.C., Forlani G., Baker M.D., Matthews E.A., Dickenson A.H., Wood J.N. (**2004**) Nociceptor-specific gene deletion reveals a major role for Na_v 1.7 (PN1) in acute and inflammatory pain. *Proc. Natl. Acad. Sci. U.S.A.* 101(34), 12706–12711.

Noda M., Shimizu S., Tanabe T., Takai T., Kayano T., Ikeda T., Takahashi H., Nakayama H., Kanaoka Y., Minamino M., Kangawa K., Matsuo H., Raftery M.A., Hirose T., Inayama S., Hayashida H., Myata T., Numa S. (**1984**) Primary structure of the Electrophorus electricus sodium channel deduced from the cDNA sequence. *Nature* 312, 121–127.

Noda M., Ikeda T., Kayano T., Suzuki H., Takeshima H., Kurasaki M., Takahashi H., Numa S. (**1986**) Existence of distinct sodium channel messenger RNAs in rat brain. *Nature* 320, 188–192.

Novakovic S.D., Tzoumake E., McGivern J.G., Haraguchi M., Sangameswaran L., Gogas K.R., Eglen R.M., Hunter J.C. (**1998**) Distribution of the tetrodotoxin-resistant sodium channel PN3 in rat sensory neurons in normal and neuropathic conditions. *J. Neurosci.* 18, 2174–2187.

Oh Y., Black J.A., Waxman S.G. (**1994**) The expression of rat brain voltage-sensitive Na$^+$ channel mRNAs in astrocytes. *Mol. Brain Res.* 23, 57–65.

Perez-Reyes, E. (**2003**) Molecular physiology of low-voltage activated T-type calcium channels. *Physiol. Rev.* 83, 117–161.

Plummer N.W., Meisler M.H. (**1999**) Evolution and diversity of mammalian sodium channel genes. *Genomics* 57, 323–331.

Satin J., Kyle J.W., Chen M., Bell P., Cribbs L.L., Fozzard H.A., Rogart R.B. (**1992**) A mutant of TTX-resistant cardiac sodium channels with TTX-sensitive properties. *Science* 256, 1202–1205.

Spafford J.D., Spencer A.N., Gallin W.J. (**1999**) Genomic organization of a voltage-gated Na$^+$ channel in a hydrozoan jellyfish: insights into the evolution of voltage-gated Na$^+$ channel genes *Receptors Channels* 6(6), 493–506.

Spampanato J., Escayg A., Meisler M.H., Goldin A.L. (**2003**) Generalized epilepsy with febrile seizures plus type 2 mutation W1204R alters voltage-dependent gating of Na(v)1.1 sodium channels. *Neuroscience* 116(1), 37–48.

Trimmer J.S., Rhodes K.J. (**2004**) Localization of voltage-gated ion channels in mammalian brain. *Annu. Rev. Physiol.* 66, 477–519.

Tzoumaka E., Tischler A.C., Sangameswaran L., Eglen R.M., Hunter J.C., Novakovic S.D. (**2000**) Differential distribution of the tetrodotoxin-sensitive rPN4/NaCH6/Scn8a sodium channel in the nervous system. *J. Neurosci. Res.* 60, 37–44.

Wallace R.H., Wang D.W., Singh R., Scheffer I.E., Georger A.L., Phillips H.A., Saar K., Rier A., Johnson E.W., Sutherland G.R., Mulley J.C. (**1998**) Febrile seizures and generalized epilepsy associated with a mutation in the Na$^+$-channel α1 subunit gene SCN1B. *Nat. Genet.* 19, 366–370.

Waxman S.G., Craner M.J., Black J.A. (**2004**) Na$^+$ channel expression along axons in multiple sclerosis and its models. *Trends Pharmacol. Sci.* 25(11), 584–591.

Westenbroek R.E., Merrick D.K., Catterall W.A. (**1989**) Differential subcellular localization of the RI and RII Na$^+$ channel subtypes in central neurons. *Neuron* 3, 695–704.

Wu L., Nishiyama K., Hollyfield J.G., Wang O. (**2002**) Localization of Nav 1.5 sodium channel protein in the mouse brain. *Neuroreport* 13(18), 2547–2551.

6.2
Small Molecule Blockers of Voltage-gated Sodium Channels

Jesús E. González, Andreas P. Termin, and Dean M. Wilson

6.2.1
Drugs that Act on Sodium Channels

Small molecule voltage-gated sodium channel (Na_Vs) blockers are used as anti-arrthymic, anesthetic, neuroprotectant, and anticonvulsant agents in man. In most cases, the agents were discovered and optimized in animal pharmacological models and it was only later that their mechanism of action on Na_Vs was eluci-dated. With the goal of identifying a local anesthetic with improved safety com-pared to cocaine, procaine was synthesized in 1905 and discovered to have useful analgesic properties. This was a milestone in the history of local anesthetics that includes the generation of a whole series of "caines" that spans over a century and are still widely used in pain management [1]. Certain examples of these "caines", including lidocaine, were found to provide protection against cardiac arrhythmia [2] via the shared Class I anti-arrythmic mechanism of blocking Na^+ currents in the heart. Lidocaine and other Na_V blockers have also been used to treat various pain conditions, including chronic neuropathic pain. In 1996 the most recent refinement, ropivacaine, emerged as a single enantiomer with differential sensory and motor blockade and an improved safety profile for acute pain management [3]. Though there is a long history of clinical use for local anesthetics, their me-chanism of action of Na_V blockade was not demonstrated until 1955–1966 [4–6].

Na_V blocking drugs are also important in treating epilepsy. The anticonvulsant phenytoin was discovered in 1937 in an effort to identify agents with decreased sedation compared to barbiturates, using an electroshock animal model compati-ble with compound screening [7]. Merritt and Putnam reasoned that the diphenyl hydantoin was a reasonable barbiturate mimic. Their success was followed by rapid demonstration of clinical efficacy in epilepsy patients, which also estab-lished that experimental animal models could be predictive of clinical efficacy. It was not until the 1980s that phenytoin was demonstrated to block neuronal Na_Vs, effects likely important for its clinical action [8, 9]. Efforts to improve upon phenytoin and other first generation anticonvulsants led to the development of lamotrigine in 1978, whose mechanism also involves Na_V block [10]. Originally launched in 1990 as an antiepileptic agent, lamotrigine (1) (and related analogs) have since been actively pursued for additional indications such as bipolar disor-der, approved in 2003, and neuropathic pain.

The molecular pharmacology of Na_Vs took a giant step forward in the 1950s when Hodgkin and Huxley demonstrated that the upstroke of the giant squid axon action potential was due to an inward sodium current that activated and in-activated in milliseconds [11]. Subsequently, Narahasi and coworkers demon-strated that the small molecule toxin from the pufferfish, tetrodotoxin (TTX, 17), selectively blocked this Na^+ current at nanomolar concentrations [12]. It was found that TTX did not affect the large outward potassium current responsible for action potential repolarization – selectivity that has played an important part in the elucidation of ion currents that contribute to the action potential; even today TTX remains one of the most important pharmacological tools for dissecting Na_V function.

To date nine Na_V subtypes have been identified and have provided a broader view of Na_Vs that includes subtypes preferentially expressed in the brain, muscle, and peripheral nervous system (see Chapter 6.1 by Krafte et al. for channel expression and structure). Interestingly, three subtypes, the cardiac ($Na_V1.5$) and two peripheral channels ($Na_V1.8$ and $Na_V1.9$) have substantially reduced affinity for TTX, thus leading to their classification as TTX resistant, or TTX-r. The remaining TTX sensitive (TTX-s) Na_V subtypes are readily blocked by TTX. $Na_V1.8$ and $Na_V1.9$, for example, are blocked at TTX concentrations of $\sim 60\,\mu M$, four orders of magnitude greater than required for the TTX-s neuronal forms. These peripheral TTX-r Na_Vs play key roles in sensory neurons and may be targets for novel analgesics [13, 14]. Except for TTX and related toxins, small molecule blockers of NaVs are generally nonselective and block all subtypes with approximately equal potency. Drugs that act via Na_V block bind to the same site, the local anesthetic receptor. The relatively recent identification of the numerous subtypes with specific distribution patterns has provided new opportunities for targeting Na_V blockers.

The medicinal chemistry of Na_V blockers has been reviewed [15], thus in this chapter we focus on recent results from existing drugs, including limitations revealed from decades of use in man, compounds in development, selected preclinical blockers, and future directions for improved and new therapeutics targeting Na_Vs.

6.2.1.1 Mechanisms of Drug Block

The blocking properties of TTX (**17**) and local anesthetics have been the foundation for the current mechanistic understanding of how small molecules block Na_Vs, and have been reviewed recently [16–18]. TTX blocks the Na^+ permeation pathway from the outside of the cell. Its receptor binding site is located on the outer pore vestibule [19] and is distinct from the local anesthetic receptor site [20]. It does not significantly traverse into the cell membrane or channel and block is not voltage-dependent. The TTX binding site is more generally described as a guanidinium site, as other structurally related toxins bind in a similar mode – a mechanism that has been compared to a cork that plugs the opening of a bottle. In contrast, local anesthetics are generally thought to act through membrane or intracellular pathways to reach its binding site. They bind to and/or stabilize closed or inactivated states and shift the equilibrium of channels states, decreasing the probability of opening. Lidocaine (**4**), for example, slows the recovery from inactivation; however, the exact mechanism of block is still unclear [17]. Many local anesthetics exhibit voltage-dependent block and cause shifts in the steady-state inactivation curves to more negative potentials upon binding, making the channel easier to inactivate.

Use-dependent or phasic block are terms used to describe the accumulation of channel block with repeated stimulation, another important property of local anesthetic block. These drugs preferentially affect the channel at a specific stage in its cycle of rest, activation and inactivation, often by delaying the recovery from the inactivated state, thereby producing a cumulative reduction of sodium currents [21]. The kinetics of block and unblock also contribute to the overall use-dependence of a given agent. When the off or recovery rates are slower than the frequency of stimulation or opening, additional block is accumulated beyond that possible from voltage-dependence alone. Use-dependent block allows these drugs to preferentially act on cells and nerves firing at different frequencies and is an important contributor to their therapeutic indices [18], since known agents demonstrate little or no selectivity for Na_V subtypes [10].

6.2.2
New Insights for Launched Compounds

6.2.2.1 Lamotrigine and Progeny
Binding studies with various mutated Na_Vs have been used to study local binding interactions as well as the induced conformational changes upon binding of lamotrigine (**1**) and its structural analogs 4030W92 (**2**) and 619C89 (**3**). An alanine scan against the S6 helix in domain III of the type IIA sodium channel has identified that mutations L1465A, I1469A decrease binding affinities of lamotrigine and three congeners, as well as the local anesthetic etidocaine. In addition, the N1466A mutant showed selective modulation of **2** and etidocaine binding. The mutated residues from the IIIS6 helix are postulated to face the lumen of the pore, suggesting a specific binding site for the pore blocking compounds examined [22]. These data add to the previous mutagenesis studies that have identified Ile-1760, Phe-1764, and Tyr-1771 as key amino acids that make up the drug receptor sites along the S6 helix of domain IV. Drug binding occurs at the interface of these S6 helices. In another study, circular dichroism was used to investigate the secondary structure of purified sodium channels from electric eel electroplax membranes with and without lamotrigine. It was found that lamotrigine induces

Lamotrigine
1

4030W92
2

619C89
3

significant increases to the helical content. A homology model based on the bacterial potassium channel was also constructed and docking of lamotrigine was consistent with previously identified residues that influenced binding. In combination with the putative preference for lamotrigine to bind the open and inactivated states, it has been proposed that more helical conformations are associated with drug block [23].

The lamotrigine congener 4030W92 (**2**) blocks Na$_V$s in rat dorsal root ganglion neurons (DRGs) in a voltage- and use-dependent manner [24]. Both slow inactivating TTX-r and fast inactivating TTX-s currents were blocked in a voltage- and use-dependent manner. Holding potential dependent IC$_{50}$s of 22–103 μM against TTX-r channels were observed, while IC$_{50}$s against TTX-s channels ranged from 5 to 37 μM, presumably through activity on a slow inactivation state of the channels. The compound blocked more potently at less negative potentials, shifted the inactivation curve 13 mV to more negative potentials, and significantly slowed the recovery from inactivation, all supporting that the molecule targets an inactivated state.

Recent pharmacology results on lamotrigine have been reported. It has shown efficacy in the rat formalin pain model [25], although both lamotrigine and 4030W92 were ineffective in the human capsaicin-induced hyperalgesia model [26]. Chronic neuropathic pain patients treated with 4030W92 exhibited transient reductions in both distribution and severity of allodynia that were not, however, maintained to day 14 of treatment [27]. Rat anxiety model results [28], coupled with substantial clinical data [29–31] indicate important utility for lamotrigine in the treatment of mood disorders. While Na$_V$ blockade has an established role in treating epilepsy, the link between sodium channel block and efficacy in these other indications is not as clear. Nevertheless, these results suggest additional therapeutic utility for other clinically known Na$_V$ blockers, as well as those in research and early development [32].

6.2.2.2 Lidocaine, Bupivacaine, Ropivacaine, and Mexiletine

The specific sodium channel binding interactions of the "-caines" lidocaine (**4**), bupivacaine (**5**), and ropivacaine (**6**), as well as the structurally similar mexiletine (**7**), have been probed through site-directed mutagenesis of heart and skeletal muscle isoforms, although a clear and complete picture of the local anesthetic (LA) binding site has yet to emerge. Recent mechanistic studies have further demonstrated that LA binding is complex and residues in both the inner and outer vestibule and can modulate the LA activity. A recent study of Na$_V$1.4 demonstrated that mutation of the externally accessible tryptophan 1531 residue in the channel pore had a profound impact on mexiletine block via the local anesthetic binding site, indeed abolishing mexiletine sensitivity in the W1531C mutant [33]. These data, coupled with those of W1531A, W1531F, and W1531Y variants, suggest an important role of this external pore vestibule residue in sensitivity to local anesthetic block, as well as ionic selectivity. Similar studies with lidocaine on K1237E and F1579A Na$_V$1.4 mutants support that local anesthetic binding to

Lidocaine
4

Bupivacaine
5

Ropivacaine
6

Mexiletine
7

the inner vestibule of the channel can inhibit the channel from entering ultra-slow inactivation (I_{US}), presumably through LA interference, with inner vestibule dependent pore closure as a proposed "foot in the door" [34]. Providing further evidence that lidocaine binding is fundamentally linked to channel gating, systematic neutralization of Na$_V$1.5 S4 basic residues demonstrated that the gating charges from domains III and IV are allosterically modulated with bound lidocaine [35].

Several reports have explored molecular hypotheses concerning toxicity, isoforms sensitivity, and described new activities of these LA and anti-arrythmic agents. The altered cardiotoxicity profiles of (R)- and (S)-bupivacaine (5) are likely not due to inhibition of L-type calcium channel [36] or selectivity of one enantiomer over the other against NaV1.5 [37]. Mexiletine (7) was reported to preferentially block the cardiac isoforms Nav1.5 over the skeletal muscle and brain type II channels. Mutation of one amino acid residue in NaV1.4 to that found in the cardiac isoforms, S251A, produced enhanced block [38]. Interestingly, bupivacaine (5) and ropivacaine (6) (but not lidocaine, 4) upregulate Na$_v$ channels and increase saxitoxin (STX, 18a) binding with 3–24 h exposure to drug [39].

The activity of these compounds has also been profiled against additional Na$_V$ isoforms, as well as in vivo correlates. For example, lidocaine (4) blocks TTX-r sodium currents in DRG neurons [40, 41], as well as the specific channels Na$_V$1.7 and Na$_V$1.8 expressed in Xenopus oocytes [42]. Lidocaine also blocks Na$_V$1.2 expressed in CHO cells with similar potency to the cardiac alpha subunit hH1 [43]. Interestingly, lidocaine increased sodium current at negative potentials and shifted the activation curves to more negative potentials. These and other studies further support the idea that local anesthetics are nonselective Na$_V$ blockers.

Lidocaine is the active pharmaceutical ingredient in emerging therapeutic products, particularly the treatment of neuropathic pain [44, 45]. Localized topical delivery of lidocaine using an adhesive patch is a valuable therapeutic in the treatment of postherpetic neuralgia [46–48] and other pain syndromes [49, 50]. Local application reduces systemic exposure and apparently attenuates the deleterious side effects of broad sodium channel block.

6.2.2.3 **Carbamazepine and Derivatives**

Carbamazepine (**8**) is an anticonvulsant used for epilepsy and trigeminal neuralgia, and Na_v blockade is a key component of its mechanism of action. Structural modification of carbamazepine has yielded additional analogs that also modulate Na_vs and demonstrate downstream *ex vivo* and *in vivo* efficacy. Most importantly, the 10-oxo-10,11-dihydro analog oxcarbazepine (**9**) is an effective clinical agent for the treatment of epilepsy, exhibiting generally improved tolerability relative to carbamazepine [51]. Oxcarbazepine is also under investigation for the treatment of neuropathic pain. Additional modifications of these structures have focused on position 10, since this and the neighboring 11 position play known roles in subsequent *in vivo* metabolism. Capitalizing on the prevalent reduction and subsequent glucoronidation of pharmacologically active 10-hydroxy metabolites, a series of ester, carbamate, and carbonate analogs at C10 have been prepared and evaluated *in vitro* using both [^{3}H]BTX binding and ^{22}Na$^+$ uptake assays in rat cortical synaptosomes, as well as in the rat maximal electric shock (MES) model of epilepsy. C-10 acetyl derivates (i.e., BIA 2-093, **10**), in particular, demonstrate unique potency and efficacy profiles that are comparable to the parent oxcarbazepine [51]. The C-10 oxime derivative BIA 2-024 (**11**) has also been investigated [52], and comparative studies have explored these agents in detail [53, 54].

Carbamazepine (**8**) and derivatives have been studied in Na_v dependent cellular models, including activity on veratridine-induced epileptiform activity in rat hippocampal CA1 pyramidal neurons [55], Na$^+$ currents from human dentate granule cells from patients with medically intractable temporal lobe epilepsy (TLE) [56], and glutamate release from rat hippocampal synaptosomes [53]. These and other studies have defined the current mechanistic understanding, particularly regarding the role of sodium and calcium channel modulation by these agents, as well as effects on adenosine receptors and the serotonergic, dopaminergic, and glutamatergic systems [54]. For example, the Na_v opener veratridine induces release of glutamate, aspartate, GABA, and dopamine from isolated rat striatal slices – effects that are reversed in a dose-dependent manner by **8–10** [57]. These findings are consistent with Na_v blockade by these agents; similar concentration and voltage dependent electrophysiological attenuation of veratridine-induced epileptiform bursting has been observed for carbamazepine [55]. Interestingly, while carbamazepine and oxcarbazepine inhibit voltage-gated calcium channels [54], **10** and **11** did not inhibit calcium channels coupled to glutamate release.

| Carbamazepine **8** | Oxcarbazepine **9** | BIA 2-093 **10** | BIA 2-024 **11** |

Clearly, the myriad of known activities of these agents are not coupled directly to structural modifications in a predictable way; the therapeutic implications of such complementary *in vitro* profiles remain unclear.

6.2.2.4 Phenytoin

Several recent studies have provided additional data that phenytoin (**12**), via Na_V blockade, is neuroprotective in models of spinal cord injury and neuroinflammation. In the rat T9 contusion spinal cord injury model (SCI), phenytoin treatment reduced the destruction of gray and white matter surrounding the lesion, and was coupled with improved performance in locomotor assessments of recovery from spinal cord injury [58]. In a similar study, dose-dependent neuroprotection was observed, presumably through indirect inhibition of lipid peroxidation (as evidenced by tissue malondialdehyde levels) and direct block of sodium influx and resulting cellular edema [59]. In the mouse experimental allergic encephalomyelitis (EAE) model, significant neuroprotection was observed in assessments of both optic nerve axon loss (50% vs. 12% in control vs. treated animals), as well as axonal loss in the dorsal corticospinal tract and dorsal column. Treated animals also exhibited both increased spinal cord compound action potentials and improved clinical scores relative to controls, suggesting significant therapeutic applications for phenytoin in conditions such spinal cord injury and multiple sclerosis [60, 61].

Phenytoin
12

6.2.2.5 Vinpocetine

The synthetic vincamine analog vinpocetine (**13**) has been used as a neuroprotectant and cognition-enhancing agent, and is presently marketed as a dietary supplement to support cerebrovascular health (ref: *Physicians' Desk Reference for Nonprescription Drugs and Dietary Supplements*, 22nd Ed., 2002). Although the compound has broad spectrum activity, including modulation of sodium channels, calcium channels, phosphodiesterases, and glutamate receptors [62], it is a potent blocker of the TTX-r sodium channel $Na_V1.8$ [63]. Using the rat $Na_V1.8$ expressed in a dorsal root ganglion derived cell line (ND7/23), vinpocetine exhibited some frequency and voltage dependent modulation of $Na_V1.8$; at relatively depolarized holding potentials ($V_h = -35\,mV$), an IC_{50} of $3.5\,\mu M$ was obtained, whereas at more hyperpolarized holding potentials ($V_h = -90\,mV$), an IC_{50} of $10.4\,\mu M$ was observed, suggesting higher affinity for the inactivated state of the channel. In another study using rat striatum synaptosomes, data suggest that the low μM ability of vinpocetine to reverse veratridine-induced Na$_v$ opening was the primary mechanism by which subsequent glutamate and aspartate release was inhibited, and thus plausibly the means by which neuroprotection occurs [62, 64]. Data obtained using 4-aminopyridine induced seizure activity in guinea pigs suggest potential utility for vinpocetine in epilepsy [65].

Vinpocetine
13

6.2.3
Challenges of Current Agents

Drugs that act via Na$_v$ blockade share common properties that are critical for their efficacy but also impose significant limitations. Importantly, they lack selectivity within the Na$_V$ family, leading to block of all subtypes with approximately equal affinity. Perhaps because of this lack of molecular selectivity, all successful drugs exhibit use-dependent block and generally show higher affinity at depolarized potentials, resulting in the preferential targeting of actively firing neurons, a key factor in the therapeutic window of existing Na$_V$ blocking drugs. While every drug has a unique therapeutic profile, current Na$_v$ blockers are generally associated with central nervous system (CNS) and cardiovascular (CV) side-effects, which are often dose-limiting. Dizziness, sedation, nausea, ataxia, and confusion are some of the specific side-effects observed for phenytoin (**12**) [66, 67] (*Physicians' Desk Reference*, 56th edition 2002, pp. 2622–2627), mexiletine (**7**) [68] (*Physicians' Desk Reference*, 56th edition 2002, pp. 1047–1049), and lidocaine (**4**) (*Physicians' Desk Reference*, 56th edition 2002, pp. 653–657). At high doses CV side-effects have been observed that may be mechanism related, e.g., bupivacaine (**5**) has been associated with prolonged QRS interval which is potentiated with lidocaine and phenytoin [69].

Biodistribution and physicochemical properties can also greatly influence efficacy and therapeutic window. Most anticonvulsants [i.e., phenytoin (**12**) and lamotrigine (**1**)] were identified using animal models such as maximal electroshock (MES) that also incorporate elements of CNS penetration. Not surprisingly, therefore, most of these agents have good CNS exposure; however, the window between therapeutic antiepileptic levels and CNS side-effects is generally small. The anti-arrythmic agents lidocaine and mexiletine also have good CNS penetration. One approach to mitigate these CNS side-effects for indications that may not require CNS exposure, perhaps certain pain syndromes, is to design blockers that do not penetrate the blood–brain barrier (BBB). This can be challenging, as decreasing brain penetration may also reduce nerve and systemic bioavailability.

Another strategy for achieving functional selectivity *in vivo* is use of the specific biophysical, state-dependent, and kinetic mechanisms of block to target certain neuronal subtypes. For example, agents that target open channel block may functionally target hyperexcitable neurons since these Na_Vs have a greater concentration of this transient state. Alternatively, it is plausible to design a blocker that binds and dissociates rapidly from the target relative to action potential frequencies that may be associated with adverse effects, such as block of $Na_V1.5$, the principal cardiac subtype. The action potential in the heart lasts for hundreds of milliseconds, which is at least 40-fold longer than neuronal action potentials; thus such a compound might not cause accumulation of block in the heart, while favoring block of high frequency firing associated with epilepsy and pain.

Identification of molecules that show molecular subtype selectivity is also an attractive approach, targeting those subtypes that are believed to mediate the desired efficacy while sparing those subtypes (like $Na_V1.5$) that may be associated with undesirable side-effects. Despite the advantages of this approach, little progress has been made in identifying small, subtype selective, drug-like molecules, which is probably due to the high degree of homology between the family members (Chapter 6.1). While existing drugs do not demonstrate subtype selectivity and there has been little or no evidence that agents that target the local anesthetic binding site can be selective, a single amino acid variation between channel subtypes can modulate sensitivity to small molecules. For example, the single amino acid Cys374 in $Na_V1.5$ [70] and the equivalent position Ser356 in $Na_V1.8$ imparts TTX resistance. In addition, Na_Vs are very large proteins, and at least six distinct drug and toxin binding sites have been identified based on binding studies [10]. Advances in ion channel screening methods and heterologous expression methods should greatly improve the prospects of identifying selective blockers.

6.2.4
Compounds in Clinical Development

6.2.4.1 **Crobenetine and Derivatives",4>**

Crobenetine (**14**), an elaborated benzomorphan, is a very potent Na_V blocker, exhibiting a K_i of 43 nM against [³H]BTX binding in rat cerebral cortex synaptosome preparations, as well as an IC_{50} of 322 nM in veratridine-induced glutamate release from rat brain slices. The compound is also active in the mouse maximal electroshock model of epilepsy, and *in vivo* efficacy was correlated with functional Na_V blockade [71]. In the same study, a series of analogs were prepared exploring (1) regiochemistry of hydroxyl substitution on the benzomorphan core (4′ was superior in [³H]BTX and glutamate release assays), (2) stereochemical preference at position 1 [(*R*) as shown ~10× more potent], (3) impact of stereochemical modifications at 2″ (not significant for Na_V potency, but improved selectivity over the L-type Ca^{2+} channel by ~10×), and (4) role of the side chain present on nitrogen (aromatic substitution 3–4 atoms distal preferred). Ultimately, the op-

Crobenetine (BIII 890 CL)
14

timized compound BIII 890 CL (crobenetine, **14**) emerged for clinical development.

An additional study supports the conclusion that the compound is a potent, use-dependent Na_V blocker with neuroprotective properties [72], consistent with its clinical development for the treatment of ischemic stroke [73]. Analgesic properties have also been observed, producing a dose dependent and statistically significant reversal of mechanical hyperalgesia and immobility in the ankle joints of rats treated with complete Freund's adjuvant (CFA), suggesting potential utility in the treatment of pain and hyperalgesia. BIII 890 CL (**14**) was also moderately superior to mexiletine (**7**) by these same measures [74].

6.2.4.2 Ralfinamide/Safinamide

Ralfinamide (**15**) and safinamide (**16**) are both under active clinical investigation, the former for the treatment of neuropathic pain, the latter for epilepsy, Parkinson's disease, and restless legs syndrome [75]. Both compounds modulate Na_V *in vitro* [76–78]. Ralfinamide exhibits voltage dependent inhibition of Na_Vs in electrophysiological studies of DRG neurons [78], with IC_{50}s dependent on preconditioning membrane potentials, showing increased TTX-r block at more depolarized potentials, i.e., $V_{pre} = -90\,mV$ (72 µM) vs. $V_{pre} = -40\,mV$ (10 µM). Affinity for TTX-s channels were comparable when channels were free from inactivation, but a modest 3-fold selectivity for TTX-r over TTX-s is claimed when depolarizing conditions were used to increase channel inactivation. The steady-state inactivation curves of both TTX-r and TTX-s channels shift towards more hyperpolarizing potentials, and block was strongly use and frequency dependent [78]. Safinamide also modulates calcium channels and selectively inhibits monoamine oxidase type B, contributing to efficacy in animal models of neuroprotection, epilepsy, and Parkinson's [79].

Ralfinamide (NW-1029)
15

Safinamide (NW-1015, PNU-151774E)
16

Modifications to **15** and **16** modulate rat brain Na_v and sigma receptor binding affinities by up to 150-fold. For example, enantiomers at the alpha carbon of the carboxamide were equipotent, whereas sigma binding was preferentially attenuated in the (S) stereoisomers. Homologation of the benzyloxy linker with additional methylene units (up to $n = 5$) generally improved Na_v potency, whereas dearylation to generate the simple methoxy analog caused a ca. 150× loss in activity. Substitution of the benzyloxy oxygen with NH was tolerated, yielding a benzylamino analog of approximately equivalent potency and efficacy in the rodent MES model. Hydrolysis of the carboxamide to the acid also appeared detrimental to activity. All analogs were inactive in the [³H]STX binding assay [76], which is consistent with drugs known to bind to a separate local anesthetic binding site. Substantial SAR (structure–activity relationship) within the series was also disclosed in an earlier study of more than 70 structural variants, including efficacy in mouse epilepsy models (MES and BIC) and rotarod safety model. From these data, safinamide (**16**) emerged as a promising development candidate [80].

Other *in vivo* studies have reinforced the efficacy of safinamide (**16**) in animal models of epilepsy, as well as adjunctive clinical application in epilepsy and Parkinson's patients [75, 81]. In healthy volunteers, dose proportional oral pharmacokinetics have been observed with a half-life of ~22 h, no clinically relevant accumulation at steady state, and good overall tolerability [79]. Ralfinamide (**15**) has shown significant efficacy in rodent pain models due to its block of Na_vs, with oral ED_{50}s of ~0.5 mg kg^{-1} in the CFA inflammatory pain model and 0.7 mg kg^{-1} in the chronic constriction injury model (CCI). In the formalin model, an ED_{50} of 10 mg kg^{-1} was observed in the second phase [82].

6.2.4.3 **TTX and Derivatives**

Well known as a potent Na_v blocker, tetrodotoxin (TTX, **17**) provides an important pharmacological tool in differentiating amongst channels resistant (TTX-r: NaV1.5, 1.8, and 1.9) or sensitive (TTX-s: NaV1.1–1.4, 1.6, 1.7) to block by TTX. While therapeutic use has been limited by toxicity of systemic administration, local administration has enabled *in vivo* evaluation of TTX in both rodents and humans.

In the SNL model of neuropathic pain, topical administration of TTX (**17**) to ligated L5 DRG neurons via implanted tubing resulted in elevation of paw withdrawal threshold. Interestingly, the concentrations required (12.5–50 nM) for

Tetrodotoxin (TTX)
17

reduction of allodynic behaviors were lower than those necessary to block action potential conduction in normal rats, possibly due to upregulation of TTX-s sodium channels in the injured test animals. It was further determined that increases to paw withdrawal threshold were highest when TTX was administered directly to the L5 DRG neurons, lesser when administered to the T11 epidural space, and insignificant when dosed i.p., suggesting a localized effect not present with systemic administration [83]. TTX-induced suppression of spinal excitatory amino acid release (Glu, Asp) in the rat plantar incision model of post-surgical pain has also been observed in test animals [84].

Rat sciatic nerve block by TTX (**17**), saxitoxin (**18a**), and structural analogs (**18b**, **18c**) have been compared *in vivo*, and marked modulation of efficacy and toxicity were observed upon structural modification [85]. Rank ordering of efficacy was generally neoSTX (**18b**) > STX (**18a**) > TTX (**17**) > dcSTX (**18c**); therapeutic indices for the saxitoxins were similar, while TTX was slightly better. Overall, changes to the saxitoxin core resulted in significant differences in the incidence and duration of block, as well as toxicity. Interestingly, the EC_{50}s for nerve block are 3–4 orders of magnitude higher than *in vitro* K_Ds of TTX-s Na_vs and are summarized in Table 6.2.1.

Saxitoxin (R^1=H, R^2=$CONH_2$) **18a**
Neosaxitoxin (R^1=OH, R^2=$CONH_2$) **18b**
Decarbamoylsaxitoxin (R^1, R^2=H) **18c**

Tab. 6.2.1 EC_{50}s for nerve block.

	$EC_{50\text{-dur60}}$[a] (μmol L^{-1})	$EC_{50\text{-max}}$[b] (μmol L^{-1})	LD_{50} (nmol kg^{-1})	Therapeutic index[c] relative to $EC_{50\text{-max}}$
TTX (**17**)	92 ± 5	76 ± 11	41 ± 2	1.82
STX (**18a**)	58 ± 3	51 ± 4	23 ± 1	1.49
neoSTX (**18b**)	34 ± 2	30 ± 1	14 ± 1	1.61
dcSTX (**18c**)	268 ± 8	202 ± 22	94 ± 5	1.59

[a] Median concentration required for thermal nociceptive block lasting 60 min.
[b] Median concentration required for maximal thermal nociceptive block. LD_{50}: median lethal dose.
[c] Therapeutic index: LD_{50}/ED_{50}.

In a preliminary open label Phase IIa assessment, severe, refractory cancer pain in humans has also been treated with intramuscular injections of TTX. Twenty-two subjects were treated with 15, 30, 45, or 60 µg TTX for four consecutive days and evaluated using the Brief Pain Inventory. In 68% of administered treatments, significant analgesia was observed, suggesting compelling therapeutic potential for Na_V blockers for the treatment of pain. Development of a subcutaneous administration regimen and broader clinical evaluations are planned [86].

6.2.5
New Blockers in Discovery or Pre-clinical Stage

New structural classes of Na_V blockers have been discovered recently using binding and functional *in vitro* cell-based screening methods (see *Expression and Analysis of Recombinant Ion Channels*, Eds. J. J. Clare and D. J. Trezise, Wiley-VCH, Weinheim, Germany, 2006. High-throughput screening technologies over the last decade have greatly improved the ability to identify novel Na_V blockers. Before these advances, most efforts focused on targeted exploration around known drugs.

6.2.5.1 Molecules Related to Known Classical Sodium Channel Blockers

Mexiletine (**7**) and its analogues have been revisited recently [87], in particular regarding the effects of compound stereochemistry on block of sodium currents in perfused skeletal muscle fibers and voltage clamp recordings. While a clear preference for one enantiomer over the other could not be established for mexiletine itself, it was subsequently demonstrated, via analysis of the phenol metabolite **20**, that the (R) configuration led to more potent Na_V block than the (S) for the mexiletine structural analog **19** [88].

Mexiletine was also used as a starting point for efforts reported by Theravance. Compound activity was assessed in rat cerebellar granule neurons using an intracellular Ca readout stimulated with veratridine and monitored on a FLIPR (US20040063760 & WO2002057215). A resulting compound (**21**) showed efficacy in SNL the model at $30 \, mg \, kg^{-1}$. *In vitro* selectivity data, stereochemical preferences, and structure–activity relationships (SARs) were not disclosed.

Boehringer-Ingelheim has reported a chemical series also containing a 2,6-disubstituted phenyl ring and pendant amine feature reminiscent of mexiletine (WO2002016308). The compounds are described as being use-dependent, displa-

19   20   21

cing BTX on TTX-sensitive sodium channels. The screening assay used veratridine to activate Na_vs and subsequent glutamate release as the readout. Compound activities were reported to be < 1000 nM. 22 is a representative example from the structural class.

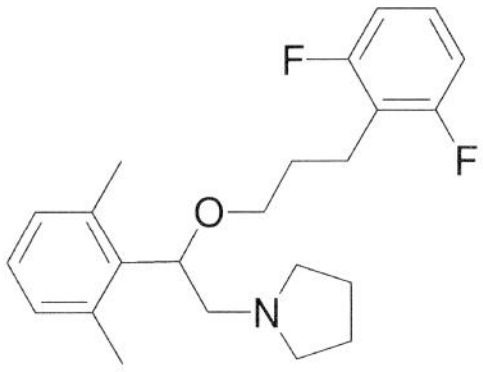

22

Lidocaine, benzocaine, and other local anesthetics contain basic tertiary amines in their structures, a critical element for their binding to Na_vs. A hybrid approach has been described in which analogues of phenytoin, which inhibited [^{3}H]BTX binding with potencies between 5 and 500 μM, were improved when a distal piperidine was incorporated into the molecule. The piperidine fragment is seen in many other Na_v/Ca_v blockers, such as flunarizine. For example, **23** has a 200 nM IC_{50} [89].

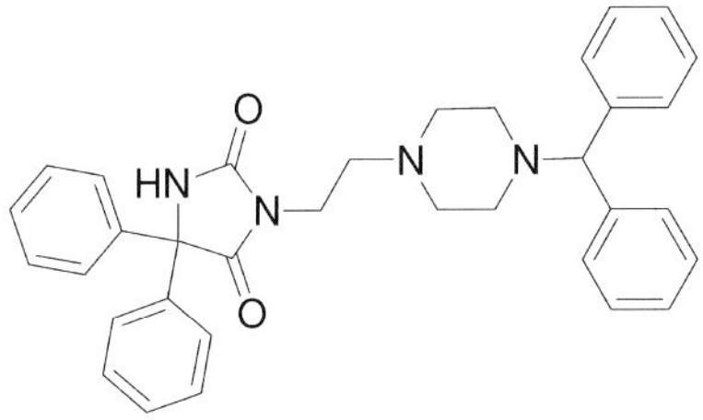

23

The high potency of **23** could be derived from the piperidine, as opposed to the phenytoin moiety, as it was shown recently that 1,1-diphenyl substituted piperidines yield sub-micromolar molecules in a [^{3}H]BTX binding assay (400 nM for **24**); further optimization led to **25** (150 nM) [90]. SAR analysis suggested that electron-withdrawing groups on R_1 (in particular F) improved the activity.

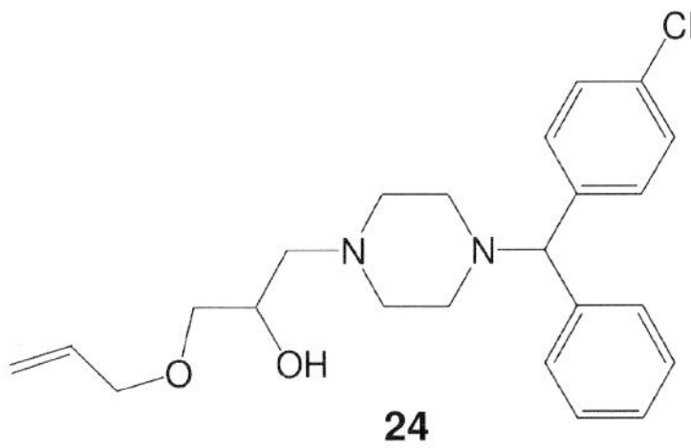

24

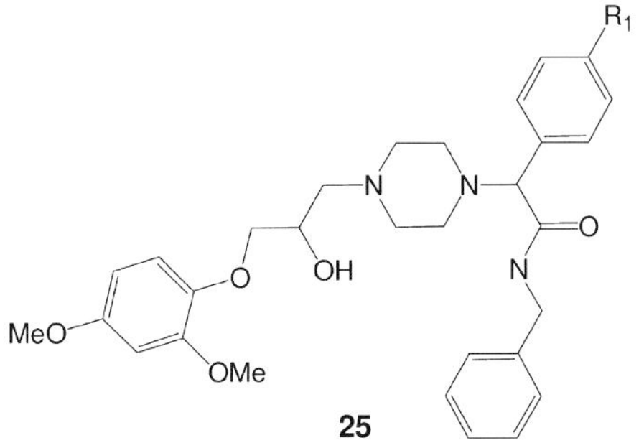

25

6.2.5.2 **Novel Structural Classes**

A novel carboline series has been reported that showed Na$_V$ block at 1.5 μM, compared to 56 μM for lidocaine (**4**). Interestingly, it was possible to replace the carboline core in **26**, thereby generating an alternative substituted imidazole class **27** that, like the piperidines **24** and **25** described above, contains basic features. Increasing the log*P* of the molecules improved potency, with the best examples showing binding assay IC$_{50}$s of ~8 and ~1000 nM activity in a veratridine-induced cytoprotection assay. The SAR described was limited to lipophilic side chains [91].

26   27

Researchers at Ionix (WO 2005/005392 & WO 2005/000309) have described two additional classes of Na$_V$ blockers, with particular emphasis on activity against Na$_V$1.8. A potent series of α-amino amides **28** were discovered with IC$_{50}$s of ~190 nM, utilizing a Na$_V$1.8 SH-SY-5Y neuroblastoma cell line and a FLIPR membrane potential assay, requiring TTX to block endogenous TTX-s Na$_V$s and 10 μM pyrethroid to prevent channel inactivation. In a second patent the aminoalkyl thiazoles were disclosed, with some of the more potent compounds exhibiting IC$_{50}$s of ~500 nM. (**29**). Although all examples described show the (S) configuration at the stereogenic center, SAR relating to stereochemistry was not discussed.

28   29

Sodium channel modulators have also been explored by Yamanouchi (WO 2004/078715 & WO 2004/011430), leading to compounds described as possessing analgesic activity in models of neuropathic pain and diabetic neuropathy, with a reduced side effect profile. Compound **30**, for example, inhibited the veratridine-induced uptake of [^{14}C]guanidine in rat brain tissue with an IC$_{50}$ of 1.1 μM. Additional structural analogs described are also shown (**31**).

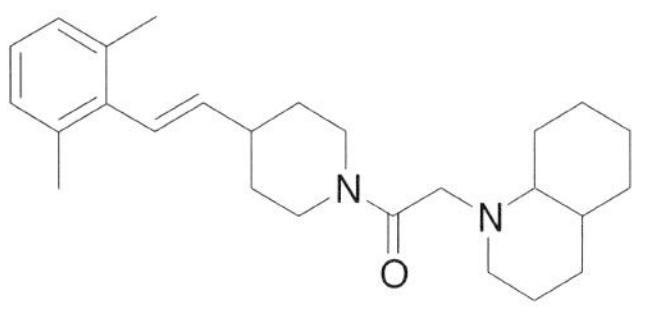

30

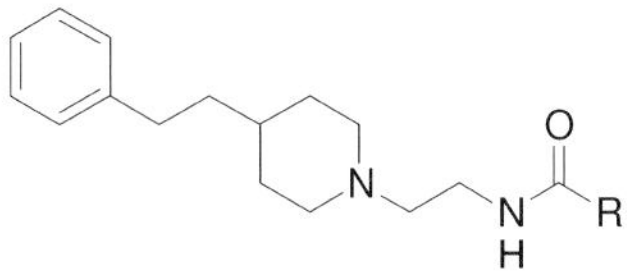

31

A distinct broad class of compounds emerging as Na_V modulators are either neutral or contain only weakly basic structural features, in contrast to the more basic class described above. In an important sub-class, structures contain lipophilic tails linked to six- or five-membered heterocycles, which in some cases have carboxamide substitutions. They are derived from V102862 (**32**), a potent, state-dependent anticonvulsant originally described by Dimmock et al. [92, 93]. Scientists at Purdue used this molecule ($IC_{50} \sim 600$ nM) as a starting point for further optimization.

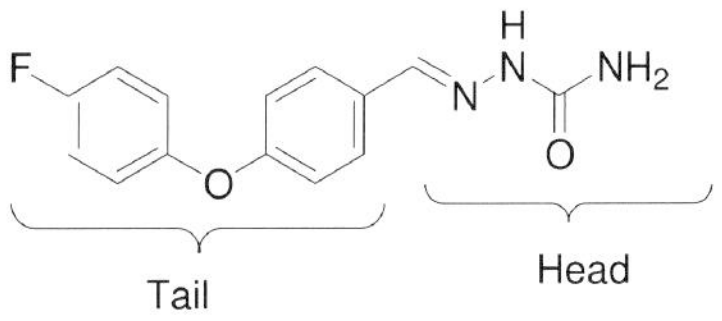

Tail

Head

V102862
32

Compound activity was tested against rat brain $Na_V1.2$ expressed in HEK-293 cells and rat DRG neurons using whole-cell patch clamp. V102862 (**32**) itself is a semicarbazide that exhibits generally low solubility and a poor metabolic profile. The series was subsequently modified to give aryloxyphenyl pyridines that maintain the spatial relationship between the aryloxyphenyl moiety and the carboxa-

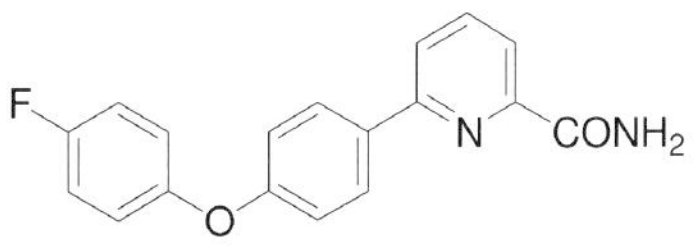

33
$K_i = 96$ nM

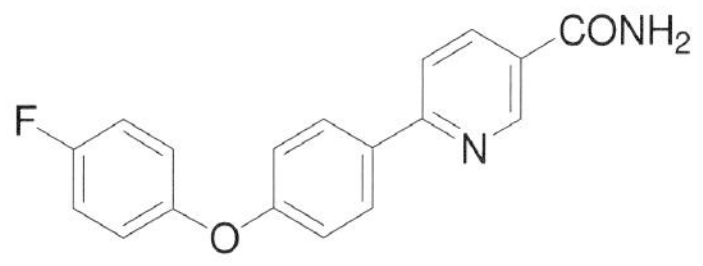

34
$K_i = 3750$ nM

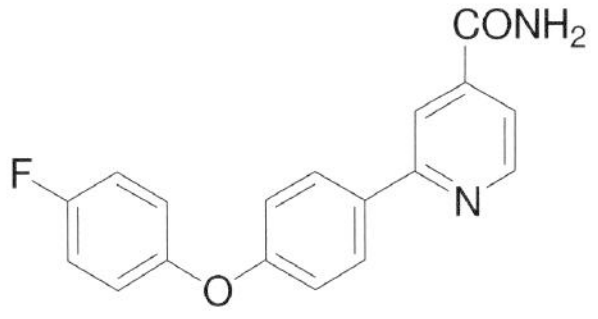

35
$K_i = 2640$ nM

mide, while offering more drug-like properties. SARs within the class were recently described, demonstrating that the regiochemistry of the pyridine nitrogen could be altered, although some loss of activity was observed relative to the optimal substitution pattern shown (**33–35**). Substitutions on the 4-position of the pyridine were tolerated, although with some loss of potency. Activity dropped significantly when the distal aryloxy group was removed, or if the carboxamide moiety was shifted to the 3- or 4-positions (**34**, **35**) [94].

The lead structure **36** (K_i = 123 nM) was tested orally *in vivo* in the SNL model of neuropathic pain, giving a minimum effective dose (MED) of 3 mg kg^{-1} in measurements of mechanical allodynia, compared to the carbamazepine MED of 100 mg kg^{-1}.

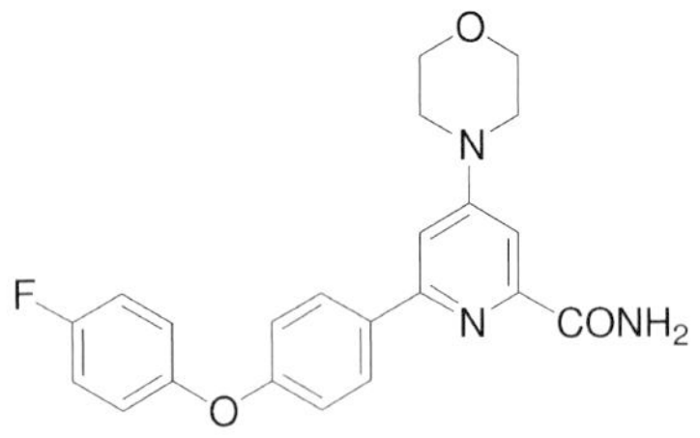

36

Other structures sharing the *p*-fluorophenoxyphenyl tail have been described [95]. Of the three representative structures **37–39**, the pyrazoles **37** were profiled in some detail. For example, the N-carboxamide substituent on **37** improved potency over the methyl version, and Na$_V$ modulation was state dependent. Compound **37** (K_i = 31 nM) was tested in the SNL model, showing efficacy at 10 mg kg^{-1} p.o. [95].

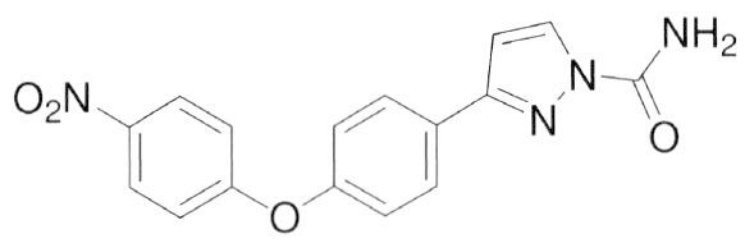

37

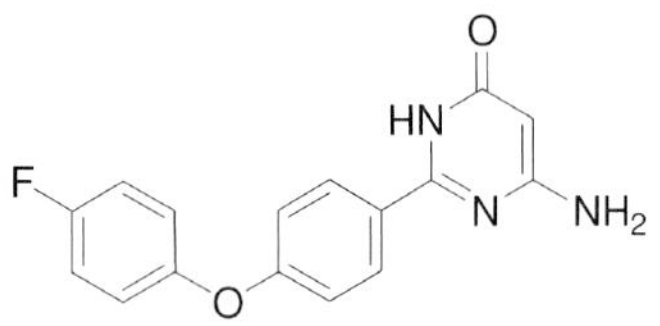

38

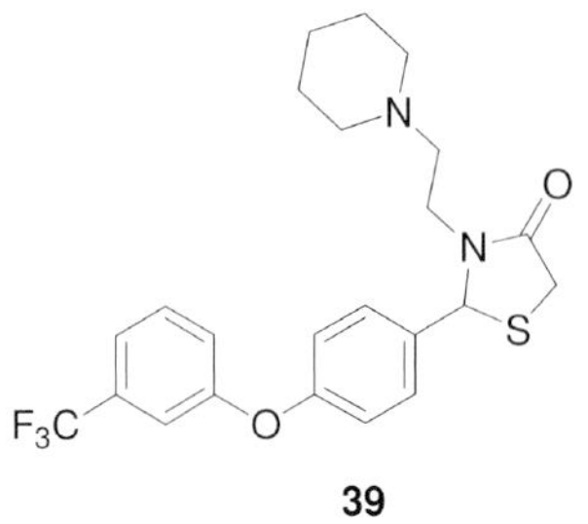

39

Additional examples incorporating additional five- and six-membered heterocycles have also been described recently by Merck (WO 2004/084824, WO 2004/083189, WO 2004/083190, WO 2004/094395). Specific data were not disclosed; compounds **40–42** are representative structures.

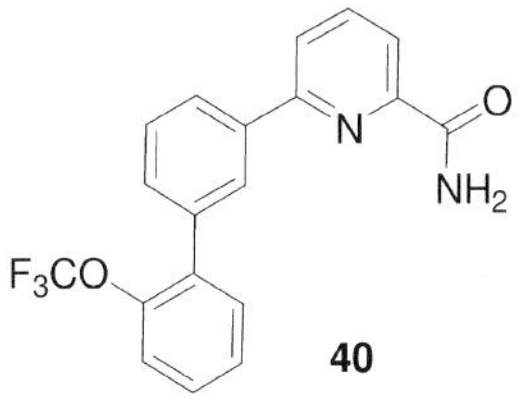

40

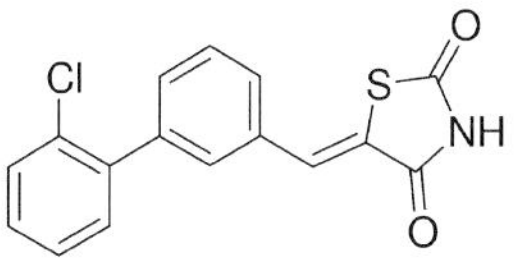

41
WO 2004/024061

42
WO 2004/092140

Researchers at Merck have also described recently cyclopentane carboxamides containing a lipophilic biaryl tail [96]. These molecules were derived from BPBTS (**43**), which exhibited an IC_{50} of 150 nM for $Na_V1.7$ in a voltage-ion-probe-reader (VIPR) membrane potential assay and a K_i of 150 nM for $Na_V1.7$ as measured by electrophysiology. Additional patch clamp experiments demonstrated that the progenitor **43** also inhibited $Na_V1.2$ and $Na_V1.5$ [97].

Modifications to **43** were systematically investigated, leading to the substituted succinyl amide linker, with the most potent example being approximately equipotent to BPBTS as measured by electrophysiology. Methylation of the amide NH or substitution on the benzylic position was detrimental, though it ultimately proved possible to replace the bithiophene moiety entirely with a *p*-trifluoromethoxyphenacyl substituent. Methylation of the sulfonamide substituent was tolerated, and while substitution at the 2-position occupied prototypically by the sulfonamide was critical, the moiety itself could be replaced. The specific optimized structure **44** showed bioavailability of 44% and was tested orally in inflammatory and neuropathic pain models. In the formalin model, the molecule showed reduction in pain behavior of 29% at 30 mg kg^{-1} compared to 33% for mexiletine (**7**) at 100 mg kg^{-1}, as well as 35% reversal in the SNL model compared to 34% for mexiletine.

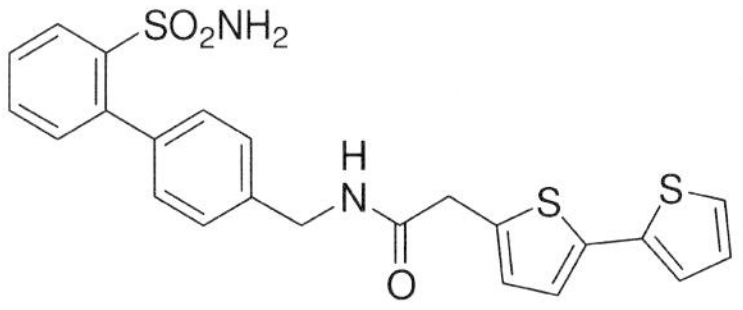

43

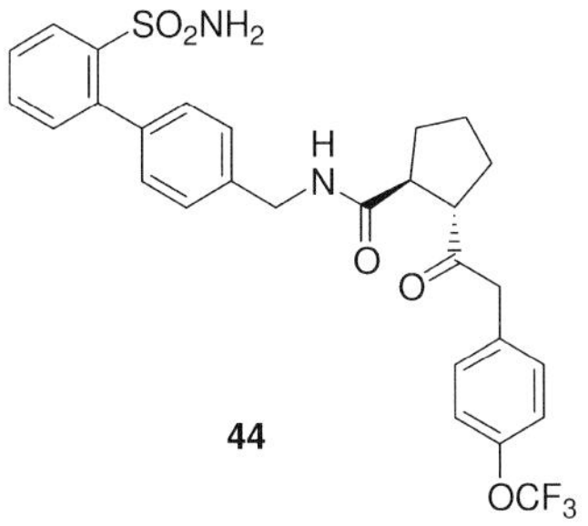

44

6.2.5.3 Other Therapeutic Agents with NaV Activity

In the last 15 years there has been accumulating evidence that some drugs developed for targets and indications not traditionally believed to be associated with Na_V block do in fact have Na_V activity that may be therapeutically relevant. One such prominent class is the tricyclic antidepressants. In particular, amitriptyline has been shown to block Na_Vs in a use-dependent manner [98]. The therapeutic blood plasma concentrations for treating pain, ~ 0.2 to $1\,\mu M$ [99], are consistent with observed Na_V blocking activity [100]. Furthermore, desipramine, which does not have potent serotonin reuptake activity, is equally efficacious in treating pain, suggesting that norepinephrine reuptake inhibition may be more important for activating the descending inhibitory pain pathway. In addition to Na_V blocking activity, amitriptyline has many other reported activities and these are also potential explanations for the analgesic activity. The atypical antidepressant fluoxetine has also demonstrated Na_V blocking activity [98].

6.2.6
Emerging Indications and Future Directions

The application of Na_v blockers for treating pain has garnered a good deal of interest. The target class has been generally validated in man through the activity of local anesthetics, antiarrhythmics, and anticonvulsants, which has helped to spur the development of multiple drug candidates currently in clinical evaluation for various pain indications, including most prominently neuropathic pain. Recent identification of individual subtypes, distinct expression patterns, and further biophysical characterization has further fueled interest in more functionally selective molecules and the promise of improved drugs. The expression patterns of Na_vs are modulated in various pain states and are believed to be attractive pain targets [13, 14]. More generally, Na_V blockers have the potential to treat other conditions where hyperactive sensory systems may contribute to symptoms, such as overactive bladder.

More nascent areas are emerging from better understanding the roles Na_Vs play in sensory systems and in neurodegenerative disease states, e.g., there have been increasing reports of Na_V dysregulation in demyelinating disease states and models [101–104], and thus Na_Vs have been proposed as drug targets in multiple sclerosis (MS) [105]. $Na_V1.8$ has been reported to be upregulated in the experimental autoimmune encephalomyelitis (EAE) animal models and in cerebellum of people afflicted with MS [106, 107]. Phenytoin [60, 61, 108] and flecainide [109], an Na_V blocking anti-arrythmic agent, protect axons in EAE animal models. It has also been proposed that Na_Vs contribute to the microglia/macrophage activation. From an EAE model, inflammatory cell infiltration was blocked 75% by phenytoin and phagocytic function was partially blocked with TTX, with $Na_V1.6$ implicated as a possible target [110]. Furthermore, flecanide was shown to also protect axonal degeneration in an experimental autoimmune neuritis (EAN) model [111], further supporting a role for Na_Vs in inflammatory demyelinating conditions.

With advances in the molecular characterization and physiological functions of Na$_v$s, combined with screening technological breakthroughs, the prospects for developing novel and more specific drug candidates are excellent. High-throughput fluorescent, flux, binding, and electrophysiological screening technologies are clearly having a significant impact in the discovery of Na$_v$ blockers, as evidenced by the increasing number of new clinical candidates and patent applications describing new classes of compounds. As the properties, function, and regulation of individual subtypes become better understood, the ability to develop more selective agents that retain the desired pharmacological activities while eliminating those undesirable will increasingly be more feasible, and will contribute a new chapter to the classical pharmacology of sodium channels.

References

1. Ruetsch, Y.A., T. Boni, A. Borgeat, From cocaine to ropivacaine: the history of local anesthetic drugs. *Curr. Top. Med. Chem.*, **2001**. 1(3), 175–182.
2. Nagle, R.E., J. Pilcher, Lignocaine for arrythmias. *Lancet*, **1968**. 1(7550), 1039.
3. Owen, M.D., L.S. Dean, Ropivacaine. *Expert Opin. Pharmacother.*, **2000**. 1(2), 325–336.
4. Hille, B., Common mode of action of three agents that decrease the transient change in sodium permeability in nerves. *Nature*, **1966**. 210(42), 1220–1222.
5. Taylor, R.E., Effect of procaine on electrical properties of squid axon membrane. *Am. J. Physiol.*, **1959**. 196(5), 1071–1078.
6. Weidmann, S., Effects of calcium ions and local anesthetics on electrical properties of Purkinje fibres. *J. Physiol.*, **1955**. 129(3), 568–582.
7. Putnam, T.J., H.H. Merritt, Experimental determination of the anticonvulsant properties of phenyl derivatives. *Science*, **1937**. 85, 525–526.
8. Willow, M., T. Gonoi, W.A. Catterall, Voltage clamp analysis of the inhibitory actions of diphenylhydantoin and carbamazepine on voltage-sensitive sodium channels in neuroblastoma cells. *Mol. Pharmacol.*, **1985**. 27(5), 549–558.
9. Worley, P.F., J.M. Baraban, Site of anticonvulsant action on sodium channels: autoradiographic and electrophysiological studies in rat brain. *Proc. Natl. Acad. Sci. U.S.A.*, **1987**. 84(9), 3051–3055.
10. Clare, J.J., et al., Voltage-gated sodium channels as therapeutic targets. *Drug Discov. Today*, **2000**. 5(11), 506–520.
11. Hodgkin, A.L., A.F. Huxley, A quantitative description of membrane current and its application to conduction and excitation in nerve. *J. Physiol.*, **1952**. 117(4), 500–544.
12. Narahashi, T., J.W. Moore, W.R. Scott, Tetrodotoxin blockage of sodium conductance increase in lobster giant axons. *J. Gen. Physiol.*, **1964**. 47, 965–974.
13. Lai, J., et al., Voltage-gated sodium channels and hyperalgesia. *Annu. Rev. Pharmacol. Toxicol.*, **2004**. 44, 371–397.
14. Wood, J.N., et al., Voltage-gated sodium channels and pain pathways. *J. Neurobiol.*, **2004**. 61(1), 55–71.
15. Anger, T., et al., Medicinal chemistry of neuronal voltage-gated sodium channel blockers. *J. Med. Chem.*, **2001**. 44(2), 115–137.
16. Catterall, W.A., Molecular mechanisms of gating and drug block of sodium channels. *Novartis Found Symp.*, **2002**. 241, 206–218; discussion 218–232.
17. Nau, C., G.K. Wang, Interactions of local anesthetics with voltage-gated Na$^+$ channels. *J. Membr. Biol.*, **2004**. 201(1), 1–8.
18. Scholz, A., Mechanisms of (local) anaesthetics on voltage-gated sodium and other ion channels. *Br. J. Anaesth.*, **2002**. 89(1), 52–61.
19. Ritchie, J.M., R.B. Rogart, The binding of saxitoxin and tetrodotoxin to excitable

tissue. *Rev. Physiol. Biochem. Pharmacol.*, **1977**. 79, 1–50.

20. Postma, S.W., W.A. Catterall, Inhibition of binding of [3H]batrachotoxinin A 20–alpha-benzoate to sodium channels by local anesthetics. *Mol. Pharmacol.*, **1984**. 25(2), 219–227.

21. Courtney, K.R., Mechanism of frequency-dependent inhibition of sodium currents in frog myelinated nerve by the lidocaine derivative GEA. *J. Pharmacol. Exp. Therapeut.*, **1975**. 195(2), 225–236.

22. Yarov-Yarovoy, V., et al., Molecular determinants of voltage-dependent gating and binding of pore-blocking drugs in transmembrane segment IIIS6 of the Na(+) channel alpha subunit. *J. Biol. Chem.*, **2001**. 276(1), 20–27.

23. Cronin, N.B., et al., Binding of the anticonvulsant drug lamotrigine and the neurotoxin batrachotoxin to voltage-gated sodium channels induces conformational changes associated with block and steady-state activation. *J. Biol. Chem.*, **2003**. 278(12), 10 675–10 682.

24. Trezise, D.J., V.H. John, X.M. Xie, Voltage- and use-dependent inhibition of Na⁺ channels in rat sensory neurones by 4030W92, a new antihyperalgesic agent. *Br. J. Pharmacol.*, **1998**. 124(5), 953–963.

25. Blackburn-Munro, G., N. Ibsen, H.K. Erichsen, A comparison of the anti-nociceptive effects of voltage-activated Na⁺ channel blockers in the formalin test. *Eur. J. Pharmacol.*, **2002**. 445(3), 231–238.

26. Wallace, M.S., S. Quessy, G. Schulteis, Lack of effect of two oral sodium channel antagonists, lamotrigine and 4030W92, on intradermal capsaicin-induced hyperalgesia model. *Pharmacol. Biochem. Behav.*, **2004**. 78(2), 349–355.

27. Wallace, M.S., et al., A multicenter, double-blind, randomized, placebo-controlled crossover evaluation of a short course of 4030W92 in patients with chronic neuropathic pain. *J. Pain*, **2002**. 3(3), 227–233.

28. Mirza, N.R., et al., Lamotrigine has an anxiolytic-like profile in the rat conditioned emotional response test of anxiety: a potential role for sodium channels? *Psychopharmacology (Berl)*, **2005**. 180 (1), 159–168.

29. Hahn, C.G., et al., The current understanding of lamotrigine as a mood stabilizer. *J. Clin. Psychiatry*, **2004**. 65(6), 791–804.

30. Hurley, S.C., Lamotrigine update and its use in mood disorders. *Ann. Pharmacother.*, **2002**. 36(5), 860–873.

31. Ketter, T.A., H.K. Manji, R.M. Post, Potential mechanisms of action of lamotrigine in the treatment of bipolar disorders. *J. Clin. Psychopharmacol.*, **2003**. 23(5), 484–495.

32. Spina, E., G. Perugi, Antiepileptic drugs: indications other than epilepsy. *Epileptic Disord*, **2004**. 6(2), 57–75.

33. Tsang, S.Y., et al., A multifunctional aromatic residue in the external pore vestibule of Na⁺ channels contributes to the local anesthetic receptor. *Mol. Pharmacol.*, **2005**. 67(2), 424–434.

34. Sandtner, W., et al., Lidocaine: a foot in the door of the inner vestibule prevents ultra-slow inactivation of a voltage-gated sodium channel. *Mol. Pharmacol.*, **2004**. 66(3), 648–657.

35. Sheets, M.F., D.A. Hanck, Molecular action of lidocaine on the voltage sensors of sodium channels. *J. Gen. Physiol.*, **2003**. 121(2), 163–175.

36. Zapata-Sudo, G., et al., Is comparative cardiotoxicity of S(-) and R(+) bupivacaine related to enantiomer-selective inhibition of L-type Ca(2+) channels? *Anesth. Analg.*, **2001**. 92(2), 496–501.

37. Nau, C., et al., Block of human heart hH1 sodium channels by the enantiomers of bupivacaine. *Anesthesiology*, **2000**. 93(4), 1022–1033.

38. Kawagoe, H., et al., Molecular basis for exaggerated sensitivity to mexiletine in the cardiac isoform of the fast Na channel. *FEBS Lett.*, **2002**. 513(2–3), 235–241.

39. Shiraishi, S., et al., Differential effects of bupivacaine enantiomers, ropivacaine and lidocaine on up-regulation of cell surface voltage-dependent sodium channels in adrenal chromaffin cells. *Brain Res.*, **2003**. 966(2), 175–184.

40. Osawa, Y., et al., The effects of class Ic antiarrhythmics on tetrodotoxin-resistant Na⁺ currents in rat sensory neurons. *Anesth Analg*, **2004**. 99(2), 464–471, table of contents.

41. Scholz, A., W. Vogel, Tetrodotoxin-resistant action potentials in dorsal root

ganglion neurons are blocked by local anesthetics. *Pain*, **2000**. 89(1), 47–52.

42. Chevrier, P., K. Vijayaragavan, M. Chahine, Differential modulation of Nav1.7 and Nav1.8 peripheral nerve sodium channels by the local anesthetic lidocaine. *Br. J. Pharmacol.*, **2004**. 142(3), 576–584.

43. Castaneda-Castellanos, D.R., et al., Lidocaine stabilizes the open state of CNS voltage-dependent sodium channels. *Brain Res. Mol. Brain Res.*, **2002**. 99(2), 102–113.

44. Mao, J., L.L. Chen, Systemic lidocaine for neuropathic pain relief. *Pain*, **2000**. 87(1), 7–17.

45. Strichartz, G.R., et al., Therapeutic concentrations of local anaesthetics unveil the potential role of sodium channels in neuropathic pain. *Novartis Found Symp.*, **2002**. 241, 189–201; discussion 202–205, 226–232.

46. Argoff, C.E., New analgesics for neuropathic pain: the lidocaine patch. *Clin. J. Pain*, **2000**. 16(2 Suppl), S62–S66.

47. Argoff, C.E., N. Katz, M. Backonja, Treatment of postherpetic neuralgia: a review of therapeutic options. *J. Pain Symptom Manage.*, **2004**. 28(4), 396–411.

48. Davies, P.S., B.S. Galer, Review of lidocaine patch 5% studies in the treatment of postherpetic neuralgia. *Drugs*, **2004**. 64(9), 937–947.

49. Burch, F., et al., Lidocaine patch 5% improves pain, stiffness, and physical function in osteoarthritis pain patients. A prospective, multicenter, open-label effectiveness trial. *Osteoarthritis Cartilage*, **2004**. 12(3), 253–255.

50. Galer, B.S., et al., Topical lidocaine patch 5% may target a novel underlying pain mechanism in osteoarthritis. *Curr. Med. Res. Opin.*, **2004**. 20(9), 1455–1458.

51. Benes, J., et al., Anticonvulsant and sodium channel-blocking properties of novel 10,11-dihydro-5H-dibenz [b,f]azepine-5-carboxamide derivatives. *J. Med. Chem.*, **1999**. 42(14), 2582–2587.

52. Learmonth, D.A., et al., Synthesis, anticonvulsant properties and pharmacokinetic profile of novel 10,11-dihydro-10-oxo-5H-dibenz/b,f/azepine-5-carboxamide derivatives. *Eur. J. Med. Chem.*, **2001**. 36(3), 227–236.

53. Ambrosio, A.F., et al., Inhibition of glutamate release by BIA 2-093 and BIA 2-024, two novel derivatives of carbamazepine, due to blockade of sodium but not calcium channels. *Biochem. Pharmacol.*, **2001**. 61(10), 1271–1275.

54. Ambrosio, A.F., et al., Mechanisms of action of carbamazepine and its derivatives, oxcarbazepine, BIA 2-093, and BIA 2-024. *Neurochem Res.*, **2002**. 27(1–2), 121–130.

55. Otoom, S.A., K.A. Alkadhi, Action of carbamazepine on epileptiform activity of the verartidine model in CA1 neurons. *Brain Res.*, **2000**. 885(2), 289–294.

56. Reckziegel, G., et al., Carbamazepine effects on Na$^+$ currents in human dentate granule cells from epileptogenic tissue. *Epilepsia*, **1999**. 40(4), 401–407.

57. Parada, A., P. Soares-da-Silva, The novel anticonvulsant BIA 2-093 inhibits transmitter release during opening of voltage-gated sodium channels: a comparison with carbamazepine and oxcarbazepine. *Neurochem. Int.*, **2002**. 40(5), 435–440.

58. Hains, B.C., et al., Sodium channel blockade with phenytoin protects spinal cord axons, enhances axonal conduction, and improves functional motor recovery after contusion SCI. *Exp. Neurol*, **2004**. 188(2), 365–377.

59. Kaptanoglu, E., et al., Blockade of sodium channels by phenytoin protects ultrastructure and attenuates lipid peroxidation in experimental spinal cord injury. *Acta Neurochir (Wien)*, **2005**. 147(4), 405–412.

60. Lo, A.C., J.A. Black, S.G. Waxman, Neuroprotection of axons with phenytoin in experimental allergic encephalomyelitis. *Neuroreport*, **2002**. 13(15), 1909–1912.

61. Lo, A.C., et al., Phenytoin protects spinal cord axons and preserves axonal conduction and neurological function in a model of neuroinflammation in vivo. *J. Neurophysiol.*, **2003**. 90(5), 3566–3571.

62. Bonoczk, P., et al., Role of sodium channel inhibition in neuroprotection: effect of vinpocetine. *Brain Res. Bull.*, **2000**. 53(3), 245–254.

63. Zhou, X., et al., Vinpocetine is a potent blocker of rat NaV1.8 tetrodotoxin-resistant sodium channels. *J. Pharmacol. Exp. Ther.*, **2003**. 306(2), 498–504.

64. Sitges, M., V. Nekrassov, Vinpocetine selectively inhibits neurotransmitter release triggered by sodium channel activation. *Neurochem. Res.*, **1999**. 24(12), 1585–1591.

65. Sitges, M., V. Nekrassov, Vinpocetine prevents 4-aminopyridine-induced changes in the EEG, the auditory brainstem responses and hearing. *Clin. Neurophysiol.*, **2004**. 115(12), 2711–2717.

66. Cohen, A.F., et al., Lamotrigine (BW430C), a potential anticonvulsant. Effects on the central nervous system in comparison with phenytoin and diazepam. *Br. J. Clin. Pharmacol.*, **1985**. 20(6), 619–629.

67. Iivanainen, M., H. Savolainen, Side effects of phenobarbital and phenytoin during long-term treatment of epilepsy. *Acta Neurol. Scand. Suppl.*, **1983**. 97, 49–67.

68. Jarvis, B., A.J. Coukell, Mexiletine. A review of its therapeutic use in painful diabetic neuropathy. *Drugs*, **1998**. 56(4), 691–707.

69. Simon, L., et al., Bupivacaine-induced QRS prolongation is enhanced by lidocaine and by phenytoin in rabbit hearts. *Anesth. Analg.*, **2002**. 94(1), 203–207, table of contents.

70. Satin, J., et al., A mutant of TTX-resistant cardiac sodium channels with TTX-sensitive properties. *Science*, **1992**. 256(5060), 1202–1205.

71. Grauert, M., et al., Synthesis and structure-activity relationships of 6,7-benzomorphan derivatives as use-dependent sodium channel blockers for the treatment of stroke. *J. Med. Chem.*, **2002**. 45(17), 3755–3764.

72. Carter, A.J., et al., Potent blockade of sodium channels and protection of brain tissue from ischemia by BIII 890 CL. *Proc. Natl. Acad. Sci. U.S.A.*, **2000**. 97(9), 4944–4949.

73. Meythaler, J., BIII-890-CL. Boehringer Ingelheim. *Curr. Opin. Investig. Drugs*, **2002**. 3(12), 1733–1735.

74. Laird, J.M., et al., Analgesic activity of a novel use-dependent sodium channel blocker, crobenetine, in mono-arthritic rats. *Br. J. Pharmacol.*, **2001**. 134(8), 1742–1748.

75. Bialer, M., et al., Progress report on new antiepileptic drugs: a summary of the Seventh Eilat Conference (EILAT VII). *Epilepsy Res.*, **2004**. 61(1–3), 1–48.

76. Pevarello, P., et al., Sodium channel activity and sigma binding of 2-aminopropanamide anticonvulsants. *Bioorg. Med. Chem. Lett.*, **1999**. 9(17), 2521–2524.

77. Salvati, P., et al., Biochemical and electrophysiological studies on the mechanism of action of PNU-151774E, a novel antiepileptic compound. *J. Pharmacol. Exp. Ther.*, **1999**. 288(3), 1151–1159.

78. Stummann, T.C., et al., The anti-nociceptive agent ralfinamide inhibits tetrodotoxin-resistant and tetrodotoxin-sensitive Na^+ currents in dorsal root ganglion neurons. *Eur. J. Pharmacol.*, **2005**. 510(3), 197–208.

79. Marzo, A., et al., Pharmacokinetics and pharmacodynamics of safinamide, a neuroprotectant with antiparkinsonian and anticonvulsant activity. *Pharmacol. Res.*, **2004**. 50(1), 77–85.

80. Pevarello, P., et al., Synthesis and anticonvulsant activity of a new class of 2-[(arylalky)amino]alkanamide derivatives. *J. Med. Chem.*, **1998**. 41(4), 579–590.

81. Bialer, M., et al., Progress report on new antiepileptic drugs: a summary of the Sixth Eilat Conference (EILAT VI). *Epilepsy Res.*, **2002**. 51(1–2), 31–71.

82. Veneroni, O., et al., Anti-allodynic effect of NW-1029, a novel Na(+) channel blocker, in experimental animal models of inflammatory and neuropathic pain. *Pain*, **2003**. 102(1–2), 17–25.

83. Lyu, Y.S., et al., Low dose of tetrodotoxin reduces neuropathic pain behaviors in an animal model. *Brain Res.*, **2000**. 871(1), 98–103.

84. Zahn, P.K., K.A. Sluka, T.J. Brennan, Excitatory amino acid release in the spinal cord caused by plantar incision in the rat. *Pain*, **2002**. 100(1–2), 65–76.

85. Kohane, D.S., et al., The local anesthetic properties and toxicity of saxitonin homologues for rat sciatic nerve block in vivo. *Reg. Anesth. Pain Med.*, **2000**. 25(1), 52–59.

86. Tetrodotoxin is safe and effective for severe, refractory cancer pain. *J. Support Oncol.*, **2004**. 2(1), 18.

87. Franchini, C., et al., Optically active mexiletine analogues as stereoselective blockers of voltage-gated Na(+) channels. *J. Med. Chem.*, **2003**. 46(24), 5238–5248.

88. Catalano, A., et al., Stereospecific synthesis of "para-hydroxymexiletine" and sodium channel blocking activity evaluation. *Chirality*, **2004**. 16(2), 72–78.

89. Zha, C., G.B. Brown, W.J. Brouillette, Synthesis and structure-activity relationship studies for hydantoins and analogues as voltage-gated sodium channel ligands. *J. Med. Chem.*, **2004**. 47(26), 6519–6528.

90. Ashwell, M.A., et al., The design, preparation and SAR of novel small molecule sodium (Na⁺) channel blockers. *Bioorg. Med. Chem. Lett.*, **2004**. 14, 2025–2030

91. Liberatore, A.M., et al., 2-Alkyl-4-arylimidazoles: structurally novel sodium channel modulators. *Bioorg. Med. Chem. Lett.*, **2004**. 14(13), 3521–3523.

92. Dimmock, J.R., et al., Anticonvulsant activities of some arylsemicarbazones displaying potent oral activity in the maximal electroshock screen in rats accompanied by high protection indices. *J. Med. Chem.*, **1993**. 36(16), 2243–2252.

93. Dimmock, J.R., et al., (Aryloxy)aryl semicarbazones and related compounds: a novel class of anticonvulsant agents possessing high activity in the maximal electroshock screen. *J. Med. Chem.*, **1996**. 39(20), 3984–3997.

94. Shao, B., et al., Phenoxyphenyl pyridines as novel state-dependent, high-potency sodium channel inhibitors. *J. Med. Chem.*, **2004**. 47(17), 4277–4285.

95. Yang, J., et al., 3-(4-Phenoxyphenyl)pyrazoles: a novel class of sodium channel blockers. *J. Med. Chem.*, **2004**. 47(6), 1547–1552.

96. Shao, P.P., et al., Novel cyclopentane dicarboxamide sodium channel blockers as a potential treatment for chronic pain. *Bioorg. Med. Chem. Lett.*, **2005**. 15(7), 1901–1907.

97. Priest, B.T., et al., A disubstituted succinamide is a potent sodium channel blocker with efficacy in a rat pain model. *Biochemistry*, **2004**. 43(30), 9866–9876.

98. Pancrazio, J.J., et al., Inhibition of neuronal Na⁺ channels by antidepressant drugs. *J. Pharmacol. Exp. Ther.*, **1998**. 284(1), 208–214.

99. Schulz, M., A. Schmoldt, Therapeutic and toxic blood concentrations of more than 800 drugs and other xenobiotics. *Pharmazie*, **2003**. 58(7), 447–474.

100. Wang, G.K., C. Russell, S.Y. Wang, State-dependent block of voltage-gated Na⁺ channels by amitriptyline via the local anesthetic receptor and its implication for neuropathic pain. *Pain*, **2004**. 110(1–2), 166–174.

101. Waxman, S.G., Acquired channelopathies in nerve injury and MS. *Neurology*, **2001**. 56(12), 1621–1627.

102. Craner, M.J., et al., Abnormal sodium channel distribution in optic nerve axons in a model of inflammatory demyelination. *Brain*, **2003**. 126(Pt 7), 1552–1561.

103. Waxman, S.G., M.J. Craner, J.A. Black, Na⁺ channel expression along axons in multiple sclerosis and its models. *Trends Pharmacol. Sci.*, **2004**. 25(11), 584–591.

104. Rasband, M.N., et al., Dysregulation of axonal sodium channel isoforms after adult-onset chronic demyelination. *J. Neurosci. Res.*, **2003**. 73(4), 465–470.

105. Waxman, S.G., Sodium channels as molecular targets in multiple sclerosis. *J. Rehabil. Res. Dev.*, **2002**. 39(2), 233–242.

106. Black, J.A., et al., Sensory neuron-specific sodium channel SNS is abnormally expressed in the brains of mice with experimental allergic encephalomyelitis and humans with multiple sclerosis. *Proc. Natl. Acad. Sci. U.S.A.*, **2000**. 97(21), 11 598–11 602.

107. Craner, M.J., et al., Temporal course of upregulation of Na(v)1.8 in Purkinje neurons parallels the progression of clinical deficit in experimental allergic encephalomyelitis. *J. Neuropathol. Exp. Neurol.*, **2003**. 62(9), 968–975.

108. Waxman, S.G., Sodium channel blockers and axonal protection in neuroinflammatory disease. *Brain*, **2005**. 128(Pt 1), 5–6.

109. Bechtold, D.A., R. Kapoor, K.J. Smith, Axonal protection using flecainide in experimental autoimmune encephalomyelitis. *Ann. Neurol.*, **2004**. 55(5), 607–616.

110. Craner, M.J., et al., Sodium channels contribute to microglia/macrophage activation and function in EAE and MS. *Glia*, **2005**. 49(2), 220–229.

111. Bechtold, D.A., et al., Axonal protection in experimental autoimmune neuritis by the sodium channel blocking agent flecainide. *Brain*, **2005**. 128(Pt 1), 18–28.

7
Potassium Channels

7.1
Potassium Channels: Overview of Molecular, Biophysical and Pharmacological Properties
Murali Gopalakrishnan, Char-Chang Shieh, and Jun Chen

7.1.1
Introduction

K^+ channels are a diverse and ubiquitous group of membrane proteins that conduct K^+ ions in or out of the cell membrane depending on electrochemical gradients. Since the first postulation of a selective K^+ permeability in excitable cells, much has been learned about the nature, diversity, architecture and function of K^+ channels. These ion channels serve a fundamental role to selectively conduct K^+ ions across the cell membrane along its electrochemical gradient at a rate of $\sim 10^6$–10^8 ions s^{-1} [1]. The concentration of K^+ ions outside the cell membrane is about 25-fold lower than that in the intracellular fluid and, consequently upon activation, an outward current due to the efflux of positively charged ions is generated by the opening of K^+ channels. Opening of K^+ channels thus offers a mechanism to counteract, dampen or restrict depolarizing activity triggered by an influx of cations (Na^+ and Ca^{2+}) or an efflux of anions (Cl^-). Activation of K^+ channels limits cell excitability whereas channel inhibition has the opposite effect. Accordingly, K^+ channels are involved in regulating a range of functions that involve setting membrane potential, dictating the duration or frequency of action potential, muscle contraction, secretion, volume regulation and cell proliferation.

Subsequent to molecular identification of the first voltage-gated K^+ channel from *Drosophila* [2], more than 200 genes encoding diverse K^+ channels have been cloned, functionally characterized, and correlated with native counterparts under physiological or pathophysiological conditions. More recently, the three-dimensional (3D) structure of several K^+ channels resolved at the atomic level has emerged, providing further information on mechanisms of K^+ channel function [3, 4]. This chapter provides a framework on general organization and properties of diverse types of K^+ channels and discusses salient aspects of modulation of K^+ channel function by ligands. Subsequent chapters provide detailed coverage of

Voltage-Gated Ion Channels as Drug Targets. Edited by D. Triggle
Copyright © 2006 WILEY-VCH Verlag GmbH & Co. KGaA, Weinheim
ISBN 3-527-31258-7

K⁺ channels of broad interest in drug discovery, including Kv1.3, Kv1.5, K_{ATP}, K_{Ca}, KCNQ and hERG.

7.1.2
Classification and General Properties

A unique set of functional determinants, in general, characterizes K⁺ channels. These include: (a) a permeation pathway or pore that conducts K⁺ ions to flow across the cell membrane, (b) a selectivity filter that specifies K⁺ as permeant ion species and (c) a gating mechanism that serves to switch between open and closed channel conformations [1]. Some K⁺ channels may be considered as ligand-gated where pore opening is coupled to the binding of an organic molecule. Other K⁺ channels are voltage-gated wherein the pore opening is coupled to the movement of a voltage sensor within the membrane electric field. Yet, other classes of K⁺ channels respond to different stimuli, including changes in intracellular Ca^{2+} concentration and G-proteins.

The ion-conducting or pore-containing subunit is generally referred to as the principal or α-subunit. The α-subunits encoded by K⁺ channel genes assemble as homomeric or heteromeric (with subfamily members) tetramers to form functional ion channels. The tripeptide sequence Gly-Tyr(Phe)-Gly is common to the pore of all K⁺ channels and constitutes the signature motif for determining K⁺ ion selectivity. As depicted in Fig. 7.1.1, K⁺ channels are classified on the basis of primary amino acid sequence of the pore-containing unit (α-subunit): (1) six transmembrane one-pore channels, (2) two transmembrane one-pore channels and (3) four transmembrane two-pore channels. Functionally, K⁺ channels can be organized as voltage-dependent K⁺ channels (Kv), Ca^{2+}-activated K⁺ channels (K_{Ca}), inward-rectifying K⁺ channels (Kir), including ATP-sensitive K⁺ channels (K_{ATP}), and two-pore K⁺ channels. Tables 7.1.1–7.1.4 summarize members of the four major families of K⁺ channels, adapted from the standardized nomenclature for K⁺ channels (Gutman, 2003 #1388) and depict amino acid sequence relationships, nomenclature, chromosomal localization, tissue expression and associated genetic disorders.

Table 7.1.1 Human voltage-gated K⁺ channels: Genes, chromosome localization, tissue distribution and genetic disorders. The human voltege-gated K⁺ channels are sorted by similarrity of amino acid sequence using Pileup program of the Wisconsin Sequence Analysis Package (Version 10.3, Accelrys Inc., San Diego, CA). Abbreviations: Adr g, adrenal gland; B, brain; BFNC, Benign familial neonatal convulsions; Co, cochlea; CTX, charybdotoxin; EA, Episodic ataxia/myokymia syndrome; eag, *ether-a-go-go* gene encoded K⁺ channel; H, heart; Herg, human *ether-a-go-go*; JLNS, Jerrell and Lange-Neilsen Syndrome; K, kidney; L, lung; Li, liver; LQT, long-QT syndrome; Lym, lymphocyte; M, muscle; Panc, pancreatic islet; Pl, placenta; Pros, prostate; PHHI, persistent hypersulinaemic hypoglycemia of infancy; R, retina; Sk m, skeletal muscle; Sm, smooth muscle; Spin, spinal cord.

▶

Name	Common name	Gene	Chromosome	Tissue expression	Genetic disorders
K$_V$11.1	Herg	KCNH2	7q35	B, H	LQT2
K$_V$11.3	erg-3	KCNH7	2q24.2	B, H	
K$_V$11.2	erg-2	KCNH6	17q23.3	B, H	
K$_V$10.1	eag-1	KCNH1	1q32	B	
K$_V$10.2	eag-2	KCNH5	14q24	B	
K$_V$12.1	Elk-1	KCNH8	3p24.3	Nerve tissue	
K$_V$12.3	Elk-3	KCNH14	17q21	B	
K$_V$12.2	Elk-2	KCNH3	12q13	B	
K$_V$7.4		KCNQ4	1p34	Hair cell, inner ear	Deafness
K$_V$7.5		KCNQ5	6q14	B, Sk m	
K$_V$7.2		KCNQ2	20q13	B, neuron	BFNC
K$_V$7.3		KCNQ3	8q24	B, neuron	BFNC
K$_V$7.1	KVLQT1	KCNQ1	11p15.5	H, Co, K, L, Pl	LQT1 (JLNS)
K$_V$4.2		KCND2	7q31–32	B	
K$_V$4.3		KCND3	1p13.2	B, H	
K$_V$4.1		KCND1	Xp11.23–11.3	H, B, L, K, Li, Pl	
K$_V$1.1		KCNA1	12p13	B, H, R, Panc	EA
K$_V$1.2		KCNA2	1p13	B, H, Panc	
K$_V$1.3		KCNA3	1p21–p13.3	Lym, B, L, Th, S	
K$_V$1.4		KCNA4	11q13.4–14.1	B, H, Panc	
K$_V$1.7		KCNA7	19q13.3	H, Panc, Sk m	
K$_V$1.6		KCNA6	12p13	B	
K$_V$1.5		KCNA5	12p13	B, H, K, L, Sk m	
K$_V$1.8		KCNA10	1p13	K, H, aorta	
K$_V$3.1		KCNC1	11p15	B, M, Lym	
K$_V$3.4		KCNC4	1p21	B, sk m	
K$_V$3.3		KCNC3	19q13.3–13.4	B, L	
K$_V$3.2		KCNC2	1pq13.3–13.4	B	
K$_V$6.1	KH2	KCNG1	20q13	B	
K$_V$6.2		KCNG2	18q22–18q23	H	
K$_V$6.3		KCNG3	2p21	B, K, L, Panc	
K$_V$9.2		KCNS2	8q22	B	
K$_V$9.3		KCNS3	2p24	L, B, artery	
K$_V$9.1		KCNS1	20q12	B	
K$_V$2.1		KCNB1	20q13.2	B, H, K, Sk m, R	
K$_V$2.2		KCNB2	8q12	Sm m	
K$_V$8.1		KCNB3	8q22.3–24.1	B	
K$_V$5.1	KH1	KCNF1	2p25	B	

Table 7.1.2 Human Ca^{2+}-activated K$^+$ channel: genes, chromosome localization and tissue distribution; for abbreviations, see Table 7.1.1 legend.

Name	Common name	Gene	Chromosome map	Tissue distribution
K$_{Ca}$2.2	SK$_{Ca}$2	KCNN2	5q21.2–q22.1	B, H, L
K$_{Ca}$2.3	SK$_{Ca}$3	KCNN3	1q21.3	B, Adr g, T cells
K$_{Ca}$2.1	SK$_{Ca}$1	KCNN1	19p13.1	B, H
K$_{Ca}$3.1	IK$_{Ca}$1	KCNN4	19q13.2	Lym, Col, Pros, Pl
K$_{Ca}$4.1	Slack	KCNMB2	9q34	Ubiquitous
K$_{Ca}$4.2	Slo2	KCNT2	1q31	
K$_{Ca}$1.1	Slo	KCNMA1	10q23.1	B, Sm, Panc
K$_{Ca}$5.1	Slo3	KCNMA3	8p11	testis

Table 7.1.3 Human inward-rectifying K$^+$ channel: genes, chromosome localization, tissue distribution and diseases; for abbreviations, see Table 7.1.1 legend.

Name	Gene	Chromosome map	Tissue expression	Genetic disorder
Kir4.1	*KCNJ10*	1q22	Glial	
Kir4.2	*KCNJ15*	21q22.2	K, L, B	
Kir1.1	*KCNJ1*	11q24	K, Panc	Bartter's
Kir6.1	*KCNJ8*	12p11.23	Ubiquitos	PHHI
Kir6.2	*KCNJ11*	11p15.1	Ubiquitos	
Kir2.1	*KCNJ2*	17q23	H, B, Sm/Sk m, K	Andersen's
Kir2.2	*KCNJ12*	17p11.1–11.2	H	
Kir2.3	*KCNJ4*	22q13.1	H, B, Sk m	
Kir2.4	*KCNJ14*	19q13	B, R	
Kir3.2	*KCNJ6*	21q22.1–22.2	Panc, cerebellum	
Kir3.4	*KCNJ5*	11q24	H, Panc	
Kir3.3	*KCJ9N*	1q21–23	H, B, Sk m	
Kir3.1	*KCNJ3*	2q24.1	H, cerebellum	
Kir5.1	*KCNJ16*	17q25	B, periphery	
Kir7.1	*KCNJ16*	2q37	Gl, K, B	

Table 7.1.4 Human two-pore K$^+$ channels: gene, chromosome localization and tissue distribution; for abbreviations, see Table 7.1 legend.

Name	Common name	Gene	Chromosome	Tissue expression
K$_{2p}$12.1	THIK2	*KCNK12*	2p22–2p23	B, K
K$_{2p}$13.1	THIK1	*KCNK13*	14q24.1–24.3	Ubiquitous
K$_{2p}$3.1	TASK1	*KCNK3*	2p23	H, B, Panc, Pl
K$_{2p}$9.1	TASK9	*KCNK9*	8q24	B, K, Li, L
K$_{2p}$15.1	TASK5	*KCNK15*	20q12	Adr g, Panc
K$_{2p}$1.1	TWIK1	*KCNK1*	1q42–q43	B, K, H
K$_{2p}$6.1	TWIK2	*KCNK6*	19q13.1	K
K$_{2p}$7.1	KCNK7	*KCNK7*	11q13	B
K$_{2p}$10.1	TREK2	*KCNK10*	14q31	B
K$_{2p}$2.1	TREK1	*KCNK2*	1q41	B
K$_{2p}$4.1	TRAAK	*KCNK4*	11q13	B, Spin, R
K$_{2p}$16.1	TALK1	*KCNK16*	6p21	Panc
K$_{2p}$17.1	TALK2	*KCNK17*	6p21	Panc, Pl, L, H, Li
K$_{2p}$5.1	TASK2	*KCNK5*	6p21	B, L
K$_{2p}$18.1	TRESK	*KCNK18*	10q26.11	Spinal cord

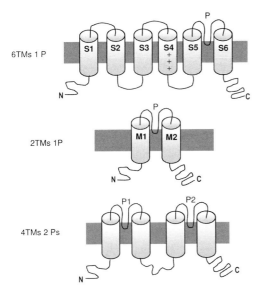

6TMs 1 P

2TMs 1P

4TMs 2 Ps

Fig. 7.1.1 Schematic representation of the primary subunits of K$^+$ channels. Shown are six transmembrane (TM)–one pore (P) (6TMS 1P) types that include voltage-gated and Ca^{2+}-activated K$^+$ channels, two transmembrane–one pore types (2TMS 1P) represented by the inward rectifiers, and four transmembrane–two pore (4TMS 2PS) channels. K$^+$ channels that belong to these subfamilies are presented in Tables 7.1.1–7.1.4. Auxiliary proteins that co-associate (or regulate) with primary subunits are summarized in Table 7.1.5.

Table 7.1.5 Human K⁺ channel auxiliary subunits: genes, chromosome localization, function, tissue distribution and diseases (for abbreviations, see Table 7.1.1 legend).

Name	Gene	Chromosome map	Associated α subunit	Functional properties	Tissue distribution	Genetic disorders
Kvβ1	KCNAB1	3q26.1	Kv1.1–Kv1.6	Inactivation, alter gating kinetics	B, H	
Kvβ2	KCNAB2	1p36.3	Kv1.1–Kv1.6	Inactivation, alter gating kinetics	B, H	
Kvβ3	KCNAB3	17p13.1	Kv1.3–Kv1.5	Inactivation, alter gating kinetics	B	
KChIP1	KCNIP1	5q35.1	Kv4.1–Kv4.3	Integral component of neuronal A-type K⁺ current; modify current density, kinetics	B	
KChIP2	KCNIP2	10q24	Kv4.2, Kv4.3	Integral component of cardiac I_{to}; modify current density, kinetics	H	Arrhythmia (?)
KChIP3	CSEN	2q21.1	Kv4.1–Kv4.3	Modify current density, kinetics	B	
KChIP4	KCNIP4	4p15.31	Kv4.1–Kv4.3	Modify current density, kinetics	B	
MinK	KCNE1	21q22.1–q22.2	Kv7.1	Constitute cardiac slow activating K⁺ current (IK_s)	K, U, H	JLNS
MiRP1	KCNE2	21q22.12	Kv11.1, Kv4.2	Modify current density, kinetics	H, Sk	
MiRP2	KCNE3	11q13–q14	Kv3.4, Kv7.4	Suppress Kv7.4 current; alter Kv3.4 kinetics	Sk, H	Periodic paralysis (?)
MiRP3	KCNE4	2q36.3	Kv7.1	Altering gating kinetics; current suppression	H, Sk, K	
MiRP4	KCNE1L	Xq22.3	–		H, Sk, B	
K$_{Ca}$β1	KCNMB1	5q34	K$_{Ca}$1.1	Increasing Ca²⁺ sensitivity, alter gating	Sm	Diastolic hypertension
K$_{Ca}$β2	KCNMB2	3q26.2–q27.1	K$_{Ca}$1.1	Fast inactivation, reduce CTX blockade	Sm, H, B	
K$_{Ca}$β3	KCNMB3	3q26.3–q27	K$_{Ca}$1.1	Partial inactivation	Testis	
K$_{Ca}$β4	KCNMB4	12q14.1–q15	K$_{Ca}$1.1	Reduce IbTx, CTX sensitivity, alter gating	B	
SUR1	SUR1	11p15.1	Kir6.2	Constitute pancreatic/neuronal K$_{ATP}$ channel	Pan, nerve	PHHI
SUR2A	SUR2A	12p12.1	Kir6.2	Constitute cardiac/skeletal K$_{ATP}$ channel	H, Sk	
SUR2B	SUR2B	12p12.1	Kir6.1, Kir6.2	Constitute smooth muscle K$_{ATP}$ channel	Sm, B	
KchAP (rat)			Kv1.2, 2.1, 2.2, 4.3	Chaperone effect, enhance current density	H, B, K	Dilated cardiomyopathy

In many cases, auxiliary subunits co-associate with the pore-forming α-subunits and could modulate expression, biophysical or pharmacological properties of the α-subunit complex. This further enhances the diversity of K^+ channels (Table 7.1.5). While the α-subunit serves as the principal binding site for a wide range of K^+ channel modulators, including organic ligands and peptides, there are notable exceptions where ligand-binding sites reside within an auxiliary subunit, as for example, K_{ATP} channels.

7.1.2.1 Voltage-gated K^+ Channels

The electrical properties of excitable cells are determined, in large part, by the voltage-gated K^+ channels. For example, the Kv channels dictate the duration of the action potential in excitable cells such as cardiac myocytes and neurons whereas in non-excitable cells they participate in processes such as volume regulation, hormone secretion or activation by mitogens. A total of 38 human Kv genes have been cloned and assigned to 12 subfamilies (Kv1 to Kv12) on the basis of sequence similarities (Table 7.1.1). K_v channels function as oligomeric proteins by tetrameric association of four α subunits, each with six transmembrane segments (S1 through S6). Both the N- and C-termini of these proteins are located intracellularly. The transmembrane segment S4 contains positively charged residues and serves as the voltage sensor. The S5-S6 domains from each of the four subunits form a functional K^+-conducting pore.

7.1.2.2 Calcium-activated K^+ Channels

The calcium-activated K^+ channels (K_{Ca}) are regulated by changes in cytosolic Ca^{2+} levels and/or membrane potential. Upon activation by intracellular Ca^{2+}, K_{Ca} channels open, resulting in hyperpolarization of the membrane potential and changes in cellular excitability. The channel activation also counteracts further increases in intracellular Ca^{2+}, thus regulating, both temporally and kinetically, the concentration of Ca^{2+}. Various functions, including the regulation of neuronal firing blood flow and cell proliferation, have been attributed to K_{Ca} channels. For example, in many central neurons, SK_{Ca} channel activity dampens excitability by contributing to the after hyperpolarization, affecting interspike intervals during a burst of action potentials as well as the duration of the burst. Initially, K_{Ca} channels have been divided on the basis of biophysical (conductance) and differential toxin sensitivities. This classification has now been complemented by the emergence of distinct genes encoding three subfamilies, viz., large conductance BK_{Ca} (KCa1.1, *slo*), small conductance SK_{Ca} (KCa2.1–2.3) and intermediate conductance IK_{Ca} (KCa3.1) channels. SK_{Ca} and IK_{Ca} channels have a similar topology to members of the voltage-gated (Kv) K^+ channel superfamily, with six transmembrane segments and the pore located between segments S5 and S6, whereas an additional transmembrane segment exists in BK_{Ca} channels (Chapter 7.4). The identification of multiple splice variants of the pore-forming α-subunit along with the various β-subunits, as in the case of BK_{Ca} channels, suggests further diversity within the K_{Ca} family. This, together with the widespread distribution throughout the CNS and in peripheral tissues, offers opportunities for discovering therapeutic agents as well as substantial challenges in the form of tissue and organ specificity.

7.1.2.3 Inward Rectifiers

The inward rectifiers (K_{ir}) are derived from four subunits, each containing two transmembrane segments (M1 and M2) equivalent to the fifth and sixth transmembrane segments (S5 and S6) of voltage-gated K^+ channels, and a segment with pore elements in between. These channels conduct K^+ currents predominantly in the inward direction [5]. This inward rectification, in some cases, is attributable to blockade of outward currents by intracellular cations such as Mg^{2+} and by the naturally occurring polyamines spermine and spermidine. There are seven Kir subfamilies (Kir1–7), which can be distinguished by their degree of rectification and responses to cellular signals. For example, the G-protein coupled channels (Kir3) are strongly rectifying and activated by $G\beta\gamma$, whereas the Kir6 family of channels are weakly rectifying and modulated by the cellular ATP:ADP ratio. Although inward rectifier channels are, in general, organized as homotetramers, complex octameric arrangements have also been described in combination with auxiliary subunits, as in the case of ATP-sensitive K^+ channels where association of Kir6 subunits with subunits of the ATP-binding cassette (ABC) family of proteins, the sulfonylurea receptors (SUR), occurs. Because the activity of K_{ATP} channels reflects the static and dynamic nature of cellular metabolism, the K_{ATP} channels are considered as sensors of intracellular metabolism, tuning the potassium permeability, and in turn the electrical activity, of the cell to its energetic balance.

7.1.2.4 Two-pore K^+ Channels

The two-pore K^+ channels are weak inward rectifiers, consisting of four transmembrane segments and two pore domains in tandem. Functional channels are dimers of subunits, thus retaining the overall tetrameric arrangement. The Gly-Tyr(Phe)-Gly residues of the K^+-signature motif are preserved in the first pore-loop of these channels, but are replaced by Gly-Phe-Gly or Gly-Leu-Gly in the second pore loop. Members of this family lack the voltage sensing S4 transmembrane domain, characteristics of voltage-gated K^+ channels, and, in general, share only moderate sequence homology outside their P regions. The channels are opened over wide voltage ranges and initially described as "background" or "baseline" K^+ channels. The TREK-type channels, TREK-1 (KCNK2), TREK-2 (KCNK10) and TRAAK (KCNK4), share many functional properties, including activation by membrane stretch and by lipids such as lysophosphatidylcholine and arachidonic acid. The TASK subgroup consists of channels that are inhibited by cellular acidosis whereas the TALK channels are activated at alkaline extracellular pH. TREK1–2 and TASK1–3 are activated by volatile general anesthetics [6, 7].

7.1.3
Auxiliary Subunits

The biophysical and pharmacological properties of the α-subunit of K$^+$ channels can be modified by the presence of diverse auxiliary subunits. As summarized in Table 7.1.5, these include distinct β-subunits that assemble with K$_V$ and K$_{Ca}$ channels, minK and minK-related peptides that coassemble with hERG and KvLQT1, and the sulfonylurea receptors SUR1 and SUR2 that associate with Kirs to form the K$_{ATP}$ channel complex. The auxiliary subunits play diverse roles, ranging from modulation of current kinetics (e.g., inactivation), modification of cellular trafficking processes, to serving as binding sites for both endogenous and exogenous ligands. Association of the SUR with Kir6.1 or Kir6.2 is necessary for ligand-dependent activation or inhibition of the K$_{ATP}$ channel. Certain α-subunits do not form functional channels by themselves, but associate with α-subunits of other subfamily members to regulate expression, biophysical and pharmacological properties. Besides auxiliary subunits, scaffolding or chaperone proteins can also interact with the primary subunits, resulting in alterations in channel trafficking and kinetic properties (for review, see, e.g., [8]). Given the diversity of K$^+$ channel subunits and the potential to vary the constituents to form α-α or α-β heteromeric channel complexes that can undergo varying degrees of post-translational processing, it is often difficult to know with precision the exact composition and state of channel complexes *in vivo*. In addition, a range of post-translational modifications that play important roles in the proper folding, assembly, and trafficking has been described, including glycosylation, methylation, ubiquitation, incorporation of fatty acids, covalent attachment of coenzymes, and phosphorylation.

7.1.4
Crystal Structure

Initial structure and function studies of K$^+$ channels using a combination of mutagenesis and biophysical approaches have revealed important domains for channel function. However, major groundbreaking events have emerged from X-ray crystallography studies, wherein atomic structures of several K$^+$ channels have been solved, including KcsA [9, 10], MthK (Jiang, 2002 #937), KirBac [11] and KvAP [12]. These bacterial channels represent major types of K$^+$ channel families such as Kir (KcsA and MthK), Ca^{2+}- (MthK) and voltage-gated K$^+$ channels (KvAP), and their structures serve as surrogates for their eukaryotic counterparts. These studies have addressed fundamental questions of K$^+$ channel function such as ion selectivity, permeation and voltage sensing, as discussed in the accompanying paragraphs.

KcsA was the first K$^+$ channel whose crystal structure was determined. It is encoded by bacterium *Streptomyces lividianswas* and gated by cytoplasmic protons. Like Kir channels, each KcsA subunit consists of two transmembrane helices (termed TM1 and TM2). Structurally, four identical subunits form a four-fold symmetric tetramer creating a central ion conduction pathway (Fig. 7.1.2). The

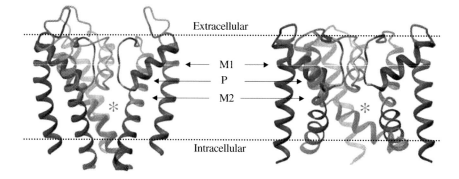

KcsA (Closed) **MthK(Open)**

Fig 7.1.2 X-ray structure of the ion conduction pores of KcsA and MthK K$^+$ channels. Three of the four subunits of each channel are depicted (facing subunit removed for clarity). The selectivity filter is colored orange and glycine-gating hinge (G99 in KcsA and G83 in MthK) is in red. A red asterisk marks the central cavity. Three helical segments are marked as M1 (outer helix), P (pore helix) and M2 (inner helix). In KcsA, the gate is formed by the bundle crossing of four M2 helices and the channel is closed. In MthK, the gate is open due to the outward bending of M2 from the central cavity. (Adapted with permission from Ref. [3].)

four TM1 (outer helices) extend from intracellular to extracellular, face lipid and line the outer border of channel, whereas four TM2 (inner helices) extend inward and cross in a bundle near the cytoplasmic surface, forming an inverted teepee-like structure. Linking the inner and outer helices in each subunit is the re-entrant pore helix and the 12 Å-long K$^+$ selectivity filter near the cytoplasmic side. Between the bundle crossing and selectivity filter is the central cavity, a 18 Å-long tunnel that widens to 10 Å near the middle of the membrane. A single K$^+$ ion is suspended in the central cavity and hydrated by surrounding water molecules. The K$^+$ ion is further stabilized by dipole forces provided by the four pore helices oriented toward the center of the central cavity. The selectivity filter is lined by the carbonyl oxygen atoms of the signature sequence and held open by structural constraints provided by pore helices. The selectivity filter can coordinate K$^+$ ions (1.33 Å) but not the smaller Na$^+$ ions (0.95 Å) [9]. In each of the four K$^+$ binding sites within the selectivity filter, eight carbonyl oxygen atoms surround the conducting K$^+$ ion such that the energy states are similar to that of a hydrated K$^+$ in the central cavity [13]. Therefore, the transition of K$^+$ between selectivity filter and central cavity is energetically barrier-less. Since more than one ion is present within the selectivity filter, the repulsion between them further favors K$^+$ movement. These findings explain elegantly two salient aspects of K$^+$ channel permeation, viz., high throughput and strict K$^+$ ion selectivity.

The ion permeation of the K$^+$ channel is controlled by opening and closing of the activation gate. KcsA and KirBac were crystallized in a closed state [9, 11] whereas the structures of MthK and KvAP were revealed in open states [12, 14].

As noted above, KcsA adopts an inverted teepee-like structure at its cytoplasmic side by bundle crossing its four inner helices and forming a 3.5 Å aperture. This aperture, lined by hydrophobic residues, can occlude K^+ from the water-filled central cavity. In KirBac, the aperture is closed by the side chains of phenylalanine residues in the four inner helices. In addition, the selectivity filter and pore-helices were aligned differently, further hindering ion flow. Contrary to KcsA and KirBac, MthK and KvAP adopt quite different conformations in their helices near the intracellular side. In MthK, each of four inner helices bends away from the central cavity at a glycine hinge, which makes the aperture as wide as 12 Å, rendering barrier-less K^+ flow between cytoplasmic side and central cavity (Fig. 7.1.2). In KvAP, the four subunits each containing the inner and outer helices is packed less tightly, resulting in a wide-open aperture, allowing free K^+ flow. Together, these structures suggest that activation gate of K^+ channels is formed by the bundle crossing of the four inner helices near the cytoplasmic site and channel opening can be accomplished by inner helix movements such as bending, tilting and rotation.

For voltage-gated channels, voltage sensor movements precede channel opening. KvAP is a voltage-gated channel and its crystal structure provided a different perspective on how the voltage sensor moves upon changes in membrane potential [12, 15]. Previous structure–function studies have suggested that the positive charges in S4 are buried in an aqueous environment and stabilized by pairing with negative charged residues. Upon change in membrane potential, the S4 domains undergo relatively small motions via lateral translation, tilting or rotation, and open the activation gate. However, the crystal structure of KvAP suggests differently, viz., instead of being localized inside the membrane, S4 helices lay parallel to the membrane at the cytoplasmic side and form a voltage-sensor paddle together with the C-terminal end of S3. Upon membrane depolarization, the paddle moves a large distance across the membrane, from inside to outside [12]. The discrepancy could arise because either the KvAP channel may have been distorted in the crystal structure (perhaps by the use of the Fab fragment) or it simply has different voltage sensor conformation and movements. Additional studies with more refined crystal structures and of different channels are expected to further clarify the mechanisms of voltage sensing.

Although crystal structures of K^+ channels are few and thus far only obtained from prokaryotes, they have started to facilitate our understanding of ion channel–drug interactions and could pave the way for further drug design approaches. For example, homology modeling using KcsA as template has guided mutagenesis studies to elucidate channel–ligand interactions [16]. A structure-based design strategy also allowed preparation of several charybdotoxin analogs with about 20-fold higher affinity to block Ca^{2+}-activated K^+ channels versus voltage-gated Kv1.3 channels [17]. Several silico screening tools for K^+ channel blockers have also emerged [18–20]. For example, natural products that block K^+ channels were identified by structure-based in silico screening [17]. Structural information, especially on ion channel–ligand complexes, could provide additional avenues for exploiting K^+ channels.

7.1.5
K⁺ Channels and Diseases

As K⁺ channels play fundamental roles in the regulation of membrane excitability, both genetic and acquired diseases involving altered functioning of neurons, smooth muscle and cardiac cells could be expected to arise subsequent to abnormalities in K⁺ channel function. Several genetically linked diseases of the cardiac, neuronal, renal and metabolic systems involving members of voltage-gated K⁺ channels, inward rectifiers and auxiliary proteins have been reviewed [21]. A well-studied example is the human KCNQ gene family where mutations in four of the five genes are associated with hereditary diseases. Mutations in either KCNQ1 or KCNE1 cause inherited long QT syndrome (LQT1 or LQT5, respectively), a condition leading to arrhythmia (Romano-Ward syndrome) as well as an associated form of deafness (Jervell and Lange-Nielsen syndrome). KCNQ2/KCNQ3 heteromultimers are thought to underlie the prototypic neuronal M-currents, and mutations in either of these genes cause inherited neonatal epilepsy (benign familial neonatal convulsion, BFNC). The KCNQ4 gene is a component of K⁺ channels described in hair cells of the cochlea and vestibular apparatus, mutations in which lead to a form of inherited progressive adult deafness (autosomal dominant non-syndromic deafness, DFNA2). Chapter 8 further elaborates on genetic and acquired channelopathies pertaining to cardiovascular system.

7.1.6
Ligands Interacting with K⁺ Channels

Considerable advances have been made in the assessment of biological activities of organic modulators and peptides interacting with K⁺ channels. Although members of the voltage-gated K⁺ channel family were the first to be cloned and characterized, medicinal chemistry efforts early were largely focused on the ATP-sensitive K⁺ channels. These efforts were pioneered by the initial discovery of nicorandil and cromakalim, which were demonstrated to relax vascular muscle with associated membrane hyperpolarization and/or K⁺ channel activation. More recently, voltage-gated K⁺ channels, including Kv1.3, Kv1.5 and KCNQ, have received prominence as drug targets whereas hERG inhibition which increases risk of drug-induced long QT effects is being considered early in the drug discovery process.

7.1.6.1 Voltage-gated K⁺ Channels

Although K_v channels served as prototypes to elucidate structure–function relationships of K⁺ channels, organic modulators, including tetraethylammonium and 4-aminopyridine are weak, and nonselective. Peptide toxins that bind with high specificity to block the channel pore or alter gating features have served as important tools for analysis of the structure–function of K⁺ channels (Section 7.1.8). More recently, the identification of the molecular components of many cardiac and neuronal voltage-gated K^+ channels has renewed medicinal chemistry efforts to identify selective openers and blockers of various channel types. Substantial efforts in the area of antiarrhythmic drug discovery have been directed towards compounds that prolong the cardiac action potential and refractoriness. For example, the Kv1.5 subunit, the major component of the cardiac ultra-rapid delayed rectifier in human atria, has been a target of heightened interest (Chapter 7.3). Because of its rapidity of activation and slow inactivation, these channels contribute significantly to repolarization in human atrium, and, consequently, specific blockade would, theoretically, overcome the shortcomings and disadvantages of currently used agents for the treatment of atrial flutter and/or atrial fibrillation. The cardiotoxicity of histamine receptor antagonists such as astemizole or terfenadine, antipsychotics such as sertindole, certain tricyclic antidepressants, antiemetics and antibiotics have been linked to their potent inhibition of hERG channels; an effect that leads to the occurrence of *torsades de pointes* in susceptible individuals [22]. Avoiding drug-induced cardiac arrhythmia, particularly via hERG channel inhibition, is a major hurdle in the optimization of new compounds (Chapter 8.21). The Kv1.3 channel expressed in T cells offers opportunities as targets for immunomodulation; the physiological roles and potential therapeutic use of selective inhibitors in immunological disorders is discussed in Chapter 7.2. The KCNQ2–5 subunit combinations, which serve as molecular correlates to different M-currents, are critical for regulating neuronal excitability and, as elaborated in Chapter 7.6, openers have the potential for the treatment of CNS disorders characterized by neuronal hyperexcitability, such as migraine, epilepsy and neuropathic pain.

7.1.6.2 Calcium-activated K^+ Channels

The therapeutic applications for BK_{Ca} channel openers have focused on stroke, epilepsy, and bladder overactivity although there is evidence for utility in the treatment of asthma, hypertension, gastric hypermotility and learning and memory. In general, most early BK_{Ca} channel openers are relatively weak agents or known to possess ancillary pharmacology that limits their utility as therapeutic agents or as probes to study the *in vivo* therapeutic relevance of BK_{Ca} channels. Benzimidazolone analogs such as NS-004 and NS-1619 stimulate BK_{Ca} activity, leading to membrane hyperpolarization. Other earlier known BK_{Ca} openers include glycosylated triterpene activators (dehydrosoyasaponin-I) and several indole diterpene blockers such as paxilline, verruculogen and penitrem A. As elaborated in Chapter 7.4, more recently reported openers of BK_{Ca} channels include aryloxindole analog BMS-204352, and certain diaryl triazolones and aminoquinolinones. Blockers of intermediate conductance channels include the antimycotic clotrimazole and related analogs such as TRAM-34 and ICA-17043 – the latter compound evaluated for treatment of sickle cell disease. SK_{Ca} channels are putative targets for indications such as cognitive dysfunction, schizophrenia and depression, and derivatives of bicuculline, dequalinium, UCL 1684 and UCL 1848 are examples of known blockers [23].

7.1.6.3 Inward Rectifiers and K$_{ATP}$ Channels

Of the seven Kir subfamilies, the ATP-sensitive K$^+$ channel has been most exploited in terms of pharmacological modulation by both agonists and antagonists. These channels are formed from four Kir6.2 pore-forming subunits, and four regulatory sulfonylurea receptor (SUR) subunits. Channel activity is modulated by voltage and by multiple ligands, including ATP and PIP$_2$, which act on the Kir6.2 subunits, as well as sulfonylureas, potassium channel openers, and Mg-nucleotides, which act on the SUR subunit. The inhibitory ATP binds to the Kir6.2 subunit, while MgATP- and ADP-activation results from interaction with the SUR subunits. Clinically well-known agents that continue to be useful in the pharmacotherapy of type II diabetes are the sulfonylureas such as glibenclamide (glyburide) that are blockers of K$_{ATP}$ channels. In contrast to the sulfonylurea analogs that interact at SUR1 and SUR2 proteins, weak and non-selective blockers such as phentolamine and cibenzoline have been suggested to directly block the Kir subunit of the K$_{ATP}$ channel. K$_{ATP}$ openers originating from various chemotypes, notably, benzopyrans, cyanoguanidines, and dihydropyridines, have been investigated for numerous therapeutic applications, prominent among which are hyperinsulinemia, hypertension, angina, cardioprotection, ventricular arrhythmia, asthma, alopecia and bladder overactivity [24]. As elaborated in Chapter 7.5, more recent medicinal chemistry efforts of K$_{ATP}$ channel modulators have been in the area of myocardial ischemia, ventricular arrhythmia, hyperinsulinemia and overactive bladder.

7.1.6.4 Two-pore K$^+$ Channels

Two-pore K$^+$ channels are thought to provide baseline regulation of membrane excitability with roles implicated in the actions of anesthetics and some neuroprotective compounds. Some of the subtypes are activated whereas others are inhibited by anesthetics. The neuroprotective agent riluzole is an activator of TREK-1 and TRAAK channels (reviewed in [6]). Knockout studies have shown that TREK-1 plays a major role in the neuroprotection against epilepsy and ischemia and that TREK-1-deficient mice display resistance to anesthetic effects [25].

7.1.7
Ligand Binding Sites

A large body of evidence pertaining to ligand binding has emerged from analysis of interactions at voltage-gated K$^+$ channels. Drugs can affect K$^+$ channel function through different mechanisms, including blockade of the ion conduction pore from the external or internal side, modulation of channel gating (opening or closing) or interactions with auxiliary subunits to modify channel function. These aspects are elaborated in the following sections.

7.1.7.1 **Interaction with the Ion-conducting Pore**

For voltage-gated K⁺ channels, the most common drug binding site is located in the central cavity. Armstrong showed that block of a voltage-gated K⁺ channel by tetraethylammonium (TEA) and its derivatives in squid giant axon required channel opening, a phenomenon referred to as open channel blockade [26, 27]. TEA and its derivatives act within the intracellular pore and impede K⁺ conduction. Once bound to the intracellular pore, blockers can either interfere with closure of the activation gate (for C10⁺, a long-chain quaternary ammonium compound) or be trapped in the channel by closure of activation gate (for C9⁺, a methylene group shorter than C10⁺). Drug trapping has also been described for cloned channels, such as TEA or decyltriethylammonium trapping by mutant Shaker K⁺ channel [28] and methanesulfonanilide (MK-499) trapping by mutant HERG channel [29]. Together, these studies support the idea that the inner pore of a channel, where drug binding occurs, is a large water-filled vestibule lined by hydrophobic residues. These predicted features of the inner vestibule were confirmed by the existence of central cavity lined by residues from the pore helices and inner helices, as revealed by recently solved crystal structures.

More recently, considerable progress has been made in understanding the molecular basis of drug binding. This was achieved by utilizing a comprehensive approach, including mutagenesis, functional characterization and molecular modeling. Typically, the ion channel of interest is altered at single or stretches of amino acids, expressed in heterologous expression systems such as *Xenopus* oocytes or mammalian cells, and compound sensitivities are assessed. Residues or regions that influence drug interactions (e.g., ligand affinity) can be identified and corroborated with homology modeling utilizing known crystal structures (i.e., KcsA) as template. These analyses often result in a model of drug–ion channel interactions that predict how well a drug docks into the channel protein by interaction with specific amino acids. One major triumph of this approach has been the elucidation of the structural basis of drug-induced long QT syndrome [30]. In an effort to identify which regions of the hERG channel are involved in high-affinity block by dofetilide, chimeras where portions of the hERG channel were replaced by the dofetilide-insensitive bovine ether-a-go-go channel (BEAG) were analyzed and a set of pore (S5-S6) residues was pinpointed as the important determinant of dofetilide binding [31]. A systematic alanine-scanning mutagenesis of the entire S6 domain and portions of the pore helix identified aromatic resides in the S6 domain (Tyr-652 and Phe-656) and pore helix (T623 and V625) critical for binding of structurally diverse drugs such as MK-499, cisapride and terfenadine [16, 32]. Homology modeling predicts these key residues face the central cavity and line the drug-binding pocket. Indeed, compounds with relatively lower hERG affinities such as vesnarinone, chloroquine and quinidine also share similar binding sites [20, 33–35]. In light of the lack of two aromatic residues in equivalent positions of other K⁺ channels such as Kv1–4, these results explain why structurally diverse drugs block hERG instead of other channels.

Drug binding sites on other voltage-gated K⁺ channels have also begun to emerge. For example, with the Kv1.3 channel, dihydrocorreolide binding occurs in the central cavity by interacting with a few residues in the S6 domain (A413, V417, A421, and V424), which form a complementary bowl-shaped binding pocket [36]. In KCNQ1, one residue in the pore helix (T312) and four residues of the S6 domain (I337, F339, F334 and A344) have been determined as the key binding residues for L-735821. Molecular docking predicts that L-735821 blocks conductance by physically precluding the occupancy of K^+ ion within the central cavity [37]. In Kv1.5, T479 and T480 in the pore helix and additional residues (V505, I508 and V512) are critical for block by S0100176 [38]. These studies point to a common theme, i.e., drug binding in the central cavity involves interaction with a few residues located in the S6 domain and pore helix.

7.1.7.2 Modification of Gating

Gating of ion channels can also greatly influence drug affinity. For example, open channel blockers require opening of the activation gate to allow access of drugs into the central cavity. Channel gating can often lead to conformational changes in drug binding sites and thus modulate drug affinity. As a result, channels with identical drug binding residues can exhibit varying sensitivities to the same compound. For example, hERG and EAG channels belong to the same gene family and share the same aromatic residues in the S6 domain (Tyr-652 and Phe-656) critical for cisapride block of hERG channels; however, EAG is >100-fold insensitive to cisapride [32]. This differential drug sensitivity has been partially explained by the presence (hERG) and absence (EAG) of inactivation gating. By repositioning of aromatic residues along the S6-helix of hERG, sensitivity of block by cisapride was reduced. Conversely, repositioning of equivalent aromatic residues in EAG channels, independent of inactivation, induced sensitivity to block by cisapride. These findings suggest that differences in gating associated positioning of S6 aromatic residues relative to the central cavity, not inactivation per se, determine drug block of hERG and EAG channels. The principle of gating-dependent drug interactions may be exploited for designing molecules with selectivity. For example, even though the same residues critical for correolide binding of Kv1.3 are present in other channels from the Kv1.x family, only Kv1.3, which possess C-type inactivation and associated conformational changes, can be selectively blocked [36]. It is also possible that channels with altered gating behavior (as, for example, under pathophysiologic conditions) may be selectively targeted.

7.1.7.3 Interaction with Auxiliary Subunits

As noted earlier, prototypical openers and blockers of K_{ATP} channels interact via the sulfonylurea receptor (SUR). It was first determined that binding of openers is critically dependent on the presence of nucleotides (MgATP), as revealed by the Mg dependence of [³H]P1075 binding to native membrane preparations and in heterologously expressed SUR subunits [39, 40]. Chimeric approaches exploiting pharmacological differences between SUR isoforms have enabled identification of residues involved in binding of openers and blockers. For example, the SUR2 isoforms confer high sensitivity to KCOs such as cromakalim and low sensitivities to sulfonylureas and, conversely, SUR1 imparts high sensitivity to sulfonylureas. Upon initial identification that the last transmembrane domain TMD2 was important in stimulation by benzopyran openers, two smaller segments were identified: the cytoplasmic loop connecting TM helices 13 and 14, and a region encompassing TM helices 16 and 17 and a short segment of nucleotide binding domain 2 (NBD2). Finally, two residues within TM helix 17 (Leu1249 and Thr1253 in SUR2A, Thr1286 and Met1290 in SUR1) were shown to be necessary and sufficient since mutation of any of these residues could confer to SUR1 a sensitivity to openers that was either partial in the case of T1286L or equivalent to wild-type SUR2A for M1290T. Further site-directed mutagenesis to investigate the role of this important residue of helix 17, SUR2A-Thr1253 (and the corresponding residues in SUR1-Met1290) in opener interactions, led to the conclusion that a threonine enables action of openers, whereas a methionine in its place prohibits it – a phenomenon that relied uniquely on side chain volume and not on shape, polarity, or hydrogen-bonding capacity of the residue [41, 42]. Interaction sites of diazoxide appear to be different and regions in SUR1 spanning TMs 6–11 and NBD1 have been implicated.

The C-terminal 42 amino acid residues that are different between SUR2A and SUR2B modulate the interaction of both NBDs with intracellular nucleotides [43]. In particular, the NBDs of SUR2x dimerize in response to ATP and nicorandil and it has been suggested that this dimerization induces the opening of the K_{ATP} channel, probably by causing a conformational change of SUR [44].

The interaction domains of glyburide and tolbutamide are positioned near cytoplasmic loops of TM12-17 of SUR1 based on the identification of segments required for high-affinity tolbutamide inhibition and [^{3}H]glyburide binding. In fact, high-affinity sulfonylurea binding was imparted on SUR2B by substituting the region separating the two KCO binding segments with the corresponding domain from SUR1. These studies show that the sulfonylurea binding pocket lies in close to the KCO binding sites within the second set (12–17 segments) of transmembrane domains, in support of pharmacological data indicating that these sites may be closely coupled. Unlike sulfonylurea analogs, the interactions of imidazolines such as phentolamine appear to reside on the Kir6.x subunit, as revealed by studies using a truncated Kir6.2 mutant that expresses in the absence of a sulfonylurea receptor. Understanding the molecular basis and mechanisms underlying opener and blocker sensitivities of SUR-Kir combinations as well as SUR-independent ligand actions will provide a basis for further exploitation of this class of K$^+$ channels, especially in the design of tissue specific compounds.

7.1.8
Peptides and Toxins

A range of neurotoxins have served as invaluable tools for identification, purification, functional characterization, tissue distribution and for probing structure and function of various ion channels, including K$^+$ channels. Since the first venom peptide isolated from *Centruroides noxius* was shown to inhibit voltage-gated K^+ channels [45], more than 80 toxins that modulate different classes of K^+ channels have been identified from venoms such as charybdotoxin (scorpions), dendrotoxins (green mamba snake), apamin and tertiapin-Q (bees), hanatoxin (spiders), APETx1, BgK, ShK (sea anemones) and (κ-M conotoxin RIIIK (marine cone snail). These peptidyl toxins affect K^+ channel function either by occluding the ion-conducting pore or by altering gating mechanism [46–48]. Furthermore, studies of toxin–ion channel interactions have revealed a membrane-access mechanism by which a voltage-sensor toxin gains access to its binding sites by partitioning into the lipid membrane [49]. Due to relatively small peptide sequences, tightly folded and stable structures, venom toxins have acted as molecular calipers to elucidate possible three-dimensional structural models of receptors that interact with toxins [50–54].

Although polypeptide toxins serve as pharmacological tools to elucidate roles of K^+ channels and their functional domains, no toxins or toxin analogs have yet been developed as therapeutic drugs. However, the utilization of toxins as leads continues to be actively investigated. For example, ShK-Dap22, derived from ShK by replacing lysine at position 22 with non-natural amino acid diaminopropionic acid, blocks Kv1.3 in T-lymphocytes [55]. Shk-Dap22 has been proposed as a useful molecule to inhibit activated T-cells as immunosuppressant, multiple sclerosis and other autoimmune diseases. Studies of toxin–channel interaction could provide invaluable information for designing potent and selective K^+ channel modulators with potential clinical utility. Indeed, the development of a synthetic equivalent of marine snail venom ω-conotoxin MVIIA, Ziconotide (PRIALT^R), a highly selective and potent N-type Ca^{2+} channel blocker for the treatment of severe chronic pain [56], provides precedent for such advancement.

7.1.9
Summary

K^+ channels continue to be embraced as drug targets for a range of central nervous system, cardiovascular and non-cardiovascular indications. Our understanding of the role of K^+ channels in various physiological processes, including neuronal signaling, vascular and nonvascular muscle contractility, cardiac pacing, auditory function, hormone secretion, immune function and cell proliferation, has been underscored by the recent flurry of discoveries linking K^+ channel mutations to various inherited disorders and the elucidation of K^+ channel structure. Molecular dissection of ion channels involved in cardiac excitability has enabled identification of drug interactions with even greater precision, and molecular counterparts such as hERG are being avoided in the development of new drugs to eliminate the risk of drug-induced QT and related cardiovascular effects. Notably, the selectivity of ligand interactions remains a key challenge since targeting drugs in an organ- or tissue-selective manner is of vital importance to have clinically meaningful benefits.

References

1. Hille, B., *Ion Channels of Excitable Membranes.* 3rd edn. **2001**, Sinauer Associates, Inc., Sunderland, MA.

2. Papazian, D.M., et al. Cloning of genomic and complementary DNA from *Shaker*, a putative potassium channel gene from Drosophila. *Science* **1987**, 237, 749–753.

3. MacKinnon, R. Potassium channels. *FEBS Lett.* **2003**, 555, 62–65.

4. MacKinnon, R. Potassium channels and the atomic basis of selective ion conduction. *Biosci. Rep.* **2004**, 24, 75–100.

5. Lu, Z. Mechanism of rectification in inward-rectifier K^+ channels. *Annu. Rev. Physiol.* **2004**, 66, 103–129.

6. Frank, N. P., Honore, E. The TREK K2P channels and their role in general anaesthesia and neuroprotection. *Trends Pharmacol. Sci.* **2004**, 25, 601–608.

7. Yost, C. Update on tandem pore (2P) domain K^+ channels. *Curr. Drug Targets* **2003**, 4, 347–351.

8. Birnbaum, S.G.V., et al. Structure and function of Kv4-family transient potassium channels. *Physiol. Rev.* **2004**, 84, 803–833.

9. Doyle, D.A., et al. The structure of the potassium channel: molecular basis of K^+ conduction and selectivity. *Science* **1998**, 280, 69–77.

10. Jiang, Y., et al. Crystal structure and mechanism of a calcium-gated potassium channel. *Nature* **2002**, 417, 515–522.

11. Kuo, A., et al. Crystal structure of the potassium channel KirBac1.1 in the closed state. *Science* **2003**, 300, 1922–1926.

12. Jiang, Y., et al. The principle of gating charge movement in a voltage-dependent K^+ channel. *Nature* **2003**, 423, 42–48.

13. Zhou, M., et al. Potassium channel receptor site for the inactivation gate and quaternary amine inhibitors. *Nature* **2001**, 411, 657–661.

14. Jiang, Y., et al. The open pore conformation of potassium channels. *Nature* **2002**, 417, 523–526.

15. Jiang, Y., et al. X-ray structure of a voltage-dependent K^+ channel. *Nature* **2003**, 423, 33–41.

16. Mitcheson, J.S., et al. A structural basis for drug-induced long QT syndrome. *Proc. Natl. Acad. Sci. U.S.A.* **2000**, 97, 12 329–12 333.

17. Liu, H., et al. Structure-based discovery of potassium channel blockers from natural products: virtual screening and electrophysiological assay testing. *Chem. Biol.* **2003**, 10, 1103–1113.

18. Bains, W., A. Basman, C. White. HERG binding specificity and binding site structure: evidence from a fragment-based evolutionary computing SAR study. *Prog. Biophys. Mol. Biol.* **2004**, 86, 205–233.

19. Cavalli, A., et al. Toward a pharmacophore for drugs inducing the long QT syndrome: insights from a CoMFA study of HERG K^+ channel blockers. *J. Med. Chem.* **2002**, 45, 3844–3853.

20. Ekins, S., et al. Three-dimensional quantitative structure-activity relationship for inhibition of human *ether-a-go-go*-related gene potassium channel. *J. Pharmacol. Exp. Ther.* **2002**, 301, 427–434.

21. Shieh, C.C., et al. Potassium channels: molecular defects, diseases, and therapeutic opportunities. *Pharmacol. Rev.* **2000**, 52, 557–594.

22. Sanguinetti, M. C., Mitcheson, J. S. Predicting drug-hERG channel interactions that cause acquired long QT syndrome. *Trends Pharmacol. Sci.* **2005**, 26, 119–124.

23. Blank T, N.I., Kye M.J., Spiess J. Small conductance Ca^{2+}-activated K^+ channels as targets of CNS drug development. *Curr. Drug Targets CNS Neurol. Disord.* **2004**, 3, 161–167.

24. Coghlan, M.J., W.A. Carroll, M. Gopalakrishnan. Recent developments in the biology and medicinal chemistry of potassium channel modulators: update from a decade of progress. *J. Med. Chem.* **2001**, 44, 1627–1653.

25. Heurteaux, C., et al. TREK-1, a K^+ channel involved in neuroprotection and general anesthesia. *EMBO J.* **2004**, 23, 2684–2695.

26. Armstrong, C.M. Inactivation of the potassium conductance and related phenomena caused by quaternary ammonium ion injection in squid axons. *J. Gen. Physiol.* **1969**, 54, 553–575.

27. Armstrong, C.M. Interaction of tetraethylammonium ion derivatives with the potassium channels of giant axons. *J. Gen. Physiol.* **1971**, 58, 413–437.

28. Holmgren, M., P.L. Smith, G. Yellen. Trapping of organic blockers by closing of voltage-dependent K^+ channels: evidence for a trap door mechanism of activation gating. *J. Gen. Physiol.* **1997**, 109, 527–535.

29. Mitcheson, J.S., J. Chen, M.C. Sanguinetti. Trapping of a methanesulfonanilide by closure of the HERG potassium channel activation gate. *J. Gen. Physiol.* **2000**, 115, 229–240.

30. Sanguinetti, M.C., et al. A mechanistic link between an inherited and an acquired cardiac arrhythmia: HERG encodes the IKr potassium channel. *Cell* **1995**, 81, 299–307.

31. Ficker, E., et al. Molecular determinants of dofetilide block of HERG K^+ channels. *Circ. Res.* **1998**, 82, 386–395.

32. Chen, J., G. Seebohm, M.C. Sanguinetti. Position of aromatic residues in the S6 domain, not inactivation, dictates cisapride sensitivity of HERG and eag potassium channels. *Proc. Natl. Acad. Sci. U.S.A.* **2002**, 99, 12 461–12 466.

33. Sanchez-Chapula, J.A., et al. Molecular determinants of voltage-dependent human ether-a-go-go related gene (HERG) K^+ channel block. *J. Biol. Chem.* **2002**, 277, 23 587–23 595.

34. Sanchez-Chapula, J.A., et al. Voltage-dependent profile of human *ether-a-go-go*-related gene channel block is influenced by a single residue in the S6 transmembrane domain. *Mol. Pharmacol.* **2003**, 63, 1051–1058.

35. Kamiya, K., et al. Open channel block of HERG K^+ channels by vesnarinone. *Mol. Pharmacol.* **2001**, 60, 244–253.

36. Hanner, M., et al. Binding of correolide to the Kv1.3 potassium channel: characterization of the binding domain by site-directed mutagenesis. *Biochemistry* **2001**, 40, 11 687–11 697.

37. Seebohm, G., et al. Pharmacological activation of normal and arrhythmia-associated mutant KCNQ1 potassium channels. *Circ. Res.* **2003**, 93, 941–947.

38. Decher, N., et al. Molecular basis for Kv1.5 channel block: conservation of drug binding sites among voltage-gated K^+ channels. *J. Biol. Chem.* **2004**, 279, 394–400.

39. Dickinson, K.E., et al., Nucleotide regulation and characteristics of potassium channel opener binding to skeletal muscle membranes. *Mol. Pharmacol.* **1997**, 52, 473–481.

40. Hambrock, A., et al. Mg^{2+} and ATP dependence of K_{ATP} channel modulator binding to the recombinant sulfonylurea receptor, SUR2B. *Br. J. Pharmacol.* **1998**, 125, 577–583.

41. Moreau, C., et al. The molecular basis of the specificity of action of K_{ATP} channel openers. *EMBO J.* **2000**, 19, 6644–6651.

42. Moreau C., et al. The size of a single residue of the sulfonylurea receptor dictates the effectiveness of K_{ATP} channel openers. *Mol. Pharmacol.* **2005**, 67, 1026–1033.

43. Yamada, M, Kurachi, Y. The nucleotide-binding domains of sulfonylurea receptor 2A and 2B play different functional roles in nicorandil-induced activation of ATP-sensitive K^+ channels. *Mol. Pharmacol.* **2004**, 65, 1198–1207.

44. Yamada, M., et al. Mutation in nucleotide-binding domains of sulfonylurea receptor 2 evokes Na-ATP-dependent activation of ATP-sensitive K^+ channels: implication for dimerization of nucleotide-binding domains to induce channel opening. *Mol. Pharmacol.* **2004**, 66, 807–816.

45. Carbone, E., et al. Selective blockage of voltage-dependent K^+ channels by a novel scorpion toxin. *Nature* **1982**, 296, 90–91.

46. Miller, C. The charybdotoxin family of K^+ channel-blocking peptides. *Neuron* **1995**, 15, 5–10.

47. Swartz, K.J., R. MacKinnon. Mapping the receptor site for hanatoxin, a gating modifier of voltage-dependent K^+ channels. *Neuron* **1997**, 18, 675–682.

48. Rodriguez de la Vega, R.C., L.D. Possani. Current views on scorpion toxins specific for K^+-channels. *Toxicon* **2004**, 43, 865–875.

49. Lee, S.Y., R. MacKinnon. A membrane-access mechanism of ion channel inhibition by voltage sensor toxins from spider venom. *Nature* **2004**, 430, 232–235.

50. Rauer, H., et al. Structural conservation of the pores of calcium-activated and voltage-gated potassium channels determined by a sea anemone toxin. *J. Biol. Chem.* **1999**, 274, 21 885–21 892.

51. Li-Smerin, Y., K.J. Swartz. Localization and molecular determinants of the Hanatoxin receptors on the voltage-sensing domains of a K⁺ channel. *J. Gen. Physiol.* **2000**, 115, 673–684.

52. Regaya, I., et al. Evidence for domain-specific recognition of SK and Kv channels by MTX and HsTx1 scorpion toxins. *J. Biol. Chem.* **2004**, 279, 55 690–55 696.

53. Frenal, K., et al. Exploring structural features of the interaction between the scorpion toxinCnErg1 and ERG K⁺ channels. *Proteins* **2004**, 56, 367–375.

54. Imredy, J.P., R. MacKinnon. Energetic and structural interactions between delta-dendrotoxin and a voltage-gated potassium channel. *J. Mol. Biol.* **2000**, 296, 1283–1294.

55. Kalman, K., et al. ShK-Dap22, a potent Kv1.3-specific immunosuppressive polypeptide. *J. Biol. Chem.* **1998**, 273, 32 697–32 707.

56. Miljanich, G.P. Ziconotide: neuronal calcium channel blocker for treating severe chronic pain. *Curr. Med. Chem.* **2004**, 11, 3029–3040.

7.2
Kv1.3 Potassium Channel: Physiology, Pharmacology and Therapeutic Indications

K. George Chandy, Heike Wulff, Christine Beeton, Peter A. Calabresi, George A. Gutman, and Michael Pennington

7.2.1
Introduction

Patch-clamp studies performed over two decades ago revealed a voltage-gated potassium channel in T lymphocytes with distinctive biophysical properties [1–6]. Initially it was called *type n* for "normal". Several blockers of this channel – verapamil (IC_{50} 6 µM), quinine (14 µM), diltiazem (60 µM), 4-aminopyridine (190 µM) and tetraethylammonium (8 mM) – suppressed mitogen-stimulated ^{3}H-thymidine and ^{3}H-leucine incorporation by human T cells at concentrations that blocked the channel [3, 7] but which were not cytotoxic [3]. The same blockers also suppressed human allogeneic mixed-lymphocyte responses, indicating a requirement for functional *type n* potassium channels for the proliferation of mitogen- and allogeneic cell-stimulated T lymphocytes [3]. T lymphocyte proliferation could only be inhibited if the blockers were added during the first 20–30 h after the stimulating signal, indicating that *type n* channels were necessary only during the early stages of lymphocyte activation [3]. In keeping with this idea, phytohemagglutinin-induced increases in intracellular calcium levels were partly inhibited by pre-treatment with 4-aminopyridine or verapamil, and, if added after the rise, the free calcium level was brought down partway to baseline [8]. *Type n* channel blockers also suppressed the production of interleukin-2 but not the expression of the interleukin-2 receptor, and exogenous interleukin-2 partially relieved inhibition [3, 9, 10]. Pharmacological studies also defined a role for this channel in volume regulation [11–13] and calcium signaling in T cells [8, 9, 14, 15]. Subsequent studies revealed the presence of *type n* channels in natural killer cells [16, 17] and B-lymphocytes [18–22], indicating that they play an important role in all three lymphocyte lineages.

The *type n* channel was shown to be the product of the Kv1.3 gene on human chromosome 1p13.3 [23–26], by demonstrating that its biophysical properties were identical to those of the cloned Kv1.3 homotetrameric channel, and by detecting Kv1.3 mRNA and protein in T cells [27–30]. Kv1.3 was later found in B-lymphocytes [31], macrophages [32, 33], microglia [34, 35], osteoclasts [36], platelets [37], oligodendrocytes [38, 39], and the olfactory bulb [40, 41]. In the brain, biochemical studies demonstrated that Kv1.3 subunits could associate with other Kv1-family subunits to form functional heteromultimeric channels [42–44]. Kv1.3 transcripts were also found in CD33$^+$ myeloid precursors in the human bone marrow and in interstitial cells of the testis (http://symatlas.gnf.org/SymAtlas/).

Many academic and industrial groups in the 1990s attempted to develop selective and potent blockers of Kv1.3 because of the channel's role in immune function (see [45–48] for recent reviews). The Merck group were the first to demonstrate that Kv1.3 blockers were immunosuppressive *in vivo*, although the blockers they used – margatoxin and correolide – were not specific for Kv1.3 [49, 50]. In 2001, Beraud's group at the University of Marseille and our group used two other Kv1.3 blockers – kaliotoxin and ShK – to ameliorate disease in a rat model of multiple sclerosis (MS) [51, 52]. Kaliotoxin was also shown to reduce T cell-mediated bone resorption in experimental periodontal disease [53]. The attractiveness of Kv1.3 as a therapeutic target was enhanced when we discovered that Kv1.3 blockers preferentially suppress the proliferation of late-stage memory T and B cells that are implicated in human autoimmune diseases (Table 7.2.1), without compromising the function of naïve and early memory lymphocytes [31, 54]. Recent reports that Kv1.3-null mice are protected from diet-induced obesity [55] and are also "super-smellers" [41] further suggest the possibility of using

Tab. 7.2.1 Phenotype of T and B lymphocytes in autoimmune disorders.

Phenotype of auto-reactive lymphocytes	Disease	Target organ(s)
T lymphocytes:		
CD28-costimulation independent	Multiple sclerosis (MS) [220]	Central nervous system
	Type-1 diabetes mellitus (T1DM) [224]	Pancreas
	Rheumatoid arthritis (RA) [372]	Joints
	Crohn's disease [389]	Digestive tract
CD45RA$^-$CD45RO$^+$	Multiple sclerosis [54, 390]	Central nervous system
	Rheumatoid arthritis [372]	Joints
	Psoriasis [226, 227]	Skin
	Pemphigus foliaceus [229],	Skin
	Pemphigus vulgaris [391]	
	Scleroderma [392, 393]	Skin
	Vitiligo [394]	Skin, mucus membranes
	Uveitis [395]	Eyes
	Crohn's disease [389]	Digestive tract
	Inclusion body myositis, Polymyositis,	Skin, skeletal muscles
	Dermatomyositis [392]	Thyroid
	Hashimoto's disease [396]	
CCR7$^-$	Multiple sclerosis [54]	Central nervous system
	Rheumatoid arthritis [372]	Joints
	Chronic Graft versus host [230]	Skin, liver, mouth, eyes
Kv1.3^{high}	Multiple sclerosis [54]	Central nervous system
B lymphocytes:		
IgD$^-$CD27$^+$ class-switched	Multiple sclerosis [244, 245]	Central nervous system
	Grave's disease, Hashimoto's disease [242]	Thyroid
	Sjögren's syndrome [241]	Salivary glands
	Systemic Lupus erythematosus (SLE) [243, 247]	Multiple organs (joints, skin, kidneys, lungs, eyes)

Kv1.3 blockers for the management of obesity, a growing international health problem, and also in the treatment of anosmia, a disease that affects significant numbers of elderly people.

In this chapter we discuss two decades of effort to develop specific and potent Kv1.3 blockers for use as therapeutics. The first two sections deal with peptide and small molecule inhibitors of the Kv1.3 channel, the third section discusses the physiological role of Kv1.3 in different cell types and the final section assesses the therapeutic potential of Kv1.3 blockers.

7.2.2
Peptide Inhibitors of Kv1.3

Scorpions, snakes, sea anemones and marine cone snails are major sources of peptide inhibitors, which target Kv1.3 channels (Table 7.2.2). In the late 1980s, venoms from these creatures were subjected to RP-HPLC separation techniques and numerous individual polypeptide components were isolated and characterized. Kv1.3 blocking peptides are primarily short peptides (<70 residues) that are stabilized by multiple disulfide bonds. Key features of these peptides such as disulfide pairing, structural motifs and binding residues have been defined using peptide synthesis, NMR and peptide sequencing. For a relatively recent review on toxin nomenclature see Ref. [56].

7.2.2.1 **Scorpion Toxins**
The first Kv1.3 channel-blocking peptide inhibitor was isolated from the venom of the scorpion *Leiurus quinquestriatus hebraeus* in the early 1980s and was named charybdotoxin (ChTX) after the mythical whirlpool of Greek legend [57–60]. Screening other scorpion venoms led to the identification of additional potent Kv1.3-blocking toxins, including agitoxin-2 [61], hongotoxin [43], margatoxin [62, 63], kaliotoxin [64], noxiustoxin [65], Pi2 [66] and OSK1 [67]. All these peptides contain 37–39 residues stabilized by three disulfide bonds [56, 68], with Cys1 paired to Cys4, Cys2 paired to Cys5 and Cys3 paired with Cys6 (Cys numbering according to their sequential positions in the sequence) (Fig. 7.2.1A and KTX in Fig. 7.2.1E). Solution structures of these peptides revealed a common structure consisting of a classical alpha helix and three folds of an antiparallel beta sheet structure [65, 69–73].

These inhibitors occlude the pore of the ion channel like a cork plugs a wine bottle (Fig. 7.2.1F). They stabilize their interaction with the channel by extending a critical lysine (position 27 in charybdotoxin) into the channel pore, and the remainder of the toxin molecule makes several secondary contacts with channel residues to further increase binding. Through the efforts of Chris Miller's group at Brandeis and Rod MacKinnon's group at Harvard, the interacting surface between scorpion toxins and the Drosophila *Shaker* potassium channel was shown to consist of a vestibule at the outer mouth of the ion conduction pathway [74–84]. During the same period we used complementary mutagenesis of Kv1.3

Tab. 7.2.2 IC$_{50}$ values of peptide and small molecule modulators of Kv1.3 in whole-cell patch-clamp.

Peptides	IC$_{50}$	Small molecules and metals	IC$_{50}$
OSK1-Lys^{16}Asp^{20} [67]	3 pM	PAP-1 [410]	2 nM
Stichodactyla helianthus toxin (ShK) [109]	11 pM	Psora-4 [159]	3 nM
Heterometrus spinnifer toxin 1 (HsTX1) [93]	12 pM	Tetraphenylporphyrin 3$^{[a]}$ [153]	20 nM
Orthochirus scrobiculosus toxin (OSK1) [67]	14 pM	Correolide C18-analog 43 [144]	37 nM
ShK-F6CA [118]	48 pM	*trans*-*N*-Propyl-carbamoyloxy-PAC [141]	50 nM
Pandinus imperator toxin 2 (Pi2) [397]	50 pM	Correolide [138]	90 nM
ShK-Dap^{22} [109]	52 pM	Sulfamidbenzamidoindane [398]	100 nM
ShK(L5) [119]	69 pM	CP-339818 [130]	150 nM
Hongotoxin (HgTX1) [43]	86 pM	WIN-17317-3 [129, 130]	200 nM
Margatoxin (MgTX) [121]	110 pM	UK-78282 [135]	200 nM
Agitoxin-2 (AgTX2) [35]	200 pM	PAC [141]	270 nM
Pandinus imperator toxin 3 (Pi3) [397]	500 pM	Khellinone dimer 2 [163]	280 nM
Kaliotoxin (KTX) [100]	650 pM	Khellinone chalcone 16 [163]	400 nM
Anuroctoxin [95]	730 pM	6-(2,5-Dimethoxyphenyl)psoralen [158]	700 nM
Noxiustoxin (NTX) [100]	1 nM	H-89 [150]	1.7 µM
Charybdotoxin (ChTX) [100]	3 nM	Resiniferatoxin [100]	3 µM
Tityustoxin-Kα (TsTX-Kα) [399]	4 nM	Phenyl-stilbene A [151]	2.9 µM
Pandinus imperator toxin 1 (Pi1) [397]	11 nM	Nifedipine [100]	5 µM
Kbot1 [400]	15 nM	Nitrendipine [125]	5 µM
Bunodosoma granulifera toxin (BgK) [86]	39 nM	Ibu-8 [156]	5 µM
Maurotoxin (MTX) [91]	150 nM	Phencyclidine (PCP) [166]	5 µM
α-Dendrotoxin (DTX) [100]	250 nM	Fluoxetine [401]	6 µM
Parabuthus toxin 3 (PbTX3) [402]	492 nM	Verapamil [3]	6 µM
Parabuthus toxin 1 (PbTX1) [403]	800 nM	H37 [157]	10 µM
ViTX [103]	2 µM	Hg^{2+} [125]	10 µM
κ-Hefutoxin 1 (κ-HfTX1) [404]	150 µM	Kokusagenine [156]	10 µM
Opisthacanthus madagascariensis toxin (OmTX3) [405]	400 µM	Quinine [3]	14 µM
		Cicutoxin [167]	18 µM
		Trifluoperazine [406]	20 µM
		Forskolin [168]	20 µM
		Capsaicin [100]	26 µM
		Diltiazem [100]	27 µM
		Progesterone [164]	30 µM
		Luteolin [119]	50 µM
		La^{3+} [125]	50 µM
		Flecainide [100]	60 µM
		K22-Y23-R11 ShK mimetic [407]	95 µM
		5-MOP [157]	101 µM
		H$_2$O$_2$ [408]	100 µM
		4-AP [100]	195 µM
		Zn^{2+}, Co^{2+} [125]	200 µM
		Melatonin [409]	1.5 mM
		Ba^{2+}, Cd^{2+} [125]	2 mM
		TEA [100]	10 mM
		Mn^{2+} [125]	20 mM

[a] Values determined in binding assays.

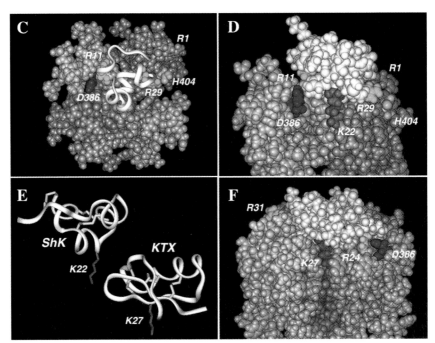

Fig. 7.2.1 (A) Sequence alignment of the scorpion toxins charybdotoxin (ChTX), margatoxin (MgTX) and kalitoxin (KTX). Disulfide bridges are shown as connecting lines. (B) Sequence and disulfide bridge pattern of the sea anemone toxin ShK. (C) Top view of ShK (white) docked into the outer vestibule of Kv1.3 (brown). All four subunits are shown. (D) Side view of ShK docked into Kv1.3. The channel subunit closest to the viewer has been removed to make the channel pore visible. ShK-Lys^{22} (red) protrudes into the pore. (E) Backbone structures of ShK and KTX with the disulfide bridges shown in yellow and the Lys residues in red. (F) Docking configuration of KTX in Kv1.3. Lys^{27} (red) occludes the pore.

and four related toxins to map the interacting surfaces of the toxins and the channel [71, 85, 86]. By knowing the NMR structures of the toxins and identifying key toxin-channel residues, we deduced the dimensions of the toxin-binding vestibule in Kv1.3 to be 28–32 Å wide at its outer margin, 28–34 Å wide at its base, and 4–8 Å deep (Fig. 7.2.1F). The deduced pore was 9–14 Å wide at its external entrance and tapered to a width of 4–5 Å at a depth of approximately 5–7 Å from the vestibule [71]. The dimensions estimated for the vestibule of the Drosophila *Shaker* channel were very similar [83, 84, 87]. These estimated dimensions turned out to be remarkably accurate when compared with the crystal structure of the bacterial potassium channel KcsA that was determined three years later [88, 89, 90].

Maurotoxin [91, 92], HsTX1 [93], Pi1 [94] and anuroctoxin [95] exemplify a second group of scorpion toxin-inhibitors of Kv1.3 that are held together by four disulfide bonds. Topology-mapping studies like those described above have yet to be performed on these four-disulfide-bonded peptides.

Scorpion toxins block other K$^+$ channels in addition to Kv1.3. Table 7.2.3 compares the blocking affinities of several potent blockers on Kv1.3 and closely related channels. Margatoxin for example, blocks Kv1.2 at picomolar concentrations and Kv1.1 at low nanomolar concentrations and as a consequence enhances intestinal peristaltic activity via Kv1.1 blockade [96]. The most potent and selective scorpion toxin inhibitor of Kv1.3 is OSK1-K^{16}D^{20}, which blocks Kv1.3 with an IC$_{50}$ of 3 pM and shows >300-fold selectivity over closely related channels. None of the other scorpion toxins is as selective for Kv1.3 [67].

Tab. 7.2.3 Selectivity of commonly used Kv1.3 blockers.

Blockers	Kv1.1	Kv1.2	Kv1.3	Kv1.5	KCa3.1	Others
Charybdotoxin	>1 µM	14 nM	3 nM	>100 nM	5 nM	KCa1.1 3 nM
Margatoxin	10 nM 144 pM[a]	520 pM 675 pM[a]	110 pM 230 pM[a]	No effect	No effect	
Kaliotoxin	41 nM	>1 µM	650 pM	>1 µM	>1 µM	
Hongotoxin	31 pM[a]	170 pM[a]	86 pM[a]		No effect	
Noxiustoxin	>25 nM	2 nM	1 nM	>25 nM	No effect	
HsTX1	7 nM	No effect	12 pM		625 nM	
Maurotoxin	No effect	110 pM	150 nM	Not tested	1 nM	
Agitoxin-2	44 pM		4 pM 200 pM		No effect	Kv1.6 44 pM
OSK1	600 pM	5.4 nM	14 pM	>1 µM	225 nM	
OSK1-K^{16}D^{20}	400 pM	2.96 nM	3 pM	>1 µM	228 nM	
Pi2			50 pM			
Anuroctoxin	Not tested	5 nM	730 pM	Not tested	No effect	
α-Dendrotoxin-I	1.1 nM 20 nM	17 nM	200 nM	>1 µM	Not tested	

Tab. 7.2.3 Selectivity of commonly used Kv1.3 blockers (continued).

Blockers	Kv1.1	Kv1.2	Kv1.3	Kv1.5	KCa3.1	Others
ShK	25 pM, 118 pM[a] 0.041 pM[a]	>1 μM	11 pM, 0.9 pM 0.25 pM[a]	No effect	28 nM	Kv1.6 200 pM Kv3.2 5 nM
ShK-F6CA	4 nM	>1 μM	48 pM	No effect	Not tested	
ShK(L5)	7 nM	48 nM	70 pM	No effect	115 nM	Kv1.6 18 nM Kv3.2 20 nM
ShK-Dap[22]	1.8 nM	39 nM	23 pM 52 pM 64 pM 110 pM	>1 μM	>1 μM	Kv1.6 10 nM
BgK	6 nM	15 nM 25 nM	10 nM 39 nM	Not tested	172 nM	
ViTX	2 μM	No effect	2 μM	Not tested		
Correolide (6)	430 nM (^{86}Rb) 21 nM[a]	700 nM (^{86}Rb) 10 nM[a]	90 nM (^{86}Rb) 10 nM[a] 110 nM	1.1 μM (^{86}Rb) 7 nM[a]	Not tested	Kv1.6 450 nM (^{86}Rb) Kv1.6 19 nM[a] Kv1.4 10 nM[a]
PAC (9)	200–400 nM	200–400 nM	149 nM	200–400 nM	Not tested	Kv1.6 200–400 nM
Psora-4 (21)	62 nM	49 nM	3 nM	8 nM	>5 μM	Kv1.4 202 nM Kv1.7 100 nM
PAP-1 (22)	65 nM	250 nM	2 nM	45 nM	10 μM	Kv1.6 62 nM Kv1.7 98 nM
UK-78282 (5)	22 μM	2.9 μM	280 nM	70 μM	> 30 μM	Kv1.4 170 nM Kv1.6 31 μM Kv3.2 127 μM
CP-339818 (4)	62 μM	14 μM	230 nM	19 μM	>500 μM	Kv1.4 300 nM Nav 10 nM
Chalcone-16 (23)	1.2 μM	>50 μM	400 nM	5.1 μM	>100 μM	Kv1.7 10 μM
Khellinone dimer-2 (24)	3.1 μM	2 μM	280 nM	1.1 μM	>100 μM	

[a] Values determined in binding assays. All other IC$_{50}$s determined by whole-cell patch-clamp or where indicated by Rb^{86}-flux.

7.2.2.2 Snake Toxins

α-Dendrotoxins, another class of K$^+$ channel blocking peptides, have been isolated from snakes of the cobra family, the black mamba *Dendroaspis polylepis* and the green mamba *D. angusticeps* [97, 98]. These peptide toxins are larger than those of scorpions, having approximately 59 residues. Their disulfide pattern is also quite different, pairing Cys1 with Cys6, Cys2 with Cys4 and Cys3 with Cys5, and essentially linking the N-terminus with the C-terminus [99]. α-Dendrotoxin blocks Kv1.3 with an IC$_{50}$ of 250 nM (Table 7.2.2), but it is a more potent inhibitor of Kv1.1 and Kv1.2 [100]. Structure–activity studies have utilized both site-directed mutagenesis and chemical synthesis to elucidate the key functional residues of α-dendrotoxin and on Kv1.1 [101], and Lys[5] and Leu[9] appear critical for activity. α-Dendrotoxin adopts a fold very similar to the pancreatic trypsin inhibitors, which places these key functional residues in close proximity, with a separation of approximately 7 Å [102].

7.2.2.3 Cone Snail Toxins

Marine cone snails also produce several potent ion channel blockers, but only one has been described recently that possesses weak Kv1.3 binding. ViTx, isolated from the venom of *Conus virgo*, is 35-residues long and held together by four disulfide bonds. This peptide represents a new superfamily of *Conus* peptides, and it is the first reported to affect vertebrate K[+] channels. Its chemical synthesis was described, but no data for the functional residues or solution structure have yet been reported [103]. Another peptide toxin, κ-M-conotoxin RIIIK from *Conus radiate*, blocks Kv1.2 with an IC_{50} of 200 nM, but has no effect on Kv1.3 or other Kv1-family channels [104].

7.2.2.4 Sea Anemone Toxins

Peptide blockers of K[+] channels were first discovered in a Caribbean sea anemone in the mid-1990s by Cuban scientists and their collaborators [105, 106]. Two peptides – ShK (from *Stichodactyla heliantus*) and BgK (from *Bunadosoma granulifera*) were found to block Kv1.3 [86, 107–110]. The initial sequence data for BgK were incorrect and slowed synthetic efforts until the native molecule was re-sequenced [108]. The ShK sequence data was reported correctly, allowing rapid identification of key functional residues using a peptide synthetic approach [107, 111, 112].

The solution structure of ShK showed it to be of a novel structural class of K[+] channel blockers (Fig. 7.2.1E) [113]. ShK contains 35 residues and is composed of two short alpha helical segments with several chain reversals. The disulfide-bonding pattern was identical to that of the larger dendrotoxins, linking Cys1 with Cys6, Cys2 with Cys4 and Cys3 with Cys5 (Fig. 7.2.1B) [114]. The solution structure of BgK, a 37-residue peptide, is similar to that of ShK with some differences in the N-terminal parts of the toxins [115]. Using an alanine scan, we identified Lys[22] and Tyr[23] in ShK as being essential for blocking Kv1.3 [86, 109, 112]. Mutagenesis studies revealed that ShK stabilized its interaction with the channel by positioning Lys[22] within the channel pore and by secondary contacts with other channel residues (Fig 7.2.1C and 7.2.1D) [86, 109]. Ala-scans of BgK revealed Lys[25], Tyr[26] and Phe[6] to be important for channel blockade [115]. By comparing the structural components of BgK with those of the other Kv1-family channel-blockers, Andre Menez's group at Saclay, France, identified a critical functional dyad – an aromatic (Tyr or Phe) or hydrophobic (Leu) residue close to the key Lys – to be an evolutionary link between Kv1-channel toxins from divergent animal species [115, 116].

ShK has been reported to block Kv1.3, with IC_{50}s ranging from 0.9 pM [117] to 9–16 pM [31, 52, 86, 109, 118, 119] to 133 pM [107] in patch-clamp experiments, and to interact with Kv1.3 with a K_i of 0.25 pM in binding studies (Tables 7.2.2 and 7.2.3) [117]. BgK is 1000-fold less potent than ShK and blocks Kv1.3 with IC_{50}s of 10–39 nM [86, 108]. ShK displays low pM affinity for Kv1.1 (IC_{50} 25 pM) in addition to Kv1.3 [109, 117–119], and also blocks Kv1.6 (IC_{50} 200 pM) and KCa3.1 (28 nM) [109, 119]. ShK has been further reported to be a "potent" inhibitor of the Kv3.2 channel in pancreatic islets [120], but in fact ShK blocks Kv3.2 (IC_{50} 5–6 nM) with 500-fold lower potency than Kv1.3 or Kv1.1 [119].

Efforts to improve the selectivity profile of ShK have benefited greatly from the use of synthetic strategies to introduce non-natural residues into the peptide. The first ShK analog with improved specificity for Kv1.3, ShK-Dap^{22}, contained a di-aminopropionic acid (DAP) in place of the critical Lys^{22}. ShK-Dap^{22} blocked Kv1.3 in the low picomolar range (IC_{50} 3–110 pM) [33, 54, 109, 117, 121, 122] and blocked Kv1.1 (IC_{50} 1.8–2.3 nM) with about 20–90-fold lower potency (Tables 7.2.2 and 7.2.3) [109,117]. The increased specificity resulted from ShK-Dap^{22} binding to Kv1.3 with a different orientation than ShK [123].

A second analog with improved specificity for Kv1.3 was discovered accidentally when we attached fluorescein-6-carboxyl through an Fmoc-Aeea-OH hydrophilic linker to the N-terminus of ShK in order to generate a fluorescent probe [118]. ShK-F6CA blocked Kv1.3 with an IC_{50} of 48 pM and exhibited 80-fold selectivity over Kv1.1 and other Kv channels (Table 7.2.2 and 7.2.3) [118]. Since the fluorescein-6-carboxyl in ShK-F6CA can exist as a restricted carboxylate or a cyclized lactone, we were not certain whether the peptide's enhanced specificity for Kv1.3 stemmed from the hydrophobicity of the fluorescein nucleus or the negative charge of F6CA. While studying this problem we developed a novel analog, ShK(L5), in which L-phosphotyrosine (*pTyr*) was attached to the N-terminus of ShK via the Fmoc-Aeea-OH hydrophilic linker [119]. ShK(L5) is 100-fold selective for Kv1.3 (IC_{50} 70 pM) over Kv1.1 (7 nM), 260-fold selective over Kv1.6 (18 nM), 280-fold selective over Kv3.2 (20 nM), 680-fold selective over Kv1.2 (48 nM), and 1600-fold selective over KCa3.1 (115 nM). Most importantly, ShK(L5) has no perceptible effect on the cardiac HERG channel (Kv11.1) [119] that is responsible for the drug-induced Long QT syndrome [124]. ShK(L5) is remarkably stable in plasma, and serum samples from ShK(L5)-treated rats exhibit the same specificity for Kv1.3 over Kv1.1 as ShK(L5). This indicates that the peptide is not modified *in vivo* by dephosphorylation of *pTyr* or cleavage of the hydrophilic linker since such modification would generate either ShK or an ShK analog less selective for Kv1.3 over Kv1.1 [119]. ShK(L5) is discussed in more detail in Section 7.2.5.

7.2.3
Small Molecules Inhibitors of Kv1.3

Several compounds were found in the early 1980s to block the *type n*/Kv1.3 channel at micromolar to millimolar concentrations, including the classical K_V channel inhibitors 4-aminopyridine and tetraethylammonium, the calcium-activated K$^+$

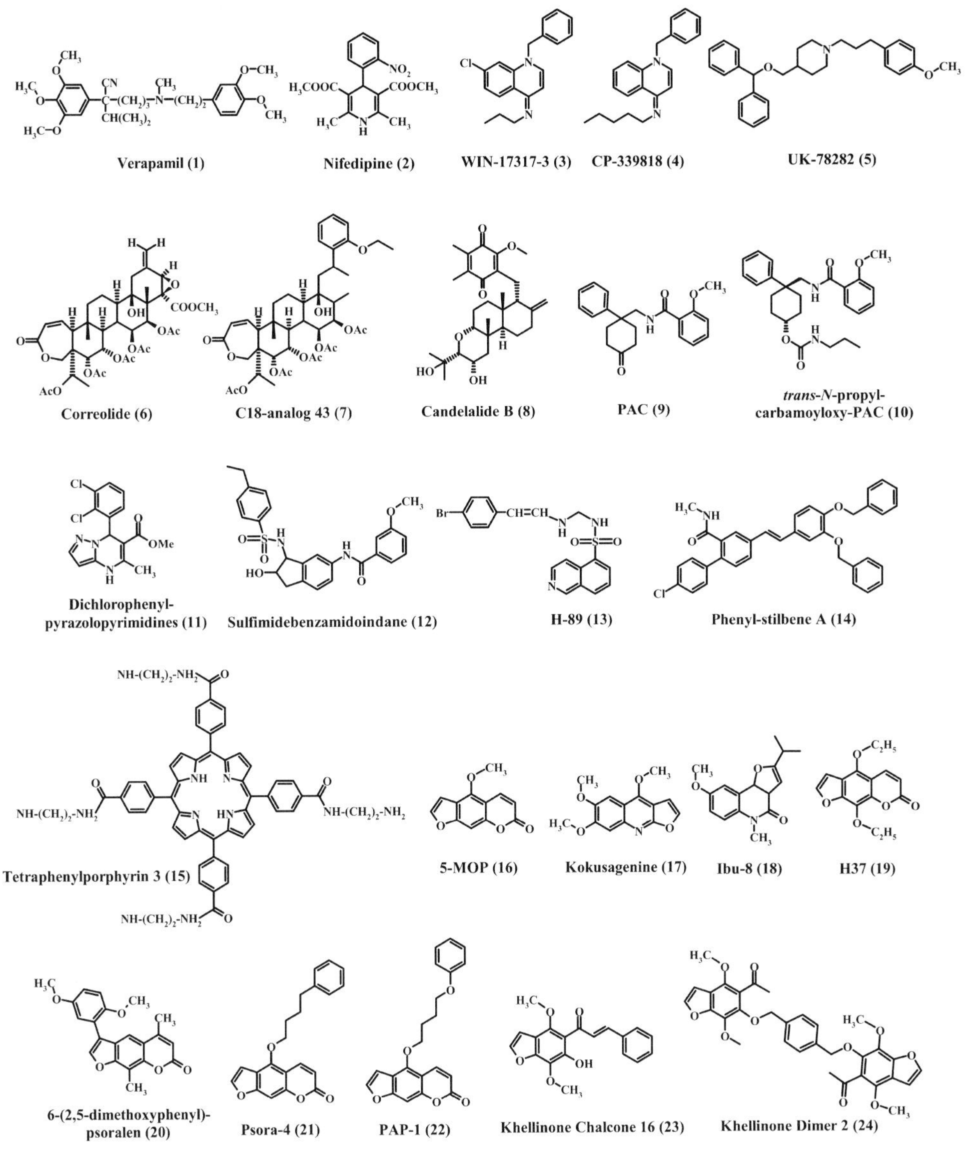

Fig. 7.2.2 Structures of small molecule Kv1.3 blockers.

channel blockers quinine and ceteidil, the phenothiazine antipsychotics chlorpromazine and trifluoperazine, the classical calcium channel inhibitors verapamil (**1**), diltiazem, nifedipine (**2**) and nitrendipine, and the β-blocker propranolol (Table 7.2.2) [1, 3, 5, 8, 125–127]. Inorganic polyvalent cations that block calcium channels in other preparations also blocked the *type n*/Kv1.3 channel in T cells – $Hg^{2+} > La^{3+} > Zn^{2+}$, $Co^{2+} > Ba^{2+}$, $Cd^{2+} > Mn^{2+}$, $Ca^{2+} > Sr^{2+} > Mg^{2+}$ (Table 7.2.2)

[125]. Since the mid-1990s efforts by scientists at pharmaceutical companies and in academia have yielded more potent and more drug-like small molecule Kv1.3 blockers (Fig. 7.2.2). These compounds fall roughly into two groups: (1) Typical combinatorial library compounds, like CP-339818 (**4**), UK-78282 (**5**), and phenyl-stilbene A (**14**) that have a relatively simple structure and low molecular weight, and are rich in nitrogen and halogen atoms. (2) Natural products or natural product-derivatives like the terpenoids correolide (**6**) and candelalide B (**8**), the psoralens and the khellinones, that are rich in oxygen atoms and have a more complex stereochemistry. Both types of small molecule inhibitors preferentially bind to a post-activation state of Kv1.3 called the C-type inactivated state. Interestingly, some of the more potent small molecule Kv1.3 blockers like the PACs, the khellinones and Psora-4 have Hill coefficients of two, indicating that two drug molecules interact with a single channel tetramer.

7.2.3.1 Dihydroquinolones

In 1995 chemists at Sterling-Winthrop identified the first nanomolar small molecule Kv1.3 blocker in a high-throughput ^{125}I-ChTX-displacement screen [128]. The lead compound, WIN-17317-3 (**3**) (1-benzyl-7-chloro-4-*n*-pentylimino-1,4-dihydroquinoline hydrochloride) blocked Kv1.3 with an IC_{50} of ~200 nM and suppressed human T cell activation (Fig. 7.2.2 and Table 7.2.4) [128, 129]. We evaluated three WIN-17317-3 analogs, CP-339818 (**4**) (1-benzyl-4-pentylimino-1,4-dihydroquinoline), CP-393223 (1-pentyl-4-pentylimino-1,4-dihydroquinoline) and CP-394322 (1-naphthyl-4-pentylimino-1,4-dihydroquinoline) to determine the mechanism of channel block [130]. All three compounds were equally potent nanomolar inhibitors of Kv1.3 in the ^{125}I-ChTX-displacement assay. Replacement of the aromatic group in N1 position with the smaller and less lipophilic allyl group (CP-393224) reduced potency 30-fold in the ^{125}I-ChTX-displacement assay, suggesting that a large lipophilic group at the N1 position was necessary for Kv1.3 blockade [130]. Since the pK_a for these compounds was ~11, they likely blocked the channel in a charged form at physiological pH. These compounds were potent in patch-clamp measurements (e.g., IC_{50} for CP-339818 ~150 nM; Hill coefficient = 1) but were ~10-fold less effective in blocking ^{86}Rb-efflux through Kv1.3 channels (IC_{50} for CP-339818 ~2300 nM) (Table 7.2.4). The difference in potency between ^{125}I-ChTX-displacement and patch-clamp assays on the one hand and ^{86}Rb-efflux on the other was determined to be because these compounds block Kv1.3 by interacting with residues exposed when the channel undergoes a conformational change during C-type inactivation. This form of channel inactivation occurs by a cooperative mechanism involving all four subunits in the tetramer and results in a dynamic rearrangement of the outer mouth of the channel [131–133]. In patch-clamp assays, CP-339818 (**4**) needed multiple channel openings to reach steady-state block [130]. The time to steady state block was lessened by biophysical manipulations that increased C-type inactivation (e.g., using a holding potential of −50 instead of −80 mV, or lengthening the depolarizing pulse duration from 200 ms to 2 s). Removal of C-type inactivation either by introducing point mutants into the outer mouth of the channel or by increasing the external K^+ ion concentration reduced the channel's affinity for CP-339818 (**4**) due to decreased availability of sites for drug binding [130]. These studies illustrated that assays using high extracellular K^+ to initiate ^{86}Rb-efflux flux may underestimate the potency of Kv1.3 blockers that bind to the C-type inactivated state.

CP-339818 (**4**) was poorly selective for Kv1.3 [130] since it also blocked the neuronal and cardiac channel Kv1.4 at nanomolar concentrations (300 nM), but it exhibited >100-fold selectivity over other Kv1-family channels (Table 7.2.3). In sub-

Tab. 7.2.4 Comparison of the pharmacological properties of commonly used Kv1.3 blockers.

	Patch-clamp	Displacement of ^{125}I-peptide binding	Inhibition of ^{86}Rb efflux	Suppression of ^{3}H-thymidine incorporation of naïve T cells	Suppression of IL2 production by naïve T cells	Suppression of ^{3}H-thymidine incorporation by T$_{EM}$ cells	Suppression of IL2 production by T$_{EM}$ cells
ShK	11 pM	118 pM 0.041 pM	10 pM	4000 pM 5000 pM 17 pM		400 pM 80 pM	
ShK(L5)	70 pM	Not done	Not done	5000 pM		80 pM	~300 pM
Margatoxin	110 pM 90 pM	240 pM 11 pM	25 pM 90 pM 110 pM	300 pM 290 pM 18 pM	13 pM 500 pM		
WIN-17317-3 (**3**)	335 nM	83 nM	Not done	Not done	800 nM		
CP-339818 (**4**)	150 nM	120 nM	2.3 μM	4.7 μM	Not done		
CP-394322	230 nM	90 nM	2100 nM				
UK-78282 (**5**)	280 nM	700 nM	400 nM	2.6 μM	3.6 μM		
Correolide (**6**)	110 nM	> 30 μM	90 nM	300 nM	44 nM		
C18-correolide analog 43 (**7**)	Not done		37 nM	12.3 nM			
PAC (**9**)	~200 nM	Not done	149 nM	1 μM			
trans-N-Propyl-carbamoyloxy PAC (**10**)	Not done	5 nM	50 nM	340 nM	Comparable to ^{3}H-thymidine but data not shown		
Psora-4 (**21**)	3 nM	Not done	Not done	600 nM	Not done	25 nM 60 nM	
PAP-1 (**22**)	2 nM	Not done	Not done	800 nM	Not done	10 nM	
Chalcone-16 (**23**)	400 nM	Not done	Not done	600 nM			

sequent studies at Merck, WIN-17317-3 (**3**) was shown to block the rat brain IIA sodium channel with a K_i of 9 nM [134]. Thus, both **3** and **4** lacked the potency and selectivity to be therapeutic drugs.

7.2.3.2 Piperidines

UK-78282 (**5**) (4-[diphenylmethoxymethyl]-1-[3-(4-methoxyphenyl)-propyl]piperidine) was discovered at Pfizer in a library screen using a high-throughput ^{86}Rb efflux assay [135]. UK-78282 blocked Kv1.3 with an IC_{50} of ~200 nM and a Hill coefficient of 1. A related analog, CP-190325, containing a benzyl moiety in place of the benzhydryl at position 4 in UK-78282, blocked Kv1.3 with 10-fold lower potency (IC_{50} 2.3 µM). Unlike the dihydroquinolones [130], UK-78282 (**5**) and CP-190325 suppressed ^{86}Rb-efflux through Kv1.3 at concentrations comparable to channel block in patch-clamp experiments (IC_{50}s of 400 nM and 1.9 µM respectively) (Fig. 7.2.2 and Table 7.2.4). Like the dihydroquinolines [130], UK-78282 (**5**) blocked the channel by interacting with residues exposed in the C-type inactivated conformation [135]. Mutations in the external vestibule of the channel that altered the rate of C-type inactivation changed the channel's sensitivity to UK-78282, and there was a direct correlation between the inactivation time constants and the IC_{50} values [135]. Despite UK-78282's ability to inhibit ^{125}I-ChTX-binding to its receptor in the external vestibule of Kv1.3, and notwithstanding the effect of external vestibular mutations on the channel's sensitivity to the drug, competition experiments demonstrated that UK-78282 (**5**) interacts with residues at the inner surface of Kv1.3, at a binding site overlapping that of verapamil [135, 136]. Based on these results we suggested that residues at the internal drug-binding site only became accessible when the channel entered the C-type inactivated state, and the drug, once bound to this internal site, allosterically interfered with ^{125}I-ChTX-binding to the external vestibule. UK-78282 (**5**) displayed 100-fold selectivity over most other Kv1-family channels with the exception of Kv1.4 (Table 7.2.3) [135]. Studies on **5** and its analogs were discontinued because the compound lacked the potency and selectivity necessary for a therapeutic.

7.2.3.3 Correolide

Scientists at Merck used a high-throughput ^{86}Rb-flux assay to identify a very different class of Kv1.3 blockers in extracts from the bark and roots of the Costa Rican tree *Spachea correa* [137]. The pentacyclic nor-triterpene correolide (**6**), containing an unusual α,β-unsaturated seven-membered lactone ring (Fig. 7.2.2), blocked Kv1.3 with a IC_{50} of 90 nM [121, 138] (Tables 7.2.2 and 7.2.4), presumably by interacting with the inactivated state of the channel [139]. Site-specific mutagenesis later defined the correolide binding-site in the intracellular cavity of Kv1.3 [140]. Molecular models showed correolide's saturated hydrocarbon surface "snuggling" into the contour of the hydrophobic surface of the S6 helix with the five polar acetyl groups facing into the water-filled cavity [140]. Unlike UK-78282 (**5**) [135] binding of correolide (**6**) to its internal site did not prevent the binding of radiolabeled peptide inhibitors to the outer vestibule [140, 141]. Correolide (**6**) was not specific for Kv1.3 (Table 7.2.3) [139]. It bound to all Kv1-family channels with equivalent affinity, and by blocking Kv1.1 channels it increased the peristaltic activity of the gastrointestinal tract [142, 143].

Prompted by correolide's limited natural supplies and its complex stereochemistry, chemists at Merck simplified the molecule by synthesizing several correolide derivatives in which the E-ring was removed [144]. The most potent of these compounds, the C18-analog 43 (**7**, Fig. 7.2.2), inhibited Kv1.3 with an IC_{50} of 37 nM in ^{86}Rb-flux assays and suppressed T cell proliferation 15-times more potently than correolide (Table 7.2.4) [144]. However, the Kv1.3-specificity of this compound has not been reported and it still requires correolide (**6**) as a starting material.

The Merck group also identified three novel diterpenoid pyrones, called candelalides A–C (**8** = candelalide B), from the fermentation broth of *Sesquicillium candelabrum*. These compounds blocked ^{86}Rb efflux through Kv1.3 channels with IC_{50}s of 3.7, 1.2 and 2.5 μM respectively [145].

7.2.3.4 Benzamides

Using ^{86}Rb-flux Merck also discovered the cyclohexyl-substituted benzamides [141, 146]. The parent compound PAC (**9**) (4-phenyl-4-[3-(2-methoxyphenyl)-3-oxo-2-azaprop-1-yl]cyclohexanone) inhibited Kv1.3 with an IC_{50} of 270 nM [141]. Further derivatization of PAC resulted in the more potent *trans*-N-propyl-carbamoyloxy PAC (**10**) (IC_{50} of 50 nM), which exhibited minimal (2–6-fold) selectivity for Kv1.3 over Kv1.1 and Kv1.2 (Table 7.2.3). Further structure–activity relationship (SAR) studies showed that modifications of the cyclohexyl ring or changes at the 2-position of the 2-methoxy phenyl ring did not affect Kv1.3 blocking activity, whereas methylation of the amide NH or substitution of a sulfonyl group in place of the amide carbonyl significantly reduced Kv1.3 blocking activity [146].

Biophysical manipulations that increased C-type inactivation (e.g., increasing the duration of the depolarizing pulse from 100 ms to 1 s) enhanced PAC's blocking efficacy, which suggested that the compound bound to the open or C-type-inactivated conformation of the channel [141]. PAC (**9**) blocked the channel when applied from the outside or to the intracellular surface. Competition studies indicated that the trans- and cis-isomers bound to the same site in the channel, and they inhibited ditritiocorreolide binding to the channel. Despite their shared binding site, the trans-isomers were more effective than the cis-isomers in inhibiting radiolabeled-peptide binding to the external vestibule of the channel and also in inhibiting ditritiocorreolide binding to Kv1.3 homotetramers versus Kv1-familiy heterotetramers in the brain [141]. One feature that distinguished PAC (**9**) and its derivatives from correolide (**6**), CP-339818 (**4**), UK-78282 (**5**) and the Kv1.3 blocking toxins was its Hill coefficient of 2, indicating that two PAC molecules bound to a single Kv1.3 tetramer [141]. Detailed binding studies with a radiolabeled PAC analog confirmed the presence of two receptor sites that display positive cooperativity [147]. In summary, the substituted benzamides represent an interesting drug-like class of compounds that might lead to therapeutic candidates if their Kv1.3-specificity can be improved.

7.2.3.5 Dichlorophenylpyrazolopyrimidines and Sulfimidebenzamidoindanes

Other small molecule Kv1.3 blockers described only in the patent literature are Bristol Myers Squibb's dichlorophenylpyrazolopyrimidines (**11**) [148] and ICAgen's sulfimidebenzamidoindanes [149]. The exemplary sulfamidebenzamidoindane (**12**, Fig. 7.2.2) blocked Kv1.3 at submicromolar concentrations in ^{86}Rb$^+$ flux assays and patch-clamp studies, and inhibited PHA-activated T cell proliferation at 10 μM. Apart from ICAgen's sulfonamide compounds, a group in Korea reported that another sulfonamide, the commonly used protein kinase inhibitor H-89 (**13**), inhibited Kv1.3 with an IC$_{50}$ of 1.7 μM [150] (Fig. 7.2.2, Table 7.2.2).

7.2.3.6 Phenyl-stilbene A

In lieu of the high-throughput screening approaches taken by pharmaceutical companies, academic groups have utilized several different strategies to identify small molecule Kv1.3 blockers. Lew and Chamberlin at the University of California, Irvine took a *de novo* design approach using Biosym/MSI's ligand design program LUDI [151]. They first generated a model of the outer vestibule of Kv1.3, focusing on the amino acids His^{404}, Gly^{380} and Asp^{386} that we had demonstrated to be important for toxin interactions with the channel (Fig. 7.2.1) [71, 152]. LUDI was then directed to suggest molecular fragments that interact with the key amino acids in this pocket. The suggested phenyl-stilbene scaffold was then varied through solid-phase combinatorial chemistry to render a library of 400 compounds, which were screened at Zeneca Pharmaceuticals (now Astra-Zeneca). Phenyl-stilbene A (**14**), the most potent compound, displaced ^{125}I-ChTX with an IC$_{50}$ of 2.9 μM [151] (Fig. 7.2.2, Table 7.2.2).

7.2.3.7 Tetraphenylporphyrins

Dirk Trauner's group at the University of California, Berkeley, took a peptidomimetic approach. His group used a tetraphenylporphyrin system (**15** in Fig. 7.2.2) as a scaffold to display four positively charged groups at the optimal distance to form salt bridges with four aspartate residues in the outer vestibule of Kv1.3, thus mimicking the high-affinity interaction of peptidic Kv1.3 blockers with the four subunits of the Kv1.3 tetramer [153]. These compounds displaced radiolabeled peptide binding to the external vestibule with EC$_{50}$s of 20–150 nM. Although the tetraphenylporphyrins might not constitute ideal drug candidates because of their charge and their relatively high molecular weight, they constitute attractive tools for the attachment of fluorophores as alternatives to fluorophore-tagged peptides or antibodies, or for the synthesis of metalloporphyrins for imaging and crystallographic studies [153].

7.2.3.8 **Psoralens**

Following up on anecdotal reports that tea prepared from leaves of *Ruta graveolens* (Garden Rue, Herb of Grace) alleviated the symptoms of MS [154] (http://herbal-remedies.com/rue.html), Eilhard Koppenhöfer's and Wolfram Hänsel's groups at the University of Kiel extracted *Ruta* and identified several K⁺ channel blocking psoralens and furoquinolines [155]. The psoralen 5-methoxypsoralen (5-MOP, **16**) reduced scotomas in single-case studies in patients with MS [154]. Hänsel's group subsequently synthesized several furoquinoline and psoralen analogs to improve potency against Kv1.3. The angular 8-methoxy-2-(1-methylethyl)-5-methyl-4,5-dihydrofuro[3,2-*c*]quinolin-4-one (Ibu-8, **18**) and the angular pyrano-quinolinone 9-methoxy-2,2,6-trimethyl-2,6-dihydro-5*H*-pyrano[3,2-*c*]quinolin-5-one are two such analogs that inhibit Kv1.3 channels with half-blocking concentrations of 5 and 10 µM, respectively [156]. The psoralen 5-MOP (**16**) served as a template for the design of H37 (**19**) [157], 6-(2,5-dimethoxyphenyl)psoralen (**20**) [158], Psora-4 (**21**) [159] and PAP-1 [410] (Fig. 7.2.2 and Table 7.2.2).

When making 5-MOP (**16**) analogs in which the methyl group at the 5-position was replaced with a series of phenylalkyl or cyclohexylalkyl substituents we found that the length of the side-chain linker had a profound effect on the potencies and Hill coefficients of these compounds. Psora-1, an analog with one CH_2 moiety in the linker, blocked Kv1.3 with a Hill coefficient of 1, whereas compounds containing 2–4 CH_2 moieties in the linker displayed Hill coefficients of 2 [159], indicating that two molecules bind per Kv1.3 tetramer. Increasing the length of the linker from one to four CH_2 moieties enhanced potency significantly, while further lengthening of the side chain reduced potency. The pharmacophore of alkoxypsoralen Kv1.3 blockers was determined to consist of a psoralen moiety attached at position-5 to a phenyl ring via an alkyl chain linker optimally containing four CH_2 groups. Replacement of the phenyl ring with an aliphatic cyclohexyl ring reduced Kv1.3-blocking potency, suggesting that the phenyl ring is required for the high-affinity interaction with Kv1.3. From these SARs, we postulated that Psora-4 (**21**) binds to Kv1.3 via two π–π electron interactions positioned ~7 Å apart (the length of the butoxy linker in Psora-4), one involving the psoralen moiety and the second involving the side chain of the phenyl ring. However, the precise binding site of Psora-4 (**21**) on Kv1.3 remains to be determined. Psora-4 (**21**), blocks Kv1.3 by interacting with residues accessible in the C-type inactivated conformation of the channel [159]. Psora-4 (**21**) is 17–70-fold selective for Kv1.3 over other Kv1-family channels with the exception of the cardiac channel Kv1.5 (IC_{50} ~8 nM) (Table 7.2.3). Because of its lack of selectivity over Kv1.5, we derivatized Psora-4 further with the aim of separating the affinities for Kv1.3 and Kv1.5. By introducing an additional oxygen atom into the side-chain linker of Psora-4, we generated PAP-1 (5-(4-phenoxybutoxy)psoralen) (22 in Figure 2) [410], which inhibits Kv1.3 with an IC_{50} of 2 nM and displays 23-fold selectivity over Kv1.5. PAP-1 is 33-125 fold selective over K1.1, Kv1.2, Kv1.4, Kv1.6 and Kv1.7 and about 1000-fold selective over more distantly related K⁺ channels like HERG, Kv2-, Kv3- and Kv4-family channels, Na⁺, Ca^{2+} and Cl⁻ channels. PAP-1 does not exhibit cytotoxic or phototoxic effects, is negative in the Ames test and affects P450-

dependent enzymes only at micromolar concentrations. Based on its logP value (a measure a compound's hydrophobicity [160, 161] of 4.0, its molecular weight of 350, and the number of hydrogen-bond donors and acceptors (donors 0, acceptors 5), the Lipinksi "rule of five" [162] further predicts that PAP-1 should be orally available. PAP-1 administered by gavage suppressed delayed-type hypersensitivity in rats [410] indicating that it could potentially be developed into an orally available immunmodulator.

7.2.3.9 Khellinones

Starting with the benzofuran khellinone (5-acetyl-4,7-dimethoxy-6-hydroxybenzo-furan), which can be regarded as a lactone ring-opened version of 5-MOP (**16**), Baell et al. synthesized two novel classes of Kv1.3 blockers: khellinone chalcones and khellinone dimers [163]. Chalcone-16 (**23**) [3-(4,7-dimethoxy-6-hydroxybenzo-furan-5-yl)-1-phenyl-3-oxopropene] and its derivatives blocked Kv1.3 with IC_{50} values of 300–800 nM and Hill coefficients of 2 [163], like the PACs and the psoralens [141, 159]. The most potent khellinone dimer (**24**) blocked Kv1.3 with an IC_{50} of 280 nM and 1:1 stoichiometry, possibly by interacting with two subunits in the Kv1.3 tetramer. The linker in the dimers has to be hydrophobic for tight binding and its optimal length is 7.2 Å. A khellinone trimer was inactive. Dimer-2 (**23**) [1,4-bis(5-acetyl-4,7-dimethoxybenzofuran-6-yloxymethyl)benzene] exhibited 10-fold selectivity over Kv1.1, Kv1.2, Kv1.5, Kv1.7 and KCa3.1, while chalcone-16 (**23**) displayed >10-fold specificity for Kv1.3 over these channels, with the exception of Kv1.1 which was only 3-fold less sensitive.

7.2.3.10 Steroids

Inhibition of Kv1.3 also seems to contribute to the immunosuppressive effect of progesterone in the placenta, where micromolar concentrations of this hormone prevent a maternal immune response against paternal antigens in the fetus [164]. Progesterone rapidly and reversibly blocked Kv1.3 (IC_{50} 28 μM) and consequently prevented Ca^{2+} signaling and NFAT-driven gene expression in T cells [164]. Other steroid hormones, with the exception of dexamethasone [165], are less effective in blocking Kv1.3 [164]. Like most other small molecule Kv1.3 blockers progesterone preferentially binds to the inactivated state of the channel.

7.2.3.11 Miscellaneous Compounds

Other more "exotic" and less potent Kv1.3 blockers include phencyclidine (PCP) [166], the poisonous principle from hemlock cicutotoxin [167], capsaicin [100], forskolin [168] and the flavanoid luteolin (Table 7.2.2), which is sold as a nutriceutical. None of these compounds is a potent or selective inhibitor of the Kv1.3 channel.

7.2.4
Physiological Role of Kv1.3 and the Effects of Kv1.3 Blockers

Kv1.3 expression has been detected in T and B lymphocytes, NK cells, macrophages, microglia, osteoclasts, platelets, adipocytes, olfactory neurons, colonic epithelia, kidney medullary cells, oligodendrocyte precursors, retina and detrusor muscle. In human T- and B-lymphocytes Kv1.3 is the only voltage-gated K^+ channel and plays a pivotal role in regulating calcium signaling, and targeting Kv1.3 may therefore have important therapeutic implications for immunological diseases. In other cells, Kv1.3 is one of many K^+ channels and blocking Kv1.3 does not have significant physiological consequences due to channel redundancy. In a third group of cells, the characterization of the channels is incomplete and/or the functional role of Kv1.3 has not been determined; these cells merit further investigation with selective Kv1.3 inhibitors and Kv1.3-specific antibodies.

7.2.4.1 K^+ Channels in T Cells
This section is divided into four sub-sections. In the first, we compare Kv1.3 expression in T cells from different species, since this is germane to the selection of animal models for the evaluation of Kv1.3 blockers. In Section 7.2.4.1.2 we discuss the important role of Kv1.3 in memory T cells, and in Sections 7.2.4.1.3 and 7.2.4.1.4 we review the functional roles of K^+ channels in T cell activation and apoptosis.

7.2.4.1.1 Comparison of Kv1.3 Expression in Human, Mouse, Rat and Minipig Naïve T Cells
The maximum Kv1.3 channel conductance in human peripheral blood T cells is about 4 nS although it varies significantly from cell to cell [1, 2, 5, 125]. It also varies in human T cell lines from 20 nS in CCRF-CEM-3A cells to 0.02 nS in CCRF-HSB2 cells [125]. In Ringer solution the mean single-channel conductance value is 12–14 pS and the unitary conductance increases in the presence of isotonic K^+-Ringer solution, paralleling the increase in whole-cell conductance [2, 5]. By dividing the maximum Kv1.3 conductance by the channel's unitary conductance, the number of Kv1.3 channels in human peripheral blood $CD4^+$ and $CD8^+$ T cells and in $CD3^+$ thymocytes is estimated to be about $250-400 \, cell^{-1}$ [2, 5, 125, 169]. Quiescent naïve human T cells also express small numbers ($\sim 10-20 \, cell^{-1}$) of the intermediate-conductance calcium-activated K^+ channel, KCa3.1 (a.k.a. IKCa1, IK1. hKCa4, SK4, KCNN4). Following activation, KCa3.1 levels are enhanced to $200-500 \, cell^{-1}$ while Kv1.3 levels do not change significantly [121]. New KCa3.1 transcripts are detected as early as 3 h after stimulation, and functional KCa3.1 channels are present in the membrane within 6 h of activation [54, 121, 170]. KCa3.1 transcription is dependent on PKC but not calcium signaling [121]. Naïve effector cells that emerge after activation express roughly equivalent numbers of Kv1.3 and KCa3.1 channels ($\sim 250-500 \, cell^{-1}$).

The channel expression pattern in mouse T cells is remarkably different from that of human T cells. The maximum K$^+$ conductance of mouse T cells is 0.16 nS compared with 4.2 nS in human T cells [6, 125, 171, 172]. Quiescent mouse CD4$^+$ cells express Kv1.3 channels, although only ~20 cell^{-1} compared with the 250–500 cell^{-1} in human CD4$^+$ T cells. They also express Kv1.1, Kv1.2 and Kv1.6 [173, 174]. Quiescent mouse CD8$^+$ T cells do not express Kv1.3 but instead express a channel (type l) with a larger single channel conductance (21 pS), which is encoded by the distantly related Kv3.1 gene [171, 172, 175–177]. The number of type l channels per quiescent mouse CD8$^+$ T cell is estimated to be ~20 cell^{-1}. Following mitogenic activation, both CD4$^+$ and CD8$^+$ upregulate Kv1.3 expression to about 200–300 cell^{-1}, the level found in human T cells, and CD8$^+$ cells simultaneously down-regulate Kv3.1 expression [6, 171, 172, 175]. Quiescent mouse T cells also express a small number of KCa3.1 channels and their numbers are augmented during activation. Repeated stimulation of mouse T cells induces a progressive increase in Kv1.3 and KCa3.1 levels with each round of activation and proliferation [178], whereas human T cells lower KCa3.1 expression and increase Kv1.3 levels when repeatedly stimulated [54]. The effectiveness of the specific KCa3.1 blocker TRAM-34 in preventing disease in mouse experimental autoimmune encephalomyelitis (EAE), a model of MS [179], highlights the importance of the KCa3.1 channel in mouse T cells, although this compound was ineffective in a rat model of EAE [52]. Mice lacking the Kv1.3 gene have a normal distribution of lymphocytes in thymus and spleen and exhibit no apparent abnormalities in thymocyte development or in the ability of their splenic T cells to proliferate [180]. This appears to be because the knockout of Kv1.3 resulted in a compensatory 50-fold increase in chloride currents [180]. Another difference between mouse and human T cells is that the membrane potential in quiescent mouse T cells is determined by Na-K-ATPase [181] whereas Kv1.3 channels are responsible for regulating the membrane potential in human T cells [49]. For all these reasons, the mouse is not considered a useful model for evaluating the immunomodulatory effects of Kv1.3 blockers.

Quiescent naïve rat T cells also differ from naïve human T cells in only expressing 10–20 Kv1.3 channels per cell [52]. ShK(L5), a highly-specific inhibitor of Kv1.3, blocks all the K$_V$ current in quiescent rat T cells, indicating that, unlike in mouse T cells, there are no other K$_V$ channels in rat T cells [119].

Naive T cells from Yucatan miniature swine express roughly the same number of Kv1.3 channels as resting human T cells, and the Kv1.3 blocker margatoxin depolarizes their membrane potential, demonstrating that Kv1.3 sets the membrane potential in minipig naïve T cells [49]. Furthermore, margatoxin suppresses mitogen-induced proliferation of minipig peripheral blood T cells as well as the mixed lymphocyte reaction [49]. Minipigs are therefore an attractive model to evaluate Kv1.3 blockers for human disease.

7.2.4.1.2 Kv1.3 in Memory T Cells

Two subsets of human memory T cells – central memory (T$_{CM}$) and effector memory (T$_{EM}$) – have been described based on the expression of the cell surface proteins CCR7, a chemokine receptor, and CD45RA, a phosphatase [182, 183]. Naïve T cells use CCR7 to gain entry into lymph nodes where they engage their specific antigen and are transformed through an activation cascade into naïve-effectors that secrete cytokines and proliferate. Naïve-effectors leave the lymph node for the site of antigenic challenge, where they carry out their protective function. Some naïve cells differentiate into long-lived T_{CM} cells that retain memory of the specific antigen, and T_{CM} cells use CCR7 to enter lymph nodes where they engage antigen and change into T_{CM}-effectors that migrate to inflamed tissues to mount a vigorous response to antigenic challenge. Naïve-effectors and T_{CM}-effectors are important protectors against pathogenic and neoplastic challenge. Repeated antigenic stimulation, as commonly occurs in a chronic infection or in autoimmune diseases, induces T cells to differentiate into T_{EM} cells that lack expression of both CCR7 and CD45RA and do not need to home to lymph nodes for antigen-induced activation. Upon activation T_{EM} cells become T_{EM}-effectors that rapidly migrate to sites of inflammation, secrete large amounts of cytokines and perform immediate effector functions. We found that 7–10 rounds of antigenic stimulation were required to convert a predominantly naïve T cell population into a mainly T_{EM} population [54].

Quiescent $CD4^+$ and $CD8^+$ T_{CM} cells, like naïve cells, express about 300 Kv1.3 and 10–20 KCa3.1 channels per cell, and upregulate KCa3.1 to ~500 channels per cell with little or no change in Kv1.3 expression when they are activated (Fig. 7.2.3A) [54, 121]. T_{EM} cells, in contrast, augment Kv1.3 to ~1500 channels per cell with no increase in KCa3.1 expression upon activation (Fig. 7.2.3A). The increase in Kv1.3 channels is detected within a few hours after activation, peaks at 15 h and remains high for a further 2–3 days [118]. The increased number of Kv1.3 channels measured by the patch-clamp correlates with the increased number of Kv1.3 tetramers in the membrane measured by flow cytometry with fluorescent ShK [118]. Addition of interleukin-2 or T cell-conditioned medium causes a decrease in Kv1.3 expression to baseline levels within 3–4 days [118]. The molecular mechanism underlying Kv1.3 upregulation has not been determined although it appears to depend on both the calcium- and PKC-mediated signaling pathways [118].

7.2.4.1.3 Role of Kv1.3 in the Activation of Naïve/T_{CM} versus T_{EM} Cells

Kv1.3's role in the activation of naïve human T cells was appreciated two decades ago when several chemically unrelated compounds were shown to suppress mitogen-induced interleukin-2 production and ^{3}H-leucine and ^{3}H-thymidine incorporation at concentrations paralleling their block of the *type n* channel [1, 3, 8]. In 1989, Carol Deutsch and her collaborators showed that ChTX, a potent Kv1.3 inhibitor from scorpion venom, suppressed mitogen-induced T cell proliferation, giving further credence to the idea that Kv1.3 was necessary for activation of naïve human T cells [184]. However, ChTX was shown to also block the intermediate-conductance calcium-activated potassium channel (KCa3.1) discovered in human T cells in the early 1990s [60, 185], and consequently it was not clear whether ChTX inhibited T cell function by blocking Kv1.3 alone, the KCa3.1 chan-

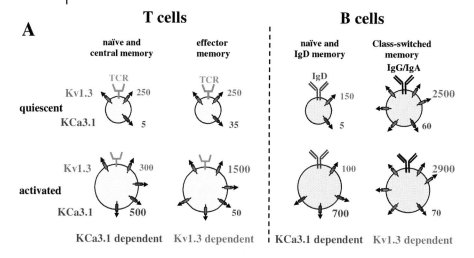

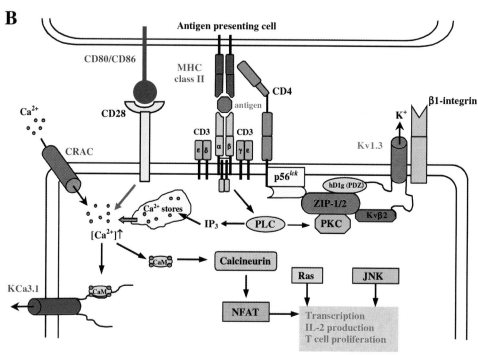

Fig. 7.2.3 (A) Average Kv1.3 and KCa3.1 expression in human T and B cells. Naïve and central memory T cells represent CCR7$^+$CD45RA$^+$ (= naïve) and CCR7$^+$CD45RA$^-$ (= T$_{CM}$) T cells in both the CD4$^+$ and the CD8$^+$ compartment. The effector memory cell represents CD4$^+$CCR7$^-$CD45RA$^-$ cells, CD8$^+$CCR7$^-$CD45RA$^-$ and CD8$^+$CCR7$^-$CD45RA$^+$ cells, which are all considered to be T$_{EM}$ cells.

For B cells the IgD$^+$ cell represents both naïve (IgD$^+$CD27$^-$) and IgD$^+$ memory cells (IgD$^+$CD27$^+$), while the class-switched memory cell represents IgD$^-$CD27$^+$ memory cells. The fourth population of human B cells (IgD$^-$CD27$^-$) is not shown. (B) Cartoon showing the involvement of Kv1.3 and KCa3.1 in the T cell activation cascade.

nel alone, or both channels together. Studies by Rader and colleagues at Monsanto with kaliotoxin (which blocks only Kv1.3) and ChTX suggested that blockade of both Kv1.3 and KCa3.1 channels were required for inhibition of mitogenesis [186]. In contrast, Leonard and his colleagues at Merck used margatoxin (which blocks Kv1.3 but not KCa3.1) to demonstrate that blockade of Kv1.3 alone inhibited the activation of human T cells [60]. Subsequent studies with several compounds that blocked Kv1.3 without affecting KCa3.1, both peptides [ShK, ShK-Dap[22], ShK(L5), NTX, AgTX2] and small molecules (WIN-17317-3, CP-339818, UK78282, correolide, Psora-4, chalcone-16, benzamides; Fig. 7.2.2), confirmed the important role played by Kv1.3 in the activation of naïve human T cells (Table 7.2.4) [9, 49, 50, 109, 119, 121, 130, 135, 141, 159, 163, 187]. These blockers were effective immunosuppressants in the first few hours of activation and their inhibitory effects on naïve human T cells was partially overcome by exogenous IL-2 [3, 9, 119]. Two groups at the end of 1992 independently demonstrated that the anti-proliferative effect of Kv1.3 blockers was due to membrane depolarization [10, 60]. A few months later, Michael Cahalan's group at UC Irvine showed that calcium oscillations in naïve human T cells depended on membrane potential [14, 15, 188, 189], while the Merck group reported that margatoxin and noxiustoxin inhibited calcium signaling at concentrations that suppress T cell activation [9]. It was only in 2000, when more selective blockers for both Kv1.3 and KCa3.1 channels were available, that the roles of Kv1.3 and KCa3.1 in T cell activation were finally understood [121, 187].

Kv1.3 channels appear to be involved in the earliest stages of T lymphocyte activation and have been suggested to participate in the formation of the "immunological synapse" at the point of contact between the T cell and the antigen-presenting cell [190–192]. In Kv1.3-transfected Jurkat T cells, a human T cell lymphoma line, the channel has been reported to lie in close to the T cell-antigen receptor complex based on FRET studies [193]. Biochemical studies suggest an explanation for this physical proximity. Kv1.3 is physically coupled via the last three residues of its C-terminus to a PDZ domain protein called hDlg (human Disc Large) or SAP97 (Synapse Associated Protein 97), which in turn is attached to the tyrosine kinase *p56lck* [194]. Other biochemical studies have shown that each N-terminus in the Kv1.3 tetramer interacts with a $K_V\beta2$ subunit [195]. In other cell systems, $K_V\beta2$ proteins have been reported to bind to adaptor proteins called Zip, which in turn are associated with protein kinase C zeta [196]. Zip is identical to the *p56lck*-associated p62-protein in T cells and the A170 protein induced by oxidative stress in macrophages [197–200]. Together, these studies suggest that Kv1.3 in naïve human T cells may be linked to *p56lck* via two sets of adaptor proteins, hDlg/SAP97 and Kvβ2-Zip. Phosphorylation of Kv1.3 by *p56lck* suppresses the current and demonstrates functional coupling of the channel and the tyrosine kinase [201–203]. Since *p56lck* is associated with CD4 and CD8 [204–206], one can envision a signaling complex consisting of Kv1.3, *p56lck*, SAP97, Zip/p62/A170, Kvβ2, PKC and either CD4 or CD8. Since CD4 and CD8 traffic to the center of the immunological synapse [190–192], it is possible that Kv1.3 and other proteins in the signaling complex also localize at the synapse. Studies in cytotoxic human

T cells transfected with Kv1.3 indicate that the channel does coalesce at the zone of contact between the cytotoxic T cell and the target cell [207]. The functional significance of this localization at the immunological synapse is not clear since T cells are small and electrically tight, and the opening of a single Kv1.3 channel should be sufficient to hyperpolarize the membrane [5, 125]. It also remains to be seen whether Kv1.3 traffics to the immune synapse in normal human CD4 and CD8 T cells, and whether Kv1.3 blockade prevents or delays synapse formation. Kv1.3 is also physically associated with β1-integrins in T cells [208] and FRET studies in melanoma cells indicate that Kv1.3 and β1-integrins are in close proximity to one another [209]. Since β1-integrins localize at the periphery of the immune synapse [210], it is tempting to speculate that Kv1.3 acts as a scaffold, linking CD4 at the center of the immunological synapse with β1-integrins at the periphery. Through its interaction with β1-integrins Kv1.3 plays a role in T cell migration, and the Kv1.3 blocker margatoxin inhibits MIP-1β-induced adhesion/migration of human T cells [208].

Figure 7.2.3(B) summarizes the current understanding of the roles of Kv1.3 and KCa3.1 in human T cells. Engagement of the T cell antigen receptor triggers Ca^{2+}-influx through voltage-independent Ca^{2+} channels called CRAC [45, 211]. The resulting increase in cytosolic Ca^{2+} causes the translocation of NFAT (nuclear factor of activated T cells) to the nucleus and the initiation of new transcription, ultimately resulting in cytokine secretion and T cell proliferation [45, 47, 48, 212, 213]. The intracellular Ca^{2+} concentration has to be high for 10–12 h to optimally activate T cells, but Ca^{2+} entry through CRAC channels diminishes as a consequence of channel inactivation brought about by membrane depolarization and a rise in Ca^{2+} in proximity to the channel [214, 215]. The opening of Kv1.3 and KCa3.1 and the resulting K^+ efflux hyperpolarizes the membrane and facilitates Ca^{2+} entry through CRAC channels. The shape and the nature of the Ca^{2+} signal regulate gene expression in response to antigenic stimulation [216, 217].

The exact role of Kv1.3 and KCa3.1 in human T cell activation depends on the relative numbers of each of these channels in a given cell. Naïve and T_{CM} cells express more Kv1.3 than KCa3.1 channels (~300 to 10 cell^{-1}). Hence, Kv1.3 blockers suppress the activation of these cells whereas the KCa3.1 blocker TRAM-34 is less effective [54, 121, 159, 187]. Naïve-effectors and T_{CM}-effectors in which KCa3.1 is upregulated rely on KCa3.1 to regulate membrane potential and calcium signaling because KCa3.1's opening is tightly linked to the intracellular calcium concentration [15, 188, 189]. Consequently, TRAM-34 suppresses the activation of naïve-effectors and T_{CM}-effectors, while Kv1.3 blockers are ineffective [54, 121, 159, 187]. T_{EM} cells upregulate Kv1.3 during activation and are therefore more sensitive to Kv1.3 blockade [ShK(L5) IC$_{50}$ ~80 pM; PAP-1 IC$_{50}$ ~10 nM] and resistant to TRAM-34. Table 7.2.3 shows the relative potencies of Kv1.3 blockers for suppression of human T cell activation. In most reports, IC$_{50}$s for block of Kv1.3 correlate well with IC$_{50}$s for suppression of T cell activation, and suggest that a substantial fraction of the Kv1.3 channels have to be blocked to affect activation [9, 49, 50, 54, 121, 159, 187]. Kv1.3 blockers are also effective in suppressing T_{EM}-mediated delayed type hypersensitivity (DTH) immune responses

[218] in rats and in mini-pigs [49–51, 119, 410]. Hypoxia-mediated inhibition of Kv1.3 channels in T cells [219], especially T_{EM} cells, may provide a mechanism to limit tissue damage at sites of inflammation.

Activated T_{EM} cells are involved in the pathophysiology of numerous autoimmune diseases such as MS [54, 220–222], type-1 diabetes mellitus (T1DM) [223, 224], rheumatoid arthritis (RA) [225], psoriasis [226, 227], pemphigus [228, 229] and chronic graft-versus-host disease [230] (Table 7.2.1). By preventing the activation and/or migration of T_{EM} cells, Kv1.3 blockers might be able to ameliorate these autoimmune diseases without compromising the function of KCa3.1-dependent naïve-effectors or T_{CM}-effectors.

7.2.4.1.4 Role of Kv1.3 in Apoptosis
Kv1.3 has been implicated in Fas-mediated apoptosis but the precise role of the channel in this process is unclear. Florian Lang and his collaborators in Tübingen reported that activation of the Fas receptor (CD95) resulted in inhibition of the Kv1.3 current in Jurkat T cells through p56^{lck} tyrosine kinase mediated phosphorylation of the channel [201]. They identified ceramide, a lipid metabolite synthesized upon Fas receptor ligation, as the messenger molecule that activates p56^{lck} [202]. The same group observed a slight but significant (10%) increase in apoptosis in Jurkat T cells if the Fas receptor was stimulated in the presence of margatoxin [201]. In contrast, Storey et al. [231] reported that Fas receptor activation resulted in a two-fold increase in Kv1.3 current amplitude in Jurkat T cells and suggested that the opening of Kv1.3 channels caused depletion of intracellular potassium and led to cell shrinkage. Their hypothesis predicts that Kv1.3 blockers would prevent apoptosis of T cells, similar to the anti-apoptotic effects of high extracellular potassium in neurons [232]. The recent demonstration that Kv1.3 is present in the inner membrane of mitochondria of T lymphocytes [233], an organelle that plays a pivotal role in apoptosis, highlights a potential role for Kv1.3 in the apoptotic process. These experiments need to be extended to memory cells to determine whether T_{EM}-effectors with elevated Kv1.3 numbers are more prone to apoptosis than other T cell subsets.

7.2.4.2 K$^+$ Channels in B Cells
Mature human B cells from tonsil and peripheral blood were reported in the late 1980s and early 1990s to express K_V and KCa currents with properties resembling those of Kv1.3 and KCa3.1 [18, 21, 22, 234–236]. Non-specific K$^+$ channel blockers suppressed anti-IgM- or PMA-stimulated proliferation of these cells [18, 21, 22, 235, 236], suggesting a requirement for K$^+$ channels for B cell proliferation. Recent studies by our group have defined Kv1.3 and KCa3.1 expression in human B cell subsets distinguished on the basis of surface expression of IgD and CD27, a member of the TNF-receptor family [237–239]. Naïve B cells express IgD, but not CD27. Following encounter of antigen and after receiving T cell help, naïve B cells acquire CD27 and differentiate into CD27$^+$IgD$^+$ memory B cells during somatic hypermutation. Replacement of surface IgD with other Ig

isotypes such as IgG or IgA during immunoglobulin class-switch yields CD27$^+$IgD$^-$ memory B cells. Class-switched IgD$^-$CD27$^+$ memory B cells are identical to a subset of CD27$^+$CD80$^+$ memory B cells that were reported to effectively present antigen to CD4$^+$ T cells without requiring pre-activation, to activate at lower thresholds and to differentiate rapidly into cells that secrete large amounts of class-switched antibodies [31, 240]. A minor population of IgD$^-$CD27$^-$ B cells exists, but the functional role of this subset is unclear [238].

In the resting state, naïve (IgD$^+$CD27$^-$) and early memory (IgD$^+$CD27$^+$) B cells express moderate numbers of Kv1.3 channels ($\sim$100 cell^{-1} in naïve and $\sim$250 cell^{-1} in IgD$^+$CD27$^+$ cells) and very few KCa3.1 channels ($\sim$5 channels per cell). Activation enhances KCa3.1 expression to 600–700 channels per cell with no change in Kv1.3 levels (Fig. 7.2.3A). In contrast, quiescent, class-switched memory B cells express high levels of Kv1.3 ($\sim$2000 channels per cell) and maintain their Kv1.3^{high} expression after activation [31]. Reflecting their channel expression patterns, proliferation of both naïve and IgD$^+$CD27$^+$ memory B cells is suppressed by KCa3.1 but not Kv1.3 blockers, whereas Kv1.3^{high} class-switched memory B cells are inhibited by Kv1.3 but not by KCa3.1 inhibitors [31].

The effects of Kv1.3 and KCa3.1 blockers on B and T cells differ in significant ways, the first major difference lying at the naïve/early memory cell stage. The proliferation of naïve and T$_{CM}$ T cells is suppressed by Kv1.3 but not KCa3.1 blockers, but once these cells have become naïve-effectors and T$_{CM}$-effectors with upregulated KCa3.1 expression, they are resistant to Kv1.3 inhibition and sensitive to KCa3.1-blockade [54, 121]. Naïve and early memory B cells, however, are sensitive to KCa3.1 blockers and resistant to Kv1.3 blockers from the outset. Differences in channel expression patterns underlie this differential sensitivity to K$^+$ channel blockers. Naïve and T$_{CM}$ cells start off with significantly more Kv1.3 than KCa3.1 channels (250–400 versus 10–20 channels per cell) and as a consequence Kv1.3 blockers suppress their proliferation during the 48 h ^{3}H-thymidine incorporation assay. Naïve and early memory B cells also start off with more Kv1.3 than KCa3.1 channels (100–250 versus 5–10), but they upregulate KCa3.1 rapidly and take considerably longer to activate than T cells ($\sim$4 days versus $\sim$2 days), resulting in their suppression by KCa3.1 but not Kv1.3 blockers. The second major difference between T and B cells lies at the late memory stage. Quiescent T$_{EM}$ cells express significantly fewer Kv1.3 channels (250 cell^{-1}) than quiescent class-switched memory B cells ($\sim$2000 cell^{-1}) and as a result are roughly 10-fold more sensitive to Kv1.3 blockers.

Class-switched memory B cells have been implicated in the pathogenesis of Sjögren's syndrome, Hashimoto's thyroiditis, Systemic Lupus Erythematosus (SLE), MS and Grave's disease (Table 7.2.1) [241–245]. In Sjögren's syndrome the number of CD27$^+$ B cells is decreased in blood, paralleling the accumulation of these cells in the inflamed salivary gland [241, 246]. In SLE, the increased number and frequency of CD27^{high} memory B cells in the blood compared with healthy controls directly correlates with disease activity and the titer of anti-DNA autoantibodies [243, 247]. Class-switched memory B cells are also a major source of pathogenic IgG autoantibodies that contribute to tissue damage in

MS [248, 249], T1DM [250] and RA [251]. Furthermore, autoantigen-specific memory B cells effectively function as antigen-presenting cells in T cell-mediated autoimmune diseases [240, 252–254] and have been reported to contribute to the phenomenon of epitope spreading in which new antigen specificities emerge during the course of disease [255–257]. Because conventional immunosuppressive therapy with cyclosporin A, azathioprine or cyclophosphamide preferentially targets CD27$^-$ naïve and early memory B cells and has no effect on CD27$^+$ memory B cells [243], Dörner et al. recently proposed that SLE flares observed in patients undergoing conventional therapy might actually be caused by the retention of CD27$^+$ memory B cells and suggested that new immunosuppressants that selectively target these cells might have immense therapeutic value [243]. Therapies that preferentially suppress class-switched CD27$^+$ Kv1.3^{high} cells without compromising the function of naïve and early memory B cells may therefore be beneficial in the therapy of autoimmune diseases.

7.2.4.3 K$^+$ Channels and Natural Killer Cells

Natural killer (NK) cells are lymphoid cells that provide host defense against tumors and viruses [258]. A subset of NK cells (CD56^{bright}CD16$^{-/dim}$) has been reported to accumulate in the joints of patients with RA and may amplify the inflammatory response [259–261]. NK cells have also been suggested to regulate autoimmune memory T cells in an antigen non-specific fashion in some MS patients [262]. Human NK cells express Kv1.3-like K$^+$ channels and non-specific blockers of K$_V$ channels were reported almost two decades ago to inhibit NK cell-mediated cytotoxicity [16, 17]. These experiments have to be repeated with Kv1.3-specific inhibitors and the role of Kv1.3 channels in NK-mediated regulation of memory T cells has to be defined.

7.2.4.4 K$^+$ Channels in Macrophages

Voltage-clamp studies were first performed on macrophages from mouse spleen cultures by Elaine Gallin in 1981 [263] and action potentials were reported in human macrophages in 1983 [264]. Subsequent patch-clamp studies revealed large- and intermediate-conductance calcium-activated K$^+$ channels, Kir2.1, Kv1.5 and Kv1.3 in monocyte-macrophages, the expression patterns depending on the species, the source of cells, the culture conditions, and the activation and differentiation status [33, 265–280].

7.2.4.4.1 Voltage-gated K$^+$ Channels

K$_V$ currents were discovered in macrophages in the early 1980s [265, 271]. Ypey and Clapham identified a delayed outwardly-rectifying K$_V$ current with properties resembling the *n type*/Kv1.3 channel within 1–4 days of culture of mouse peritoneal macrophages [271] and suggested that K$^+$ channel expression in macrophages might change according to their activation status. In humans, only 5% of peripheral blood monocyte-derived-macrophages express Kv1.3-like currents whereas roughly 50% of alveolar macrophages exhibit large Kv1.3-like currents [272, 273]. In FACS-sorted human alveolar macrophages Kv1.3 mRNA is the only K_V channel mRNA present, indicating that the K_V current in these cells is indeed Kv1.3 [32]. Margatoxin depolarized the membrane potential of human alveolar macrophages, demonstrating that Kv1.3 sets the membrane potential in these cells, but it failed to influence Fc-receptor mediated phagocytosis or IL-1β secretion [32]. Margatoxin, agitoxin-2 and ChTX inhibited chemotactic migration of human monocyte-derived macrophages through Boyden chambers and through an artificial blood–brain barrier consisting of transwell membranes coated with fibronectin, a layer of human astrocytes and a layer of brain microvascular endothelial cells [280]. Margatoxin was also shown to suppress macrophage-colony-stimulating factor (M-CSF)-stimulated [^{3}H]thymidine incorporation and expression of inducible nitric oxide synthase (iNOS) in mouse bone marrow-derived macrophages [33]. Inhibition of Kv1.3 in macrophages may contribute to the effects of Kv1.3 blockers *in vivo*.

In vitro exposure of macrophages or macrophage-like cell lines to different stimuli changes their K^+ channel expression pattern. Phorbol ester-induced differentiation of THP-1 human monocytes into macrophage-like cells is accompanied by down-regulation of the Kv1.3 current and Kv1.3 mRNA [281]. In both human monocyte-derived-macrophages and mouse bone marrow-derived-macrophages, lipopolysaccharide (LPS) causes a 4–5-fold increase in Kv1.3 expression [33, 274]. Interleukin-2 has been reported to increase the frequency of K_V currents in human monocyte-derived-macrophages [274] and GM-CSF (granulocyte/macrophage colony-stimulating factor) increases expression of Kv1.3 in mouse bone marrow-derived-macrophages [277].

Mouse macrophages also express Kv1.5 [33]. Of all human tissues analyzed, Kv1.5 transcripts are most abundantly expressed in human dendritic cells, followed by liver and heart (http://symatlas.gnf.org/SymAtlas/). The functional role of Kv1.5 in macrophages remains to be determined. Psora-4 blocks Kv1.5 and Kv1.3 at low nanomolar concentrations [159] and it would be interesting to evaluate its effect on macrophage function.

7.2.4.4.2 **Calcium-activated K^+ Channels**
Large-conductance K_{Ca} channels (a.k.a. KCa1.1, BK, KCNMA1, Slo, Maxi-K_{Ca}) were first described by Elaine Gallin's group at the NIH in human monocyte-derived-macrophages and in the murine macrophage cell line J774.1 [265, 266, 268–270]. KCa1.1 channels were also reported in phorbol ester-differentiated human monocytic leukemia THP-1 cells [276]. A later study reported that treatment of human macrophages with LPS enhanced KCa1.1 current density 2.5-fold [275]. Paxilline, a KCa1.1 blocker, suppressed LPS-induced IL6 and TNF-α production by human monocyte-derived-macrophages [275, 282, 283]. Mouse and human macrophages were also reported to express KCa3.1 channels [284] and these channels were shown to open during calcium oscillations induced by extracellular ATP [279] or when HIV engages chemokine receptors CCR5 and CXCR4 during entry into macrophages [278]. However, the effect of TRAM-34

or of other selective KCa3.1 blockers on monocyte-macrophage function has not been evaluated.

7.2.4.4.3 Inward-rectifier K⁺ Channels

Elaine Gallin was the first to identify an inwardly rectifying K^+ conductance (K_{ir}) in macrophages [266, 285, 286], and studies by various other groups later confirmed this finding [33, 276, 277, 287, 288]. The gene encoding this channel was identified to be Kir2.1 (a.k.a. IRK1, KCNJ2) by Lily Jan and colleagues at UCSF [289]. Phorbol ester-induced differentiation of THP-1 cells into macrophages is accompanied by an upregulation of Kir2.1 expression [276]. K_{ir} currents in cell-attached patches of phorbol ester-differentiated macrophages from the human pro-myelocytic cell line HL60 are reported to be inhibited by M-CSF within 1 minute of exposure, indicating rapid channel modulation via intracellular mediators [287]. However, other groups found that M-CSF-cultured mouse macrophages express K_{ir} currents (~300 Kir2.1 channels per cell) [33, 277], and LPS-stimulation decreases K_{ir} levels ~6-fold [33]. The reasons for these differences are unclear. The lack of specific Kir2.1 inhibitors has precluded studies to define the role of this channel in monocyte-macrophage function, although Ba^{2+}, a non-selective blocker of this channel, is known to suppress LPS-induced ^{3}H-thymidine incorporation and expression of iNOS [33]. In lieu of selective inhibitors, it may be feasible to overexpress dominant-negative isoforms of Kir2.1 [290] in macrophages to ascertain this channel's role in the monocyte-macrophage lineage.

7.2.4.5 Kv1.3 in Microglia

Microglia are brain-resident macrophages and their K^+ channel expression resembles that of macrophages in many ways. Depending on the culture conditions, the length of the culture and the added stimuli, the reported K^+ channels include Kir2.1, Kv1.3, Kv1.5, HERG, KCa2.3 and KCa3.1.

7.2.4.5.1 Voltage-gated K⁺ Channels

Korotzer and Cotman provided the first evidence for voltage-gated K^+ currents in microglia in 1992 [291]. They identified a K_V current in rat microglia that was activated by depolarization to potentials more positive than -40 mV, and the rates of activation and deactivation showed a voltage-dependence similar to other delayed rectifier currents. In the same year, Norenberg and colleagues reported that agents (e.g., bacterial LPS and interferon-γ) that promote microglial differentiation induced K_V currents in cultured rat microglia with properties similar to Kv1.3 [292]. A year later the same group identified Kv1.3 mRNA by RT-PCR in rat microglia [293]. The HIV-1 regulatory protein Tat was also reported to induce a Kv1.3-like current in rat microglia [294]. Ramified (process-bearing; resting) and ameboid (non-process-bearing; activated) mouse microglia were likewise reported to express Kv1.3 currents after culture with several agents: GM-CSF, interferon-γ, astrocyte-conditioned medium, transforming growth-factor-beta (TGF-β), pituitary adenylate cyclase activating polypeptide or beta-amyloid peptide [277,

295–299]. TGF-β-mediated upregulation of Kv1.3 expression was enhanced by pre-treatment with the p38 MAP kinase inhibitor SB203580 or the phosphatidy-linositol 3-OH (PI3) kinase inhibitor wortmannin, suggesting that p38 MAP kinase and PI3-kinase negatively regulate the expression of Kv1.3 in microglia [299]. Other negative-regulators of Kv1.3 were prostaglandin E2 [300] and arachidonic acid [301]. The Kv1.3 inhibitors agitoxin-2 and ChTX suppressed BrdU incorporation by microglia, highlighting a requirement for functional Kv1.3 channels in microglial proliferation [302]. Agitoxin-2 was also shown to partially inhibit the phorbol ester-stimulated respiratory burst (an indicator of microglial activation) by attenuating the rise in intracellular calcium that is required for the respiratory burst. Kv1.3 has been proposed to both initiate and then limit the respiratory burst because src-family tyrosine kinases that get activated during the respiratory burst phosphorylate Kv1.3 and thus reduce the Kv1.3 current, resulting in depolarization and decreased electron flux [303]. In cultured rat microglia, Kv1.3, *src* and the scaffolding protein PSD-95 form a signaling complex [303].

Kv1.5 was discovered in 1997 by immunohistochemistry in rat microglia cultured with LPS [304]. The same group found Kv1.5$^+$OX-42$^+$ microglia accumulating at sites of LPS injection into the rat brain preceding the expression of iNOS; Kv1.3 immunoreactivity was not detected [305]. A year later, Lyanne Schlichter's group in Toronto used a tissue printing technique on hippocampal rat brain slices to obtain microglia close to their *in vivo* condition and found Kv1.5 but not Kv1.3 currents [302]. After 10 days in culture, the tissue-printed microglia began expressing Kv1.3 channels in place of Kv1.5 and became highly proliferative. Immunostaining confirmed this Kv1.5 to Kv1.3 switch as cells shifted from a non-proliferating state into proliferating cells.

Kv11.1 (a.k.a. HERG, erg1, KCNH2, HERG-b) and Kv11.2 (a.k.a. erg-2) are present in the MLS-9 cell line derived from rat primary microglial cultures and co-assemble with *Src* tyrosine kinase in these cells [306, 307]. The functional role of these channels in microglia remains to be determined.

7.2.4.5.2 Calcium-activated K$^+$ Channels

Transcripts of KCa2.2 (a.k.a. SK2, KCNN2), KCa2.3 (SK3, KCNN3) and KCa3.1 were detected in rat microglia, and the presence of KCa2.3 protein was confirmed by immunostaining; no antibodies for KCa2.2 and KCa3.1 were available at that time [35]. The SK$_{Ca}$ channel blocker apamin and the KCa3.1 blockers ChTX and clotrimazole suppressed the phorbol ester-triggered respiratory burst, suggesting that both types of channels are involved in this process [35]. Microglial activation and migration induced by lysophosphatidic acid (LPA) caused the activation of KCa3.1 currents and the KCa3.1 blockers ChTX (also blocks Kv1.3) and clotrimazole (also blocks cytochrome p450 enzymes) partially inhibited microglial migration [308, 309]. Similar studies need to repeated with more specific KCa3.1 blockers.

Given the involvement of Kv1.3 and KCa3.1 in microglia proliferation [302], migration [308] and the respiratory burst [35], both channels have been proposed as therapeutic targets for the suppression of detrimental microglia functions that exacerbate brain damage during stroke and acute brain injury [35, 302] or HIV-related neuropathy [294]. However, this hypothesis remains to be tested in animal models since it is currently hotly debated whether microglia activation is beneficial or detrimental during ischemic stroke and traumatic brain injury.

7.2.4.5.3 Inward Rectifier K$^+$ Channels

Kettenman and colleagues were the first to identify a 30 pS inward-rectifying K$^+$ conductance (K_{ir}) in mouse microglia [310]. Studies by other groups confirmed the presence of K_{ir} currents in ramified and ameboid cultured microglia from rats and mice [291, 311, 312]. Ramified microglia in acutely isolated rat brain slices displayed little or no voltage-dependent currents but, within 12 h of nerve axotomy, microglia expressed a prominent K_{ir} current like cultured cells [313]. These K_{ir} currents had properties similar to Kir2.1, and Kir2.1 mRNA was detected by RT-PCR in mouse microglia [312]. Interferon-γ and protein kinase C were both reported to enhance K_{ir} current density in rat microglia [314]. Ba^{2+}, a non-selective blocker of Kir2.1, suppressed colony-stimulated-factor 1-induced proliferation of rat microglia, suggesting a requirement for this channel in microglial function [34].

In summary, several K$^+$ channels are involved in microglial function and it is therefore difficult to predict whether Kv1.3 blockers will adversely affect microglial function if used as therapeutics for immune modulation.

7.2.4.6 K$^+$ Channels in Oligodendrocytes

Oligodendrocytes are essential for remyelination of neurons. While there seems to be general agreement that non-dividing mature oligodendrocytes express very small outward K$^+$ currents, the reports about Kv1.3 in proliferative oligodendrocyte precursors vary, depending on the species that was used. Transcripts of Kv1.1–Kv1.6 are detected in mouse oligodendrocyte precursors, but at the protein level only Kv1.4, Kv1.5 and Kv1.6 are present in most cells, whereas Kv1.3 and Kv1.2 are undetectable [38]. The predominant K_V current in these cells appears to be Kv1.5 [38]. In contrast, oligodendrocyte precursors from rats upregulate Kv1.3 and Kv1.5 expression during the G1 phase of the cell cycle and then down-regulate these proteins as cells mature [39]. Overexpression of Kv1.3 in these cells augments activation while Kv1.3 inhibitors suppress proliferation but not differentiation [39, 315]. In brain tissue slice cultures from the subventricular zone where oligodendrocytes originate, the Kv1.3 inhibitors margatoxin and agitoxin-2 inhibited proliferation of oligodendrocyte precursors and pre-oligoendrocytes present in abundance in this region [39]. The inward rectifier channel, Kir4.1, is also expressed by oligodendrocyte precursors and mice lacking Kir4.1 exhibit hypomyelination, indicating that Kir4.1 has an important role in precursor function [316]. Oligodendrocyte progenitor cells are not seen in chronic MS lesions in humans and do not appear to contribute to remyelination in these lesions [317, 318]. The potentially harmful effects of Kv1.3 blockers due to the suppression of oligodendrocyte precursors may therefore be limited if these compounds are used as therapeutics for MS.

7.2.4.7 Kv1.3 in the Olfactory Bulb

Kv1.3 is expressed by neurons but as heteromultimers with Kv1.1, Kv1.2 and Kv1.4 subunits [43]. Highly selective blockers of homotetrameric Kv1.3 channels are therefore unlikely to block Kv1.3-containing heteromultimers and adversely affect neuronal function.

However, one region of the brain where homotetrameric Kv1.3 channels are present are the olfactory bulb and olfactory cortex [319] and K1.3 has been shown to carry 60–80% of the voltage-dependent K^+ current in cultured rat olfactory bulb neurons [40, 320]. In olfactory neurons, activation of the insulin receptor kinase by insulin [40, 321] or neurotrophin B (TrkB) by the brain-derived neurotrophic factor (BDNF) [322] results in tyrosine phosphorylation of the Kv1.3 channel and in acute reduction of the Kv1.3 current [323]. Mice lacking the Kv1.3 gene exhibit a "Super-Smeller" phenotype with 1000 to 10 000-fold lower threshold for the detection of odors and an increased ability to discriminate between different odorants [41]. Absence of Kv1.3 causes a compensatory increase in the expression of a non-inactivating K^+ current in olfactory neurons along with structural changes in the glomerular layer of the olfactory bulb. These results suggest a signal transduction role for Kv1.3 in the response of olfactory bulb neurons to sensory inputs. It remains to be seen whether Kv1.3 blockers will increase the sense of smell.

7.2.4.8 Other Cell Types

7.2.4.8.1 K$^+$ Channels in Adipocytes

K_V channels were described in brown fat rat cells by Pam Pappone's group at UC Davis in 1989 [324, 325]. These currents are resistant to ChTX, but are sensitive to verapamil, quinine, nifedipine, 4-aminopyridine and tetraethylammonium. Similar K_V currents were described in white fat cells from rats and humans [326, 327]. In recent studies, Kv1.3 protein was detected by Western blot analysis in white and brown fat tissue from mice [55] and the Kv1.3 inhibitor margatoxin was reported to facilitate translocation of the glucose transporter, GLUT4, to the plasma membrane of adipocytes and thus improve insulin sensitivity [328]. While Kv1.3 protein has been detected in adipocytes, the electrophysiological properties of K_V currents in brown and white adipocytes do not appear to fit the biophysical and pharmacological characteristics of a Kv1.3 homotetramer (e.g., they are ChTX-resistant and do not exhibit use-dependence). Moreover, adipocytes express only low levels of Kv1.3 transcripts along with low levels of Kv1.1–Kv1.7 mRNAs (http://symatlas.gnf.org/SymAtlas/), suggesting that the functional channel may consist of a heteromultimer of Kv1-containing subunits. An apamin-sensitive SK_{Ca} channel has also been reported in brown fat cells [324, 325]. Non-specific Kv channel blockers suppressed proliferation of cultured brown fat rat cells whereas apamin, the SK_{Ca} blocker, had no effect [324, 325].

7.2.4.8.2 K$^+$ Channels in Epithelia
Epithelia from rabbit colon and kidneys were reported to express Kv1.3, Kv1.2 and Kv1.1 protein [329–331]. In the colon, Kv1.3 protein was identified exclusively at the basolateral membrane of crypt cells, whereas in the kidney it was found in medullary interstial cells and the distal loop of Henle [330]. Kv1.3 and Kv1.1 protein was also identified at the basolateral surfaces of rat inner medullary cells where it was co-localized with Na^+-K^+-ATPase [331]. Immunostaining also revealed robust Kv1.3 expression in about a third of human colon cancer specimens and moderate expression in another 60% of tumors [332]. Electrophysiological analysis of these cells and evaluation of the effects of Kv1.3-inhibitors on function need to be undertaken.

7.2.4.8.3 K$^+$ Channels in the Bladder
Kv1.3 and Kv1.6 proteins were detected by immunostaining in human detrusor smooth muscle, and patch-clamp experiments revealed a K_V current that was blocked by 3,4-diaminopyridine [333]. Correolide (a blocker of all Kv1-family channels) and agitoxin-2 (a blocker of Kv1.3, Kv1.1, Kv1.6) enhanced the amplitude of spontaneous contractions but did not change the frequency of contractions [333]. The effects of selective Kv1.3 blockers on bladder function remain to be determined.

7.2.4.8.4 Kv1.3 in the Retina
Muller cells are the principal glia of the retina. A margatoxin- and ChTX-sensitive K_V current was identified in human retinal Muller cells, indicating that the channel contains Kv1.3 and/or Kv1.2 subunits [334], and both LPA and serum caused an increase in expression of the K_V channel. In the mouse retina, Kv1.3 protein was detected along with Kv1.1, Kv1.2, Kv1.4, Kv2.1 and Kv4.2 [335]. Kv1.2 and Kv1.3 immunoreactivity was detected in the axon and the postsynaptic membrane of the rod ribbon synapse [336]. Additional studies on K_V channels in cells of the retina are clearly warranted.

7.2.4.8.5 Kv1.3 in Osteoclasts
Osteoclasts, specialized cells responsible for bone resorption, share a common origin with cells of the monocyte/macrophage lineage [337]. In 1989, Ravesloot et al. first described that freshly isolated chicken osteoclasts expressed three distinct types of potassium currents, two K_V and one K_{ir} current [338]. Stephen Sims' group at the University of Western Ontario, Canada, later identified Kv1.3 and Kir2.1 mRNA in mouse and rat osteoclasts and characterized K$^+$ currents in these cells as exhibiting properties closely resembling Kv1.3 and Kir2.1 [36, 339]. An intermediate-conductance calcium-activated potassium channel with properties resembling KCa3.1 was later identified in rabbit osteoclasts and impli-

cated in osteoclast-spreading and bone resorption [340]. Human osteoclasts from deciduous teeth undergoing root resorption were shown to express a large-conductance K_{Ca} channel and a K_{ir} channel (presumably Kir2.1) [341]. ChTX, a blocker of Kv1.3, KCa3.1 and KCa1.1 channels, depolarized the osteoclast membrane potential by 5 to 10 mV, suggesting that one or more of these channels regulated the membrane potential of these cells. Experiments with inhibitors specific for each of these channels are required to identify the channel responsible for the ChTX-mediated membrane depolarization. Osteoclast ion channels have been proposed as novel therapeutic targets for antiresorptive drugs [342]. Kaliotoxin, a Kv1.3 inhibitor, effectively prevents bone resorption in experimental periodontal disease in rats, although disease amelioration was indicated to be mediated by the suppression of T_{EM} cells specific for *Actinobacillus actinomycetemcomitans* rather than to an effect on osteoclasts [53]. Future studies with specific Kv1.3 inhibitors may determine the therapeutic importance of this channel in osteoclasts and in osteoclast-mediated bone resorption.

7.2.4.8.6 K$^+$ Channels and Platelets

Maruyama identified a Kv1.3-like current in human, rabbit and rat platelets in 1987, and he estimated the number of K_V channels per platelet to be about 325, corresponding to a density of 25 μm^{-2} [37]. A ChTX-sensitive K_V current (i.e., either Kv1.2 or Kv1.3) was described in human platelets three years later and the channel was reported to play a major role in setting the resting membrane potential of these cells [343]. A KCa3.1-like current was also identified in human platelets and was shown to be important in regulating the membrane potential during calcium signaling [344, 345]. Platelets need to be reinvestigated with the newer tools available for the study of K$^+$ channels, and the functional roles of these channels in platelet aggregation and other platelet functions should be determined.

7.2.4.8.7 Kv1.3 in the Heart

Kv1.3 transcripts have been identified in human, rat, mouse and ferret cardiac tissue (http://symatlas.gnf.org/SymAtlas/) [346], although immunostaining has not been done to establish the presence of Kv1.3 protein in the heart. Kv1.3 is not a component of the major cardiac K$^+$ currents (I_{KR}, I_{KS}, I_{KUR}, I_{TO}) and its blockade is therefore unlikely to affect cardiac function, as evidenced by the lack of detectable electrocardiographic changes in rats administered a pharmacological dose of ShK(L5) [119]. It would be prudent, however, to look closely for cardiac toxicity when evaluating Kv1.3 blockers as therapeutics.

7.2.5
Disease Indications

This section assesses the therapeutic potential of Kv1.3 blockers for the treatment of immunological diseases, obesity, type-2 diabetes mellitus and anosmia.

7.2.5.1 **Autoimmune Diseases**

Autoimmune disorders are a broad class of diseases that affect virtually every tissue in the body and are a health threat to several hundred million people globally. T_{EM} cells contribute to inflammatory tissue damage in several autoimmune diseases, including MS, T1DM, RA, psoriasis and chronic graft-versus-host disease (Table 7.2.1). Class-switched CD27$^+$IgD$^-$ memory B cells are also implicated in the pathogenesis of several autoimmune diseases (Table 7.2.1). Therapies that suppress the function of these disease-causing autoreactive T_{EM} or class-switched memory B cells while sparing naïve-effectors, T_{CM}-effectors, naïve B cells and CD27$^+$IgD$^+$ B cells might ameliorate autoimmune diseases and at the same time allow patients to mount immune responses against most pathogenic agents.

Kv1.3 blockers would suppress the function of both CD4$^+$ and CD8$^+$ T_{EM} cells, while naïve/T_{CM} cells would escape inhibition by upregulating the KCa3.1 channel. As an unwanted side effect, suppression of all activated T_{EM} cells could adversely affect the individual's ability to overcome chronic infections. However, the inhibitory effects of Kv1.3 blockers can be overcome by strong antigenic stimuli, and pathogens typically induce robust immune responses. T_{EM} cells specific for pathogens should therefore be able to over-ride Kv1.3 blockade and effectively challenge invading pathogens. Conversely, autoreactive cells display low- to moderate-affinities for autoantigens and are therefore likely to be suppressed by Kv1.3 blockers. Another advantage of Kv1.3-based therapy is that it can be stopped when infections develop. Kv1.3 blockers suppress class-switched memory B cells with 10-fold lower potency than T_{EM} cells. Depending on the type of autoimmune disease, it may therefore be feasible to use low doses of Kv1.3 blockers to shut down T_{EM} cells alone and higher doses to suppress both T_{EM} and class-switched memory B cells without impacting other immune cell subsets.

7.2.5.1.1 **Multiple Sclerosis (MS)**

MS is a chronic inflammatory autoimmune disease of the central nervous system [347–350] that affects about 0.4 million people in the US and 2 million worldwide. Active MS lesions contain inflammatory infiltrates consisting of T lymphocytes, a few B cells, plasma cells, activated macrophages and microglia. This pathology is similar to that found in EAE in rodents, a model of MS. EAE can be induced in susceptible mouse and rat strains and in primates by immunization with either whole CNS tissue or purified proteins or peptides from CNS myelin or glia cells. The central role of T cells in EAE was demonstrated by the induction of EAE in naive animals following adoptive transfer of myelin antigen-specific T cells [351]. Although myelin antigen-specific T cells are also present in the blood of healthy controls, their activation state in MS patients is different [352]. Myelin-reactive T cells from MS patients secrete larger amounts of IL-2, IFN-γ and TNF-α than T cells from controls, and exhibit features of chronically stimulated memory T cells [220, 221, 353, 354]. Myelin basic protein (MBP)-specific encephalitogenic rat T cell lines [52] and short-term myelin antigen-specific T cell lines from the peripheral blood of patients with MS are predominantly Kv1.3^{high} activated CCR7$^-$CD45RA$^-$ T_{EM} cells [54]. T cells in active MS brain lesions are almost exclusively CCR7⁻ T_{EM} cells [355] and express high levels of Kv1.3 [411]. In contrast, naïve CCR7⁺ T cells predominate in brain infiltrates in mouse EAE [356, 357]. Unsurprisingly, since CCR7⁺ naïve-effectors and T_{CM}-effectors express KCa3.1, the KCa3.1 inhibitor TRAM-34 effectively ameliorated EAE in mice [179]. Mouse EAE may in general be more amenable to therapeutic interventions than MS because non-specifically recruited cells expressing many early activation markers predominate in the brain in mouse EAE and these are relatively easy to suppress. Differentiated T_{EM} cells that predominate in the brain in human MS patients are harder to shut down.

Present therapies for MS include the interferon-based drugs (Avonex®, Betaseron®, Rebif®), Copaxone® (glatiramer acetate) and the antineoplastic agent mitoxantrone (Novantrone). These "disease-modifying drugs" have significantly improved the management of autoimmune diseases during the last decade, but they broadly and indiscriminately modulate the entire immune system and they are not completely effective [358]. Mitoxantrone is only approved for patients with secondary progressive or worsening relapsing-remitting MS because of its toxicity. Interferon-β and glatiramer acetate ameliorate disease in about 50–70% of patients, of whom 20–30% experience sustained improvements, and about 20–40% of patients develop neutralizing antibodies to interferon-β that diminish drug efficacy over time. Tysabri® (Antegren® or natalizumab), a humanized monoclonal antibody against VLA-4, effectively reduced the number of active lesions and decreased the relapse rate in MS patients [359, 360], but two patients, who were also receiving other immune modulators, developed progressive multifocal leukoencephalopathy and one died. Cladribine (2-chloro-2-deoxyadenosine), Laquinimod and anti-CD25 antibodies (daclizumab) have recently been reported to be effective in MS therapy [361–363]. Of note is the failure of a vaccination trial for MS that had to be terminated because patients worsened after receiving altered myelin antigen-peptide ligand [364, 365].

A Kv1.3-based therapy would differ from existing therapies for MS by preferentially targeting T_{EM} cells. In proof-of-concept animal studies, the Kv1.3 blockers ShK and ShK(L5) effectively prevented disease development and reduced disease severity in a EAE model where disease was induced by the adoptive transfer of MBP-specific CD4⁺CD45RC⁻Kv1.3^high T_{EM} cells into Lewis rats [52, 119]. ShK(L5) currently constitutes the most suitable Kv1.3 blocker for further preclinical and possibly clinical development. It exhibits a high degree of specificity for Kv1.3, achieves steady state plasma levels of 300 pM following a single subcutaneous injection (10 μg kg⁻¹), exhibits no perceptible *in vitro* toxicity, is negative in the Ames test, has no effect on the heart rate and other cardiac parameters as measured by continuous EKG monitoring, and did not alter hematological and clinical chemistry parameters in a subchronic toxicity study [119]. Administration of a single subcutaneous injection of 1000 μg kg⁻¹ ShK(L5) or five repeated daily injections of 600 μg kg⁻¹ day⁻¹ caused no obvious signs of toxicity in normal rats, indicating that the therapeutic safety index is in excess of 100 [119]. The therapeutic safety index was 75 in animals with EAE, probably because the peptide crosses the compromised blood–brain barrier and at high doses blocks neuronal

Kv1.1 channels (IC_{50} for Kv1.1 block = 7 nM). Drug-discovery programs focused on Kv1.3 are ongoing at many BioPharma companies, and the hope is that a suitable therapeutic will emerge in the foreseeable future.

7.2.5.1.2 Type-1 Diabetes Mellitus (T1DM), Rheumatoid Arthritis (RA) and Psoriasis

Kv1.3 blockers should also be useful for the treatment of other autoimmune diseases mediated by T_{EM} cells, including T1DM, RA and psoriasis (Table 7.2.1). T1DM is caused by T cell-mediated autoimmune destruction of insulin-producing pancreatic beta cells [366, 367]. The central role of T cells in the pathogenesis of T1DM is supported by the demonstration that islet-infiltrating T cells from diabetic mice or rats can transfer disease into healthy animals. Such "adoptive transfer" has also unintentionally been demonstrated in humans when bone marrow was transplanted from diabetic donors to non-diabetic relatives [368, 369]. Intensive work over the last two decades has identified insulin, glutamic acid decarboxylase (GAD65 and GAD67) and islet antigen-2 (IA-2) as target autoantigens for both T cells and B cells in T1DM [253, 367, 370]. Although T cells responsive to these antigens are also present in healthy controls, circulating autoantigen reactive T cells from T1DM patients are predominantly of a memory phenotype and produce pro-inflammatory Th1 cytokines, indicating that they have undergone repeated activation *in vivo* and are critically involved in pathogenis [224, 371]. Kv1.3-based therapy to suppress anti-insulin and anti-GAD65-specific T_{EM} cells might have a use as an adjunct to islet cell transplantation (to prevent autoimmune rejection of transplanted islets) or could be used in conjunction with therapies designed to regenerate islet cells.

The Arthritis Foundation estimates that there are 2.1 million people with RA in the United States and about 6 million people affected with this disease in the world. RA is a chronic inflammatory disease that primarily affects peripheral joints but also commonly features systemic inflammation. T_{EM} cells seem to be critically involved in RA pathogenesis. Their percentage is increased in the peripheral blood of one-third of patients with RA [372] and they represent the majority of infiltrating T cells in the synovial fluid [225]. Peripheral blood $CD4^+$ T cells in RA patients have shorter telomeres [373]. This reflects increased replication *in vivo*, i.e., consistent with these cells being chronically activated memory cells. Biologic-response-modifiers such as the inhibitors of TNF-α (Etanercept, Adalimumab, Infliximab) have been reported to provide relief in RA, as did the COX-2 inhibitor Vioxx before it was withdrawn from the market. Kv1.3 blockers may have use in RA in combination with TNF-α-inhibitors or possibly as a stand-alone drug.

In psoriasis, an inflammatory skin disease, infiltration of $CD8^+$ cutaneous lymphocyte antigen$^+$ memory T cells into the epidermis is thought to be the key pathological event [227, 374–376]. $CD45RO^+CD8^+$ memory T cells have also been identified in the synovial fluid from patients with psoriatic arthritis [377]. The telomere lengths of $CD8^+CD28^+$ T cells from patients with psoriasis are shorter than in healthy controls, which is indicative of chronic stimulation *in vivo* [378]. Three autoantigens – keratin 13 (K13), heterogeneous nuclear ribonucleoprotein-A1 (hnRNP-A1), and a protein called FLJ00294 – that are recognized by T and B cells have been isolated from psoriatic plaques [379]. Methotrexate and cyclosporin A are effective therapies in psoriasis, but their significant side effect profiles preclude their long-term use. The TNF-α inhibitor Etanercept, the LFA-1 blocker efalizumab, and alefacept, a blocker of the CD2-LFA3 interaction, are effective in psoriasis [380]. Kv1.3 blockers, by targeting T_{EM} cells may enhance the therapeutic effectiveness of the biological disease modifiers and possibly have therapeutic use on their own.

7.2.5.1.3 Bone Resorption in Periodontal Disease

Periodontitis is a form of gum disease characterized by damage to the bone and connective tissue that supports the teeth. The disease is initiated by bacterial infection although memory cells have been implicated in bone resorption and tissue damage [381–384]. The Kv1.3 blocker kaliotoxin reduced bone resorption in rats induced by the transfer of T-cells directed against *Actinobacillus actinomycetemcomitans*, presumably by suppressing gingival-infiltrating T_{EM} cells [53], although the peptide could also have affected osteoclasts [53]. Selective Kv1.3 blockers may prevent bone loss in periodontitis.

7.2.5.2 Transplant Rejection and Chronic GvHD

Alloantigen-specific T cells play a crucial role in transplant rejection, and the precursor frequency of these cells is high (20–30%) in both the naïve and memory T cell pools of healthy individuals [385]. During a mixed lymphocyte reaction, CCR7 is rapidly lost from 50–60% of CD4$^+$ and CD8$^+$ T cells, suggesting that strong allogenic stimulation is effective in generating alloantigen-specific T_{EM} cells [385]. T_{EM} cells may therefore participate in the early stages of transplant rejection [385]. T_{EM} cells are blamed for transplant rejection that occurs in patients who have undergone drastic T-cell depletion with anti-CD52 antibody (alemtuzumab) because T_{EM} cells are remarkably resistant to anti-CD52 antibody [386]. The T_{EM} pool expands in the first few months after depletion and can constitute as much as 95% of CD4$^+$ cells in the peripheral blood of T-cell-depleted patients [386]. They also represent the majority of cells in the failed renal graft [386]. Kv1.3 blockers with their ability to shut down T_{EM} cells might be useful as an adjunct to alemtuzumab therapy by preventing the post-depletion expansion of T_{EM} cells and thus late-stage graft rejection. By attenuating calcium responses, Kv1.3 blockers may reduce the dose of cyclosporin A required for immunosuppression and thereby diminish the toxic side effects of this compound. Kv1.3 blockers may also synergize with rapamycin, FK-506 and mycophenolate mofetil.

Chronic graft-versus-host disease (GvHD) is the most common problem affecting long-term survivors of allogenic hematopoietic stem transplantation. Although bone marrow transplantation cures many patients with otherwise incurable diseases, chronic GvHD occurs in about 30% of recipients of transplants from well-matched siblings and in 60–70% of recipients of transplants from unmatched donors [387]. Chronic GvHD typically develops several months after transplantation and most commonly affects the skin, liver, mouth and eyes of the patient. In humans with severe chronic GvHD the percentage of CCR7$^-$ T$_{EM}$ cells in the peripheral blood is significantly increased in comparison to healthy controls and correlates with the severity of the syndrome [230]. A recent study in mice showed that the transfer of allogenic CD8$^+$ effector memory T cells from mice that had previously received a bone marrow transplant into secondary recipients caused virulent GvHD [388]. There is a clear scientific rationale to explore the effectiveness of Kv1.3 blockers as therapy for chronic GvHD.

7.2.5.3 Obesity

Examination of the Kv1.3$^{-/-}$ mice revealed that the channel is involved in body weight regulation [55]. Kv1.3$^{-/-}$ mice gained less weight on a high-fat diet than Kv1.3$^{+/+}$ littermates. Kv1.3 has been proposed as a therapeutic target for the improvement of insulin sensitivity and weight reduction in type-2 diabetes in humans. Additional studies with Kv1.3 inhibitors are merited.

7.2.5.4 Anosmia

Disorders of smell (anosmia) or taste affect about 4 million Americans. The sense of smell deteriorates with age, causing many elderly people to lose their appetite and become malnourished. Kv1.3$^{-/-}$ knockout mice exhibit a superior sense of smell [41], and pharmacological blockade of Kv1.3 might have the same functional effect as genetic knockout of the channel. Kv1.3 blockers could therefore have use in the treatment of anosmia.

7.2.6
Conclusions

Although much work remains to be done to develop either a peptidic or a small molecule Kv1.3 blocker into a clinically useful drug, great progress has been made in identifying Kv1.3 selective compounds and in evaluating Kv1.3 as a new drug target. Similar to cyclosporin and FK506, Kv1.3 blockers were initially thought to suppress the entire immune response by preventing the rise in intracellular calcium during T cell activation. However, our recent findings that K$^+$ channel expression changes during T and B cell differentiation from dominance of KCa3.1 in naïve and early memory cells to dominance of Kv1.3 in T$_{EM}$ cells and class-switched memory B cells have shown that Kv1.3 blockers are much more specific and selectively suppress the function of late-stage memory cells without affecting the function of naïve and early memory cells, which rely on KCa3.1. Since autoreactive "late" memory T and B cells play a critical role in the pathogenesis of autoimmune diseases, transplant rejection and chronic graft-versus-host disease, it therefore appeared logical to propose Kv1.3 blockade as a new therapeutic principle for the treatment of these diseases.

Ideally, autoimmune therapy should selectively target autoreactive responses without compromising essential immune functions. Antigen-specific vaccination strategies are therefore actively pursued for many autoimmune diseases. However, as mentioned above, vaccination trials for MS had to be terminated because MS worsened after patients received an altered myelin antigen-peptide ligand. Vaccination further requires a complete knowledge of the autoantigens against which the misguided immune response is directed, and in many autoimmune diseases the antigens have been only insufficiently characterized or new antigens are generated during the disease through epitope spreading, making it difficult to design an effective vaccine. Targeting T_{EM} cells with Kv1.3 blockers therefore seems to constitute a "good compromise" between the currently not realizable ideal of antigen-specific therapy and general immunosuppression. In proof of this concept Kv1.3 blockers can prevent T_{EM} cell mediated DTH reactions in rats and mini-pigs and treat EAE and inflammatory bone resorbtion in rats without exhibiting any obvious side effects. So, in summary, Kv1.3 blockade constitutes a promising therapeutic principle for the treatment of autoimmune diseases such as MS, T1DM, RA, psoriasis and lupus and for the prevention of transplant rejection and chronic graft-versus-host disease.

References

1. DeCoursey, T.E., Chandy, K.G., Gupta, S., Cahalan, M.D. Voltage-gated K⁺ channels in human T lymphocytes: a role in mitogenesis? *Nature* **1984**, *307*, 465–468.

2. Matteson, D.R., Deutsch, C. K channels in T lymphocytes: a patch clamp study using monoclonal antibody adhesion. *Nature* **1984**, *307*, 468–471.

3. Chandy, K.G., DeCoursey, T.E., Cahalan, M.D., McLaughlin, C., Gupta, S. Voltage-gated potassium channels are required for human T lymphocyte activation. *J. Exp. Med.* **1984**, *160*, 369–385.

4. Fukushima, Y., Hagiwara, S., Henkart, M. Potassium current in clonal cytotoxic T lymphocytes from the mouse. *J. Physiol.* **1984**, *351*, 645–656.

5. Cahalan, M.D., Chandy, K.G., DeCoursey, T.E., Gupta, S. A voltage-gated potassium channel in human T lymphocytes. *J. Physiol. (Lond.)* **1985**, *358*, 197–237.

6. Decoursey, T.E., Chandy, K.G., Gupta, S., Cahalan, M.D. Mitogen induction of ion channels in murine T lymphocytes. *J. Gen. Physiol.* **1987**, *89*, 405–420.

7. Sabath, D.E., Monos, D.S., Lee, S.C., Deutsch, C., Prystowsky, M.B. Cloned T-cell proliferation and synthesis of specific proteins are inhibited by quinine. *Proc.*

Natl. Acad. Sci. U.S.A. **1986**, *83*, 4739–4743.

8. Chandy, K.G., DeCoursey, T.E., Cahalan, M.D., Gupta, S. Electroimmunology: the physiologic role of ion channels in the immune system. *J. Immunol.* **1985**, *135*, 787s–791s.

9. Lin, C.S., Boltz, R.C., Blake, J.T., Nguyen, M., Talento, A., Fischer, P.A., Springer, M.S., Sigal, N.H., Slaughter, R.S., Garcia, M.L., Kaczorowski, G., Koo, G. Voltage-gated potassium channels regulate calcium-dependent pathways involved in human T lymphocyte activation. *J. Exp. Med.* **1993**, *177*, 637–645.

10. Freedman, B.D., Price, M.A., Deutsch, C.J. Evidence for voltage modulation of IL-2 production in mitogen-stimulated human peripheral blood lymphocytes. *J. Immunol.* **1992**, *149*, 3784–3794.

11. Cahalan, M.D., Lewis, R.S. Role of potassium and chloride channels in volume regulation by T lymphocytes. *In Cell Physiology of Blood*, R. Gunn, J. Parker, (eds.) Rockefeller University Press, New York, **1988**, pp. 282–301.

12. Deutsch, C., Chen, L.Q. Heterologous expression of specific K⁺ channels in T lymphocytes: functional consequences for volume regulation. *Proc. Natl. Acad. Sci. U.S.A.* **1993**, *90*, 10036–10040.

13. Lee, S.C., Price, M., Prystowsky, M.B., Deutsch, C. Volume response of quiescent and interleukin 2-stimulated T-lymphocytes to hypotonicity. *Am. J. Physiol.* **1988**, *254*, C286–296.

14. Hess, S.D., Oortgiesen, M., Cahalan, M.D. Calcium oscillations in human T and natural killer cells depend upon membrane potential and calcium influx. *J. Immunol.* **1993**, *150*, 2620–2633.

15. Verheugen, J.A., Vijverberg, H.P. Intracellular Ca^{2+} oscillations and membrane potential fluctuations in intact human T lymphocytes: role of K^+ channels in Ca^{2+} signaling. *Cell Calcium* **1995**, *17*, 287–300.

16. Schlichter, L., Sidell, N., Hagiwara, S. Potassium channels mediate killing by human natural killer cells. *Proc. Natl. Acad. Sci. U.S.A.* **1986**, *83*, 451–455.

17. Sidell, N., Schlichter, L.C., Wright, S.C., Hagiwara, S., Golub, S.H. Potassium channels in human NK cells are involved in discrete stages of the killing process. *J. Immunol.* **1986**, *137*, 1650–1658.

18. Sutro, J.B., Vayuvegula, B.S., Gupta, S., Cahalan, M.D. Voltage-sensitive ion channels in human B lymphocytes. *Adv. Exp. Med. Biol.* **1989**, *254*, 113–122.

19. Gollapudi, S.V., Vayuvegula, B.S., Thadepalli, H., Gupta, S. Effect of K^+ channel blockers on anti-immunoglobulin-induced murine B cell proliferation. *J. Clin. Lab. Immunol.* **1988**, *27*, 121–125.

20. Amigorena, S., Choquet, D., Teillaud, J.L., Korn, H., Fridman, W.H. Ion channels and B cell mitogenesis. *Mol. Immunol.* **1990**, *27*, 1259–1268.

21. Amigorena, S., Choquet, D., Teillaud, J.L., Korn, H., Fridman, W.H. Ion channel blockers inhibit B cell activation at a precise stage of the G1 phase of the cell cycle. Possible involvement of K^+ channels. *J. Immunol.* **1990**, *144*, 2038–2045.

22. Brent, L.H., Butler, J.L., Woods, W.T., Bubien, J.K. Transmembrane ion conductance in human B lymphocyte activation. *J. Immunol.* **1990**, *145*, 2381–2398.

23. Stuhmer, W., Ruppersberg, J.P., Schroter, K.H., Sakmann, B., Stocker, M., Giese, K.P., Perschke, A., Baumann, A., Pongs, O. Molecular basis of functional diversity of voltage-gated potassium channels in mammalian brain. *EMBO J.* **1989**, *8*, 3235–3244.

24. Chandy, K.G., Williams, C.B., Spencer, R.H., Aguilar, B.A., Ghanshani, S., Tempel, B.L., Gutman, G.A. A family of three mouse potassium channel genes with intronless coding regions. *Science* **1990**, *247*, 973–975.

25. Swanson, R., Marshall, J., Smith, J.S., Williams, J.B., Boyle, M.B., Folander, K., Luneau, C.J., Antanavage, J., Oliva, C., Buhrow, S.A., et al. Cloning and expression of cDNA and genomic clones encoding three delayed rectifier potassium channels in rat brain. *Neuron* **1990**, *4*, 929–939.

26. Folander, K., Douglass, J., Swanson, R. Confirmation of the assignment of the gene encoding Kv1.3, a voltage-gated potassium channel (KCNA3) to the proximal short arm of human chromosome 1. *Genomics* **1994**, *23*, 295–296.

27. Grissmer, S., Dethlefs, B., Wasmuth, J.J., Goldin, A.L., Gutman, G.A., Cahalan, M.D., Chandy, K.G. Expression and chromosomal localization of a lymphocyte K^+ channel gene. *Proc. Natl. Acad. Sci. U.S.A.* **1990**, *87*, 9411–9415.

28. Douglass, J., Osborne, P.B., Cai, Y.C., Wilkinson, M., Christie, M.J., Adelman, J.P. Characterization and functional expression of a rat genomic DNA clone encoding a lymphocyte potassium channel. *J. Immunol.* **1990**, *144*, 4841–4850.

29. Attali, B., Romey, G., Honore, E., Schmid-Alliana, A., Mattei, M.G., Lesage, F., Ricard, P., Barhanin, J., Lazdunski, M. Cloning, functional expression, and regulation of two K^+ channels in human T lymphocytes. *J. Biol. Chem.* **1992**, *267*, 8650–8657.

30. Cai, Y.C., Osborne, P.B., North, R.A., Dooley, D.C., Douglass, J. Characterization and functional expression of genomic DNA encoding the human lymphocyte type n potassium channel. *DNA Cell. Biol.* **1992**, *11*, 163–172.

31. Wulff, H., Knaus, H.G., Pennington, M., Chandy, K.G. K^+ channel expression during B-cell differentiation: implications for immunomodulation and autoimmunity. *J. Immunol.* **2004**, *173*, 776–786.

32. Mackenzie, A.B., Chirakkal, H., North, R.A. Kv1.3 potassium channels in human alveolar macrophages. *Am. J. Physiol. Lung Cell Mol. Physiol.* **2003**, *85*, L862–868.

33. Vicente, R., Escalada, A., Coma, M., Fuster, G., Sanchez-Tillo, E., Lopez-Iglesias, C., Soler, C., Solsona, C., Celada, A., Felipe, A. Differential voltage-dependent K$^+$ channel responses during proliferation and activation in macrophages. *J. Biol. Chem.* **2003**, *278*, 46 307–46 320.

34. Schlichter, L.C., Sakellaropoulos, G., Ballyk, B., Pennefather, P.S., Phipps, D.J. Properties of K$^+$ and Cl$^-$ channels and their involvement in proliferation of rat microglial cells. *Glia* **1996**, *17*, 225–236.

35. Khanna, R., Roy, L., Zhu, X., Schlichter, L.C. K$^+$ channels and the microglial respiratory burst. *Am. J. Physiol. Cell. Physiol.* **2001**, *280*, C796–806.

36. Arkett, S.A., Dixon, J., Yang, J.N., Sakai, D.D., Minkin, C., Sims, S.M. Mammalian osteoclasts express a transient potassium channel with properties of Kv1.3. *Receptors Channels* **1994**, *2*, 281–293.

37. Maruyama, Y. A patch-clamp study of mammalian platelets and their voltage-gated potassium current. *J. Physiol.* **1987**, *391*, 467–845.

38. Schmidt, K., Eulitz, D., Veh, R.W., Kettenmann, H., Kirchhoff, F. Heterogeneous expression of voltage-gated potassium channels of the shaker family (Kv1) in oligodendrocyte progenitors. *Brain Res.* **1999**, *843*, 145–160.

39. Chittajallu, R., Chen, Y., Wang, H., Yuan, X., Ghiani, C.A., Heckman, T., C.J., M., Gallo, V. Regulation of Kv1 subunit expression in oligodendrocyte progentitor cells and their role in G1/S phase progression of the cell cycle. *Proc. Natl. Acad. Sci. U.S.A.* **2002**, *99*, 2350–2355.

40. Fadool, D.A., Levitan, I.B. Modulation of olfactory bulb neuron potassium current by tyrosine phosphorylation. *J. Neurosci.* **1998**, *18*, 6126–6137.

41. Fadool, D.A., Tucker, K., Perkins, R., Fasciani, G., Thompson, R.N., Parsons, A.D., Overton, J.M., Koni, P.A., Flavell, R.A., Kaczmarek, L.K. Kv1.3 channel gene-targeted deletion produces "Super-Smeller Mice" with altered glomeruli,

interacting scaffolding proteins, and biophysics. *Neuron* **2004**, *41*, 389–404.

42. Helms, L.M., Felix, J.P., Bugianesi, R.M., Garcia, M.L., Stevens, S., Leonard, R.J., Knaus, H.G., Koch, R., Wanner, S.G., Kaczorowski, G.J., Slaughter, R.S. Margatoxin binds to a homomultimer of K(V)1.3 channels in Jurkat cells. Comparison with K(V)1.3 expressed in CHO cells. *Biochemistry* **1997**, *36*, 3737–3744.

43. Koschak, A., Bugianesi, R.M., Mitterdorfer, J., Kaczorowski, G.J., Garcia, M.L., Knaus, H.G. Subunit composition of brain voltage-gated potassium channels determined by hongotoxin-1, a novel peptide derived from Centruroides limbatus venom. *J. Biol. Chem.* **1998**, *273*, 2639–2644.

44. Coleman, S.K., Newcombe, J., Pryke, J., Dolly, J.O. Subunit composition of Kv1 channels in human CNS. *J. Neurochem.* **1999**, *73*, 849–859.

45. Cahalan, M.D., Wulff, H., Chandy, K.G. Molecular properties and physiological roles of ion channels in the immune system. *J. Clin. Immunol.* **2001**, *21*, 235–252.

46. Chandy, K.G., Cahalan, M.D., Pennington, M., Norton, R.S., Wulff, H., Gutman, G.A. Potassium channels in T lymphocytes: toxins to therapeutic immunosuppressants. *Toxicon* **2001**, *39*, 1269–1276.

47. Wulff, H., Beeton, C., Chandy, K.G. Potassium channels as therapeutic targets for autoimmune disorders. *Curr. Opin. Drug Discovery Develop.* **2003**, *6*, 640–647.

48. Chandy, K.G., Wulff, H., Beeton, C., Pennington, M., Gutman, G.A., Cahalan, M.D. Potassium channels as targets for specific immunomodulation. *Trends Pharmacol. Sci.* **2004**, *25*, 280–289.

49. Koo, G.C., Blake, J.T., Talento, A., Nguyen, M., Lin, S., Sirotina, A., Shah, K., Mulvany, K., Hora, D., Jr., Cunningham, P., Wunderler, D.L., McManus, O.B., Slaughter, R., Bugianesi, R., Felix, J., Garcia, M., Williamson, J., Kaczorowski, G., Sigal, N.H., Springer, M.S., Feeney, W. Blockade of the voltage-gated potassium channel Kv1.3 inhibits immune responses in vivo. *J. Immuno. l* **1997**, *158*, 5120–5128.

50. Koo, G.C., Blake, J.T., Shah, K., Staruch, M.J., Dumont, F., Wunderler, D., Sanchez, M., McManus, O.B., Sirotina-Meisher, A., Fischer, P., Boltz, R.C., Goetz, M.A., Baker, R., Bao, J., Kayser, F., Rupprecht, K.M., Parsons, W.H., Tong, X.C., Ita, I.E., Pivnichny, J., Vincent, S., Cunningham, P., Hora, D., Jr., Feeney, W., Kaczorowski, G., et al. Correolide and derivatives are novel immunosuppressants blocking the lymphocyte Kv1.3 potassium channels. *Cell Immunol.* **1999**, *197*, 99–107.

51. Beeton, C., Barbaria, J., Giraud, P., Devaux, J., Benoliel, A., Gola, M., Sabatier, J., Bernard, D., Crest, M., Beraud, E. Selective blocking of voltage-gated K$^+$ channels improves experimental autoimmune encephalomyelitis and inhibits T cell activation. *J. Immunol.* **2001**, *166*, 936–944.

52. Beeton, C., Wulff, H., Barbaria, J., Clot-Faybesse, O., Pennington, M., Bernard, D., Cahalan, M.D., Chandy, K.G., Beraud, E. Selective blockade of T lymphocyte K$^+$ channels ameliorates experimental autoimmune encephalomyelitis, a model for multiple sclerosis. *Proc. Natl. Acad. Sci. U.S.A.* **2001**, *98*, 13 942–13 947.

53. Valverde, P., Kawai, T., Taubman, M.A. Selective blockade of voltage-gated potassium channels reduces inflammatory bone resorption in experimental periodontal disease. *J. Bone Miner. Res.* **2004**, *19*, 155–164.

54. Wulff, H., Calabresi, P.A., Allie, R., Yun, S., Pennington, M., Beeton, C., Chandy, K.G. The voltage-gated Kv1.3 K$^+$ channel in effector memory T cells as new target for MS. *J. Clin. Invest.* **2003**, *111*, 1703–1713.

55. Xu, J., Koni, P.A., Wang, P., Li, G., Kaczmarek, L., Wu, Y., Li, Y., Flavell, R.A., Desir, G.V. The voltage-gated potassium channel Kv1.3 regulates energy homeostasis and body weight. *Hum. Mol. Genet.* **2003**, *12*, 551–559.

56. Tytgat, J., Chandy, K.G., Garcia, M.L., Gutman, G.A., Martin-Eauclaire, M.F., van der Walt, J.J., Possani, L.D. A unified nomenclature for short-chain peptides isolated from scorpion venoms: alpha-KTx molecular subfamilies. *Trends Pharmacol. Sci.* **1999**, *20*, 444–447.

57. Miller, C., Moczydlowski, E., Latorre, R., Phillips, M. Charybdotoxin, a protein inhibitor of single Ca^{2+}-activated K$^+$ channels from mammalian skeletal muscle. *Nature* **1985**, *313*, 316–368.

58. Sands, S.B., Lewis, R.S., Cahalan, M.D. Charybdotoxin blocks voltage-gated K$^+$ channels in human and murine T lymphocytes. *J. Gen. Physiol.* **1989**, *93*, 1061–1074.

59. Deutsch, C., Price, M., Lee, S., King, V., Garcia, M. Characterization of high affinity binding sites for charybdotoxin in human T lymphocytes. Evidence for association with the voltage-gated K$^+$ channel. *J. Biol. Chem.* **1991**, *266*, 3668–3674.

60. Leonard, R., Garcia, M., Slaughter, R., Reuben, J. Selective blockers of voltage-gated K$^+$ channels depolarize human T lymphocytes: mechanism of the antiproliferative effect of charybdotoxin. *Proc. Natl. Acad. Sci. U.S.A.* **1992**, *89*, 10 094–10 098.

61. Garcia, M.L., Garcia-Calvo, M., Hidalgo, P., Lee, A., MacKinnon, R. Purification and characterization of three inhibitors of voltage-dependent K$^+$ channels from Leiurus quinquestriatus var. hebraeus venom. *Biochemistry* **1994**, *33*, 6834–6839.

62. Garcia-Calvo, M., Leonard, R.J., Novick, J., Stevens, S.P., Schmalhofer, W., Kaczorowski, G.J., Garcia, M.L. Purification, characterization, and biosynthesis of margatoxin, a component of Centruroides margaritatus venom that selectively inhibits voltage-dependent potassium channels. *J. Biol. Chem.* **1993**, *268*, 18 866–18 874.

63. Bednarek, M., Bugianesi, R., Leonard, R., Felix, J. Chemical synthesis and structure-function studies of margatoxin, a potent inhibitor of voltage-dependent potassium channel in human T lymphocytes. *Biochem. Biophys. Res. Commun.* **1994**, *198*, 619–625.

64. Crest, M., Jacquet, G., Gola, M., Zerrouk, H., Benslimane, A., Rochat, H., Mansuelle, P., Martin-Eauclaire, M.F. Kaliotoxin, a novel peptidyl inhibitor of neuronal BK-type Ca^{2+}-activated K$^+$ channels characterized from Androctonus mauretanicus mauretanicus venom. *J. Biol. Chem.* **1992**, *267*, 1640–1647.

65. Dauplais, M., Gilquin, B., Possani, L., Gurrola-Briones, G., Roumestand, C.,

Menez, A. Determination of the three-dimensional solution structure of nox-iustoxin: analysis of structural differences with related short-chain scorpion toxins. *Biochemistry* **1995**, *34*, 16 563–16 573.

66. Peter, M., Jr., Varga, Z., Hajdu, P., Gaspar, R., Jr., Damjanovich, S., Horjales, E., Possani, L.D., Panyi, G. Effects of toxins Pi2 and Pi3 on human T lymphocyte Kv1.3 channels: the role of Glu7 and Lys24. *J. Membr. Biol.* **2001**, *179*, 13–25.

67. Mouhat, S., Visan, V., Ananthakrishnan, S., Wulff, H., Andreotti, N., Grissmer, S., Darbon, H., De Waard, M., Sabatier, J.M. K⁺ channel types targeted by synthetic OSK1, a toxin from Orthochirus scrobiculosus scorpion venom. *Biochem. J.* **2005**, *385*, 95–104.

68. Gimenez-Gallego, G., Navia, M.A., Reuben, J.P., Katz, G.M., Kaczorowski, G.J., Garcia, M.L. Purification, sequence, and model structure of charybdotoxin, a potent selective inhibitor of calcium-activated potassium channels. *Proc. Natl. Acad. Sci. U.S.A.* **1988**, *85*, 3329–3333.

69. Bontems, F., Roumestand, C., Boyot, P., Gilquin, B., Doljansky, Y., Menez, A., Toma, F. Three-dimensional structure of natural charybdotoxin in aqueous solution by 1H-NMR. Charybdotoxin possesses a structural motif found in other scorpion toxins. *Eur. J. Biochem.* **1991**, *196*, 19–28.

70. Johnson, B., Stevens, S., Williamson, J. Determination of the three-dimensional structure of margatoxin by ¹H, ¹³C, ¹⁵N triple-resonance nuclear magnetic resonance spectroscopy. *Biochemistry* **1994**, *33*, 15 061–15 070.

71. Aiyar, J., Withka, J.M., Rizzi, J.P., Singleton, D.H., Andrews, G.C., Lin, W., Boyd, J., Hanson, D.C., Simon, M., Dethlefs, B., Lee, C.-L., Hall, J.E., Gutman, G.A., Chandy, K.G. Topology of the pore-region of a K⁺ channel revealed by the NMR-derived structures of scorpion toxins. *Neuron* **1995**, *15*, 1169–1181.

72. Krezel, A.M., Kasibhatla, C., Hidalgo, P., MacKinnon, R., Wagner, G. Solution structure of the potassium channel inhibitor agitoxin 2: caliper for probing channel geometry. *Protein Sci.* **1995**, *4*, 1478–1489.

73. Pragl, B., Koschak, A., Trieb, M., Obermair, G., Kaufmann, W.A., Gerster, U., Blanc, E., Hahn, C., Prinz, H., Schutz, G., Darbon, H., Gruber, H.J., Knaus, H.G. Synthesis, characterization, and application of cy-dye- and alexa-dye-labeled hongotoxin(1) analogues. The first high affinity fluorescence probes for voltage-gated K⁺ channels. *Bioconjug. Chem.* **2002**, *13*, 416–425.

74. MacKinnon, R., Miller, C. Mutant potassium channels with altered binding of charybdotoxin, a pore-blocking peptide inhibitor. *Science* **1989**, *245*, 1382–1385.

75. MacKinnon, R., Heginbotham, L., Abramson, T. Mapping the receptor site for charybdotoxin, a pore-blocking potassium channel inhibitor. *Neuron* **1990**, *5*, 767–771.

76. Goldstein, S.A., Miller, C. A point mutation in a Shaker K⁺ channel changes its charybdotoxin binding site from low to high affinity. *Biophys. J.* **1992**, *62*, 5–7.

77. Park, C.S., Miller, C. Interaction of charybdotoxin with permeant ions inside the pore of a K⁺ channel. *Neuron* **1992**, *9*, 307–313.

78. Goldstein, S.A., Miller, C. Mechanism of charybdotoxin block of a voltage-gated K⁺ channel. *Biophys. J.* **1993**, *65*, 1613–1619.

79. Goldstein, S.A., Pheasant, D.J., Miller, C. The charybdotoxin receptor of a Shaker K⁺ channel: peptide and channel residues mediating molecular recognition. *Neuron* **1994**, *12*, 1377–1388.

80. Stocker, M., Miller, C. Electrostatic distance geometry in a K⁺ channel vestibule. *Proc. Natl. Acad. Sci. U.S.A.* **1994**, *91*, 9509–9513.

81. Stampe, P., Kolmakova-Partensky, L., Miller, C. Intimations of K⁺ channel structure from a complete functional map of the molecular surface of charybdotoxin. *Biochemistry* **1994**, *33*, 443–450.

82. Naranjo, D., Miller, C. A strongly interacting pair of residues on the contact surface of charybdotoxin and a Shaker K⁺ channel. *Neuron* **1996**, *16*, 123–130.

83. Ranganathan, R., Lewis, J.H., MacKinnon, R. Spatial localization of the K⁺ channel selectivity filter by mutant cycle-based structure analysis. *Neuron* **1996**, *16*, 131–139.

84. Gross, A., MacKinnon, R. Agitoxin foot-printing the shaker potassium channel pore. *Neuron* **1996**, *16*, 399–406.

85. Aiyar, J., Rizzi, J.P., Gutman, G.A., Chandy, K.G. The signature sequence of voltage-gated potassium channels projects into the external vestibule. *J. Biol. Chem.* **1996**, *271*, 31 013–31 016.

86. Rauer, H., Pennington, M., Cahalan, M., Chandy, K.G. Structural conservation of the pores of calcium-activated and voltage-gated potassium channels determined by a sea anemone toxin. *J. Biol. Chem.* **1999**, *274*, 21 885–21 892.

87. Hidalgo, P., MacKinnon, R. Revealing the architecture of a K$^+$ channel pore through mutant cycles with a peptide inhibitor. *Science* **1995**, *268*, 307–310.

88. Doyle, D.A., Morais Cabral, J., Pfuetzner, R.A., Kuo, A., Gulbis, J.M., Cohen, S.L., Chait, B.T., MacKinnon, R. The structure of the potassium channel: molecular basis of K$^+$ conduction and selectivity. *Science* **1998**, *280*, 69–77.

89. MacKinnon, R., Cohen, S.L., Kuo, A., Lee, A., Chait, B.T. Structural conservation in prokaryotic and eukaryotic potassium channels. *Science* **1998**, *280*, 106–109.

90. Legros, C., Pollmann, V., Knaus, H., Farrell, A., Darbon, H., Bougis, P., Martin-Eauclaire, M., Pongs, O. Generating a high affinity scorpion toxin receptor in KcsA-Kv1.3 chimeric potassium channels. *J. Biol. Chem.* **2000**, *275*, 16 918–16 924.

91. Kharrat, R., Mabrouk, K., Crest, M., Darbon, H., Oughideni, R., Martin-Eauclaire, M.F., Jacquet, G., el Ayeb, M., Van Rietschoten, J., Rochat, H., Sabatier, J.M. Chemical synthesis and characterization of maurotoxin, a short scorpion toxin with four disulfide bridges that acts on K$^+$ channels. *Eur. J. Biochem.* **1996**, *242*, 491–498.

92. Blanc, E., Sabatier, J., Kharrat, R., Meunier, S., el Ayeb, M., Van Rietschoten, J., Darbon, H. Solution structure of maurotoxin, a scorpion toxin from Scorpio maurus, with high affinity for voltage-gated potassium channels. *Proteins* **1997**, *29*, 321–333.

93. Lebrun, B., Romi-Lebrun, R., Martin-Eauclaire, M.F., Yasuda, A., Ishiguro, M., Oyama, Y., Pongs, O., Nakajima, T. A four-disulphide-bridged toxin, with high affinity towards voltage-gated K$^+$ channels, isolated from Heterometrus spinnifer (Scorpionidae) venom. *Biochem. J.* **1997**, *328*, 321–327.

94. Peter, M.J., Hajdu, P., Varga, Z., Damjanovich, S., Possani, L., Panyi, G., Gaspar, R.J. Blockage of human T lymphocyte Kv1.3 channels by Pi1, a novel class of scorpion toxin. *Biochem. Biophys. Res. Commun.* **2000**, *278*, 34–37.

95. Bagdany, M., Batista, C.V., Valdez-Cruz, N.A., Somodi, S., Rodriguez de la Vega, R.C., Licea, A.F., Varga, Z., Gaspar, R., Possani, L.D., Panyi, G. Anuroctoxin, a new scorpion toxin of the {alpha}-KTx 6 subfamily is highly selective for Kv1.3 over IKCa1 ion channels of human T lymphocytes. *Mol. Pharmacol.* **2005**, *67*, 1034–1044.

96. Suarez-Kurtz, G., Vianna-Jorge, R., Pereira, B.F., Garcia, M.L., Kaczorowski, G.J. Peptidyl inhibitors of shaker-type Kv1 channels elicit twitches in guinea pig ileum by blocking Kv1.1 at enteric nervous system and enhancing acetylcholine release. *J. Pharmacol. Exp. Ther.* **1999**, *289*, 1517–1522.

97. Strydom, D.J. Snake venom toxins. Purification and properties of low-molecular-weight polypeptides of Dendroaspis polylepis polylepis (black mamba) venom. *Eur. J. Biochem.* **1976**, *69*, 169–176.

98. Benishin, C.G., Sorensen, R.G., Brown, W.E., Krueger, B.K., Blaustein, M.P. Four polypeptide components of green mamba venom selectively block certain potassium channels in rat brain synaptosomes. *Mol. Pharmacol.* **1988**, *34*, 152–159.

99. Harvey, A.L., Anderson, A.J. Dendrotoxins: snake toxins that block potassium channels and facilitate neurotransmitter release. *Pharmacol. Ther.* **1985**, *31*, 33–55.

100. Grissmer, S., Nguyen, A.N., Aiyar, J., Hanson, D.C., Mather, R.J., Gutman, G.A., Karmilowicz, M.J., Auperin, D.D., Chandy, K.G. Pharmacological characterization of five cloned voltage-gated K$^+$ channels, types Kv1.1, 1.2, 1.3, 1.5, and 3.1, stably expressed in mammalian cell lines. *Mol. Pharmacol.* **1994**, *45*, 1227–1234.

101. Tytgat, J., Debont, T., Carmeliet, E., Daenens, P. The alpha-dendrotoxin footprint on a mammalian potassium channel. *J. Biol. Chem.* **1995**, *270*, 24 776–24 781.

102. Gasparini, S., Danse, J.M., Lecoq, A., Pinkasfeld, S., Zinn-Justin, S., Young, L.C., de Medeiros, C.C., Rowan, E.G., Harvey, A.L., Menez, A. Delineation of the functional site of alpha-dendrotoxin. The functional topographies of dendrotoxins are different but share a conserved core with those of other Kv1 potassium channel-blocking toxins. *J. Biol. Chem.* **1998**, *273*, 25 393–25 403.

103. Kauferstein, S., Huys, I., Lamthanh, H., Stocklin, R., Sotto, F., Menez, A., Tytgat, J., Mebs, D. A novel conotoxin inhibiting vertebrate voltage-sensitive potassium channels. *Toxicon* **2003**, *42*, 43–52.

104. Ferber, M., Al-Sabi, A., Stocker, M., Olivera, B.M., Terlau, H. Identification of a mammalian target of kappaM-conotoxin RIIIK. *Toxicon* **2004**, *43*, 915–921.

105. Aneiros, A., Garcia, I., Martinez, J.R., Harvey, A.L., Anderson, A.J., Marshall, D.L., Engstrom, A., Hellman, U., Karlsson, E. **(1993)** A potassium channel toxin from the secretion of the sea anemone Bunodosoma granulifera. Isolation, amino acid sequence and biological activity. *Biochim. Biophys. Acta* **1993**, *1157*, 86–92.

106. Castaneda, O., Sotolongo, V., Amor, A.M., Stocklin, R., Anderson, A.J., Harvey, A.L., Engstrom, A., Wernstedt, C., Karlsson, E. Characterization of a potassium channel toxin from the Caribbean Sea anemone Stichodactyla helianthus. *Toxicon* **1995**, *33*, 603–613.

107. Pennington, M., Mahnir, V., Krafte, D., Zaydenberg, I., Byrnes, M., Khaytin, I., Crowley, K., Kem, W. Identification of three separate binding sites on ShK toxin, a potent inhibitor of voltage-dependent potassium channels in human T-lymphocytes and rat brain. *Biochem. Biophys. Res. Commun.* **1996**, *219*, 696–701.

108. Cotton, J., Crest, M., Bouet, F., Alessandri, N., Gola, M., Forest, E., Karlsson, E., Castaneda, O., Harvey, A., Vita, C., Menez, A. A potassium-channel toxin from the sea anemone Bunodosoma granulifera, an inhibitor for Kv1 channels. Revision of the amino acid sequence, disulfide-bridge assignment, chemical synthesis, and biological activity. *Eur. J. Biochem.* **1997**, *244*, 192–202.

109. Kalman, K., Pennington, M.W., Lanigan, M.D., Nguyen, A., Rauer, H., Mahnir, V., Paschetto, K., Kem, W.R., Grissmer, S., Gutman, G.A., Christian, E.P., Cahalan, M.D., Norton, R.S., Chandy, K.G. ShK-Dap[22], a potent Kv1.3-specific immunosuppressive polypeptide. *J. Biol. Chem.* **1998**, *273*, 32 697–32 707.

110. Alessandri-Haber, N., Lecoq, A., Gasparini, S., Grangier-Macmath, G., Jacquet, G., Harvey, A., de Medeiros C, Rowan E.G., Gola M., Menez A.M.C. Mapping the functional anatomy of BgK on Kv1.1, Kv1.2, and Kv1.3. Clues to design analogs with enhanced selectivity. *J. Biol. Chem.* **1999**, *274*, 35 653–35 661.

111. Pennington, M., Byrnes, M., Zaydenberg, I., Khaytin, I., de Chastonay, J., Krafte, D., Hill, R., Mahnir, V., Volberg, W., Gorczyca, W., Kem, W. Chemical synthesis and characterization of ShK toxin: a potent potassium channel inhibitor from a sea anemone. *Int. J. Pept. Protein Res.* **1995**, *346*, 354–358.

112. Pennington, M., Mahnir, V., Khaytin, I., Zaydenberg, I., Byrnes, M., Kem, W. An essential binding surface for ShK toxin interaction with rat brain potassium channels. *Biochemistry* **1996**, *35*, 16 407–16 411.

113. Tudor, J.E., Pallaghy, P.K., Pennington, M.W., Norton, R.S. Solution structure of ShK toxin, a novel potassium channel inhibitor from a sea anemone. *Nat. Struct. Biol.* **1996**, *3*, 317–320.

114. Pohl, J., Hubalek, F., Byrnes, M.E., Nielsen, K.R., Woods, A., Pennington, M.W. Assignment of the three disulfide bonds in ShK toxin: A potent potassium channel inhibitor from the sea anemone Stichodactyla helianthus. *Lett. Int. Peptide Sci.* **1995**, *1*, 291–297.

115. Dauplais, M., Lecoq, A., Song, J., Cotton, J., Jamin, N., Gilquin, B., Roumestand, C., Vita, C., de Medeiros, C.L.C., Rowan, E.G., Harvey, A.L., Menez, A. On the convergent evolution of animal toxins. Conservation of a diad of functional residues in potassium channel-blocking toxins with unrelated structures. *J. Biol. Chem.* **1997**, *272*, 4302–4309.

116. Gasparini, S., Gilquin, B., Menez, A. Comparison of sea anemone and scorpion toxins binding to Kv1 channels: an example of convergent evolution. *Toxicon* **2004**, *43*, 901–908.

117. Middleton, R.E., Sanchez, M., Linde, A.R., Bugianesi, R.M., Dai, G., Felix, J.P., Koprak, S.L., Staruch, M.J., Bruguera, M., Cox, R., Ghosh, A., Hwang, J., Jones, S., Kohler, M., Slaughter, R.S., McManus, O.B., Kaczorowski, G.J., Garcia, M.L. Substitution of a single residue in Stichodactyla helianthus peptide, ShK-Dap[22], reveals a novel pharmacological profile. *Biochemistry* **2003**, *42*, 13698–13707.

118. Beeton, C., Wulff, H., Singh, S., Bosko, S., Crossley, G., Gutman, G.A., Cahalan, M.D., Pennington, M., Chandy, K.G. A novel fluorescent toxin to detect and investigate Kv1.3-channel up-regulation in chronically activated T lymphocytes. *J. Biol. Chem.* **2003**, *278*, 9928–9937.

119. Beeton, C., Pennington, M.W., Wulff, H., Singh, S., Nugent, D., Crossley, G., Khaytin, I., Calabresi, P.A., Chen, C.Y., Gutman, G.A., Chandy, K.G. Targeting effector memory T cells with a selective peptide inhibitor of Kv1.3 channels for therapy of autoimmune diseases. *Mol. Pharmacol.* **2005**, *67*, 1369–13681.

120. Yan, L., Herrington, J., Goldberg, E., Dulski, P.M., Bugianesi, R.M., Slaughter, R.S., Banerjee, P., Brochu, R.M., Priest, B.T., Kaczorowski, G.J., Rudy, B., Garcia, M.L. ShK, a pharmacological tool for studying Kv3.2 channels. *Mol. Pharmacol.* **2005**, *67*, 1513–1521.

121. Ghanshani, S., Wulff, H., Miller, M.J., Rohm, H., Neben, A., Gutman, G.A., Cahalan, M.D., Chandy, K.G. Up-regulation of the IKCa1 potassium channel during T-cell activation: Molecular mechanism and functional consequences. *J. Biol. Chem.* **2000**, *275*, 37137–37149.

122. Yao, X., Liu, W., Tian, S., Rafi, H., Segal, A.S., Desir, G.V. Close association of the N terminus of Kv1.3 with the pore region. *J. Biol. Chem.* **2000**, *275*, 10859–10863.

123. Lanigan, M.D., Kalman, K., Lefievre, Y., Pennington, M.W., Chandy, K.G., Norton, R.S. Mutating a critical lysine in ShK toxin alters its binding configuration in the pore-vestibule region of the voltage-gated potassium channel, Kv1.3. *Biochemistry* **2002**, *41*, 11963–11971.

124. Roden, D.M. Drug-induced prolongation of the QT interval. *New Engl. J. Med.* **2004**, *350*, 1013–1022.

125. DeCoursey, T.E., Chandy, K.G., Gupta, S., Cahalan, M.D. Voltage-dependent ion channels in T-lymphocytes. *J. Neuroimmunol.* **1985**, *10*, 71–95.

126. Deutsch, C., Krause, D., Lee, S.C. Voltage-gated potassium conductance in human T lymphocytes stimulated with phorbol ester. *J. Physiol. (Lond.)* **1986**, *372*, 405–423.

127. Lee, S.C., Sabath, D.E., Deutsch, C., Prystowsky, M.B. Increased voltage-gated potassium conductance during interleukin 2-stimulated proliferation of a mouse helper T lymphocyte clone. *J. Cell. Biol.* **1986**, *102*, 1200–1208.

128. Michne, W., Guiles, J., Treasurywala, A., Castonguay, L., Weigelt, C., Oconnor, B., Volberg, W., Grant, A., Chadwick, C., Krafte, D. Novel inhibitors of potassium ion channels on human T lymphocytes. *J. Med. Chem.* **1995**, *38*, 1877–1883.

129. Hill, R.J., Grant, A.M., Volberg, W., Rapp, L., Faltynek, C., Miller, D., Pagani, K., Baizman, E., Wang, S., Guiles, J.W., et al. WIN 17317-3: novel nonpeptide antagonist of voltage-activated K$^+$ channels in human T lymphocytes. *Mol. Pharmacol.* **1995**, *48*, 98–104.

130. Nguyen, A., Kath, J.C., Hanson, D.C., Biggers, M.S., Canniff, P.C., Donovan, C.B., Mather, R.J., Bruns, M.J., Rauer, H., Aiyar, J., Lepple-Wienhues, A., Gutman, G.A., Grissmer, S., Cahalan, M.D., Chandy, K.G. Novel nonpeptide agents potently block the C-type inactivated conformation of Kv1.3 and suppress T cell activation. *Mol. Pharmacol.* **1996**, *50*, 1672–1679.

131. Panyi, G., Sheng, Z., Deutsch, C. C-type inactivation of a voltage-gated K$^+$ channel occurs by a cooperative mechanism. *Biophys. J.* **1995**, *69*, 896–903.

132. Ogielska, E.M., Zagotta, W.N., Hoshi, T., Heinemann, S.H., Haab, J., Aldrich, R.W. Cooperative subunit interactions in C-type inactivation of K channels. *Biophys. J.* **1995**, *69*, 2449–2457.

133. Levy, D.I., Deutsch, C. Recovery from C-type inactivation is modulated by extracellular potassium. *Biophys. J.* **1996**, *70*, 798–805.

134. Wanner, S., Glossmann, H., Knaus, H., Baker, R., Parsons, W., Rupprecht, K., Brochu, R., Cohen, C., Schmalhofer, W., Smith, M., Warren, V., Garcia, M., Kaczorowski, G. WIN 17317-3, a new high-affinity probe for voltage-gated sodium channels. *Biochemistry* **1999**, *38*, 11 137–11 146.

135. Hanson, D.C., Nguyen, A., Mather, R.J., Rauer, H., Koch, K., Burgess, L.E., Rizzi, J.P., Donovan, C.B., Bruns, M.J., Canniff, P.C., Cunningham, A.C., Verdries, K.A., Mena, E., Kath, J.C., Gutman, G.A., Cahalan, M.D., Grissmer, S., Chandy, K.G. UK-78,282, a novel piperidine compound that potently blocks the Kv1.3 voltage-gated potassium channel and inhibits human T cell activation. *Br. J. Pharmacol.* **1999**, *126*, 1707–1716.

136. Rauer, H., Grissmer, S. Evidence for an internal phenylalkylamine action on the voltage-gated potassium channel Kv1.3. *Mol. Pharmacol.* **1996**, *50*, 1625–1634.

137. Goetz, M.A., Hensens, O.D., Zink, D.L., Borris, R.P., Morales, F., Tamayo-Castillo, G., Slaughter, R.S., Felix, J., Ball, R.G. Potent nor-triterpenoid blockers of the voltage-gated potassium channel Kv1.3 from Spachea correa. *Tetrahedron Lett.* **1998**, *39*, 2895–2898.

138. Felix, J.P., Bugianesi, R.M., Schmalhofer, W.A., Borris, R., Goetz, M.A., Hensens, O.D., Bao, J.M., Kayser, F., Parsons, W.H., Rupprecht, K., Garcia, M.L., Kaczorowski, G.J., Slaughter, R.S. Identification and biochemical characterization of a novel nortriterpene inhibitor of the human lymphocyte voltage-gated potassium channel, Kv1.3. *Biochemistry* **1999**, *38*, 4922–4930.

139. Hanner, M., Schmalhofer, W.A., Green, B., Bordallo, C., Liu, J., Slaughter, R.S., Kaczorowski, G.J., Garcia, M.L. Binding of correolide to Kv1 family potassium channels. *J. Biol. Chem.* **1999**, *274*, 25 237–25 244.

140. Hanner, M., Green, B., Gao, Y.-D., Schmalhofer, W., Matyskiela, M., Durand, D.J., Felix, J.P., Linde, A.-R., Bordallo, C., Kaczorowski, G.J., Kohler, M., Garcia, M.L. Binding of correolide to the Kv1.3 potassium channel: charcaterization of the binding domain by site-directed mutagenesis. *Biochemistry* **2001**, *40*, 11 687–11 697.

141. Schmalhofer, W.A., Bao, J., McManus, O.B., Green, B., Matyskiela, M., Wunderler, D., Bugianesi, R.M., Felix, J.P., Hanner, M., Linde-Arias, A.-R., Ponte, C.G., Velasco, L., Koo, G., Staruch, M.J., Miao, S., Parsons, W.H., Rupprecht, K., Slaughter, R.S., Kaczorowski, G.J., Garcia, M.L. Identification of a new class of inhibitors of the voltage-gated potassium channel, Kv1.3, with immunosuppressant properties. *Biochemistry* **2002**, *18*, 7781–7794.

142. Vianna-Jorge, R., Oliveira, C., Garcia, M., Kaczorowski, G., Suarez-Kurtz, G. Correolide, a nor-triterpenoid blocker of shaker-type Kv1 channels elicits twitches in guinea-pig ileum by stimulating the enteric nervous system and enhancing neurotransmitter release. *Br. J. Pharmacol.* **2000**, *131*, 772–778.

143. Vianna-Jorge, R., Oliveira, C.F., Garcia, M.L., Kaczorowski, G.J., Suarez-Kurtz, G. Shaker-type Kv1 channel blockers increase the peristaltic activity of guinea-pig ileum by stimulating acetylcholine and tachykinins release by the enteric nervous system. *Br. J. Pharmacol.* **2003**, *138*, 57–62.

144. Bao, J., Miao, S., Kayser, F., Kotliar, A.J., Baker, R.K., Doss, G.A., Felix, J.P., Bugianesi, R.M., Slaughter, R.S., Kaczorowski, G.J., Garcia, M.L., Ha, S.N., Castonguay, L., Koo, G.C., Shah, K., Springer, M.S., Staruch, M.J., Parsons, W.H., Rupprecht, K.M. Potent Kv1.3 inhibitors from correolide-modification of the C18 position. *Bioorg. Med. Chem. Lett.* **2005**, *15*, 447–451.

145. Singh, S.B., Zink, D.L., Dombrowski, A.W., Dezeny, G., Bills, G.F., Felix, J.P., Slaughter, R.S., Goetz, M.A. Candelalides A–C: novel diterpenoid pyrones from fermentations of Sesquicillium candelabrum as blockers of the voltage-gated potassium channel Kv1.3. *Org. Lett.* **2001**, *3*, 247–250.

146. Miao, S., Bao, J., Garcia, M.L., Goulet, J.L., Hong, X.J., Kaczorowski, G., Kayser, F., Koo, G.C., Kotliar, A., Schmalhofer,

W., Shah, K., Sinclair, P.J., Slaughter, R.S., Springer, M.S., Staruch, M.J., Tsou, N.N., Wong, F., Parsons, W.H., Rupprecht, K. Benzamide derivatives as blockers of the Kv1.3 ion channel. *Bioorg. Med. Chem. Lett.* **2003**, *13*, 1161–1164.

147. Schmalhofer, W.A., Slaugther, R.S., Matyskiela, M., Felix, J.P., Tang, Y.S., Rupprecht, K., Kaczorowski, G.J., Garcia, M.L. Di-substituted cyclohexyl derivatives bind to two identical sites with positive cooperativity on the voltage-gated potassium channel Kv1.3. *Biochemistry* **2003**, *42*, 4733–4743.

148. Atwal, K.S., Vaccaro, W., Lloyd, J., Finlay, H., Lin, Y., Bhandaru, R.S. *WO 01/40231 A1*, **2001**.

149. Castle, N.A., Hollinshead, S.P., Hughes, P.F., Mendoza, G.S., Wilson, J.W., Amato, G.S., Beaudoin, S., Gross, M., McNaughton-Smith, G. Potassium channel inhibitors. *U.S. Patent 6083986*, **2000**.

150. Choi, J., Choi, B.H., Hahn, S.J., Yoon, S.H., Min, D.S., Jo, Y., Kim, M. Inhibition of Kv1.3 channels by H-89 (N-[2-(p-bromocinnamylamino)ethyl]-5-isoquinolinesulfonamide) independent of protein kinase A. *Biochem. Pharmacol.* **2001**, *61*, 1029–1032.

151. Lew, A., Chamberlin, A.R. Blockers of human T cell Kv1.3 potassium channels using de novo ligand design and solid-phase parallel combinatorial chemistry. *Bioorg. Med. Chem. Lett.* **1999**, *9*, 3267–3272.

152. Rauer, H., Lanigan, M.D., Pennington, M.W., Aiyar, J., Ghanshani, S., Cahalan, M.D., Norton, R.S., Chandy, K.G. Structure-guided transformation of charybdotoxin yields an analog that selectively targets Ca^{2+}-activated over voltage-gated K^+ channels. *J. Biol. Chem.* **2000**, *275*, 1201–1208.

153. Gradl, S.N., Felix, J.P., Isacoff, E.Y., Garcia, M.L., Trauner, D. Protein surface recognition by rational design: nanomolar ligands for potassium channels. *J. Am. Chem. Soc.* **2003**, *125*, 12668–12669.

154. Bohuslavizki, H.K., Hinck-Kneip, C., Kneip, A., Koppenhofer, E., Reimers, A. Reduction of MS-related scotoma by a new class of potassium channel blockers from Ruta graveolens. *Neuroopthalmol.* **1993**, *13*, 191–198.

155. Bohuslavizki, K.H., Hansel, W., Kneip, A., Koppenhofer, E., Niemoller, E., Sanmann, K. Mode of action of psoralens, benzofurans, acridones and coumarins on the ionic currents in intact myelinated nerve fibres and its significance in demyelinating diseases. *Gen. Physiol. Biophys.* **1994**, *13*, 309–328.

156. Butenschon, I., Moller, K., Hansel, W. Angular methoxy-subsituted furo- and pyranoquinolinones as blockers of the voltage-gated potassium channel Kv1.3. *J. Med. Chem.* **2001**, *44*, 1249–1256.

157. Wulff, H., Rauer, H., During, T., Hanselmann, C., Ruff, K., Wrisch, A., Grissmer, S., Hansel, W. Alkoxypsoralens, novel nonpeptide blockers of Shaker-type K^+ channels: synthesis and photoreactivity. *J. Med. Chem.* **1998**, *41*, 4542–4549.

158. Wernekenschnieder, A., Korner, P., Hansel, W. 3-Alkyl- and 3-aryl-7H-furo[3,2-g]chromen-7-ones as blockers of the voltage-gated potassium channel Kv1.3. *Pharmazie* **2004**, *59*, 319–320.

159. Vennekamp, J., Wulff, H., Beeton, C., Calabresi, P.A., Grissmer, S., Hansel, W., Chandy, K.G. Kv1.3 blocking 5-phenylalkoxypsoralens: a new class of immunomodulators. *Mol. Pharmacol.* **2004**, *65*, 1364–1373.

160. Hansch, C., Anderson, S.M. The effect of intramolecular hydrophobic bonding on partition coefficients. *J. Org. Chem.* **1967**, *32*, 2583–2586.

161. Leo, A. Octanol/water partition coefficients. In *Handbook of Property Estimation Methods for Chemicals* (R.S. Boethling, D. Mackay, eds), Lewis Publishers, Boca Raton, FA 89-114, **2000**.

162. Lipinski, C.A., Lombardo, F., Dominy, B.W., Feeney, P.J. Experimental and computational approaches to estimate solubility and permeability in drug discovery and development settings. *Adv. Drug Delivery Rev.* **1997**, *23*, 3–25.

163. Baell, J.B., Gable, R.W., Harvey, A.J., Toovey, N., Herzog, T., Hansel, W., Wulff, H. Khellinone derivatives as blockers of the voltage-gated potassium channel Kv1.3: synthesis and immunosuppressive activity. *J. Med. Chem.* **2004**, *47*, 2326–2336.

164. Ehring, G.R., Kerschbaum, H.H., Eder, C., Neben, A.L., Fanger, C.M., Khoury,

R.M., Negulescu, P.A., Cahalan, M.D. A nongenomic mechanism for progesterone-mediated immunosuppression: inhibition of K^+ channels, Ca^{2+} signaling, and gene expression in T lymphocytes. *J. Exp. Med.* **1998**, *188*, 1593–1602.

165. Lampert, A., Muller, M.M., Berchtold, S., Lang, K.S., Palmada, M., Dobrovinskaya, O., Lang, F. Effect of dexamethasone on voltage-gated K^+ channels in Jurkat T-lymphocytes. *Pflugers. Arch.* **2003**, *447*, 168–174.

166. Fiorica-Howells, E., Gambale, F., Horn, R., Osses, L., Spector, S. Phencyclidine blocks voltage-dependent potassium currents in murine thymocytes. *J. Pharmacol. Exp. Ther.* **1990**, *252*, 610–615.

167. Strauss, U., Wittstock, U., Teuscher, E., Jung, S., Mix, E. Cicutotoxin from Cicuta virosa - a new and potent potassium channel blocker in T lymphocytes. *Biochem. Biophys. Res. Commun.* **1996**, *219*, 332–336.

168. Krause, D., Lee, S.C., Deutsch, C. Forskolin effects on the voltage-gated K^+ conductance of human T cells. *Pflugers Arch* **1998**, *12*, 133–140.

169. Schlichter, L., Sidell, N., Hagiwara, S. K channels are expressed early in human T-cell development. *Proc. Natl. Acad. Sci. U.S.A.* **1986**, *83*, 5625–5629.

170. Logsdon, N.J., Kang, J., Togo, J.A., Christian, E.P., Aiyar, J. A novel gene, hKCa4, encodes the calcium-activated potassium channel in human T lymphocytes. *J. Biol. Chem.* **1997**, *272*, 32723–32726.

171. Chandy, K.G., DeCoursey, T.E., Fischbach, M., Talal, N., Cahalan, M.D., Gupta, S. Altered K^+ channel expression in abnormal T lymphocytes from mice with the lpr gene mutation. *Science* **1986**, *233*, 1197–1200.

172. Decoursey, T.E., Chandy, K.G., Gupta, S., Cahalan, M.D. Two types of potassium channels in murine T lymphocytes. *J. Gen. Physiol.* **1987**, *89*, 379–404.

173. Freedman, B.D., Fleischmann, B.K., Punt, J.A., Gaulton, G., Hashimoto, Y., Kotlikoff, M.I. Identification of Kv1.1 expression by murine $CD4^-CD8^-$ thymocytes. A role for voltage-dependent K^+ channels in murine thymocyte development. *J. Biol. Chem.* **1995**, *270*, 22406–22411.

174. Liu, Q.-H., Fleischmann, B.K., Hondowitcz, B., Maier, C.C., Turka, L.A., Yui, K., Kotlikoff, M.I., Wells, A.D., Freedman, B.D. Modulation of Kv channel expression and function by TCR and costimulatory signals during peripheral $CD4^+$ lymphocyte differentiation. *J. Exp. Med.* **2002**, *196*, 897–909.

175. Lewis, R.S., Cahalan, M.D. Subset-specific expression of potassium channels in developing murine T lymphocytes. *Science* **1988**, *239*, 771–775.

176. Ghanshani, S., Pak, M., McPherson, J.D., Strong, M., Dethlefs, B., Wasmuth, J.J., Salkoff, L., Gutman, G.A., Chandy, K.G. Genomic organization, nucleotide sequence, and cellular distribution of a Shaw-related potassium channel gene, Kv3.3, and mapping of Kv3.3 and Kv3.4 to human chromosomes 19 and 1. *Genomics* **1992**, *12*, 190–196.

177. Grissmer, S., Ghanshani, S., Dethlefs, B., McPherson, J.D., Wasmuth, J.J., Gutman, G.A., Cahalan, M.D., Chandy, K.G. The Shaw-related potassium channel gene, Kv3.1, on human chromosome 11, encodes the type l K^+ channel in T cells. *J. Biol. Chem.* **1992**, *267*, 20971–20979.

178. Beeton, C., Chandy, K.G. Potassium channels, memory T cells and multiple sclerosis. *Neuroscientist* **2005**, *11*, 550–562.

179. Reich, E.P., Cui, L., Yang, L., Pugliese-Sivo, C., Golovko, A., Petro, M., Vassileva, G., Chu, I., Nomeir, A.A., Zhang, L.K., Liang, X., Kozlowski, J.A., Narula, S.K., Zavodny, P.J., Chou, C.C. Blocking ion channel KCNN4 alleviates the symptoms of experimental autoimmune encephalomyelitis in mice. *Eur. J. Immunol.* **2005**, *35*, 1027–1036.

180. Koni, P.A., Khanna, R., Chang, M.C., Tang, M.D., Kaczmarek, L.K., Schlichter, L.C., Flavella, R.A. Compensatory anion currents in Kv1.3 channel-deficient thymocytes. *J. Biol. Chem.* 2003, **278**, 39443–39451.

181. Ishida, Y., Chused, T.M. Lack of voltage sensitive potassium channels and generation of membrane potential by sodium potassium ATPase in murine T lymphocytes. *J. Immunol.* **1993**, *151*, 610–620.

182. Sallusto, F., Lenig, D., Forster, R., Lipp, M., Lanzavecchia, A. Two subsets of memory T lymphocytes with distinct homing potentials and effector functions. *Nature* 1999, *401*, 708–712.

183. Geginat, J., Sallusto, F., Lanzavecchia, A. Cytokine-driven proliferation and differentiation of human naive, central memory, and effector memory CD4$^+$ T cells. *J. Exp. Med.* 2001, *194*, 1711–1719.

184. Price, M., Lee, S.C., Deutsch, C. Charybdotoxin inhibits proliferation and interleukin 2 production in human peripheral blodd lymphocytes. *Proc. Natl. Acad. Sci. U.S.A.* 1989, *86*, 10171–10175.

185. Grissmer, S., Nguyen, A.N., Cahalan, M.D. Calcium-activated potassium channels in resting and activated human T lymphocytes. Expression levels, calcium dependence, ion selectivity, and pharmacology. *J. Gen. Physiol.* 1993, *102*, 601–630.

186. Rader, R.K., Kahn, L.E., Anderson, G.D., Martin, C.L., Chinn, K.S., Gregory, S.A. T cell activation is regulated by voltage-dependent and calcium-activated potassium channels. *J. Immunol.* 1996, *156*, 1425–1430.

187. Wulff, H., Miller, M.J., Haensel, W., Grissmer, S., Cahalan, M.D., Chandy, K.G. Design of a potent and selective inhibitor of the intermediate-conductance Ca^{2+}-activated K$^+$ channel, IKCa1: A potential immunosuppressant. *Proc. Natl. Acad. Sci. U.S.A.* 2000, *97*, 8151–8156.

188. Verheugen, J.A., Le Deist, F., Devignot, V., Korn, H. Enhancement of calcium signaling and proliferation responses in activated human T lymphocytes. Inhibitory effects of K$^+$ channel block by charybdotoxin depend on the T cell activation state. *Cell Calcium* 1997, *21*, 1–17.

189. Fanger, C.M., Rauer, H., Neben, A.L., Miller, M.J., Rauer, H., Wulff, H., Rosa, J.C., Ganellin, C.R., Chandy, K.G., Cahalan, M.D. Calcium-activated potassium channels sustain calcium signaling in T lymphocytes. *J. Biol. Chem.* 2001, *276*, 12249–12256.

190. Grakoui, A., Bromley, S.K., Sumen, C., Davis, M.M., Shaw, A.S., Allen, P.M., Dustin, M.L. The immunological synapse: a molecular machine controlling T cell activation. *Science* 1999, *285*, 221–217.

191. Davis, D.M. Assembly of the immunological synapse for T cells and NK cells. *Trends Immunol.* 2000, *23*, 356–363.

192. Davis, D.M., Dustin, M.L. What is the importance of the immunological synapse? *Trends Immunol.* 2004, *25*, 323–327.

193. Panyi, G., Bagdany, M., Bodnar, A., Vamosi, G., Szentesi, G., Jenei, A., Matyus, L., Varga, S., Waldmann, T.A., Gaspar, R., Damjanovich, S. Colocalization and nonrandom distribution of the Kv1.3 potassium channel and CD3 molecules in the plasma membrane of human T lymphocytes. *Proc. Natl. Acad. Sci. U.S.A.* 2003, *100*, 2592–2597.

194. Hanada, T., Lin, L., Chandy, K.G., Oh, S.S., Chishti, A.H. Human homologue of the Drosophila discs large tumor suppressor binds to p56lck tyrosine kinase and Shaker type Kv1.3 potassium channel in T lymphocytes. *J. Biol. Chem.* 1997, *272*, 26899–26904.

195. McCormack, T., McCormack, K., Nadal, M.S., Vieira, E., Ozaita, A., Rudy, B. The effects of shaker beta-subunits on the human lymphocyte K$^+$ channel Kv1.3. *J Biol Chem* 1999, *274*, 20123–20126.

196. Gong, J., Xu, J., Bezanilla, M., van Huizen, R., Derin, R., Li, M. Differential stimulation of PKC phosphorylation of potassium channels by ZIP1 and ZIP2. *Science* 1999, *285*, 1565–1569.

197. Joung, I., Strominger, J.L., Shin, J. Molecular cloning of a phosphotyrosine-independent ligand of the p56lck SH2 domain. *Proc. Natl. Acad. Sci. U.S.A.* 1996, *93*, 5991–5995.

198. Ishii, T., Yanagawa, T., Kawane, T., Yuki, K., Seita, J., Yoshida, H., Bannai, S. Murine peritoneal macrophages induce a novel 60-kDa protein with structural similarity to a tyrosine kinase p56lck-associated protein in response to oxidative stress. *Biochem. Biophys. Res. Commun.* 1996, *226*, 456–460.

199. Park, Y.C., Jun, C.D., Kang, H.S., Kim, H.D., Kim, H.M., Chung, H.T. Intracellular Ca^{2+} pool depletion is linked to the induction of nitric oxide synthesis in murine peritoneal macrophages. *Biochem. Mol. Biol. Int.* 1995, *36*, 949–955.

200. Puls, A., Schmidt, S., Grawe, F., Stabel, S. Interaction of protein kinase C zeta with ZIP, a novel protein kinase C-binding protein. *Proc. Natl. Acad. Sci. U.S.A.* **1997**, *94*, 6191–6196.

201. Szabo, I., Gulbins, E., Apfel, H., Zhang, X., Barth, P., Busch, A., K, S., Pongs, O., Lang, F. Tyrosine phosphorylation-dependent suppression of a voltage-gated K⁺ channel in T lymphocytes upon Fas stimulation. *J. Biol. Chem.* **1996**, *271*, 20 465–20 469.

202. Gulbins, E., Szabo, I., Baltzer, K., Lang, F. Ceramide-induced inhibition of T lymphocyte voltage-gated potassium channel is mediated by tyrosine kinases. *Proc. Natl. Acad. Sci. U.S.A.* **1997**, *94*, 7661–7666.

203. Dellis, O., Bouteau, F., Guenounou, M., Rona, J.P. HIV-1 gp160 decreases the K⁺ voltage-gated current from Jurkat E6.1 T cells by up-phosphorylation. *FEBS Lett.* **1999**, *443*, 187–191.

204. Veillette, A., Bookman, M.A., Horak, E.M., Bolen, J.B. The CD4 and CD8 T cell surface antigens are associated with the internal membrane tyrosine-protein kinase p56lck. *Cell* **1988**, *55*, 301–308.

205. Barber, E.K., Dasgupta, J.D., Schlossman, S.F., Trevillyan, J.M., Rudd, C.E. The CD4 and CD8 antigens are coupled to a protein-tyrosine kinase (p56lck) that phosphorylates the CD3 complex. *Proc. Natl. Acad. Sci. U.S.A.* **1989**, *86*, 3277–3281.

206. Veillette, A., Bookman, M.A., Horak, E.M., Samelson, L.E., Bolen, J.B. Signal transduction through the CD4 receptor involves the activation of the internal membrane tyrosine-protein kinase p56lck. *Nature* **1989**, *338*, 257–259.

207. Panyi, G., Vamosi, G., Bacso, Z., Bagdany, M., Bodnar, A., Varga, Z., Gaspar, R., Matyus, L., Damjanovich, S. Kv1.3 potassium channels are localized in the immunological synapse formed between cytotoxic and target cells. *Proc. Natl. Acad. Sci. U.S.A.* **2004**, *101*, 1285–1290.

208. Levite, M., Cahalon, L., Peretz, A., Hershkoviz, R., Sobko, A., Ariel, A., Desai, R., Attali, B., Lider, O. Extracellular K⁺ and opening of voltage-gated potassium channels activate T cell integrin function: physical and functional association between Kv1.3 channels and beta1 integrins. *J. Exp. Med.* **2000**, *191*, 1167–1176.

209. Artym, V.V., Petty, H.R. Molecular proximity of the Kv1.3 voltage-gated potassium channel and beta1-integrins on the plasma membrane of melanoma cells: effects of cell adherence and channel blockers. *J. Gen. Physiol.* **2002**, *120*, 29–37.

210. Sims, T.N., Dustin, M.L. (2002) The immunological synapse: integrins take the stage. *Immunol. Rev.* **186**, 100–117.

211. Lewis, R.S., Cahalan, M.D. Potassium and calcium channels in lymphocytes. *Annu. Rev. Immunol.* **1995**, *13*, 623–653.

212. Dolmetsch, R.E., Lewis, R.S., Goodnow, C.C., Healy, J.I. Differential activation of transcription factors induced by Ca²⁺ response amplitude and duration. *Nature* **1997**, *386*, 855–858.

213. Lewis, R.S. Calcium signaling mechanisms in T lymphocytes. *Annu. Rev. Immunol.* **2001**, *19*, 497–521.

214. Zweifach, A., Lewis, R.S. Mitogen-regulated Ca²⁺ current of T lymphocytes is activated by depletion of intracellular Ca²⁺ stores. *Proc. Natl. Acad. Sci. U.S.A.* **1993**, *90*, 6295–6299.

215. Zweifach, A., Lewis, R.S. Rapid inactivation of depletion-activated calcium current (ICRAC) due to local calcium feedback. *J. Gen. Physiol.* **1995**, *105*, 209–226.

216. Dolmetsch, R.E., Xu, K., Lewis, R.S. Calcium oscillations increase the efficiency and specificity of gene expression. *Nature* **1998**, *392*, 933–936.

217. Feske, S., Giltnane, J., Dolmetsch, R., Staudt, L.M., Rao, A. Gene regulation mediated by calcium signals in T lymphocytes. *Nat. Immunol.* **2001**, *2*, 316–324.

218. Soler, D., Humphreys, T.L., Spinola, S.M., Campbell, J.J. CCR4 versus CCR10 in human cutaneous TH lymphocyte trafficking. *Blood* **2003**, *101*, 1677–1682.

219. Conforti, L., Petrovic, M., Mohammad, D., Lee, S., Ma, Q., Barone, S., Filipovich, A.H. Hypoxia regulates expression and activity of Kv1.3 channels in T lymphocytes: a possible role in T cell proliferation. *J. Immunol.* **2003**, *170*, 695–702.

220. Lovett-Racke, A.E., Trotter, J.L., Lauber, J., Perrin, P.J., June, C.H., Racke, M.K. Decreased dependence of myelin basic protein-reactive T cells on CD28-mediated costimulation in multiple sclerosis patients: a marker of activated/memory T cells. *J. Clin. Invest.* **1998**, *101*, 725–730.

221. Markovic-Plese, S., Cortese, I., Wandinger, K.P., McFarland, H.F., Martin, R. CD4⁺CD28⁻ costimulation-independent T cells in multiple sclerosis. *J. Clin. Invest.* **2001**, *108*, 1185–1194.

222. Calabresi, P.A., Yun, S.H., Allie, R., Whartenby, K.A. Chemokine receptor expression on MBP-reactive T cells: CXCR6 is a marker of IFNgamma-producing effector cells. *J. Neuroimmunol.* **2002**, *127*, 96–105.

223. Atkinson, M.A., Maclaren, N.K. The pathogenesis of insulin-dependent diabtes mellitus. *New Eng. J. Med.* **1994**, *331*, 1428–1436.

224. Viglietta, V., Kent, S.C., Orban, T., Hafler, D.A. GAD65-reactive T cells are activated in patients with autoimmune type 1a diabetes. *J. Clin. Invest.* **2002**, *109*, 895–903.

225. Ezawa, K., Yamamura, M., Matsui, H., Ota, Z., Makino, H. Comparative analysis of CD45RA- and CD45RO-positive CD4+T cells in peripheral blood, synovial fluid, and synovial tissue in patients with rheumatoid arthritis and osteoarthritis. *Acta Med. Okayama* **1997**, *51*, 25–31.

226. Friedrich, M., Krammig, S., Henze, M., Docke, W.D., Sterry, W., Asadullah, K. Flow cytometric characterization of lesional T cells in psoriasis: Intracellular cytokine and surface antigen expression indicates an activated, memory effector/type 1 immunophenotype. *Arch. Dermatol. Res.* **2000**, *292*, 519–521.

227. Vissers, W.H., Arndtz, C.H., Muys, L., Van Erp, P.E., de Jong, E.M., van de Kerkhof, P.C. Memory effector (CD45RO⁺) and cytotoxic (CD8⁺) T cells appear early in the margin zone of spreading psoriatic lesions in contrast to cells expressing natural killer receptors, which appear late. *Br. J. Dermatol.* **2004**, *150*, 852–859.

228. Lin, M.S., Swartz, S.J., Lopez, A., Ding, X., Fernandez-Vina, M.A., Stastny, P., Fairley, J.A., Diaz, L.A. Development and characterization of desmoglein-3 specific T cells from patients with pemphigus vulgaris. *J. Clin. Invest.* **1997**, *99*, 31–40.

229. Lin, M.S., Fu, C.L., Aoki, V., Hans-Filho, G., Rivitti, E.A., Moraes, J.R., Moraes, M.E., Lazaro, A.M., Giudice, G.J., Stastny, P., Diaz, L.A. Desmoglein-1-specific T lymphocytes from patients with endemic pemphigus foliaceus (fogo selvagem). *J. Clin. Invest.* **2000**, *105*, 207–213.

230. Yamashita, K., Choi, U., Woltz, P.C., Foster, S.F., Sneller, M.C., Hakim, F.T., Fowler, D.H., Bishop, M.R., Pavletic, S.Z., Tamari, M., Castro, K., Barrett, A.J., Childs, R.W., Illei, G.G., Leitman, S.F., Malech, H.L., Horwitz, M.E. Severe chronic graft-versus-host disease is characterized by a preponderance of CD4⁺ effector memory cells relative to central memory cells. *Blood* **2004**, *103*, 3986–3988.

231. Storey, N.M., Gomez-Angelats, M., Bortner, C.D., Armstrong, D.L., Cidlowski, J.A. Stimulation of Kv1.3 potassium channels by death receptors during apoptosis in Jurkat T lymphocytes. *J. Biol. Chem.* **2003**, *278*, 33 319–33 326.

232. Yu, S.P., Yeh, C., Strasser, U., Tian, M., Choi, D.W. NMDA receptor-mediated K⁺ efflux and neuronal apoptosis. *Science* **1999**, *284*, 336–369.

233. Szabo, I., Bock, J., Jekle, A., Soddemann, M., Adams, C., Lang, F., Zoratti, M., Gulbins, E. A novel potassium channel in lymphocyte mitochondria. *J. Biol. Chem.* **2005**, *280*, 12 790–12 798.

234. Mahaut-Smith, M.P., Schlichter, L.C. Ca²⁺-activated K⁺ channels in human B lymphocytes and rat thymocytes. *J. Physio.l (Lond.)* **1989**, *415*, 69–83.

235. Partiseti, M., Choquet, D., Diu, A., Korn, H. Differential regulation of voltage- and calcium-activated potassium channels in human B lymphocytes. *J. Immunol.* **1992**, *148*, 3361–3368.

236. Partiseti, M., Korn, H., Choquet, D. Pattern of potassium channel expression in proliferating B lymphocytes depends upon the mode of activation. *J. Immunol.* **1993**, *151*, 2462–2470.

237. Klein, U., Rajewsky, K., Kuppers, R. Human immunoglobulin (Ig)M⁺IgD⁺ peripheral blood B cells expressing CD27 cell surface antigen carry somatically mutated variable region genes: CD27 as a general marker for somatically mutated (memory) B cells. *J. Exp. Med.* **1998**, *188*, 1679–1689.

238. Agematsu, K., Hokibara, S., Nagumo, H., Komiyama, A. CD27: a memory B-cell

marker. *Immunol. Today* **2000**, *21*, 204–206.

239. Shi, Y., Agematsu, K., Ochs, H.D., Sugane, K. Functional analysis of human memory B-cell subpopulations: IgD⁺CD27⁺ B cells are crucial in secondary immune response by producing high affinity IgM. *Clin. Immunol.* **2003**, *108*, 128–137.

240. Bar-Or, A., Oliveira, E.M., Anderson, D.E., Krieger, J.I., Duddy, M., O'Connor, K.C., Hafler, D.A. Immunological memory: contribution of memory B cells expressing costimulatory molecules in the resting state. *J. Immunol.* **2001**, *167*, 5669–5677.

241. Hansen, A., Odendahl, M., Reiter, K., Jacobi, A.M., Feist, E., Scholze, J., Burmester, G.R., Lipsky, P.E., Dorner, T. Diminished peripheral blood memory B cells and accumulation of memory B cells in the salivary glands of patients with Sjogren's syndrome. *Arthritis Rheum.* **2002**, *46*, 2160–2171.

242. Leyendeckers, H., Voth, E., Schicha, H., Hunzelmann, N., Banga, P., Schmitz, J. Frequent detection of thyroid peroxidase-specific IgG⁺ memory B cells in blood of patients with autoimmune thyroid disease. *Eur. J. Immunol.* **2002**, *32*, 3126–3132.

243. Dorner, T., Lipsky, P.E. Correlation of circulating CD27high plasma cells and disease activity in systemic lupus erythematosus. *Lupus* **2004**, *13*, 283–289.

244. Archelos, J.J., Storch, M.K., Hartung, H.P. The role of B cells and autoantibodies in multiple sclerosis. *Ann. Neurol.* **2000**, *47*, 694–706.

245. Corcione, A., Casazza, S., Ferretti, E., Giunti, D., Zappia, E., Pistorio, A., Gambini, C., Mancardi, G.L., Uccelli, A., Pistoia, V. Recapitulation of B cell differentiation in the central nervous system of patients with multiple sclerosis. *Proc. Natl. Acad. Sci. U.S.A.* **2004**, *101*, 11 064–11 069.

246. Bohnhorst, J.O., Thoen, J.E., Natvig, J.B., Thompson, K.M. Significantly depressed percentage of CD27⁺ (memory) B cells among peripheral blood B cells in patients with primary Sjogren's syndrome. *Scand. J. Immunol.* **2001**, *54*, 421–427.

247. Jacobi, A.M., Odendahl, M., Reiter, K., Bruns, A., Burmester, G.R., Radbruch, A., Valet, G., Lipsky, P.E., Dorner, T. Correlation between circulating CD27high plasma cells and disease activity in patients with systemic lupus erythematosus. *Arthritis Rheum.* **2003**, *48*, 1332–1342.

248. O'Connor, K.C., Bar-Or, A., Hafler, D.A. The neuroimmunology of multiple sclerosis: possible roles of T and B lymphocytes in immunopathogenesis. *J. Clin. Immunol.* **2001**, *21*, 81–92.

249. Berger, T., Rubner, P., Schautzer, F., Egg, R., Ulmer, H., Mayringer, I., Dilitz, E., Deisenhammer, F., Reindl, M. Antimyelin antibodies as a predictor of clinically definite multiple sclerosis after a first demyelinating event. *New Engl. J. Med.* **2003**, *349*, 139–145.

250. Atkinson, M.A., Eisenbarth, G.S. Type 1 diabetes: new perspectives on disease pathogenesis and treatment. *Lancet* **2001**, *358*, 221–220.

251. Dorner, T., Burmester, G.R. The role of B cells in rheumatoid arthritis: mechanisms and therapeutic targets. *Curr. Opin. Rheumatol.* **2003**, *15*, 246–252.

252. Serreze, D.V., Fleming, S.A., Chapman, H.D., Richard, S.D., Leiter, E.H., Tisch, R.M. B lymphocytes are critical antigen-presenting cells for the initiation of T cell-mediated autoimmune diabetes in nonobese diabetic mice. *J. Immunol.* **1998**, *15*, 3912–3918.

253. Yoon, J.W., Jun, H.S. Cellular and molecular pathogenic mechanisms of insulin-dependent diabetes mellitus. *Ann. New York Acad. Sci.* **2001**, *928*, 200–211.

254. Serreze, D.V., Silveira, P.A. The role of B lymphocytes as key antigen-presenting cells in the development of T-cell mediated autoimmune type-1 diabetes. *Curr. Dir. Autoimmun.* **2003**, *6*, 212–227.

255. Vanderlugt, C.L., Neville, K.L., Nikcevich, K.M., Eagar, T.N., Bluestone, J.A., Miller, S.D. Pathologic role and temporal appearance of newly emerging autoepitopes in relapsing experimental autoimmune encephalomyelitis. *J. Immunol.* **2000**, *164*, 670–678.

256. Jaume, J.C., Parry, S.L., Madec, A.M., Sonderstrup, G., Baekkeskov, S. Suppressive effect of glutamic acid decarboxylase 65-specific autoimmune B lymphocytes on processing of T cell determinants located within the antibody epitope. *J. Immunol.* **2002**, *169*, 665–6672.

257. Vanderlugt, C.L., Miller, S.D. Epitope spreading in immune-mediated diseases: implications for immunotherapy. *Nat. Rev.* **2002**, *2*, 85–95.

258. Moretta, A., Bottino, C., Mingari, M.C., Biassoni, R., Moretta, L. What is a natural killer cell? *Nat. Immunol.* **2002**, *3*, 6–8.

259. Stewart-Akers, A.M., Cunningham, A., Wasko, M.C., Morel, P.A. Fc gamma R expression on NK cells influences disease severity in rheumatoid arthritis. *Genes Immun.* **2004**, *5*, 521–529.

260. Pridgeon, C., Lennon, G.P., Pazmany, L., Thompson, R.N., Christmas, S.E., Moots, R.J. Natural killer cells in the synovial fluid of rheumatoid arthritis patients exhibit a CD56bright,CD94bright,CD158-negative phenotype. *Rheumatology (Oxford)* **2003**, *42*, 870–878.

261. Dalbeth, N., Gundle, R., Davies, R.J., Lee, Y.C., McMichael, A.J., Callan, M.F. CD56bright NK cells are enriched at inflammatory sites and can engage with monocytes in a reciprocal program of activation. *J. Immunol.* **2004**, *173*, 6418–6426.

262. Takahashi, K., Aranami, T., Endoh, M., Miyake, S., Yamamura, T. The regulatory role of natural killer cells in multiple sclerosis. *Brain* **2004**, *127*, 1917–1927.

263. Gallin, E.K. Voltage clamp studies in macrophages from mouse spleen cultures. *Science* **1981**, *214*, 458–460.

264. McCann, F.V., Cole, J.J., Guyre, P.M., Russell, J.A. Action potentials in macrophages derived from human monocytes. *Science* **1983**, *219*, 991–993.

265. Gallin, E.K. Calcium- and voltage-activated potassium channels in human macrophages. *Biophys. J.* **1984**, *46*, 821–825.

266. Gallin, E.K., Sheehy, P.A. Differential expression of inward and outward potassium currents in the macrophage-like cell line J774.1. *J. Physiol. (Lond.)* **1985**, *369*, 475–499.

267. Gallin, E.K. Ionic channels in leukocytes. *J. Leukoc. Biol.* **1986**, *39*, 241–254.

268. Gallin, E.K., McKinney, L.C. Patch-clamp studies in human macrophages: single-channel and whole-cell characterization

of two K^+ conductances. *J. Membr. Biol.* **1988**, *103*, 55–66.

269. McKinney, L.C., Gallin, E.K. Effect of adherence, cell morphology, and lipopolysaccharide on potassium conductance and passive membrane properties of murine macrophage J774.1 cells. *J. Membr. Biol.* **1990**, *116*, 47–56.

270. Judge, S.I., Montcalm-Mazzilli, E., Gallin, E.K. IKir regulation in murine macrophages: whole cell and perforated patch studies. *Am. J. Physiol.* **1994**, *267*, C1691–1698.

271. Ypey, D.L., Clapham, D.E. Development of a delayed outward-rectifying K^+ conductance in cultured mouse peritoneal macrophages. *Proc. Natl. Acad. Sci. U.S.A.* **1984**, *81*, 3083–3087.

272. Nelson, D.J., Jow, B., Jow, F. Whole-cell currents in macrophages: I. Human monocyte-derived macrophages. *J. Membr. Biol.* **1990**, *17*, 29–44.

273. Nelson, D.J., Jow, B., Popovich, K.J. Whole-cell currents in macrophages: II. Alveolar macrophages. *J. Membr. Biol.* **1990**, *117*, 45–55.

274. Nelson, D.J., Jow, B., Jow, F. Lipopolysaccharide induction of outward potassium current expression in human monocyte-derived macrophages: lack of correlation with secretion. *J. Membr. Biol* .**1992**, *125*, 207–218.

275. Blunck, R., Scheel, O., Muller, M., Brandenburg, K., Seitzer, U., Seydel, U. New insights into endotoxin-induced activation of macrophages: involvement of a K^+ channel in transmembrane signaling. *J. Immunol.* **2001**, *166*, 1009–10015.

276. DeCoursey, T.E., Kim, S.Y., Silver, M.R., Quandt, F.N. Ion channel expression in PMA-differentiated human THP-1 macrophages. *J. Membr. Biol.* **1996**, *152*, 141–157.

277. Eder, C., Fischer, H.G. Effects of colony-stimulating factors on voltage-gated K^+ currents of bone marrow-derived macrophages. *Naunyn Schmiedebergs Arch. Pharmacol.* **1997**, *355*, 198–202.

278. Liu, Q.H., Williams, D.A., McManus, C., Baribaud, F., Doms, R.W., Schols, D., De Clercq, E., Kotlikoff, M.I., Collman, R.G., Freedman, B.D. HIV-1 gp120 and chemokines activate ion channels in primary macrophages through CCR5 and CXCR4

stimulation. *Proc. Natl. Acad. Sci. U.S.A.* **2000**, *97*, 4832–4827.

279. Hanley, P.J., Musset, B., Renigunta, V., Limberg, S.H., Dalpke, A.H., Sus, R., Heeg, K.M., Preisig-Muller, R., Daut, J. Extracellular ATP induces oscillations of intracellular Ca^{2+} and membrane potential and promotes transcription of IL-6 in macrophages. *Proc. Natl. Acad. Sci. U.S.A.* **2004**, *101*, 9479–9484.

280. Chung, I., Zelivyanskaya, M., Gendelman, H.E. Mononuclear phagocyte biophysiology influences brain transendothelial and tissue migration: implication for HIV-1-associated dementia. *J. Neuroimmunol.* **2002**, *122*, 40–54.

281. DeCoursey, T.E., Cherny, V.V. Voltage-activated proton currents in human THP-1 monocytes. *J. Membr. Biol.* **1996**, *152*, 131–140.

282. Seydel, U., Scheel, O., Muller, M., Brandenburg, K., Blunck, R. A K^+ channel is involved in LPS signaling. *J. Endotoxin Res.* **2001**, *7*, 243–247.

283. Muller, M., Scheel, O., Lindner, B., Gutsmann, T., Seydel, U. The role of membrane-bound LBP, endotoxin aggregates, and the MaxiK channel in LPS-induced cell activation. *J. Endotoxin Res.* **2003**, *9*, 181–186.

284. Eder, C., Klee, R., Heinemann, U. Pharmacological properties of Ca^{2+}-activated K^+ currents of ramified murine brain macrophages. *Naunyn Schmiedebergs Arch. Pharmacol.* **1997**, *356*, 233–239.

285. Gallin, E.K., Livengood, D.R. Inward rectification in mouse macrophages: evidence for a negative resistance region. *Am. J. Physiol.* **1981**, *241*, C9–17.

286. Gallin, E.K. Electrophysiological properties of macrophages. *Fed. Proc.* **1984**, *43*, 2385–2389.

287. Wieland, S.J., Chou, R.H., Gong, Q.H. Macrophage-colony-stimulating factor (CSF-1) modulates a differentiation-specific inward-rectifying potassium current in human leukemic (HL-60) cells. *J. Cell. Physiol.* **1990**, *142*, 643–651.

288. Banati, R.B., Hoppe, D., Gottmann, K., Kreutzberg, G.W., Kettenmann, H. A subpopulation of bone marrow-derived macrophage-like cells shares a unique ion channel pattern with microglia. *J. Neurosci. Res.* **1991**, *30*, 593–600.

289. Kubo, Y., Baldwin, T.J., Jan, Y.N., Jan, L.Y. Primary structure and functional expression of a mouse inward rectifier potassium channel. *Nature* **1993**, *362*, 127–133.

290. Diaz, R.J., Zobel, C., Cho, H.C., Batthish, M., Hinek, A., Backx, P.H., Wilson, G.J. Selective inhibition of inward rectifier K^+ channels (Kir2.1 or Kir2.2) abolishes protection by ischemic preconditioning in rabbit ventricular cardiomyocytes. *Circ. Res.* **2004**, *95*, 325–332.

291. Korotzer, A.R., Cotman, C.W. Voltage-gated currents expressed by rat microglia in culture. *Glia* **1992**, *6*, 81–88.

292. Norenberg, W., Gebicke-Haerter, P.J., Illes, P. Inflammatory stimuli induce a new K^+ outward current in cultured rat microglia. *Neurosci. Lett.* **1992**, *147*, 171–174.

293. Norenberg, W., Appel, K., Bauer, J., Gebicke-Haerter, P.J., Illes, P. Expression of an outwardly rectifying K^+ channel in rat microglia cultivated on teflon. *Neurosci. Lett.* **1993**, *160*, 69–72.

294. Visentin, S., Renzi, M., Levi, G. Altered outward-rectifying K^+ current reveals microglial activation induced by HIV-1 Tat protein. *Glia* **2001**, *33*, 181–190.

295. Eder, C., Fischer, H.G., Hadding, U., Heinemann, U. Properties of voltage-gated potassium currents of microglia differentiated with granulocyte/macrophage colony-stimulating factor. *J. Membr. Biol.* **1995**, *147*, 137–146.

296. Eder, C., Fischer, H.G., Hadding, U., Heinemann, U. Properties of voltage-gated currents of microglia developed using macrophage colony-stimulating factor. *Pflugers Arch.* **1995**, *430*, 526–533.

297. Eder, C., Heinemann, U. Proton modulation of outward K^+ currents in interferon-gamma-activated microglia. *Neurosci. Lett.* **1996**, *206*, 101–104.

298. Ichinose, M., Asai, M., Sawada, M., Sasaki, K., Oomura, Y. Induction of outward current by orexin-B in mouse peritoneal macrophages. *FEBS Lett.* **1998**, *440*, 51–54.

299. Schilling, T., Eder, C. Effects of kinase inhibitors on TGF-beta induced upregulation of Kv1.3 K^+ channels in brain macrophages. *Pflugers Arch.* **2003**, *447*, 312–315.

300. Caggiano, A.O., Kraig, R.P. Prostaglandin E2 and 4-aminopyridine prevent the lipopolysaccharide-induced outwardly rectifying potassium current and interleukin-1beta production in cultured rat microglia. *J. Neurochem.* **1998**, *70*, 2357–2368.

301. Visentin, S., Levi, G. Arachidonic acid-induced inhibition of microglial outward-rectifying K$^+$ current. *Glia* **1998**, *22*, 1–10.

302. Kotecha, S.A., Schlichter, L.C. A Kv1.5 to Kv1.3 switch in endogenous hippocampal microglia and a role in proliferation. *J. Neurosci.* **1999**, *19*, 10 680–10 693.

303. Cayabyab, F., Khanna, R., Jones, O., Schlichter, L. Suppression of the rat microglia Kv1.3 current by src-family tyrosine kinases and oxygen/glucose deprivation. *Eur. J. Neurosci.* **2000**, *12*, 1949–1960.

304. Pyo, H., Chung, S., Jou, I., Gwag, B., Joe, E.H. Expression and function of outward K$^+$ channels induced by lipopolysaccharide in microglia. *Mol. Cells* **1997**, *7*, 610–614.

305. Jou, I., Pyo, H., Chung, S., Jung, S.Y., Gwag, B.J., Joe, E.H. Expression of Kv1.5 K$^+$ channels in activated microglia in vivo. *Glia* **1998**, *24*, 408–414.

306. Zhou, W., Cayabyab, F.S., Pennefather, P.S., Schlichter, L.C., DeCoursey, T.E. HERG-like K$^+$ channels in microglia. *J. Gen. Physiol.* **1998**, *111*, 781–794.

307. Cayabyab, F.S., Schlichter, L.C. Regulation of an ERG K$^+$ current by Src tyrosine kinase. *J. Biol. Chem.* **2002**, *277*, 13 673–13 681.

308. Schilling, T., Stock, C., Schwab, A., Eder, C. Functional importance of Ca^{2+}-activated K$^+$ channels for lysophosphatidic acid-induced microglia migration. *Eur. J. Neurosci.* **2004**, *19*, 1469–1474.

309. Schilling, T., Lehmann, F., Ruckert, B., Eder, C. Physiological mechanisms of lysophosphatidylcholine-induced de-ramification of murine microglia. *J. Physiol.* **2004**, *557*, 105–120.

310. Kettenmann, H., Hoppe, D., Gottmann, K., Banati, R., Kreutzberg, G. Cultured microglial cells have a distinct pattern of membrane channels different from peritoneal macrophages. *J. Neurosci. Res.* **1990**, *26*, 278–287.

311. Chung, S., Jung, W., Lee, M.Y. Inward and outward rectifying potassium currents set membrane potentials in acti-

vated rat microglia. *Neurosci. Lett.* **1999**, *262*, 121–124.

312. Schilling, T., Quandt, F.N., Cherny, V.V., Zhou, W., Heinemann, U., Decoursey, T.E., Eder, C. Upregulation of Kv1.3 K$^+$ channels in microglia deactivated by TGF-beta. *Am. J. Physiol. Cell. Physiol.* **2000**, *279*, C1123–1134.

313. Boucsein, C., Kettenmann, H., Nolte, C. Electrophysiological properties of microglial cells in normal and pathologic rat brain slices. *Eur. J. Neurosci.* **2002**, *12*, 2049–2058.

314. Visentin, S., Levi, G. Protein kinase C involvement in the resting and interferon-gamma-induced K$^+$ channel profile of microglial cells. *J. Neurosci. Res.* **1997**, *47*, 233–241.

315. Vautier, F., Belachew, S., Chittajallu, R., Gallo, V. Shaker-type potassium channel subunits differentially control oligodendrocyte progenitor proliferation. *Glia* **2004**, *48*, 337–345.

316. Neusch, C., Rozengurt, N., Jacobs, R.E., Lester, H.A., Kofuji, P. Kir4.1 potassium channel subunit is crucial for oligodendrocyte development and in vivo myelination. *J. Neurosci.* **2001**, *21*, 5429–5438.

317. Schonrock, L.M., Kuhlmann, T., Adler, S., Bitsch, A., Bruck, W. Identification of glial cell proliferation in early multiple sclerosis lesions. *Neuropathol. Appl. Neurobiol.* **1998**, *24*, 320–330.

318. Wolswijk, G. Oligodendrocyte survival, loss and birth in lesions of chronic-stage multiple sclerosis. *Brain* **2000**, *123*, 105–115.

319. Kues, W.A., Wunder, F. Heterogeneous expression patterns of mammalian potassium channel genes in developing and adult rat brain. *Eur. J. Neurosci.* **1992**, *4*, 1296–1308.

320. Fadool, D.A., Holmes, T.C., Berman, K., Dagan, D., Levitan, I.B. Tyrosine phosphorylation modulates current amplitude and kinetics of a neuronal voltage-gated potassium channel. *J. Neurophysiol.* **1997**, *78*, 1563–1573.

321. Fadool, D.A., Tucker, K., Phillips, J.J., Simmen, J.A. Brain insulin receptor causes activity-dependent current suppression in the olfactory bulb through multiple phosphorylation of Kv1.3. *J. Neurophysiol.* **2000**, *83*, 2332–2348.

322. Tucker, K., Fadool, D.A. Neurotrophin modulation of voltage-gated potassium channels in rat through TrkB receptors is time and sensory experience dependent. *J. Physiol.* **2002**, *542*, 413–429.

323. Colley, B., Tucker, K., Fadool, D.A. Comparison of modulation of Kv1.3 channel by two receptor tyrosine kinases in olfactory bulb neurons of rodents. *Receptors Channels* **2004**, *10*, 25–36.

324. Lucero, M.T., Pappone, P.A. Voltage-gated potassium channels in brown fat cells. *J. Gen. Physiol.* **1989**, *93*, 451–472.

325. Pappone, P.A., Ortiz-Miranda, S.I. Blockers of voltage-gated K channels inhibit proliferation of cultured brown fat cells. *Am. J. Physiol.* **1993**, *264*, C1014–1019.

326. Ramirez-Ponce, M.P., Mateos, J.C., Carrion, N., Bellido, J.A. Voltage-dependent potassium channels in white adipocytes. *Biochem. Biophys. Res. Commun.* **1996**, *223*, 250–256.

327. Ramirez-Ponce, M.P., Mateos, J.C., Bellido, J.A. Human adipose cells have voltage-dependent potassium currents. *J. Membr. Biol.* **2003**, *196*, 129–134.

328. Xu, J., Wang, P., Li, Y., Li, G., Kaczmarek, L.K., Wu, Y., Koni, P.A., Flavell, R.A., Desir, G.V. The voltage-gated potassium channel Kv1.3 regulates peripheral insulin sensitivity. *Proc. Natl. Acad. Sci. U.S.A.* **2004**, *101*, 3112–3127.

329. Volk, K.A., Husted, R.F., Pruchno, C.J., Stokes, J.B. Functional and molecular evidence for Shaker-like K$^+$ channels in rabbit renal papillary epithelial cell line. *Am. J. Physiol.* **1994**, *267*, F671–678.

330. Grunnet, M., Rasmussen, H., Hay-Schmidt, A., Klaerke, D.A. The voltage-gated potassium channel subunit, Kv1.3, is expressed in epithelia. *Biochim. Biophys. Acta* **2003**, *1616*, 85–94.

331. Escobar, L.I., Martinez-Tellez, J.C., Salas, M., Castilla, S.A., Carrisoza, R., Tapia, D., Vazquez, M., Bargas, J., Bolivar, J.J. A voltage-gated K$^+$ current in renal inner medullary collecting duct cells. *Am. J. Physiol. Cell. Physiol.* **2004**, *286*, C965–974.

332. Abdul, M., Hoosein, N. Voltage-gated potassium ion channels in colon cancer. *Oncol. Rep.* **2002**, *9*, 961–964.

333. Davies, A.M., Batchelor, J.P., Eardleu, I., Beech, D.J. potassium channel KV 1 subunit expression and function in human detrusor muscle. *J. Urol.* **2002**, *167*, 1881–1886.

334. Kusaka, S., Kapousta-Bruneau, N., Green, D.G., Puro, D.G. Serum-induced changes in the physiology of mammalian retinal glial cells: role of lysophosphatidic acid. *J. Physiol.* **1998**, *506*, 445–458.

335. Klumpp, D.J., Song, E.J., Pinto, L.H. Identification and localization of K$^+$ channels in the mouse retina. *Vis. Neurosci.* **1995**, *12*, 1177–1790.

336. Klumpp, D.J., Song, E.J., Ito, S., Sheng, M.H., Jan, L.Y., Pinto, L.H. The Shaker-like potassium channels of the mouse rod bipolar cell and their contributions to the membrane current. *J. Neurosci.* **1995**, *15*, 5004–5013.

337. Lorenzo, J. Interactions between immune and bone cells: new insights with many remaining questions. *J. Clin. Invest.* **2000**, *106*, 749–752.

338. Ravesloot, J.H., Ypey, D.L., Vrijheid-Lammers, T., Nijweide, P.J. Voltage-activated K$^+$ conductances in freshly isolated embryonic chicken osteoclasts. *Proc. Natl. Acad. Sci. U.S.A.* **1989**, *86*, 6821–6825.

339. Sims, S.M., Kelly, M.E., Dixon, S.J. K$^+$ and Cl$^-$ currents in freshly isolated rat osteoclasts. *Pflugers Arch.* **1991**, *419*, 358–370.

340. Espinosa, L., Paret, L., Ojeda, C., Tourneur, Y., Delmas, P.D., Chenu, C. Osteoclast spreading kinetics are correlated with an oscillatory activation of a calcium-dependent potassium current. *J. Cell. Sci.* **2002**, *115*, 3837–3848.

341. Weidema, A.F., Dixon, S.J., Sims, S.M. Electrophysiological characterization of ion channels in osteoclasts isolated from human deciduous teeth. *Bone* **2000**, *27*, 5–11.

342. Komarova, S.V., Dixon, S.J., Sims, S.M. Osteoclast ion channels: potential targets for antiresorptive drugs. *Curr. Pharm. Des.* **2001**, *7*, 637–654.

343. Mahaut-Smith, M.P., Rink, T.J., Collins, S.C., Sage, S.O. Voltage-gated potassium channels and the control of membrane potential in human platelets. *J. Physiol. (Lond.)* **1990**, *428*, 723–735.

344. Fine, B.P., Hansen, K.A., Salcedo, J.R., Aviv, A. Calcium-activated potassium channels in human platelets. *Proc. Soc. Exp. Biol. Med.* **1989**, *192*, 109–113.

345. Mahaut-Smith, M.P. Calcium-activated potassium channels in human platelets. *J. Physiol. (Lond.)* **1995**, *484*, 15–24.

346. Brahmajothi, M.V., Morales, M.J., Liu, S., Rasmusson, R.L., Campbell, D.L., Strauss, H.C. In situ hybridization reveals extensive diversity of K$^+$ channel mRNA in isolated ferret cardiac myocytes. *Circ. Res.* **1996**, *78*, 1083–1089.

347. Hohlfeld, R., Wekerle, H. (2001) Immunological update on multiple sclerosis. *Curr. Opin. Neurol.* **2001**, *14*, 299–304.

348. Noseworthy, J.H., Lucchinetti, C., Rodriguez, M., Weinshenker, B.G. Multiple sclerosis. *N. Engl. J. Med.* **2000**, *343*, 938–952.

349. Steinman, L. Multiple sclerosis: a two-stage disease. *Nat. Immunol.* **2001**, *2*, 762–764.

350. Lucchinetti, C.W.B., Noseworthy, J. Multiple sclerosis: recent developments in neuropathology, pathogenesis, magnetic resonance imaging studies and treatment. *Curr. Opin. Neurol.* **2001**, *14*, 259–269.

351. Ben-Nun, A., Cohen, I.R. Experimental autoimmune encephalomyelitis (EAE) mediated by T cell lines: process of selection of lines and characterization of the cells. *J. Immunol.* **1982**, *129*, 303–308.

352. Zhang, J., Markovic-Plese, S., Lacet, B., Raus, J., Weiner, H.L., Hafler, D.A. Increased frequency of interleukin 2-responsive T cells specific for myelin basic protein and proteolipid protein in peripheral blood and cerebrospinal fluid of patients with multiple sclerosis. *J. Exp. Med.* **1994**, *179*, 973–984.

353. Allegretta, M., Nicklas, J.A., Sriram, S., Albertini, R.J. T cells responsive to myelin basic protein in patients with multiple sclerosis. *Science* **1990**, *247*, 718–721.

354. Scholz, C., Anderson, D.E., Freeman, G.J., Hafler, D.A. Expansion of autoreactive T cells in multiple sclerosis is independent of exogenous B7 costimulation. *J. Immunol.* **1998**, *160*, 1532–1538.

355. Kivisakk, P., Mahad, D.J., Callahan, M.K., Sikora, K., Trebst, C., Tucky, B., Wujek, J., Ravid, R., Staugaitis, S.M., Lassmann, H., Ransohoff, R.M. Expression of CCR7 in multiple sclerosis: Implications for CNS immunity. *Ann. Neurol.* **2004**, *55*, 627–638.

356. Cross, A.H., Cannella, B., Brosnan, C.F., Raine, C.S. Homing to central nervous system vasculature by antigen-specific lymphocytes. I. Localization of ^{14}C-labeled cells during acute, chronic, and relapsing experimental allergic encephalomyelitis. *Lab. Invest.* **1990**, *63*, 162–170.

357. Alt, C., Laschinger, M., Engelhardt, B. Functional expression of the lymphoid chemokines CCL19 (ELC) and CCL 21 (SLC) at the blood-brain barrier suggests their involvement in G-protein-dependent lymphocyte recruitment into the central nervous system during experimental autoimmune encephalomyelitis. *Eur. J. Immunol.* **2002**, *32*, 2133–2144.

358. Kernich, C.A. Current treatments for multiple sclerosis. *Neurologist* **2005**, *11*, 137–138.

359. Tubridy, N., Behan, P.O., Capildeo, R., Chaudhuri, A., Forbes, R., Hawkins, C.P., Hughes, R.A., Palace, J., Sharrack, B., Swingler, R., Young, C., Moseley, I.F., MacManus, D.G., Donoghue, S., Miller, D.H. The effect of anti-alpha4 integrin antibody on brain lesion activity in MS. The UK Antegren Study Group. *Neurology* **1999**, *53*, 466–472.

360. Miller, D.H., Khan, O.A., Sheremata, W.A., Blumhardt, L.D., Rice, G.P., Libonati, M.A., Willmer-Hulme, A.J., Dalton, C.M., Miszkiel, K.A., O'Connor, P.W. A controlled trial of natalizumab for relapsing multiple sclerosis. *New Engl. J. Med.* **2003**, *348*, 15–23.

361. Bartosik-Psujek, H., Belniak, E., Mitosek-Szewczyk, K., Dobosz, B., Stelmasiak, Z. Interleukin-8 and RANTES levels in patients with relapsing-remitting multiple sclerosis (RR-MS) treated with cladribine. *Acta Neurol. Scand.* **2004**, *109*, 390–392.

362. Polman, C., Barkhof, F., Sandberg-Wollheim, M., Linde, A., Nordle, O., Nederman, T. Treatment with laquinimod reduces development of active MRI lesions in relapsing MS. *Neurology* **2005**, *64*, 987–991.

363. Bielekova, B., Richert, N., Howard, T., Blevins, G., Markovic-Plese, S., McCartin, J., Frank, J.A., Wurfel, J., Ohayon, J., Waldmann, T.A., McFarland, H.F., Martin, R. Humanized anti-CD25 (daclizumab) inhibits disease activity in multiple sclerosis patients failing to respond to

interferon beta. *Proc. Natl. Acad. Sci. U.S.A.* **2004**, *101*, 8705–8708.

364. Kappos, L., Comi, G., Panitch, H., Oger, J., Antel, J., Conlon, P., Steinman, L. Induction of a non-encephalitogenic type 2 T helper-cell autoimmune response in multiple sclerosis after administration of an altered peptide ligand in a placebo-controlled, randomized phase II trial. The Altered Peptide Ligand in Relapsing MS Study Group. *Nat. Med.* **2000**, *6*, 1176–1182.

365. Bielekova, B., Goodwin, B., Richert, N., Cortese, I., Kondo, T., Afshar, G., Gran, B., Eaton, J., Antel, J., Frank, J.A., McFarland, H.F., Martin, R. Encephalitogenic potential of the myelin basic protein peptide (amino acids 83-99) in multiple sclerosis: results of a phase II clinical trial with an altered peptide ligand. *Nat. Med.* **2000**, *6*, 1167–1175.

366. Gepts, W. Pathologic anatomy of the pancreas in juvenile diabetes mellitus. *Diabetes* **1965**, *14*, 619–633.

367. Roep, B.O. The role of T-cells in the pathogenesis of Type 1 diabetes: from cause to cure. *Diabetologia* **2003**, *46*, 305–321.

368. Lampeter, E.F., Homberg, M., Quabeck, K., Schaefer, U.W., Wernet, P., Bertrams, J., Grosse-Wilde, H., Gries, F.A., Kolb, H. Transfer of insulin-dependent diabetes between HLA-identical siblings by bone marrow transplantation. *Lancet* **1993**, *341*, 1243–1244.

369. Lampeter, E.F., McCann, S.R., Kolb, H. Transfer of diabetes type 1 by bone-marrow transplantation. *Lancet* **1998**, *351*, 568–569.

370. Roep, B.O., Arden, S.D., de Vries, R.R., Hutton, J.C. T-cell clones from a type-1 diabetes patient respond to insulin secretory granule proteins. *Nature* **1990**, *345*, 632–634.

371. Arif, S., Tree, T.I., Astill, T.P., Tremble, J.M., Bishop, A.J., Dayan, C.M., Roep, B.O., Peakman, M. Autoreactive T cell responses show proinflammatory polarization in diabetes but a regulatory phenotype in health. *J. Clin. Invest.* **2004**, *113*, 451–463.

372. Fasth, A.E., Cao, D., van Vollenhoven, R., Trollmo, C., Malmstrom, V. CD28nullCD4$^+$ T cells – characterization

of an effector memory T-cell population in patients with rheumatoid arthritis. *Scand. J. Immunol.* **2004**, *60*, 199–208.

373. Wagner, U.G., Koetz, K., Weyand, C.M., Goronzy, J.J. Perturbation of the T cell repertoire in rheumatoid arthritis. *Proc. Natl. Acad. Sci. U.S.A.* **1998**, *95*, 14 447–14 452.

374. Lin, W.J., Norris, D.A., Achziger, M., Kotzin, B.L., Tomkinson, B. Oligoclonal expansion of intraepidermal T cells in psoriasis skin lesions. *J. Invest. Dermatol.* **2001**, *117*, 1546–1553.

375. Kupper, T.S. Immunologic targets in psoriasis. *New Engl. J. Med.* **2003**, *349*, 1987–1990.

376. Gudjonsson, J.E., Johnston, A., Sigmundsdottir, H., Valdimarsson, H. Immunopathogenic mechanisms in psoriasis. *Clin. Exp. Immunol.* **2004**, *135*, 1–8.

377. Ross, E.L., D'Cruz, D., Morrow, W.J. Localized monocyte chemotactic protein-1 production correlates with T cell infiltration of synovium in patients with psoriatic arthritis. *J. Rheumatol.* **2000**, *27*, 2432–2443.

378. Wu, K., Higashi, N., Hansen, E.R., Lund, M., Bang, K., Thestrup-Pedersen, K. Telomerase activity is increased and telomere length shortened in T cells from blood of patients with atopic dermatitis and psoriasis. *J. Immunol.* **2000**, *165*, 4742–4727.

379. Jones, D.A., Yawalkar, N., Suh, K.Y., Sadat, S., Rich, B., Kupper, T.S. Identification of autoantigens in psoriatic plaques using expression cloning. *J. Invest. Dermatol.* **2004**, *123*, 93–100.

380. Winterfield, L.S., Menter, A., Gordon, K., Gottlieb, A. Psoriasis treatment: current and emerging directed therapies. *Ann. Rheum. Dis.* **2005**, *64*, 87–92.

381. Yamazaki, K., Nakajima, T., Aoyagi, T., Hara, K. Immunohistological analysis of memory T lymphocytes and activated B lymphocytes in tissues with periodontal disease. *J. Periodontal. Res.* **1993**, *28*, 324–334.

382. Seymour, G.J., Taubman, M.A., Eastcott, J.W., Gemmell, E., Smith, D.J. CD29 expression on CD4$^+$ gingival lymphocytes supports migration of activated memory T lymphocytes to diseased periodontal tissue. *Oral Microbiol. Immunol.* **1997**, *12*, 129–134.

383. Taubman, M.A., Kawai, T. Involvement of T-lymphocytes in periodontal disease and in direct and indirect induction of bone resorption. *Crit. Rev. Oral Biol. Med.* **2001**, *12*, 125–135.

384. Aroonrerk, N., Pichyangkul, S., Yongvanitchit, K., Wisetchang, M., Sa-Ard-Iam, N., Sirisinha, S., Mahanonda, R. Generation of gingival T cell lines/clones specific with Porphyromonas gingivalis pulsed dendritic cells from periodontitis patients. *J Periodontal Res.* **2003**, *38*, 262–268.

385. Nikolaeva, N., Uss, E., van Leeuwen, E.M., van Lier, R.A., ten Berge, I.J. Differentiation of human alloreactive CD4$^+$ and CD8$^+$ T cells in vitro. *Transplantation* **2004**, *78*, 815–824.

386. Pearl, J.P., Parris, J., Hale, D.A., Hoffmann, S.C., Bernstein, W.B., McCoy, K.L., Swanson, S.J., Mannon, R.B., Roederer, M., Kirk, A.D. Immunocompetent T-cells with a memory-like phenotype are the dominant cell type following antibody-mediated T-cell depletion. *Am. J. Transplant* **2005**, *5*, 465–474.

387. Bhushan, V., Collins, R.H., Jr. Chronic graft-vs-host disease. *J. Am. Med. Assoc.* **2003**, *290*, 2599–2603.

388. Zhang, Y., Joe, G., Hexner, E., Zhu, J., Emerson, S.G. Alloreactive memory T cells are responsible for the persistence of graft-versus-host disease. *J. Immunol.* **2005**, *174*, 3051–3058.

389. Garcia de Tena, J., Manzano, L., Leal, J.C., San Antonio, E., Sualdea, V., Alvarez-Mon, M. Active Crohn's disease patients show a distinctive expansion of circulating memory CD4$^+$CD45RO$^+$CD28null T cells. *J. Clin. Immunol.* **2004**, *24*, 185–196.

390. Burns, J., Bartholomew, B., Lobo, S. Isolation of myelin basic protein-specific T-cells predominantly from the memory T-cell compartment in multiple sclerosis. *Ann. Neurol.* **1999**, *45*, 33–39.

391. Nishifuji, K., Amagai, M., Kuwana, M., Iwasaki, T., Nishikawa, T. Detection of antigen-specific B cells in patients with pemphigus vulgaris by enzyme-linked immunospot assay: requirement of T cell collaboration for autoantibody produc-

tion. *J. Invest. Dermatol.* **2000**, *114*, 88–94.

392. De Bleecker, J.L., Engel, A.G. Immunocytochemical study of CD45 T cell isoforms in inflammatory myopathies. *Am. J. Pathol.* **1995**, *146*, 1178–1187.

393. Fiocco, U., Rosada, M., Cozzi, L., Ortolani, C., De Silvestro, G., Ruffatti, A., Cozzi, E., Gallo, C., Todesco, S. Early phenotypic activation of circulating helper memory T cells in scleroderma: correlation with disease activity. *Ann. Rheum. Dis.* **1993**, *52*, 272–277.

394. Mahmoud, F., Abul, H., Haines, D., Al-Saleh, C., Khajeji, M., Whaley, K. Decreased total numbers of peripheral blood lymphocytes with elevated percentages of CD4$^+$CD45RO$^+$ and CD4$^+$CD25$^+$ of T-helper cells in non-segmental vitiligo. *J. Dermatol.* **2002**, *29*, 68–73.

395. Ohta, K., Norose, K., Wang, X.C., Ito, S., Yoshimura, N. Abnormal naive and memory T lymphocyte subsets in the peripheral blood of patients with uveitis. *Curr. Eye Res.* **1997**, *16*, 650–655.

396. Gessl, A., Waldhausl, W. Elevated CD69 expression on naive peripheral blood T-cells in hyperthyroid Graves' disease and autoimmune thyroiditis: discordant effect of methimazole on HLA-DR and CD69. *Clin. Immunol. Immunopathol.* **1998**, *87*, 168–175.

397. Peter, M.J., Varga, Z., Panyi, G., Bene, L., Damjanovich, S., Pieri, C., Possani, L., Gaspar, R.J. Pandinus imperator scorpion venom blocks voltage-gated K$^+$ channels in human lymphocytes. *Biochem. Biophys. Res. Commun.* **1998**, *242*, 621–625.

398. Baker, R., Chee, J., Bao, J., Garcia, M.L., Kaczorowski, G., Kotliar, A., Kayser, F., Liu, C., Miao, S., Rupprecht, K.M., Parsons, W.H., Schmalhofer, W., Liverton, N., Clairborne, C.F., Claremon, D.A., W.J., T., Wayne, J. Carbocyclic potassium channel inhibitors. *PCT Int. Appl. WO 0025770*, **2000**.

399. Rodrigues, A.R., Arantes, E.C., Monje, F., Stuhmer, W., Varanda, W.A. Tityustoxin-K(alpha) blockade of the voltage-gated potassium channel Kv1.3. *Br. J. Pharmacol.* **2003**, *139*, 1180–1186.

400. Mahjoubi-Boubaker, B., Crest, M., Khalifa, R.B., Ayeb, M.E., Kharrat, R. Kbot1, a

three disulfide bridges toxin from Buthus occitanus tunetanus venom highly active on both SK and Kv channels. *Peptides* **2004**, *25*, 637–645.

401. Choi, J.S., Hahn, S.J., Rhie, D.J., Yoon, S.H., Jo, Y.H., Kim, M.S. Mechanism of fluoxetine block of cloned voltage-activated potassium channel Kv1.3. *J. Pharmacol. Exp. Ther.* **1999**, *291*, 1–6.

402. Huys, I., Dyason, K., Waelkens, E., Verdonck, F., van Zyl, J., du Plessis, J., Muller, G.J., van der Walt, J., Clynen, E., Schoofs, L., Tytgat, J. Purification, characterization and biosynthesis of parabutoxin 3, a component of Parabuthus transvaalicus venom. *Eur. J. Biochem.* **2002**, *269*, 1854–1865.

403. Huys, I., Olamendi-Portugal, T., Garcia-Gomez, B.I., Vandenberghe, I., Van Beeumen, J., Dyason, K., Clynen, E., Zhu, S., van der Walt, J., Possani, L.D., Tytgat, J. A subfamily of acidic alpha-K$^+$ toxins. *J. Biol. Chem.* **2004**, *279*, 2781–2789.

404. Nirthanan, S., Pil, J., Abdel-Mottaleb, Y., Sugahara, Y., Gopalakrishnakone, P., Joseph, J.S., Sato, K., Tytgat, J. Assignment of voltage-gated potassium channel blocking activity to kappa-KTx1.3, a non-toxic homologue of kappa-hefutoxin-1, from Heterometrus spinifer venom. *Biochem. Pharmacol.* **2005**, *69*, 669–678.

405. Chagot, B., Pimentel, C., Dai, L., Pil, J., Tytgat, J., Nakajima, T., Corzo, G., Darbon, H., Ferrat, G. An unusual fold for potassium channel blockers: NMR structure of three toxins from the scorpion Opisthacanthus madagascariensis. *Biochem. J.* **2005**, *388*, 263–271.

406. Teisseyre, A., Michalak, K. The voltage- and time-dependent blocking effect of trifluoperazine on T lymphocyte Kv1.3 channels. *Biochem. Pharmacol.* **2003**, *65*, 551–561.

407. Baell, J.B., Harvey, A.J., Norton, R.S. Design and synthesis of type-III mimetics of ShK toxin. *J. Comput. Aided Mol. Des.* **2002**, *16*, 245–262.

408. Szabo, I., Nilius, B., Zhang, X., Busch, A.E., Gulbins, E., Suessbrich, H., Lang, F. Inhibitory effects of oxidants on n-type K$^+$ channels in T lymphocytes and Xenopus oocytes. *Pflugers Arch.* **1997**, *433*, 626–632.

409. Varga, Z., Panyi, G., Peter, M., Jr., Pieri, C., Csecsei, G., Damjanovich, S., Gaspar, R., Jr. Multiple binding sites for melatonin on Kv1.3. *Biophys. J.* **2001**, *80*, 1280–1297.

410. Schmitz, A., Sankaranarayanan, A., Azam, P., Schmidt-Lassen, K., Homerick, D., Hansel, W., Wulff, H. Design of PAP-1, a selective small molecule Kv1.3 blocker, for the suppression of effector memory T cells in autoimmune diseases. *Mol. Pharmacol.* **2005**, *68*, 1254–1270.

411. Rus, H., Pardo, C.A., Hu, L., Darrah, E., Cudrici, C., Niculescu, T., Niculescu, F., Mullen, K.M., Allie, R., Guo, L., Wulff, H., Beeton, C., Judge, S.I., Kerr, D.A., Knaus, H.G., Chendy, K.G., Calbresi, P.A. The voltage-gated potassium channel Kv1.3 is highly expressed on inflammatory infiltrates in multiple sclerosis brain. *Proc. Nal. Acad. Sci. USA* **2005**, *102*, 11094–11099.

7.3
Drugs Active at Kv1.5 Potassium Channels [1]

Stefan Peukert and Heinz Gögelein

7.3.1
Structure of the Kv1.5 Channel

The Kv1.5 channel belongs to the superfamily of voltage-gated potassium channels, which all consist of four pore-forming subunits, each subunit containing six transmembrane segments (S1–S6). The S4 segment contains positively charged residues at every third position and serves as the principal voltage sensor. The linker between the S5 and S6 segments, the pore loop, is highly conserved in all K⁺ selective channels and defines the ion selectivity (Fig. 7.3.1) [2, 3].

The bacterial potassium channel KcsA, a two transmembrane domain channel is the first potassium channel with a known crystal structure [4]. This channel shows 54% sequence homology with the corresponding pore-forming domains (S5–S6) of Kv channels and therefore provides insight into the ion selectivity and the pore structure of voltage-gated potassium channels in general. The pore is constructed of an inverted tepee, with the selectivity filter held at its wide end. The selectivity filter is formed by the Thr-Val-Gly-Tyr-Gly part of the pore loops of the four subunits. The carbonyl oxygens of this peptide backbone are directed inwards and constrained in an optimal geometry to coordinate a dehydrated K⁺ ion. The remainder of the pore is wider and lined with hydrophobic amino acids to form a water-filled cavity. The channel can exist in a closed and open state, changing from one conformation to the other by bending the four inner helixes S6, which form a bundle in the closed state, around the "gating hinge" glycine by approximately 30° [5]. A new crystal structure of K$_v$AP, a voltage-dependent K⁺ channel from the bacterium *Aeropyrum pernix*, helped to elucidate the mechanism of pore opening [6]. This channel, which is closely related in amino acid sequence and electrophysiological properties to eukaryotic K$_v$ channels, contains a central ion-conduction pore surrounded by voltage sensors. It is proposed that in response to a change in the membrane voltage the voltage sensor paddles could move across the membrane, which in turn open the pore [6].

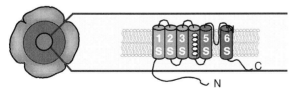

Fig. 7.3.1 Basic architecture of the Kv1.5 channel.

7.3.2
Pharmacological Significance of the Kv1.5 Channel

7.3.2.1 **Kv1.5 and the Ultrarapid Component of the Delayed Rectifier**
Potassium Current, IK$_{ur}$
The shaker related Kv1.5 clone was identified in various tissues like heart, aorta, skeletal muscle and to a lesser extent in brain [7]. Kv1.5 protein is present in the sarcolemma of human atrial cells [8]. Comparison of the biophysical and pharmaceutical properties of the native currents in human atrial cells with currents obtained from expression of cloned Kv1.5 channels provided the first evidence that Kv1.5 is the molecular basis for the ultrarapid potassium current, IK$_{ur}$ [9]. In addition, a study with antisense oligonucleotides directed against Kv1.5 mRNA demonstrated specific reduction of IK$_{ur}$ density in cultured human atrial myocytes, confirming the involvement of Kv1.5 in this current [10]. In contrast to IK$_r$ and IK$_s$, the ultrarapid delayed rectifier IK$_{ur}$ has been found only in human atrial cells but not in ventricles [11–13]. However, notably, the Kv1.5 channel protein was detected by immunohistochemical techniques in both human atrium and ventricle [8], contrary to the functional current IK$_{ur}$, which only exists in atrial myocytes. This apparent discrepancy might be explained by different β-subunit compositions that might render the ventricular channel non-functional.

7.3.2.2 **Kv1.5 as a Drug Target in Atrial Fibrillation**
Atrial fibrillation, the most frequent cardiac arrhythmia, is caused, or at least maintained, by so-called re-entrant wavelets. In healthy myocardium the cardiac impulse that originates at the pacemaker cells of the sino-atrial node is propagated through the atria and ventricles in a co-ordinated and unidirectional fashion. In diseased myocardium, however, it may happen that the excitation wave front circles around anatomical barriers. This can lead to self-propagating irregular excitation circles (fibrillation) when the cells are no longer refractory as a circling wave front reaches its origin.

A well-established principle to extinguish or prevent such re-entry circuits consists in the increase of the myocardial refractoriness by prolongation of the action potential duration (APD). Many different ionic currents contribute to the cardiac action potential of human atrial myocytes (Fig. 7.3.2). While inward sodium currents initiate the action potential by rapid depolarization of the cell, sequential calcium influx is responsible for the electromechanical coupling leading to contraction of the cardiac muscle cells. Finally, various outward potassium currents are responsible for repolarizing the cell back to its resting potential before a new excitation can occur. The most prominent depolarizing potassium currents are the so-called transient outward current I$_{to}$, which is responsible for early repolarization, the ultrarapid potassium current IK$_{ur}$, which plays a role during early and plateau phase repolarization, and finally the rapid and the slow components of the delayed rectifier current, IK$_r$ and IK$_s$ respectively, which are active during the plateau phase and the final repolarization. In addition, other potassium cur-

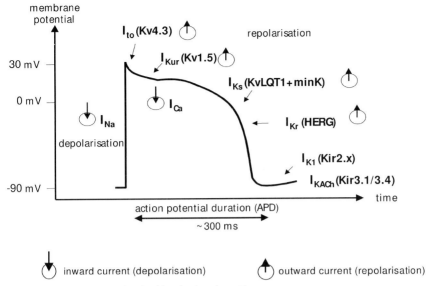

Fig. 7.3.2 Ionic currents involved in shaping the action potential of atrial myocytes. (Reprinted from Ref. [1] with permission from Bentham Science Publishers Ltd.)

rents such as IK_1, IK_{ACh}, and IK_{ATP} are involved in shaping the action potential [14].

Owing to the selectivity of the functional IK_{ur} to the atrium, the Kv1.5 channel should be a promising target for the development of new atrial selective anti-arrhythmic drugs.

During the last decade, several publications have pointed to the potential advantages that a selective Kv1.5 channel blocker could have in the treatment of AF [15–17].

Due to the lack of selective Kv1.5 blockers, two research groups have calculated the effect of Kv1.5 channel blockade in mathematical models of the human atrial action potential [18–20]. Nygren et al. [18, 19] concluded that reduction of IK_{ur} can produce a substantial prolongation of the APD in the atria. In contrast, Courtemanche et al. [20] predicted that selective inhibition of IK_{ur} produces no net change of the action potential duration in atria of healthy individuals. This apparently paradoxical result is explained by the assumption that inhibition of the early depolarizing current IK_{ur} disturbs the current balance during the plateau phase, leading to increased activation of the L-type Ca-current, which can drive the plateau potential well above 0 mV. The more positive plateau voltage would then result in increased activation of mid- and late-depolarizing currents (IK_r, IK_s) and thus compensate the increase in plateau height and duration. However, a completely different result is obtained when the action potential of patients with chronic AF is considered. Chronic AF leads to a remodeling of the heart, resulting in changed ionic currents and shortening of the APD, moreover adaptation of the

APD to frequency is lost [21–23]. Especially, $I_{Ca,L}$ and I_{to} are reduced in cells from AF patients while there are conflicting data about the effect of AF on IK_{ur} density [24]. If these depressed currents are simulated in the mathematical model, selective inhibition of IK_{ur} is able to lengthen the APD by 12%. In summary, although both models result in slightly different conclusions, both support the usefulness of Kv1.5 blockers as new potential drugs for AF, at least under the pathological conditions of chronic atrial fibrillation. Interestingly, these models predict stronger prolongation of the APD when additional K^+ currents, such as the IK_r or I_{to}, are inhibited. Therefore, Kv1.5 could be considered as one of several potassium channel targets in the pursuit of a new anti-arrhythmic drug.

7.3.3
Known Drugs with Activity on Kv1.5

For a surprisingly large number of known drugs, inhibition of Kv1.5 or IK_{ur} has been described in the literature. This section summarizes the pharmacological blockade of Kv1.5 or IK_{ur}, expressed either as IC_{50} value or percent inhibition at a given concentration, and reports the species and cell type.

7.3.3.1 Anti-arrhythmic Drugs with Activity on Kv1.5
For several known anti-arrhythmics (Table 7.3.1 and Fig. 7.3.3), a blockade of Kv1.5, in addition to other mechanisms, has been described.

Tab. 7.3.1 Pharmacological blockade of Kv1.5/IK_{ur} by known anti-arrhythmic drugs.

Drug	Kv1.5 (IC_{50}, µM)	Ikur (IC_{50})	Species	Ref.
Ambasilide (**1**)		34–45 µM	HAM	25
Amiodarone (**2**)	22.9		CHO	26
Azimilide (**3**)		39% @ 100 µM	HAM	27
Bepridil (**4**)	6.6		HEK	28
Chromanol 293B (**5**)		30.9 µM	HAM	29
Clofilium (**6**)	0.14		CHO	30
Diltiazem (**7**)	29.2		Ltk	31
Dronedarone (**8**)	1.0		CHO	26
Flecainide (**9**)	101		HAM	32
HMR-1556 (**10**)		36% @ 10 µM	Xo	33
SSR149744C (**11**)	2.7		CHO	26
Propafenone (**12**)	4.4		Ltk	34
Quinidine (**13**)		5 µM	HAM	34
d-Sotalol (**14**)		0.5–1 mM	HAM	35
Verapamil (**15**)	3.2		HAM	36

HAM = human atrial myocytes, CHO = Chinese hamster ovarian cells, Ltk = mice fibroblasts, Xo = *Xenopus* oocytes, HEK = human embryonic kidney 293 cells.

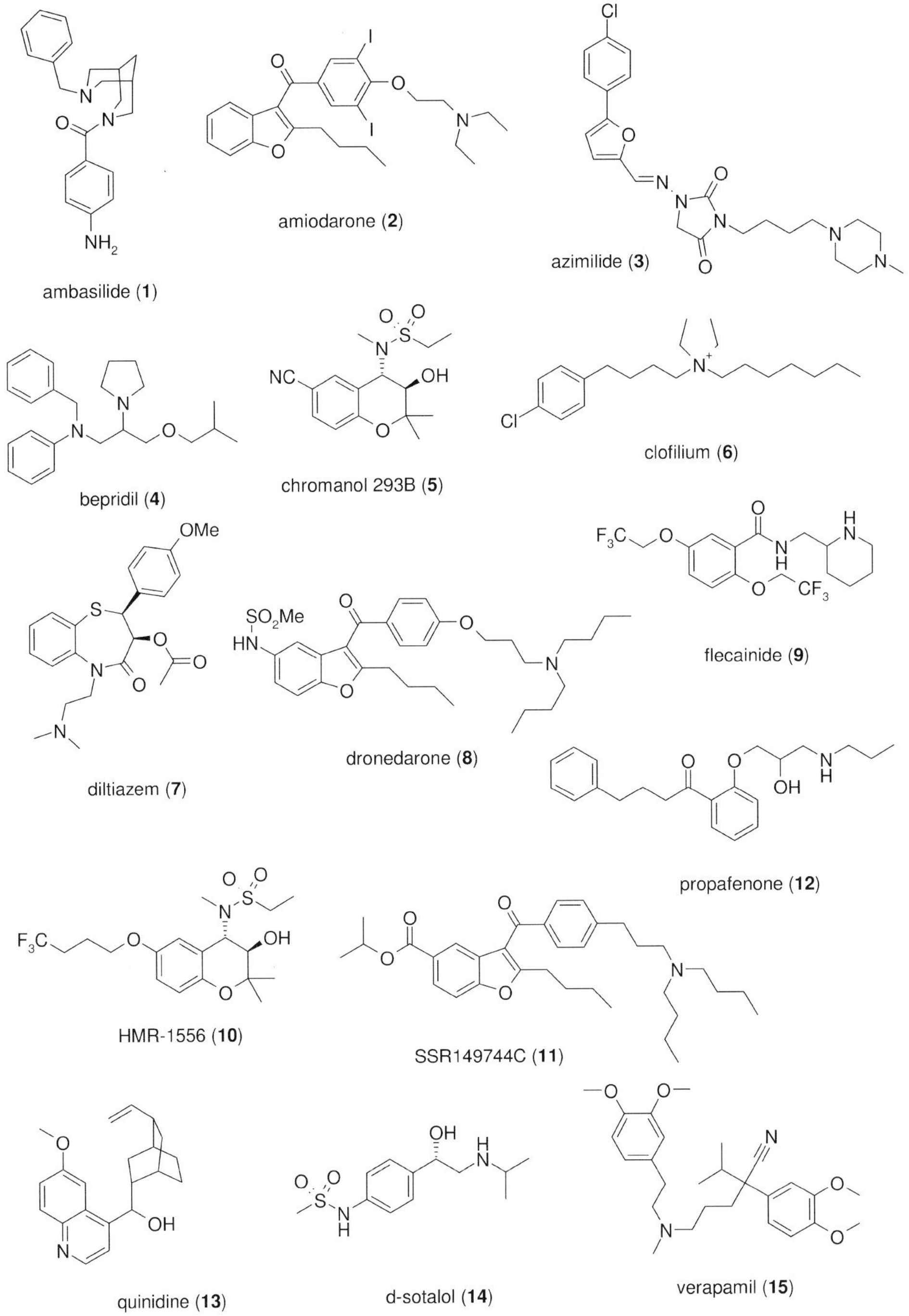

amiodarone (**2**)

ambasilide (**1**)

azimilide (**3**)

bepridil (**4**)

chromanol 293B (**5**)

clofilium (**6**)

diltiazem (**7**)

dronedarone (**8**)

flecainide (**9**)

propafenone (**12**)

HMR-1556 (**10**)

SSR149744C (**11**)

quinidine (**13**)

d-sotalol (**14**)

verapamil (**15**)

Fig. 7.3.3 Known anti-arrhythmic agents acting on Kv1.5/IK$_{ur}$.

The new class III drug ambasilide (**1**) inhibits IK_{ur} in addition to stronger effects on IK_r and IK_s; similarly, the old class III drug d-sotalol shows weak, but measurable inhibition of IK_{ur} [25, 26]. Azimilide (**3**), a mixed IK_r and IK_s inhibitor, inhibits multiple cardiac potassium currents in human atrial myocytes, including IK_{ur} [27]. The multichannel blocker bepridil (**4**) inhibited hKv1.5 with an IC_{50} of 6.6 µM, indicating a partial block of IK_{ur} under clinical conditions [28]. Similarly, the unselective class III quaternary ammonium compound clofilium (**6**) exhibits Kv1.5 blockade at submicromolar concentrations [30]. The class Ic anti-arrhythmic agent propafenone (**12**) and its main metabolite 5-hydroxy-propafenone also block IK_{ur} at therapeutical concentrations besides other channels [34], whereas another class Ic anti-arrhythmic agent, flecainide (**9**), shows only weak inhibitory activity [32]. The three anti-arrhythmic drugs amiodarone (**2**), dronedarone (**8**), and SSR149744C (**11**) from Sanofi-Synthelabo (now Sanofi-Aventis) possess pharmacological effects of class I–IV agents and all inhibited Kv1.5, with IC_{50}s of 22.9, 1.0 and 2.7 µM, respectively [26]. Chromanol 293B (**5**) and HMR-1556 (**10**), two inhibitors of the slow component of the delayed rectifier K^+ current (IK_s), are weak blockers of the Kv1.5 channel [29, 33]. Diltiazem (**7**) and verapamil (**15**), two widely used calcium antagonists, inhibit IK_{ur} at clinically relevant concentrations [31, 36]. In addition, quinidine (**13**), the oldest anti-arrhythmic agent, inhibits IK_{ur} in human atrial myocytes at clinically relevant concentrations [32].

This list demonstrates that many anti-arrhythmic drugs show at least some degree of Kv1.5 inhibitory activity, suggesting that their anti-arrhythmic activity may, in part, be related to their Kv1.5 blocking activity.

7.3.3.2 **Other Drugs with Activity on Kv1.5**

In addition to these anti-arrhythmics, drugs for other indications involving a range of different mode of actions have also been reported to possess Kv1.5 blocking activities (Table 7.3.2, Fig. 7.3.4). The local anesthetics benzocaine (**16**) and bupivacaine (**17**), two potent sodium channel blockers, exhibit in addition Kv1.5 channel inhibition [37, 38]. The class of type 1 angiotensin II receptor antagonists (**18–21**) shows, consistently, inhibition of Kv1.5, although losartan transiently increases Kv1.5 current by 8% at 1 µM [39–41]. This class of compounds is one of the few reported Kv1.5 blockers with acidic functionality and might not bind in the channel pore [40]. Clotrimazole (**22**) and ketoconazole (**23**), two cytochrome P450 inhibitors, showed direct inhibition of Kv1.5, although to very different degrees [42, 43].

Pfizer has filed a use patent application [50] for the treatment of atrial fibrillation with the marketed antihistamine loratadine (**24**) [51]. Using patch-clamp electrophysiology, the drug blocked Kv1.5 currents in HEK 293 cells stably expressing the hKv1.5 channel with an IC_{50} of 0.8 µM, whereas no effect was observed at 10 µM on the HERG channel current [44]. Other antihistamines, rupatadine (**25**) and terfenadine (**26**), also showed Kv1.5 inhibition in the low micromolar range [45, 46]. Papaverine (**27**), which has been used as vasodilator, showed effects

Tab. 7.3.2 Pharmacological blockade of Kv1.5/IK$_{ur}$ by known anti-arrhythmic drugs.

Drug	Main mode of action	Kv1.5 (IC$_{50}$, μM)	IKur (IC$_{50}$, μM)	Species	Ref.
Benzocaine (16)	Na channel blocker	901		Ltk	37
Bupivacaine (17)	Na channel blocker	7.4 (R enantiomer)		Ltk	38
Candesartan (18)	AT$_1$ blocker	21% @ 0.1 μM		CHO	39
Eprosartan (19)	AT$_1$ blocker	14% @ 1 μM		CHO	39
Irbesartan (20)	AT$_1$ blocker	22% @ 0.1 μM		Ltk	40
Losartan (21)	AT$_1$ blocker	9% @ 1 μM		Ltk	41
Clotrimazole (22)	Cytochrome P450 inhibitor	1.99		RPV-M	42
Ketoconazole (23)	Cytochrome P450 inhibitor		107	Xo	43
Loratadine (24)	Antihistamine	0.8		HEK	44
Rupatadine (25)	Antihistamine & PAF antagonist	2.4		Ltk	45
Terfenadine (26)	Antihistamine	1.1		Ltk	46
Papaverine (27)	Cholinoceptor antagonist	43.4		Ltk	47
U50,488H (28)	κ-Opioid receptor agonist		3.3	HAM	48
Erythromycin (29)	Bacteria protein biosynthesis	26		HEK	49

HAM = human atrial myocytes, CHO = Chinese hamster ovarian cells, Ltk = mice fibroblasts, HEK = human embryonic kidney cells, Xo = oocytes of *Xenopus laevis*, RPV-M = rabbit portal vein myocytes, PAF = platelet activating factor.

on hKv1.5, as did U50,488H (28), a selective κ-opioid receptor agonist in human atrial myocytes [47, 48]. Finally, erythromycin (29), a macrolide antibiotic, exhibited activity on the Kv1.5 channel at clinically relevant concentrations [49].

Based on the broad abundance of Kv1.5 blockade among marketed drugs it might be concluded that blockade of Kv1.5 *per se* does not seem to be detrimental. The multitude of compound classes suggests that the Kv1.5 channel presents a relatively promiscuous drug target.

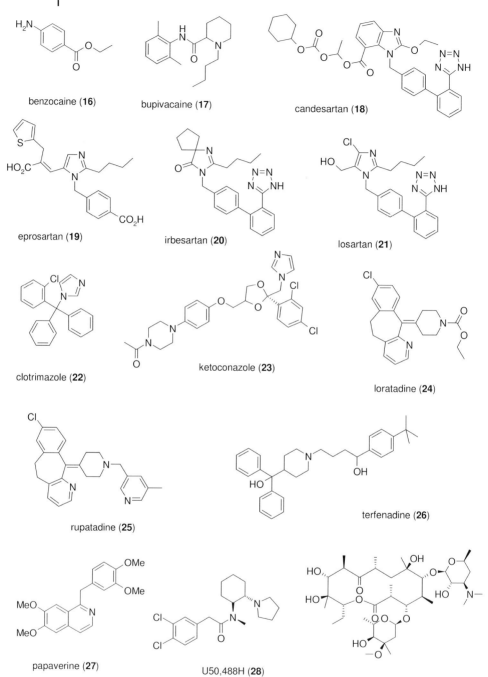

benzocaine (**16**)

bupivacaine (**17**)

candesartan (**18**)

eprosartan (**19**)

irbesartan (**20**)

losartan (**21**)

clotrimazole (**22**)

ketoconazole (**23**)

loratadine (**24**)

rupatadine (**25**)

terfenadine (**26**)

papaverine (**27**)

U50,488H (**28**)

erythromycin (**29**)

Fig. 7.3.4 Known drugs with Kv1.5 blocking activity as an additional effect.

7.3.4
Structural Classes of New Kv1.5 Channel Blockers, their Structure–Activity Relationship and Pharmacology

To find more selective Kv1.5 channel blockers as new drugs, in particular for the treatment of atrial arrhythmias, many pharmaceutical companies have been working in this field during the last several years. From the many patent applications and scientific publications in this area the structure–activity relationship and pharmacology of these compounds acting on the Kv1.5 channel will be discussed in the following sections.

7.3.4.1 Pyridazinones and Phosphine Oxides
In 1998, two use patent applications on Kv1.5 blockers for the treatment of cardiac arrhythmias were published by Merck [52, 53]. The claimed compounds consist of pyridazinones such as **30** and phosphine oxides such as **31** (Fig. 7.3.5). The pyridazinones had previously been filed as immunosuppressants [54]. For compound **31**, anti-arrhythmic effects in a canine model of atria flutter were shown. Intravenous application of **31** at doses between 1 and 10 mg kg^{-1} led to increases of the atrial effective refractory period by 10% and termination of the arrhythmia [53]. No effect on atrioventricular or ventricular function was seen. However, the relevance of the used dog model is under debate due to contradictory results about the role of Kv1.5 in the dog atrium. In 1996, Nattel proposed that the IK$_{ur}$ in the canine atria seems to be carried by the Kv3.1 rather than the Kv1.5 channel [55], whereas in 2003 a study by Fedida et al. suggested that in canine atria, as in other species (including human), Kv1.5 protein is highly expressed and contributes to IK$_{ur}$ [56].

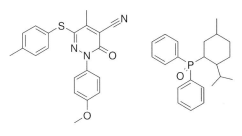

30
(Merck)

31
(Merck)

Fig. 7.3.5 Structural classes of Kv1.5 blockers: Pyridazinones and phosphine oxides.

7.3.4.2 Chromanes, Indanes and Related Compounds
In recent years, Nissan has worked intensively on the chromane class and published many new compounds as potential anti-arrhythmics (e.g., **32–39**, Fig. 7.3.6) [57–61]. An intensively characterized lead compound from these series is the benzopyran **32** (example I-21 in [57]), which has been reported in the litera-

ture as NIP-142 (free amine) and NIP-141 (hydrochloride of NIP-142). NIP-142 selectively prolongs the effective refractory period in guinea pig atrium, but not in the ventricle [62]. It was shown to effectively terminate vagal stimulation induced atrial fibrillation and Y-shaped incision induced flutter in the dog [63, 64]. To explain its atrial selective anti-arrhythmic activity, several researchers investigated the modes of action of NIP-141/142. It was found that NIP-142 inhibited hKv1.5 expressed in HEK293 cells with an IC_{50} of 4.75 μM [65]. Experiments in human atrial myocytes indicated that NIP-141 inhibited not only IK_{ur}, with an IC_{50} of 5.3 μM, but also the transient outward current (I_{to}) with an IC_{50} of 16.3 μM, but it did not affect sodium or HERG currents [66, 67]. A recent study in dog showed that the prolongation of atrial ERP with NIP-142 was greater in the presence of vagal stimulation than in the control state, indicating an effect of NIP-142 on the acetylcholine-induced potassium current IK_{ACh} [68, 69]. In summary, NIP-141/142 is a new class III anti-arrhythmic that blocks IK_{ur}, I_{to} and IK_{Ach}, resulting in atrial selective action potential prolongation without affecting conduction velocity and ventricular function.

The chromane scaffold, next to the tetrahydroquinoline scaffold, was also utilized by Bristol-Myers Squibb in the synthesis of compounds such as **37** [70]. However, no biological data were provided in this patent application.

In 1998, Icagen and Eli Lilly published structurally related new indanes such as compound **38** as inhibitors of Kv1.5 and Kv1.3 channels [71]. 1 μM of **38** was described to inhibit IK_{ur} and APD in isolated human myocytes by >50%. In addition, an immunosuppressive effect of **38** was shown at 10 μM in a lymphocyte proliferation assay.

In a subsequent patent application by Icagen the indane ring was expanded to a tetrahydronaphthalene ring, leading to the potent Kv1.5 blocker **39** (IC_{50} 0.05 μM), and many analogues [72]. In further applications Icagen elaborated the indane scaffold and disclosed compounds such as **40** and **41** [73, 74]. The specified compounds inhibited Kv1.5 by 66% in a rubidium efflux assay and by 46% expressed in Chinese hamster ovary (CHO) cells at a concentration of 0.1 μM. Recently, Icagen published two new patent applications with compounds such as **42** and **43** [75, 76]. The compounds from these applications are claimed to have the advantage of a slower dissociation rate from the channel than previously known compounds. The slow off-rate of these open channel blockers allows little recovery from block during interpulse intervals and thus results in inhibition of the peak current, which is expected to lead to optimal Kv1.5 block under physiological conditions. Compound **42** inhibited the peak current in Chinese hamster ovary cells stably expressing hKv1.5 by 70% at 0.1 μM.

Hoffmann-La Roche used the indane **38** from Icagen as template for a *de novo* design of new Kv1.5 blockers (cf. Section 7.3.5.2) [77]. This study resulted in compound **44**, which inhibited Kv1.5 with an IC_{50} of 0.47 μM in CHO cells expressing hKv1.5.

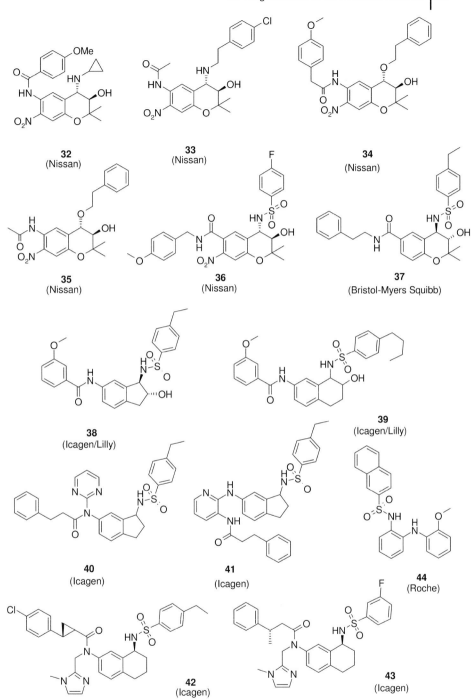

Fig. 7.3.6 Structural classes of Kv1.5 blockers: Chromanes, indanes and related compounds.

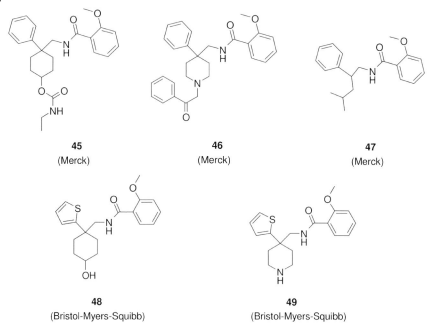

Fig. 7.3.7 Structural classes of Kv1.5 blockers: Cyclohexanes, piperidines and related structures.

7.3.4.3 Cyclohexanes and Piperidines and Related Structures

Merck has published three patent applications on new blockers of Kv1.3 and Kv1.5 (e.g., **45–47**; Fig. 7.3.7) [78–80]. The main focus of these applications seems to be the development of Kv1.3 blockers for the treatment of various diseases involving the immune system but the use of the compounds as Kv1.5 blockers for the treatment of arrhythmia is also claimed.

Knowing the Merck applications, researchers at Bristol-Myers Squibb further elaborated on these scaffolds and published in two patent applications more than 1000 novel cycloalkyl derivatives and piperidines [81, 82]. These compounds show significant structural similarity with the previously disclosed structures, as exemplified by comparison of **45** with **48** and **46** with **49**. The compounds are stated to be inhibitors of potassium channels and particularly of the Kv1.5 channel and to be useful for the treatment of IK$_{ur}$-associated disorders, including atrial fibrillation. However, no biological data are given.

7.3.4.4 Bisaryls

In 2001, Aventis published 2′-substituted 1,1′-biphenyl-2-carboxamides (e.g., **50**, Fig. 7.3.8) as new Kv1.5 blockers for the treatment of atrial arrhythmias [83]. In subsequent patent applications it was shown that the phenyl rings from the

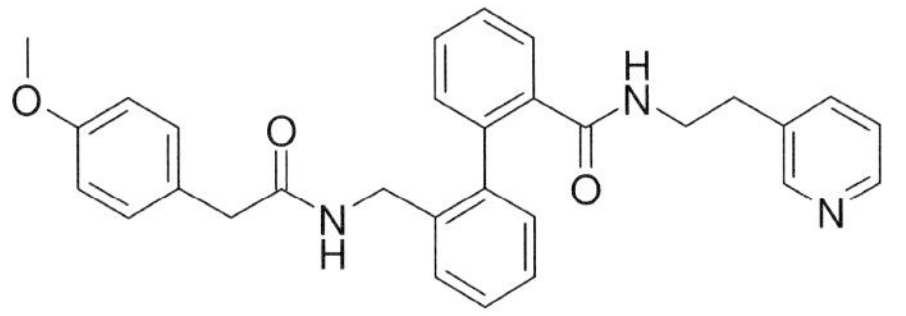

Fig. 7.3.8 Structural classes of Kv1.5 blockers: Bisaryl compounds (all by Aventis).

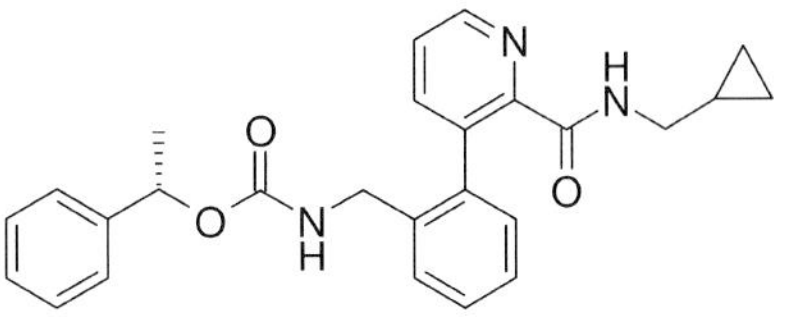

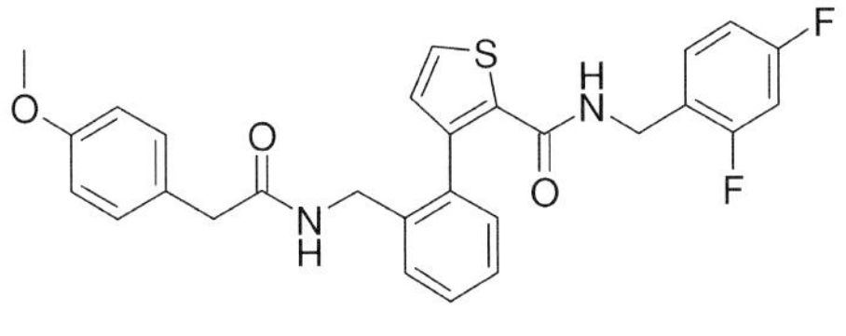

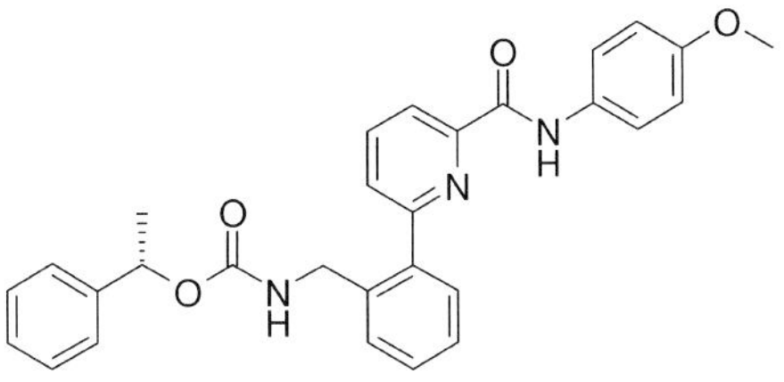

biphenyl scaffold could be replaced by six-membered N-heterocycles such as **51** [84] or five-membered heterocycles such as **52** [85] without losing Kv1.5 blocking activity. Furthermore, besides ortho,ortho-substituted bisaryls, ortho,meta-substituted compounds such as **53** were also active [86].

Variation of the amide side chain in the o,o-disubstituted biphenyls allowed a broad range of different amines, as shown in examples **54a–e** (Fig. 7.3.9). Aliphatic, aromatic, heteroaromatic, basic and amide functionalized side chains were all tolerated. In general, amides were more potent than the corresponding esters, as illustrated with esters **54f** and **54g**, which show a loss in activity compared to the corresponding amides **54c** and **54a**, respectively. The linkage to the other ortho substituent was of minor importance. As seen by comparison of compounds **54h–l** there is no significant difference between carbamate **54h**, sulfonamide **54i**, carbon amide **54j**, urea **54k** and amine **54l**. Of greater influence however, is the structure of the linked lipophilic end group. A very small residue such as the methyl carbamate **54n** leads to weaker activity. The best substituent in this position was the methyl substituted benzyl group of example **54o**, having an IC_{50} of 0.16 µM in *Xenopus* oocytes and 90 nM in CHO cells.

Whilst compounds with (S)-stereochemistry at the methyl branch such as compounds **54o** or **54q** were generally very potent, the corresponding (R)-enantiomers such as **54p** or **54r** were 3- to 20-fold weaker in activity in most cases. In addition to the above-mentioned o,o-disubstituted biphenyls that do not form stable atropisomers, some tetrasubstituted analogues have been described [87]. The ortho,ortho-tetrasubstituted compounds (R)-**55a**/(S)-**55a** and (R)-**55b**/(S)-**55b** form stable atropisomers that could be separated by chiral HPLC, but there was no significant difference in activity between the optical antipodes.

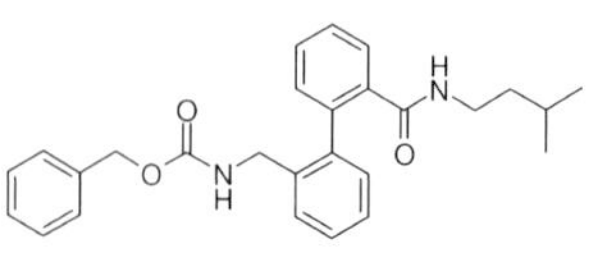

54a (0.8 µM)

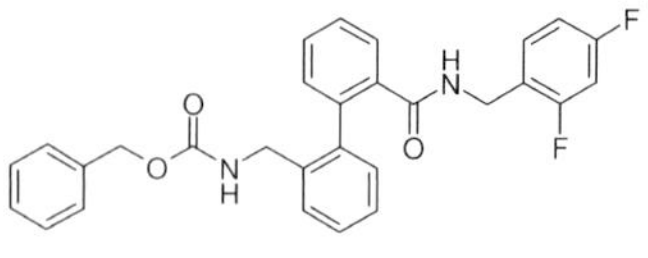

54b (1.2 µM)

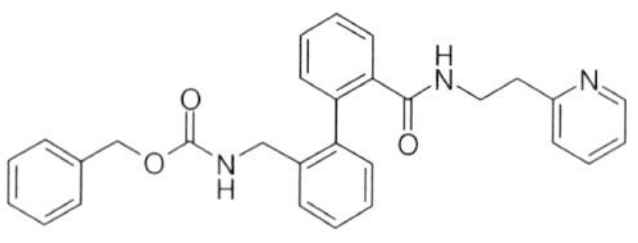

54c (0.7 µM)

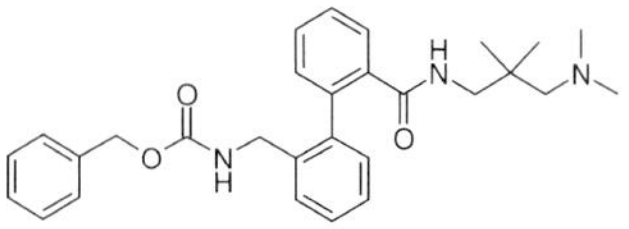

54d (2.2 µM)

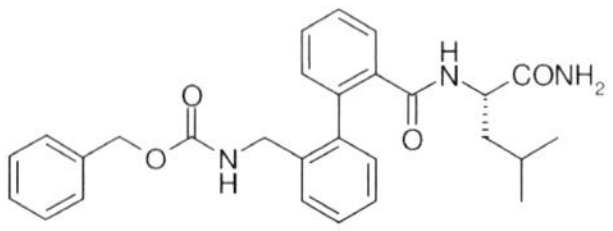

54e (0.3 µM)

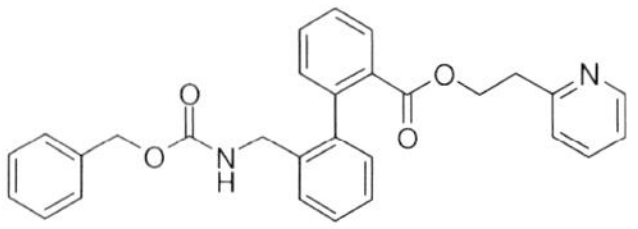

54f (3.5 µM)

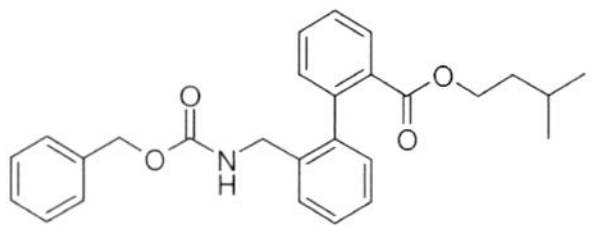

54g (8% inhibition at 10 µM)

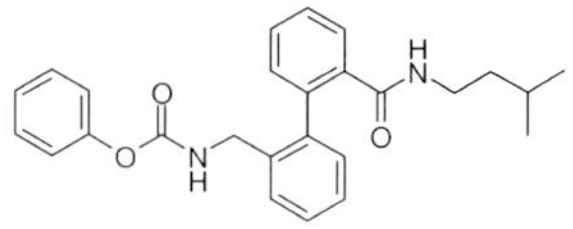

54h (3.3 µM)

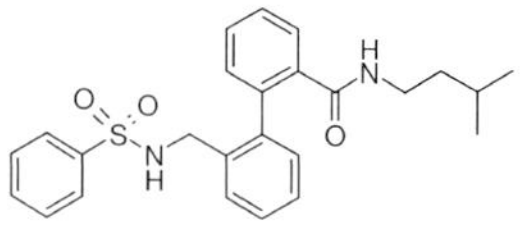

54i (2.6 µM)

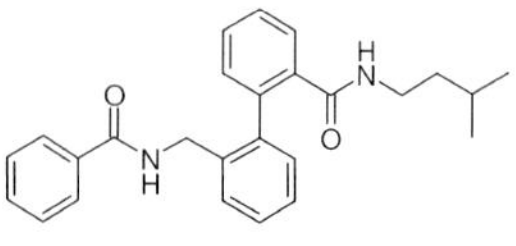

54j (4.1 µM)

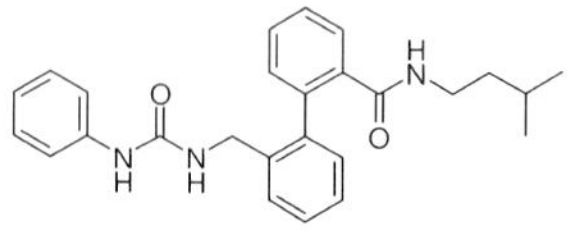

54k (4.5 µM)

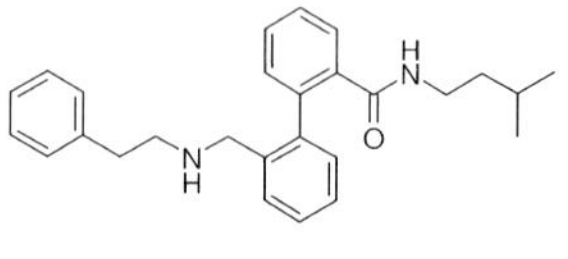

54l (4.0 µM)

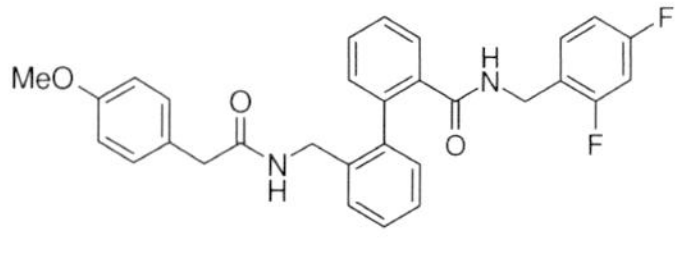

54m (1.2 µM)

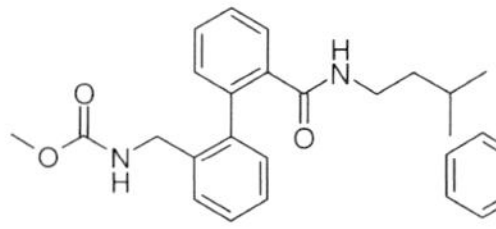

54n (7.1 µM)

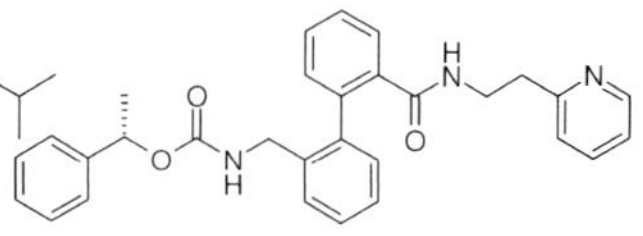

54o (0.16 µM)

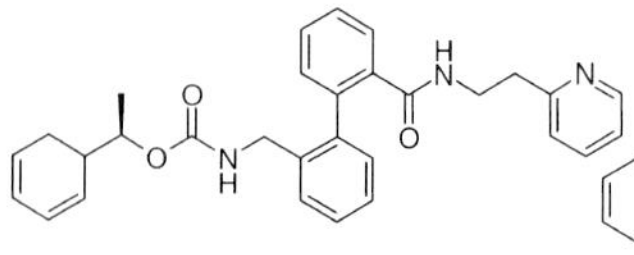

54p (2.4 µM)

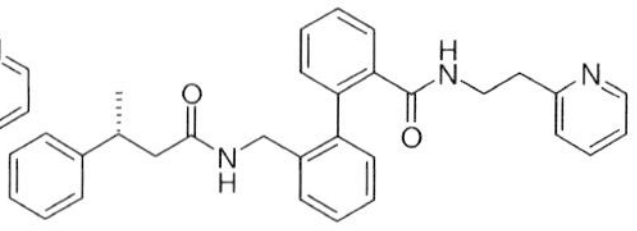

54q (0.3 µM)

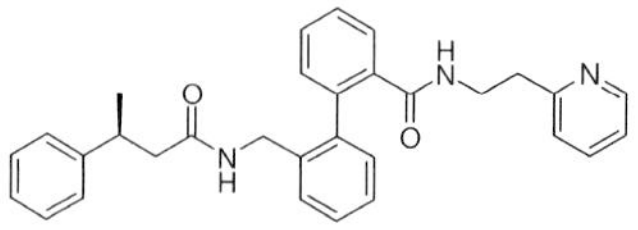

54r (1.0 µM)

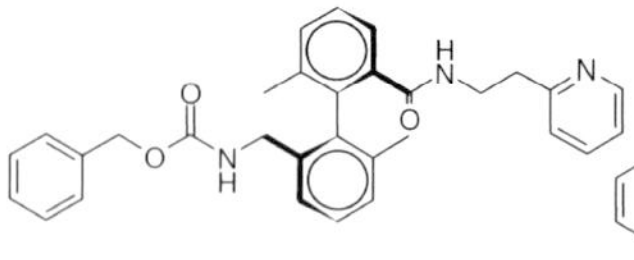

(R)-55a (1.7 µM)

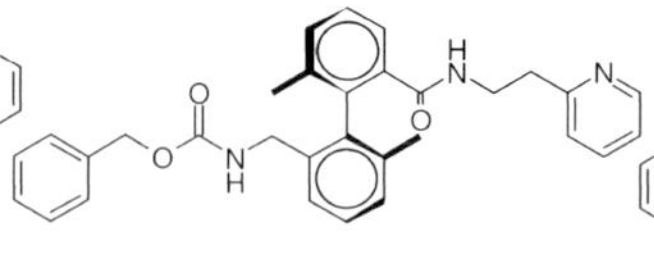

(S)-55a (0.7 µM)

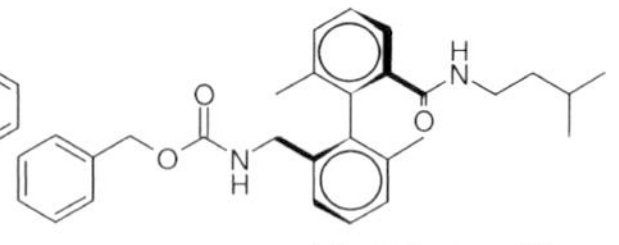

(R)-55b (1 µM)

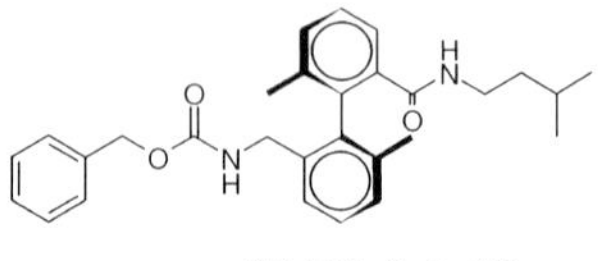

(S)-55b (1.3 µM)

Fig. 7.3.9 SAR for selected biphenyls (IC_{50} for human Kv1.5 measured in *Xenopus laevis* oocytes).

7.3.4.4.1 **Electrophysiology and Pharmacology of SS9947 and S20951**

For several of the bisaryls discussed above, additional biological data have been published. The biphenyl carbamate **54c** has been reported as S9947 [88]. S9947 inhibited the human Kv1.5-channel expressed in CHO cells with an IC_{50} of 417 nM and blocked the sustained current (I_{sus}) in human atrial myocytes with an IC_{50} of 71 nM while it had no effect on HERG or sodium currents. This data indicate that blockade of I_{sus} in human atrial cells occurs at lower drug concentration than inhibition of the hKv1.5 channel in an expression system. Furthermore, this compound increased action potential duration in rat ventricular myocytes (which contain Kv1.5, contrary to human ventricular cells) at both slow and fast pacing rates with an IC_{50} of 1 µM. As a consequence of this rate-independence, S9947 could be expected to show better efficacy at high rates than presently available class III anti-arrhythmics, which often exhibit a reverse rate-dependency.

The two biphenyls **54c** (S9947) and **54m** (S20951) have been investigated in a new pig model in comparison with other class III drugs [89]. Both S9947 and S20951 showed a marked prolongation of atrial refractoriness with no apparent effect on ventricular repolarization. In contrast to this atrial selective effect of the two IK_{ur} blockers, typical IK_r blockers such as dofetilide, dl-sotalol (cf. **14**, Fig. 7.3.3), azimilide (**3**, Fig. 7.3.3) and ibutilide also exhibited substantial increases of the QTc-interval. Whereas the IK_r blockers were more potent in the right than in the left atrium, the two IK_{ur} blockers showed the opposite profile and were highly active in preventing left atrial tachycardias. At a dosage of 3 mg kg^{-1}, S20951 and S9947 decreased the left atrial vulnerability by 100% and 82%, respectively, without affecting the QT-interval. These experiments indicate that IK_{ur} blockers have the potential to become a new type of atrial selective drugs for the termination of AF and prevention of its recurrence.

7.3.4.4.2 **Electrophysiology and Pharmacology of AVE0118**

As a third member from the bisaryl class, compound **50**, known as AVE0118 in the literature, has been investigated in detail.

In vitro **Investigations**

Investigations in cell cultures expressing cardiac ion channels revealed that the compound inhibits the human isoform of the Kv1.5 channel but, in addition, it significantly blocks the Kv4.3, which is co-expressed together with the regulatory subunit KChIP2.2. [90]. In both channels the time of inactivation is accelerated, indicating open channel block (Fig. 7.3.10). At higher concentrations also the HERG channel is blocked (Table 7.3.3). Moreover, studies in isolated cardiomyocytes showed that the atrium-specific acetylcholine-activated K$^+$ channel is blocked, whereas other cardiac ion channels are affected only at very high concentrations. Thus, the compound cannot be considered as a selective Kv1.5 channel blocker but rather as a blocker of multiple K$^+$ channels. However, in contrast to other drugs that are on the market or under development for atrial fibrillation, AVE0118 inhibits K$^+$ channels, which are activated in the early phase of repolarization (I_{sus}, I_{to}, IK_{ACh}), in contrast to the "late" K$^+$ channels IK_r and IK_s.

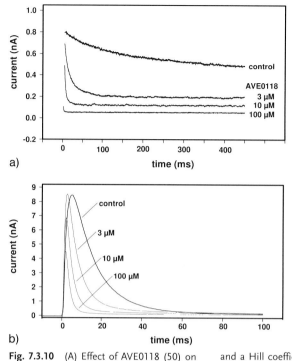

Fig. 7.3.10 (A) Effect of AVE0118 (50) on whole-cell currents in CHO cells, expressing hKv1.5 channels. Original recordings obtained by voltage pulses from the holding potential of −30 to +20 mV for 450 ms at different concentrations of AVE0118 (as indicated). To show the peak current amplitudes more clearly, the up-strokes of the currents at time zero have been eliminated. Evaluation of the current at the end of the voltage pulse yielded an IC_{50} of 1.1 μM and a Hill coefficient of 0.6. (B) Effect of AVE0118 on whole-cell currents in CHO cells, expressing Kv4.3 + KChIP2.2. Original current traces obtained by voltage pulses from the holding potential of −50 to −10 mV for 200 ms. Concentrations of AVE0118 are indicated in the figure. Evaluation of the integral current yielded an IC_{50} of 3.4 μM and a Hill coefficient of 0.9. (Reprinted from Ref. [90] with permission of Springer Verlag, Heidelberg.)

Tab. 7.3.3 Inhibition of ion channels by AVE0118.

Ion channel or current	Tissue	IC_{50} or % inhibition at 10 μM
hKv1.5	CHO cells	1.1 μM
hKv4.3 + hKChIP2.2	CHO cells	3.4 μM
HERG	CHO cells	8.4 μM
IK_{ACh}	Pig left atrial myocytes	4.5 μM
IK_{s}	Guinea pig ventricular myocytes	10%
IK_{1}	Guinea pig ventricular myocytes	0%
IK_{ATP}	Guinea pig ventricular myocytes	28%
I_{L-Ca}	Pig left atrial myocytes	22%

AVE0118 was investigated in isolated right atrial trabeculae obtained from patients with atrial fibrillation or being in sinus rhythm by means of the microelectrode technique [91]. In cells from patients with sinus rhythm, AVE0118 (6 μM) significantly elevated the plateau potential and caused a slight shortening of the APD_{90}. In atrial fibrillation, an increase in plateau potential was also observed next to a significant increase of the APD, especially in the early phase of repolarization.

Investigations in Anesthetized Pig

It was recently demonstrated that investigations of atrial effective refractory period (AERP) and left atrial vulnerability in anesthetized pigs reflects the clinical experience of some conventional drugs [92]. Only amiodarone (**2**, Fig. 7.3.3), which was shown to prevent significantly recurrence of AF in patients, was able to prevent the inducibility of AF, whereas dofetilide, flecainide (**9**, Fig. 7.3.3) and propafenone (**12**, Fig. 7.3.3) were ineffective. In the absence of drugs, the refractory period was shorter in the left than in the right atria. Amiodarone prolonged AERP in both the left and right atrium, whereas dofetilide was mainly prolonging right AERP, and flecainide and propafenone were mainly prolonging left AERP.

AVE0118 prolonged left AERP significantly more than the right one and potently prevented left atrial vulnerability (LAV) in this pig model [93]. At a dose of $0.5 \, mg \, kg^{-1}$ AVE0118 inhibited left atrial vulnerability completely whereas dofetilide inhibited LAV at a dose of $0.01 \, mg \, kg^{-1}$ (the typical dose given in humans) with less than 20%. Recordings of the monophasic action potentials (MAP) of the left atrium revealed that AVE0118 prolonged the MAP duration mainly in the early phase of repolarization (APD_{50}) whereas dofetilide prolongs the late phase (APD_{90}). Prolongation of the left and right AERP by AVE0118 was independent of the basic cycle length in the range 240–400 ms.

Conscious Goats

AVE0118 was investigated in conscious goats in sinus rhythm as well as with chronic atrial fibrillation by the group of Allessie [94]. In goats with sinus rhythm, AVE0118 caused marked prolongation of the AERP, which was less at short (200 ms) than long (400 ms) cycle length. However, in goats with electrically remodeled atria, prolongation of the AERP by AVE0118 was considerably more pronounced and nearly independent of the pacing cycle length. In contrast, prolongation of the AERP by dofetilide was less in normal and was nearly absent in remodeled atria (Fig. 7.3.11).

In goats, atrial vulnerability induced by an extra stimulus is low in animals with sinus rhythm but high in goats with remodeled atria. AVE0118 (at $3 \, mg \, kg^{-1} \, h^{-1}$) reduced the inducibility of AF in remodeled atria effectively to 32%. Importantly, AVE0118 did not prolong the QTc time at dosages up to $5 \, mg \, kg^{-1}$ and had no effect on atrial conduction velocity. Thus, the effect of AVE0118 both in goats with sinus rhythm and with chronic atrial fibrillation is atrium selective and is superior in its therapeutic action to drugs blocking the IK_r channels.

Dofetilide

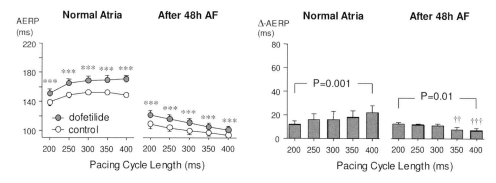

AVE 0118

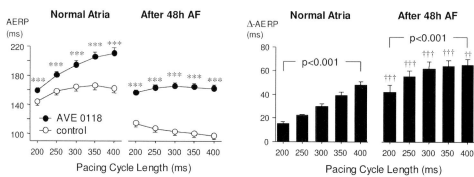

Fig. 7.3.11 Effect of dofetilide ($20 \mu g \, kg^{-1} \, h^{-1}$) and AVE0118 ($3 \, mg \, kg^{-1}$) on atrial refractoriness in goats at different pacing rates before and after 48 hours of AF. In contrast to dofetilide, the class III action of AVE0118 was sig- nificantly enhanced in remodeled atria. ***$P < 0.001$ vs. no drug; ††$P < 0.01$, ††† $P < 0.001$ vs. normal atria. (Reprinted from Ref. [94] with permission of the publisher Lippincott Williams & Wilkins.)

7.3.4.5 Anthranilic Acid Amides

In 2002, Aventis published three patent applications with anthranilic acid amides as blockers of the Kv1.5 channel for the treatment of arrhythmias (Fig. 7.3.12) [95–97].

The discovery of the compound class was facilitated by employing a pharmacophore-based virtual screening approach (Section 7.3.5.3) [98]. The SAR of this compound class was investigated by variations of the three building blocks: amines, anthranilic acids and sulfonyl chlorides (Fig. 7.3.13). Systematic investigation of the amide substituents revealed that in accordance with the pharmacophore model at least one hydrophobic contact is needed at this part of the compound for good activity. The distance of the hydrophobic moiety matters for potency (**59a** vs. **59b**) with an optimum for the phenyl group at the benzylic position. Further increase of activity is obtained by small α-substituents such as ethyl

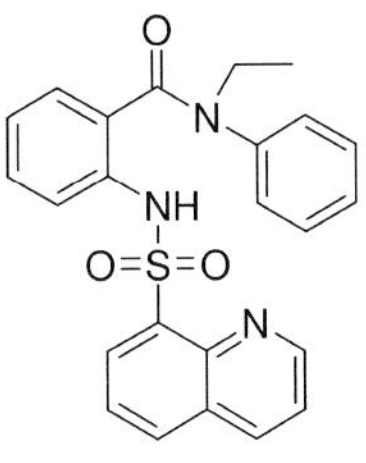

Fig. 7.3.12 Structural classes of Kv1.5 blockers:
Anthranilic acid amides (all Aventis).

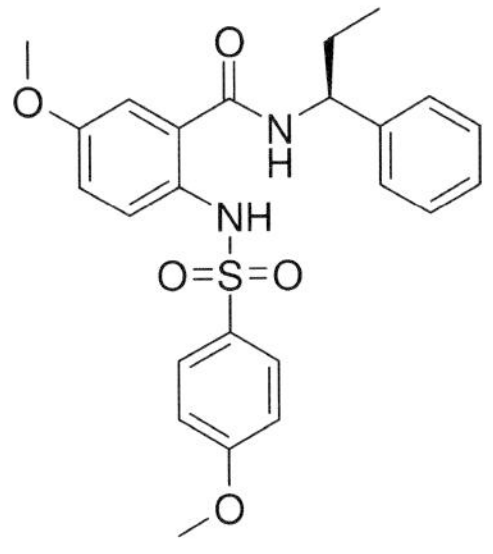

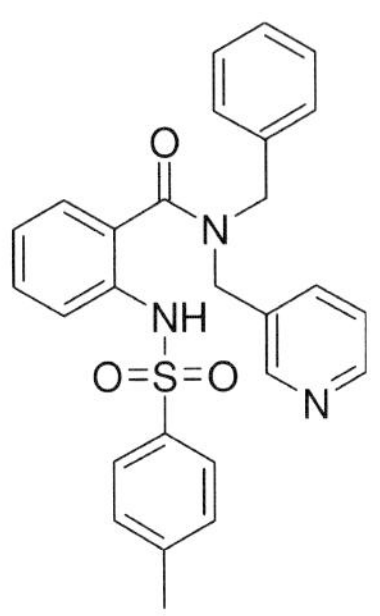

(**59c, d**) with a preference for the S-stereochemistry at this position. Next to these secondary amides, tertiary amides were potent that carried a heterocyclic substituent (**59e**) or a cycloalkyl substituent (**59f**) in addition to the benzyl group. Smaller substituents such as the ethyl group in **59g** result in a decrease of potency. The benzyl group can be readily replaced by heterocyclic substituents (e.g., as in **59h**) without significant loss of activity.

The sulfonyl substituent can be a substituted phenyl group, such as in **59c** or **59i**, an 8-quinolinyl substituent (**59j**) or a more bulky aliphatic group, as in **59k**. Reduced steric bulk at this position resulted in lower potency, as seen by comparison of **59k** and **59l**.

The acidity of the sulfonamide functionality correlated with the inhibitory activity of the compounds, as demonstrated by comparison of **59m–o**. Replacement of the acidic hydrogen by a methyl group (**59p**) resulted in further reduction of activity. These results suggest that the hydrogen might be involved in an intramolecular hydrogen bond with the carbonyl oxygen, which stabilizes a favorable active conformation.

7.3.4.6 Tetrahydroindolone Semicarbazones

Procter & Gamble disclosed in 2003 a series of tetrahydroindolone-based semicarbazones as potent Kv1.5 inhibitors (Fig. 7.3.14) [99]. The compounds show remarkable selectivity towards a number of ion channels, e.g., **60** inhibited the Kv1.5 channel with a IC_{50} of 125 nM and showed weak activity on IK_r (IC_{50} = ~40 µM), Kv1.3 (IC_{50} = ~18 µM) and the L-type Ca channel (IC_{50} > 30 µM), which is typical for this whole compound class. The structure–activity relationship revealed that substitution at the pyrrole ring is crucial. A methyl/ethyl combination at carbon C2/3 is best, methyl/methyl or ethyl/ethyl is also tolerated whereas the lack of substituents at this position or larger alkyl substituents leads to a loss of activity. N-methyl and N–H pyrrole compounds were similarly potent. Substitution of the annelated pyrrole ring to thiophene, pyrazole, or

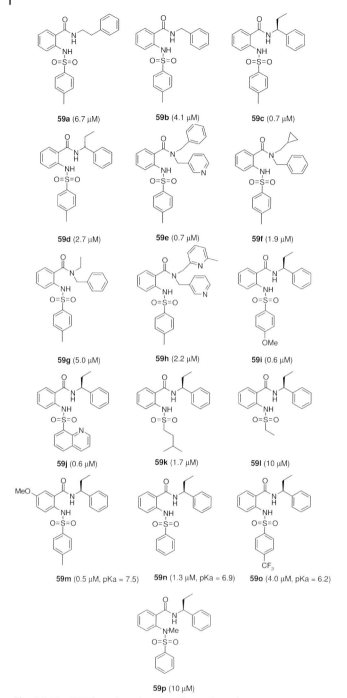

59a (6.7 μM) **59b** (4.1 μM) **59c** (0.7 μM)

59d (2.7 μM) **59e** (0.7 μM) **59f** (1.9 μM)

59g (5.0 μM) **59h** (2.2 μM) **59i** (0.6 μM)

59j (0.6 μM) **59k** (1.7 μM) **59l** (10 μM)

59m (0.5 μM, pKa = 7.5) **59n** (1.3 μM, pKa = 6.9) **59o** (4.0 μM, pKa = 6.2)

59p (10 μM)

Fig. 7.3.13 SAR for selected anthranilic acid amides
(IC$_{50}$ for human Kv1.5 measured in *Xenopus* oocytes).

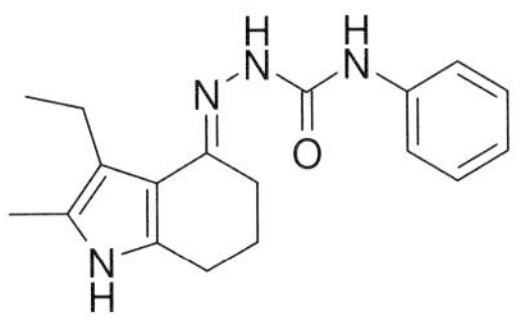

60 (125 nM) 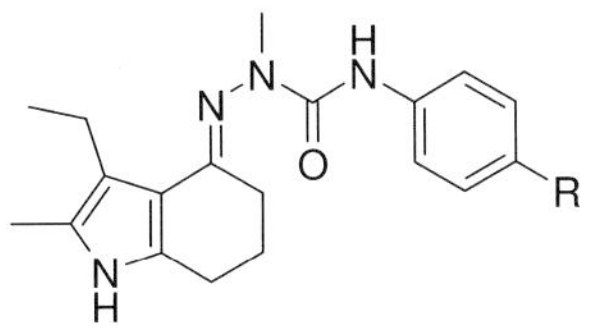   **61a** R = H   (127 nM)   **62** (67 nM)
                         b R = Cl   (127 nM)

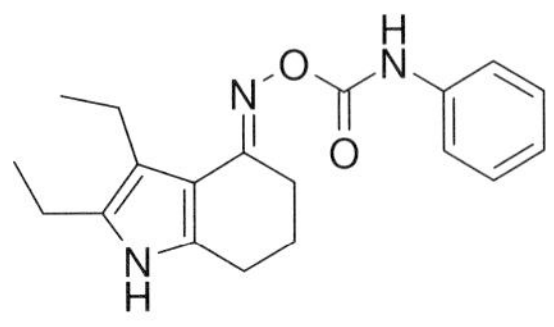

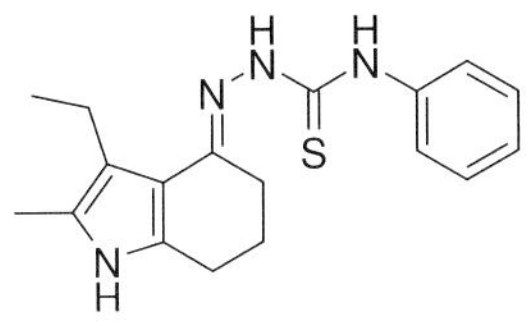

63 (~200nM) 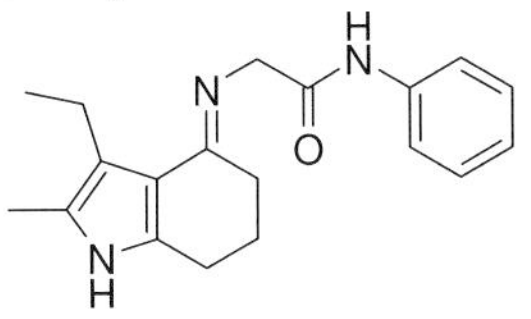   **64** (not active)

Fig. 7.3.14 SAR for selected compounds from the tetrahydroindolone series.

pyrimidine resulted in inactive compounds. Some modifications in the linker were allowed (e.g., **60** to **63**) whereas a change of the nitrogen to a methylene unit (**64**) was not tolerated at all.

Despite the high potency of **61b** on the Kv1.5 channel this compound showed only moderate efficacy in an *in vivo* pig model. At 10 mg kg^{-1} a 10% increase in the right atrium AERP was recorded with no effect on the ventricle [99].

7.3.4.7 Miscellaneous Structures

In 1999, a second patent application by Icagen and Eli Lilly claimed many thiazo-lidinone and tetrahydrothiazinone derivatives for the treatment of cardiac arrhyth-mias [100]. The most potent example (**65**, Fig. 7.3.15) inhibited Kv1.5 currents with an IC$_{50}$ of 0.2 µM.

In 2001, Bristol-Myers Squibb claimed novel heterocyclic dihydropyrimidines (e.g., **66**) as Kv1.5 blockers [101]. However, no biological data were disclosed.

Aventis published indanylcarbamoyl-substituted benzenesulfonamides as new Kv1.5 channel blockers [102]. For example, compound **67** inhibited human Kv1.5 channels expressed in *Xenopus* oocytes with an IC$_{50}$ of 2.8 µM and in CHO cells with an IC$_{50}$ of 0.4 µM.

Recently, Merck claimed isoquinolinones as new Kv1.5 blockers for the treat-ment of atrial arrhythmias [103]. As preferred example, **68** is disclosed but no bio-logical data are presented.

Cardiome has published a patent application on imidazo[1,2-*a*]pyridine ether compounds (e.g., **69**) for the treatment of various diseases, including cardiac arrhythmia [104]. The compounds are claimed to inhibit the Kv1.5 and Kv4.2 channels in addition to sodium channels.

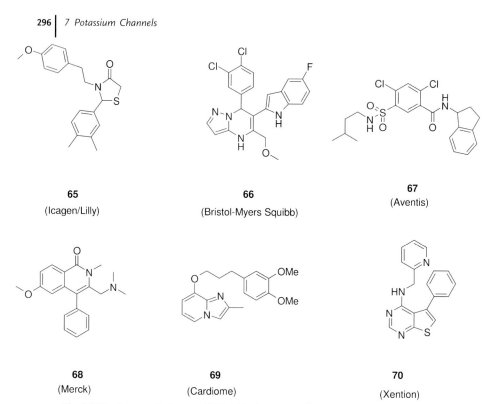

65

(Icagen/Lilly)

66

(Bristol-Myers Squibb)

67

(Aventis)

68

(Merck)

69

(Cardiome)

70

(Xention)

Fig. 7.3.15 Structural classes of Kv1.5 blockers: Miscellaneous structures.

Most recently, Xention disclosed new thienopyrimidines as potent Kv1.5 blockers, such as **70**, which inhibited the human Kv1.5 with an IC_{50} of 30 nM in a rubidium efflux assay and inhibited Kv1.5 in an automated whole cell patch clamp assay with 76% at 1 μM [105].

7.3.5
Strategies in Lead Identification for Kv1.5 Blockers

Despite the lack of structural information, on the one hand, and the unavailability of high-throughput assays on the other hand until very recently, voltage-gated potassium channels have been successfully pursued targets in the pharmaceutical industry. The Kv1.5 channel can serve as an excellent example of chemists' creativity to come up with multiple methods for lead generation in this environment.

7.3.5.1 Privileged Structures
The term "privileged structures" was first introduced in 1988 by Evans and co-workers, describing their design of benzodiazepine-based cholecystekinin-1 (CCK1) antagonists [106]. Evans et al. refer to a finding by Chang et al. [107],

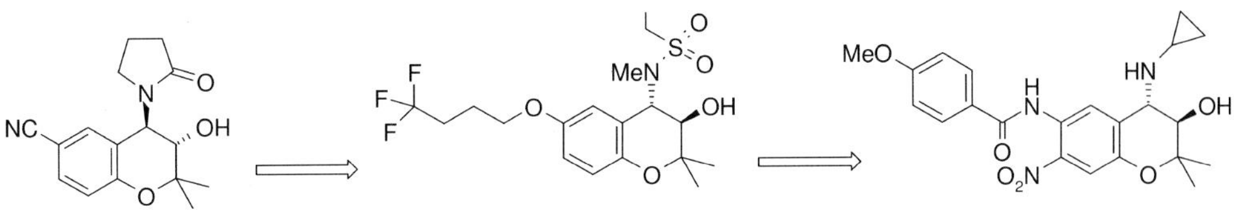

cromakalim (**71**)
(K_{ATP} opener)

HMR-1556 (**10**)
(IKs blocker)

NIP-142 (**32**)
(Kv1.5 blocker)

Fig. 7.3.16 Chromanes as potassium channel modulators.

who discovered that the previously described analgesic tifluadom, a κ-opioid benzodiazepine agonist, also acts as a peripheral CCK receptor antagonist. This documented activity of a single compound on two different target proteins of the same gene family (GPCRs) implies that a single molecular framework can provide ligands for various receptors. Therefore, modifying such structures could be a viable alternative in the search for new ligands [106]. This concept has been used in particular for identifying GPCR modulators [108]. However, there exist compelling examples in the area of ion channel modulators.

Chromanes have long been prominent scaffolds in the synthesis of potassium channel modulators (Fig. 7.3.16). Starting from the K_{ATP} channel opener (–)-cromakalim (**71**), a peripheral vasodilator, the selective IK_s channel blocker HMR1556 (**10**) was developed later [109, 110]. A structurally related compound is the Kv1.5 blocker NIP-142 (**32**). Interestingly, both K-channel blockers (**10** and **32**) have the opposite absolute configuration than the K-channel opener cromakalim (**71**). As shown in Section 7.3.4.2, the chromane scaffold then served as a starting point for numerous Kv1.5 inhibitors.

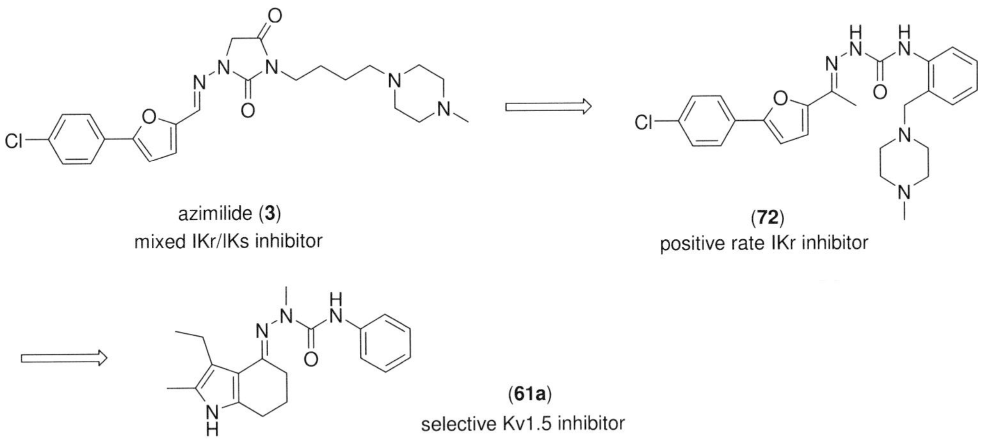

azimilide (**3**)
mixed IKr/IKs inhibitor

(**72**)
positive rate IKr inhibitor

(**61a**)
selective Kv1.5 inhibitor

Fig. 7.3.17 Imine-N-imidazolidines and semicarbazones as potassium channel modulators.

Another example is the work by scientists at Procter & Gamble, a company long involved in research on anti-arrhythmic drugs acting on ion channels. Originating from azimilide (**3**), a mixed IK$_r$ and IK$_s$ blocker with an imine-N-imidazolidine structural element, positive rate dependent IK$_r$ blockers such as **72** were developed by opening up the imidazolidine to a semicarbazone (Fig 7.3.17). Further modifications on the substituents of the semicarbazone core resulted in compound **61a**, which was devoid of any significant IK$_r$ blockade [99].

7.3.5.2 Similarity Searches and Rescaffolding

Known inhibitors for a target can serve as an excellent starting point in the search for new chemical compounds. Using similarity algorithms helps to find *de novo* core structures with improved physicochemical, pharmacokinetic and pharmacological properties in comparison to the parent compound. Starting from the indane compound **38** (Fig. 7.3.6), a micromolar inhibitor of the Kv1.5 channel, scientists at Aventis searched for novel inhibitors within the company inventory using a 2D similarity search [111]. Seventy-five compounds with a similarity coefficient of at least 0.8 were screened for inhibition of IK$_{ur}$. Using this protocol the *m*-sulfonylbenzamide (**73**, Fig. 7.3.18) and the naphthalene derivative **74** that blocked Kv1.5 with moderate potency were discovered. However, because of chemical instability and insufficient physical properties these compounds were not considered a lead but subjected to a rescaffolding approach that resulted in the 1,1'-disubstituted biphenyls **75** (Section 7.3.4.4) [87].

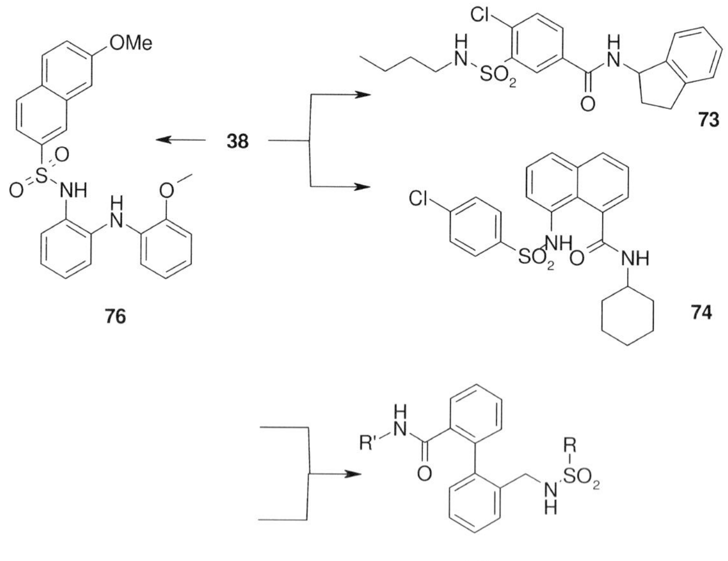

Fig. 7.3.18 Similarity searches and rescaffolding on Kv1.5 blockers.

Independently, the enantiomer of the above-mentioned indane **38** was used by Hoffman-La Roche as template for a *de novo* design of new Kv1.5 blockers using an evolutionary algorithm (Fig 7.3.18) [77]. This approach is based on the principle of topological pharmacophore similarity and uses molecular fragments derived from compounds listed in the Derwent World Drug Index (WDI) to construct virtual structures. The rational for employing building blocks based on known drugs is that reassembly of fragments potentially produces "druglike" structures. The designed structures exhibited similar 2D distribution of generalized atom types (hydrogen-bond donors/acceptors, positive/negative charge, lipophilic) with the template structure. Chemical synthesis and biological testing of the highest ranked structures resulted in the new Kv1.5 inhibitor **76**.

7.3.5.3 Pharmacophore-based Lead Identification

Once sufficient structure–activity data for a target are available pharmacophore modeling and subsequent virtual screening can be pursued to identity new leads. This approach was taken by computational chemists at Aventis employing seven Kv1.5 blockers from the bisaryl (e.g., **54o**, Fig. 7.3.9) and *m*-sulfonylbenzamide series (e.g., **67**, Fig. 7.3.15) to identity a pharmacophore model [98]. The derived pharmacophore consists of three hydrophobic centers in a triangular arrangement (Fig. 7.3.19a). As illustrated in Fig. 7.3.19(b), one phenyl of the bisaryl core and the central phenyl moiety of the other compound (**67**) matches the middle hydrophobic centre while, within each structure class, both ends of the side chains correspond to the remaining hydrophobic centers of the model. The model is consistent with the SAR previously observed for bisaryl Kv1.5 blockers [87]. Strikingly, many of the other compounds described in this chapter also share this motif of three hydrophobic residues and some of them fit quite well into the arrangement of Fig. 7.3.19(a).

This pharmacophore model served as a query to screen an in-house data bank of 423 Kv1.5 blockers, which resulted in the retrieval of 58% of the known Kv1.5 blockers. This rate of false negatives is typically satisfactory for lead identification. Accordingly, the researchers ran a 3D search within the Aventis inventory and identified 4234 virtual hits, 1975 of them passed various filters selecting for "drug-

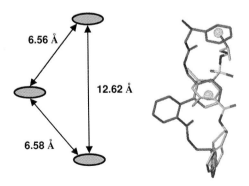

Fig. 7.3.19 (a) Pharmacophore model and (b) matching of bisaryl compound **54o** and *m*-sulfonylbenzamide **67** with pharmacophore model. (Reprinted from Ref. [98], copyright 2004, with permission of Elsevier.)

6.56 Å

12.62 Å

6.58 Å

like" compounds. After hierarchical cluster analysis, 27 clusters resulted, of which representatives from 18 were actually available for *in vitro* screening. This approach yielded the series of anthranilic amides described in Section 7.3.4.5.

7.3.5.4 Structure-based Lead Identification

The method of structure-based lead identification became available for ion channels only recently since the groundbreaking discovery of the first crystal structure of a bacterial KcsA channel [4]. However, no experimentally determined 3D structure is currently available for a Kv1.5 channel. Therefore, the same group of researchers at Aventis used the published crystal structure of KscA as a template for homology modeling of the Kv1.5 pore domain as each of the four subunits forming the Kv1.5 pore share 49% of sequence identity with its KscA counterpart [112]. The minimized homology model (Fig. 7.3.20a) served as input for a binding site identification procedure, based purely on geometrical criteria. This led to the identification of two relevant putative binding sites: a large and shallow site located on the extra-cellular surface and an internal cavity underneath the selectivity

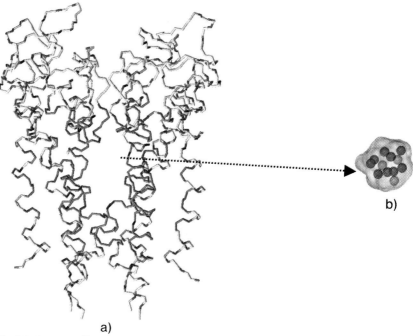

a)

b)

Fig. 7.3.20 (a) Backbone representation of the homology model of the Kv1.5 pore. Oxygen atoms have been omitted for clarity. The backbone of the amino acids forming the extracellular site is colored in yellow, while the backbone of the amino acids forming the internal site is in magenta. (b) Protein-based pharmacophore derived from the GRID analysis of the internal site. H bond donor, acceptor and hydrophobic features are displayed in blue, red and orange. The yellow region corresponds to the excluded volume defined by the protein. (Reprinted with permission from Ref. [112]. Copyright 2005 American Chemical Society.)

filter. The second was considered more drugable and characterized by three hydrophobic sites, eight hydrogen bond donor sites and one hydrogen bond acceptor site. These were combined with an excluded volume, defined by the walls of the internal cavity, resulting in the channel pore-based pharmacophoric query shown in Fig. 7.3.20(b).

The protein-derived pharmacophore served as query to the corporate inventory, yielding 244 molecules for *in vitro* screening, 19 of which (7.8%) were found actives. Five of them, belonging to five different chemical classes, exhibited IC_{50} values under $10 \mu M$ (structures not disclosed).

As most of the virtual screening methods described in this chapter were performed within one company using the same compound inventory as a base, a comparison is very instructive: The structure-based strategy proved to be more effective than two ligand-based virtual screening methods, in terms of hit rate and number of identified chemotypes. However, the two ligand-based approaches enabled the identification of compounds that have not been found by the structure-based protocol. Therefore, the different approaches can be seen as complementary, providing medicinal chemists with different starting points for optimization.

7.3.6
Selectivity against other Ion Channels

Most compounds discussed in this chapter are not selective for Kv1.5. In many cases blockade of Kv1.3 of the same order of magnitude as for Kv1.5 has been reported. For example, the bisaryl compounds **54c** and **54o** (Fig. 7.3.9) inhibit the Kv1.5 channel with an IC_{50} of 0.7 and $0.16 \mu M$ and at the same time block the Kv1.3 channel with IC_{50}s of 2.3 and $0.5 \mu M$, respectively [87].

Also blockade of Kv4.2/Kv4.3 channels are frequent side activities. This low selectivity towards different members of the Kv-channel family is not surprising if one considers the described pharmacophore model, which mainly consists of unspecific lipophilic interactions with no or little hints for specific interactions of the blockers with the Kv1.5 channel protein. Semicarbazones from Procter & Gamble are a remarkable exception (Section 7.3.4.6), showing a roughly 100-fold selectivity towards Kv1.3 [99].

In some cases Kv1.5 blockers were also active on the acetylcholine activated potassium current IK_{Ach} [69, 90].

Selectivity toward the HERG channel is typically much more marked, partly because the structural similarity between the channels is less distinct and partly because the compounds were optimized not to show inhibitory activity on the IK_r current. Table 7.3.4 summarizes some of the literature results.

The question arises of whether the limited selectivity often observed in anti-arrhythmic drugs is detrimental or beneficial. Since several different ionic currents contribute to the cardiac action potential of human atrial myocytes, a blocker that affects more than one of these currents would probably be much more effective than a selective blocker of just one current. However, to exploit the advantage of atrial selectivity of the IK_{ur} current, the blocking activity on this current should

Tab. 7.3.4 Inhibition of the HERG and Kv1.3 channel in comparison to Kv1.5.

Compound	HERG (inhibition at 10 μM, %)	Kv1.3 (IC$_{50}$, μM)	Kv1.5 (IC$_{50}$, μM)
54c [87]	7	2.3	0.7
54o [87]	11	0.5	0.16
59l [98]	6		0.7
59h [98]	11		2.2
60 [99]	40 μM	18	0.125
62 [99]	>50 μM	23	0.067

clearly dominate over the activity on currents that are also present in the ventricles such as IK$_r$ or I$_{to}$. An attractive combination might be a block of IK$_{ur}$ and IK$_{ACh}$, since the acetylcholine-induced potassium current, similar to IK$_{ur}$, was also reported to be present in atrial cells with little or no activity in ventricular myocytes [14, 113–115]. A combination of different mechanisms of cardiac action, preferably atrial selective, might also lead to lower required doses for the desired anti-arrhythmic effect on the atria and therefore reduce the risk of non-cardiac side effects, such as immunosuppression via Kv1.3 block (Chapter 7.2).

7.3.7
Structural Basis for Kv1.5 Channel Block

As no structural data for Kv1.5 channel-inhibitors complexes are currently available, all binding hypothesis so far are based on computational modeling and site-directed mutagenesis studies. The anti-arrhythmic and local anesthetic drugs quinidine (**13**, Fig. 7.3.3), benzocaine (**16**, Fig. 7.3.4), and bupivacaine (**17**, Fig. 7.3.4) have been reported to block Kv1.5 by interaction with residues in the S6 domain that forms the pore region [38, 116–120]. In particular, bupivacaine has been the object of numerous studies (e.g., [38, 121]). Electrophysiological and site-directed mutagenesis studies of Kv1.5 channel block show that bupivacaine binds to the channel pore from the intramolecular side in a stereoselective manner and predominantly in the open state involving the hydroxyl groups of threonine in positions 479 and 507, combined with hydrophobic interactions at positions Leu-510 and Val-514 of the channel protein [38]. A computational modeling study of binding of R(+) bupivacaine to the open-state of Kv1.5 was based on the crystal structures of the closed bacterial ion channel KcsA and the open-state of the calcium-gated potassium channel MthK [121]. The multiple alignment between the KcsA, MthK, and Kv1.5 sequence shows a good match for amino acids that code the internal pore region, including the short pore α-helix, selectivity filter, and part of the internal α-helixes that precede the PVP (proline-valine-proline) sequence (Fig. 7.3.21).

```
                    P            SF                                    S6
KcsA        65/  ALWWSVETATTVGYGDLYPVTLWGRLVAVVVMVAGITSFGLVTAALATWF........VGREQE
MthK       364/  SLYWTFVTIATVGYGDYSPSTPLGMYFTVTLIVLGIGTFAVAVERLLEFL........INREQM
hKv1.5     470/  AFWWAVVTMTTVGYGDMRPITVGGKIVGSLCAIAGVLTIALPVPVIVSNFNYFYHRETDHEEPA
                      TT-479/480                      G-504   PVP-511/513
```

Fig. 7.3.21 Sequence alignment of the internal pore region for the bacterial channels KcsA and MthK and the human channel Kv1.5. P and SF stand for the short pore α-helix and the selectivity filter in the KcsA template, respectively. S6 corresponds to the internal cavity-forming helices. TVGYG, G, and PVP are the highly conserved sequences for mammalian Kv channels. Amino acids mentioned in the text are marked in red. In the bottom line some sequence numbering is given for orientation.

From existing electrophysiological data, a PVP-bend model that introduces the PVP kink (prolines generally show α-helix breaking properties) into the otherwise KcsA-like architecture of the channel, thereby opening up the pore, seems most supported. From the two predicted binding sites for (+)-bupivacaine in the internal pore, the first is in the upper part of the internal cavity, and the second binding region lies in the vicinity of the highly conserved PVP sequence in the S6 segment. Most results favor the PVP bending hinge as bupivacaine binding site (Fig. 7.3.22). The binding site does not seem to include any side-chains that can make specifically strong interactions with bupivacaine.

A completely different inhibitor, S0100176 (**59e**, Fig. 7.3.13), an anthranilic acid amide was used in a Ala-scanning mutagenesis study of the Kv1.5 pore residues combined with modeling of the Kv1.5 open-state pore and docking of the inhibi-

Type 6   view from the internal entrance   side view

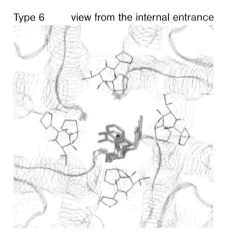
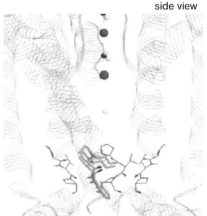

Fig. 7.3.22 A preferred mode of binding of (R)-(+)-bupivacaine to the PVP-bend Kv1.5 homology model. Selected residues explicitly indicate dimensions of the internal pore. Two channels subunits are clipped from the side view for greater clarity. The side view also shows positions of two K$^+$ ions and two waters in the selectivity filter. The position of the cavity centre in the side view is marked by a small circle. (Reprinted from Ref. [121], copyright 2003, with permission from Elsevier.)

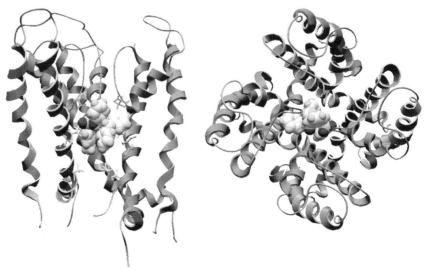

Fig. 7.3.23 Side and top view of the homology model of the Kv1.5 pore region with **59e** docked into the putative binding site. In the side view, two units are omitted for clarity. Amino acid side chains of Thr-479 and 480 (green), Val-505 and 512 (orange) and Ile-508 (yellow) are all less than 4.5 Å from the compound. (Reprinted from Ref. [122] with permission from The American Society for Biochemistry and Molecular Biology.)

tor into this channel pore model [122]. Electrophysiology on the mutants and docking of the compound into the pore indicate that next to hydrophobic residues in the S6 segment near the PVP region (Val-505, Ile-508, Val-512) residues in the pore loop (Thr-479, Thr-480) could constitute drug binding sites (Fig. 7.3.23).

In summary, an emerging theme from these and other studies is that a few residues, located near the selectivity filter and the PVP motif in the S6 domain, form the binding sites for drugs that block the open state of Kv1.5.

References

1. Parts of this book chapter have been published in: Brendel, J., Peukert, S. Blockers of the Kv.15 channel for the treatment of atrial arrhythmias. *Curr. Med. Chem. – C&HA*, **2003**, *1*, 273–287.

2. Ashcroft M. A. *Voltage gated K+ channels, Ion Channels and Disease*, Academic Press, London, **2000**, p. 97.

3. Durell, S. R., Hao, Y., Guy, H. R. Structural models of the transmembrane region of voltage-gated and other K+ channels in open, closed, and inactivated conformations. *J. Struct. Biol.*, **1998**, *121*, 263.

4. Doyle, D. A., Cabral, J. M., Pfuetzner, R. A., Kuo, A., Gulbis, J. M., Cohen, S. L., Chait, B. T., MacKinnon, R. The structure of the potassium channel: Molecular basis of K+ conduction and selectivity, *Science*, **1998**, *280*, 69.

5. Jiang, Y., Lee, A. Chen. J., Cadene, M., Chait, B. T., McKinnon, R. The open pore conformation of potassium channels. *Nature*, **2002**, *417*, 523–526.

6. Jiang, Y., Lee, A., Chen. J., Ruta, V., Cadene, M., Chait, B. T., McKinnon, R. X-ray structure of a voltage-dependent K+ channel. *Nature*, **2003**, *423*, 33–41.

7. Roberds, S. L., Tamkun, M. M. Cloning and tissue-specific expression of five voltage-gated potassium channel cDNAs expressed in rat heart. *Proc. Natl. Acad. Sci. U.S.A.* **1991**, *88*, 1798–1802.

8. Mays, D. J., Foose, J. M., Philipson, L. H., Tamkun, M. M. Localization of the Kv1.5 K+ channel protein in explanted cardiac tissue. *J. Clin. Invest.* **1995**, *96*, 282–292.

9. Fedida, D., Wible, B., Fermini, B., Faust, F., Nattel, S., Brown, A. M. Identity of a novel delayed rectifier current from human heart with a cloned K+ channel current. *Circ. Res.*, **1993**, *73*, 210.

10. Feng, J., Wible, B., Li, G.-R., Wang, Z., Nattel, S., Antisense oligodeoxynucleotides directed against Kv1.5 mRNA specifically inhibit iltrarapid delayed rectifier K+ current in cultured adult human atrial myocytes. *Circ. Res.*, **1997**, *80*, 572.

11. Wang, Z., Fermini, B., Nattel, S. Sustained depolarization-induced outward current in human atrial myocytes. *Circ. Res.*, **1993**, *73*, 1061.

12. Li, G.-R., Feng, J., Yue, L., Carrier, M., Nattel, S. Evidence for two components of delayed rectifier K^+ current in human ventricular myocytes. *Circ. Res.*, **1996**, *78*, 689.

13. Amos, G. J., Wettwer, E. , Metzger, F., Li, Q., Himmel, H. M., Ravens, U. Differences between outward currents of human atrial and subepicardial ventricular myocytes. *J. Physiol. (London)*, **1996**, *491*, 31.

14. Nattel, S. The molecular and ionic specificity of anti-arrhythmic drug actions. *J. Cardiovasc. Electrophysiol.*, **1999**, *10*, 272.

15. Feng, J., Xu, D., Wang, Z., Nattel, S. Ultrarapid delayed rectifier current inactivation in human atrial myocytes: properties and consequences. *Am. J. Physiol.*, **1998**, *275*, H1717.

16. Nattel, S., Yue, L., Wang, Z., Cardiac ultrarapid delayed rectifiers. *Cell Physiol. Biochem.*, **1999**, *9*, 217.

17. Schaffer, P., Pelzmann, B., Bernhart, E., Lang, P., Lökebö, J. E., Mächler, H. Rigler, B., Koidl, B. Estimation of outward currents in isolated human atrial myocytes using inactivation time course analysis. *Pflügers Arch. - Eur. J. Physiol.*, **1998**, *436*, 457.

18. Nygren, A., Leon, L. J., Giles, W. R., Simulations of the human atrial action potential. *Phil. Trans. R. Soc. London A*, **2001**, *359*, 1111.

19. Nygren, A., Fiset, C., Firek, L., Clark, J. W., Lindblad, D. S., Clark, R. B., Giles, W. R. Mathematical model of an adult human atrial cell. *Circ. Res.*, **1998**, *82*, 63.

20. Courtemanche, M., Ramirez, R. J., Nattel, S. Ionic targets for drug therapy and atrial fibrillation-induced electrical remodeling: insights from a mathematical model. *Cardiovasc. Res.*, **1999**, *42*, 477.

21. Wijffels, M. C., Kirchhof, C. J., Dorland, R., Allessie, M. A. Atrial fibrillation begets atrial fibrillation. A study in awake chronically instrumented goats. *Circulation* **1995**, *92*, 1954–1968.

22. Daoud, E. G., Bogun, F., Goyal, R., et al. Effect of atrial fibrillation on atrial refractoriness in humans. *Circulation* **1996**, *94*, 1600–1606.

23. Yue, L., Feng, J., Gaspo, R., Li, G. R., Wang, Z., Nattel, S. Ionic remodeling underlying action potential changes in a canine model of atrial fibrillation. *Circ. Res.* **1997**, *81*, 512–525.

24. Bosch, R. F., Nattel, S. Cellular electrophysiology of atrial fibrillation. *Cardiovasc. Res.*, **2002**, *54*, 259.

25. Koidl, B., Flaschberger, P., Schaffer, P. et al. Effects of the class III anti-arrhythmic drug ambasilide on outward currents in human atrial myocytes. *Naunyn-Schmiedeberg's Arch. Pharmacol.* **1996**, *353*, 226–232.

26. Gautier, P., Guillemare, E., Djanjighian, L. et al. In vivo and in vitro characterization of the novel anti-arrhythmic agent SSR149744C. *J. Cardiovasc. Pharmacol.* **2004**, *44*, 244–257.

27. Chen, F., Esmailian, F., Sun, W. et al. Azimilide inhibits multiple cardiac potassium currents in human atrial myocytes. *J. Pharmacol. Exp. Ther.* **2002**, *7*, 255–262.

28. Kobayashi, S., Reien, Y., Ogura, T., Saito, T., Masuda, Y., Nakaya, H. Inhibitory effect of bepridil on hKv1.5 channel current: comparison with amiodarone and E-4031. *Eur. J. Pharmacol.* **2001**, *430*, 149.

29. Du, X.-L., Lau, C.-P., Chiu, S.-W., Tse, H.-F., Gerlach, U., Li, G.-R. Effects of chromanol 293B on transient outward and ultra-rapid delayed rectifier potassium currents in human atrial myocytes. *J. Mol. Cell. Cardiol.*, **2003**, *35*, 293–300.

30. Malayev, A. A., Nelson, D. J., Philipson, L. H. Mechanism of clofilium block of the human Kv1.5 delayed rectifier potassium channel. *Mol. Pharmacol.* **1995**, *47*, 198–205.

31. Cabellero, R., Gomez, R., Nunez, L., Moreno, I., Tamargo, J., Delpon, E. Diltiazem inhibits hKv1.5 and Kv4.3 currents at therapeutic concentrations. *Cardiovas. Res.* **2004**, *64*, 457–466.

32. Wang, Z., Fermini, B., Nattel, S. Effects of flecainide, quinidine, and 4-aminopyridine on transient outward and ultrarapid delayed rectifier currents in human atrial myocytes. *J. Pharmacol. Exp. Ther.*, **1995**, *272*, 184.

33. Goegelein, H., Brueggemann, A., Gerlach, U. et al. Inhibition of IK_s channels by HMR1556. *Naunyn Schmiedebergs Arch. Pharmacol.* **2000**, *362*, 480–488.

34. Franqueza, L., Valenzuela, C., Delpón, E., Longobardo, M., Caballero, R., Tamargo, J. Effects of propafenone and 5-hydroxypropafenone on hKv1.5 channels. *Br. J. Pharmacol.*, **1998**, *125*, 969.

35. Feng, J., Wang, Z., Li, G., Nattel, S. Effects of class III anti-arrhythmic drugs on transient outward and ultra-rapid delayed rectifier currents in human atrial myocytes. *J. Pharmacol. Exp. Ther.* **1997**, *281*, 384–392.

36. Gao, Z., Lau, C.-P., Chiu, S.-W., Li, G.-R. Inhibition of ultra-rapid delayed rectifier K+ current by verapamil in human atrial myocytes. *J. Mol. Cell. Cardiol.* **2004**, *36*, 257–263.

37. Delpon, E., Caballero, R., Valenzuela, C. et al. Benzocaine enhances and inhibits the K+ current through a human cardiac cloned channel (Kv1.5). *Cardiovasc. Res.* **1999**, *42*, 510–520.

38. Franqueza, L., Longobardo, M., Vicente, J. et al. Molecular determinants of stereoselective bupivacaine block of hKv1.5 channels. *Circ. Res.* **1997**, *81*, 1053–1064.

39. Cabellero, R., Delpon, E., Valenzuela, C. et al. Direct effects of candesartan and

eprosartan on human cloned potassium channels involved in cardiac repolarization. *Mol. Pharmacol.* **2001**, *296*, 573–583.

40. Moreno, I., Caballero, R. Gonzalez, T., et al. Effects of irbesartan on cloned potassium channels involved in human cardiac repolarization. *J. Pharmacol. Exp. Ther.* **2003**, *304*, 862–873.

41. Caballero, R., Delpon, E., Valenzuela, C. et al. Losartan and its metabolite E3174 modify cardiac delayed rectifier K+ currents. *Circulation* **2000**, *101*, 1199–1205.

42. Iftinca, M., Waldron, G. J., Triggle, C. R., Cole, W. C. State-dependent block of rabbit vascular smooth muscle delayed rectifier and Kv1.5 channels by inhibitors of cytochrome P450-dependent enzymes. *J. Pharmacol. Exp. Ther.* **2001**, *298*, 718.

43. Dumaine, R., Roy, M., Brown, A. Blockade of HERG and Kv1.5 by ketoconazole. *J. Pharmacol. Exp. Ther.* **1998**, *286*, 727–735.

44. Lacerda, A. E., Roy, M. L., Lewis, E. W., Rampe, D. Interactions of the nonsedating antihistamine loratadine with a Kv1.5-type potassium channel cloned from human heart. *Mol. Pharmacol.* **1997**, *52*, 314.

45. Caballero, R., Valanzuela, C., Longobardo, M., Tamargo, J., Delpon, E. Effects of rupatadine, a new dual antagonist of histamine and platelet-activating factor receptors, on human cardiac Kv1.5 channels. *Br. J. Pharmacol.* **1999**, *128*, 1071–1081.

46. Delpon, E., Valenzuela, C., Tamargo, J. Blockade of cardiac potassium channels by antihistamines. *Drug Safety* **1999**, *21(Suppl. 1)*, 11–18.

47. Choe, H., Lee, Y.-K., Lee, Y.-T. et al. Papaverine blocks hKv1.5 channel currents and human atrial ultrarapid delayed rectifier K+ currents. *J. Pharmacol. Exp. Ther.* **2002**, *304*, 706–712.

48. Xiao, G.-S., Zhou, J.-J., Cheung, Y.-F., Li, G.-R., Wong, T.-M. Effects of U50,488H on transient outward and ultra-rapid delayed rectifier K+ currents in young human atrial myocytes. *Eur. J. Pharmacol.* **2003**, *473*, 97–103.

49. Rampe, D., Murawsky, M. K. Blockade of the human cardiac K+ channel Kv1.5 by the antibiotic erythromycin. *Naunyn*

Schmiedebergs Arch. Pharmacol., **1997**, *355*, 743.

50. Pfizer, *EP 0968715*, Loratadine for use as an anti-arrhythmic.

51. Delpon, E, Valenzuela, C., Gay, P., Franqueza, L., Snyders, D. J., Tamargo, J., Block of human cardiac Kv1.5 channels by loratadine: voltage-, time- and use-dependent block at concentrations above therapeutic levels. *Cardiovasc. Res.*, **1997**, *35*, 341.

52. Merck & Co., Methods of treating or preventing cardiac arrhythmia, *WO9818475*, **1998**.

53. Merck & Co., Methods of treating or preventing cardiac arrhythmia, *WO 9818476*, **1998**.

54. Merck & Co., 2,6-Diaryl pyridazinones with immunosuppressant activity, *WO 9625936*, **1996**.

55. Yue, L., Feng, J., Li, G.-R., Nattel, S. Characterization of an ultrarapid delayed rectifier potassium channel involved in canine atrial repolarization. *J. Physiol.*, **1996**, *496*, 647.

56. Fedida, D., Eldstrom, J., Hesketh, J.C. et al. Kv1.5 is an important component of repolarizing K+ current in canine atrial myocytes. *Circ. Res.*, **2003**, *93*, 744–751.

57. Nissan, Chroman derivatives, *WO 9804542*, **1998**.

58. Nissan, Chroman derivatives, *WO 0058300*, **2000**.

59. Nissan, 4-Oxybenzopyran derivatives, *WO 0125224*, **2001**.

60. Nissan, 4-Oxybenzopyran derivatives, *WO 0121609*, **2001**.

61. Nissan, 4-Benzopyran derivatives, *WO 0121610*, **2001**.

62. Matsuda, T., Saito, T., Itokawa, M., Yamashita, T., Tsuruzoe, N., Tanaka, Y., Tanaka, H., Shigenobu, K., Atrial selective prolongation by NIP-142 of refractory period and action potential duration in guinea-pig myocardium. *Jpn. J. Pharmacol.*, **2001**, *85 (Suppl. 1)*, 48P.

63. Nagasawa, H., Fujiki, A., Usui, M., Tani, M., Inoue, H., Yamashita, T., Fujikura, N., *Jpn. Circ. J.*, **1999**, *63 (Suppl. 1)*, 361P.

64. Nagasawa, H., Fujiki, A., Inoue, H., Fujikura, N. *Jpn. Circ. J.*, **2000**, *64 (Suppl. 1)*, 606P.

65. Matsuda, T., Masumiya, H., Tanaka, N., Yamashita, T., Tsuruzoe, N., Tanaka, Y., Tanaka, H., Shigenoba, K. Inhibition by a novel anti-arrhythmic agent, NIP-142, of cloned human cardiac K+ channel Kv1.5 current. *Life Sci.*, **2001**, *68*, 2017.

66. Seki, A., Hagiwara, N., Kasanuki, H. *Jpn. J. Electrocardiol.*, **1999**, *19*, 595P.

67. Seki, A., Hagiwara, N., Kasanuki, H., Effects of NIP-141 on K-currents in human atrial myocytes. *J. Cardiovasc. Pharmacol.*, **2002**, *39*, 29.

68. Nagasawa, H., Fujiki, A., Fujikura, N., Matsuda, T., Yamashita, T., Inoue, H., Effects of a novel class III anti-arrhythmic agent, NIP-142, on canine atrial fibrillation and flutter. *Circ. J.*, **2002**, *66*, 185.

69. Matsuda, T., Ishimaru, S., Saito, T., Aikawa, T., Hashimoto, N., Yamashita, T., Tsuruzoe, N., Tanaka, Y., Tanaka, H., Shigenobu, K. Effect of NIP-142 on carbachol-induced myocardial action potential shortening and human GIRK1/4 channel current *Jpn. J. Pharmacol.*, **2002**, *88 (Suppl. 1)*, 260P.

70. Bristol-Myers Squibb, Potassium channel inhibitors and method, *WO 0012077*, **2000**.

71. Icagen, Eli Lilly, Potassium channel inhibitors, *WO 9804521*, **1998**.

72. Icagen, Potassium channel inhibitors, *WO 9937607*, **1999**.

73. Icagen, Potassium channel inhibitors, *WO 0146155*, **2001**.

74. Icagen, Potassium channel inhibitors, *WO 02060874*, **2002**.

75. Icagen, Potassium channel inhibitors, *WO 0208183*, **2002**.

76. Icagen, Potassium channel inhibitors, *WO 0208191*, **2002**.

77. Schneider, G., Clément-Chomienne, O., Hilfiger, L., Schneider, P., Kirsch, S., Böhm, H.-J., Neidhart, W., Virtual screening for bioactive molecules by evolutionary de novo design. *Angew. Chem., Int. Ed..*, **2000**, *39*, 4130.

78. Merck & Co., Heterocyclic potassium channel inhibitors, *WO 0025786*, **2000**.

79. Merck & Co., Carbocyclic potassium channel inhibitors, *WO 0025770*, **2000**.

80. Merck & Co., Benzamide potassium channel inhibitors, *WO 0025774*, **2000**.

81. Bristol-Myers Squibb, Cycloalkyl inhibitors of potassium channel function, *WO 03063797*, **2003**.

82. Bristol-Myers Squibb, Heterocyclic inhibitors of potassium channel function, *WO 03088908*, **2003**.

83. Aventis Pharma Deutschland GmbH, 2'-Substituted 1,1'-biphenyl-2-carbonamides, method for the production thereof, use thereof as a medicament and pharmaceutical preparations containing said compounds, *WO 0125189*, **2001**.

84. Aventis Pharma Deutschland GmbH, Preparation of ortho-substituted nitrogen containing bisaryl compounds as potassium channel blockers, *WO 02046162*, **2002**.

85. Aventis Pharma Deutschland GmbH, Preparation of [(aminomethyl)phenyl]-furan- and -thiophenecarboxamides as potassium channel blockers, *WO 02048131*, **2002**.

86. Aventis Pharma Deutschland GmbH, Preparation of ortho-substituted and meta-substituted bisaryl compounds as potassium channel blockers, *WO 02044137*, **2002**.

87. Peukert, S., Brendel, J., Pirard, B., Brüggemann, A., Below, P., Kleemann, H.-W., Hemmerle, H., Schmidt, W. Identification, synthesis and activity of novel blockers of the voltage-gated potassium channel Kv1.5. *J. Med. Chem.*, **2003**, *46*, 486–498.

88. Bachmann, A., Gutcher, I., Kopp, K., Brendel, J., Bosch, R. F., Busch, A. E., Goegelein, H. Characterization of a novel Kv1.5 channel blocker in *Xenopus* oocytes, CHO cells, human and rat cardiomyocytes. *Naunyn Schmiedebergs. Arch. Pharmacol.*, **2001**, *364*, 472.

89. Knobloch, K., Brendel, J., Peukert, S., Rosenstein, B., Busch, A. E., Wirth, K. J., Electrophysiological and anti-arrhythmic effects of the novel IK_{ur} blockers, S9947 and S20951, on left vs. right pig atrium *in vivo* in comparison with the IK_r blockers dofetilide, azimilide, d,l-sotalol and ibutilide. *Naunyn Schmiedebergs Arch. Pharmacol.*, **2002**, *366*, 482.

90. Goegelein, H., Brendel, J., Steinmeyer. K., et al. Effects of the atrial anti-arrhythmic drug AVE0118 on cardiac ion channels. *Naunyn Schmiedebergs Arch. Pharmacol.* **2004**, *370*, 183–192.

91. Wettwer, E., Hála, O., Christ, T., et al. Role of IK_{ur} in controlling action potential shape and contractility in the human atrium: influence of chronic atrial fibrillation. *Circulation* **2004**, *110*, 2299–2306.

92. Wirth, K.J., Knobloch, K. Differential effects of dofetilide, amiodarone, and class Ic drugs on left and right atrial refractoriness and left atrial vulnerability in pigs. *Naunyn Schmiedebergs Arch. Pharmacol.* **2001**, *363*, 166–174.

93. Wirth, K.J., Paehler, T., Rosenstein, B., et al. Atrial effects of the novel K⁺-channel-blocker AVE0118 in anesthetized pigs. *Cardiovasc. Res.* **2003**, *60*, 298–306.

94. Blaauw, Y., Gögelein, H., Tieleman, R. G., Van Hunnik. A., Schotten. U., Allessie, M. A. "Early" class III drugs for the treatment of atrial fibrillation - Efficacy and atrial selectivity of AVE0118 in remodeled atria of the goat. *Circulation*, **2004**, *110*, 1717–1724.

95. Aventis Pharma Deutschland GmbH, Use of anthranilic acid amides as potassium channel blockers for the treatment of arrhythmia, *WO 02087568*, **2002**.

96. Aventis Pharma Deutschland GmbH, Use of anthranilic acid amide derivates as potassium channel blockers for the treatment and prevention of potassium channel-mediated diseases including arrhythmia, *WO 02088073*, **2002**.

97. Aventis Pharma Deutschland GmbH, New anthranilic acid amide derivatives are potassium channel blockers – useful for the treatment and prevention of arrhythmia, supraventricular arrhythmia and atrial fibrillation and flutter, *WO 02100825*, **2002**.

98. Peukert, S., Brendel, J., Pirard, B., Struebing, C., Kleemann, H.-W., Boehme, T., Hemmerle, H. Pharmacophore-based search, synthesis, and biological evaluation of anthranilic amide as novel blockers of the Kv1.5 channel. *Bioorg. Med. Chem. Lett.* **2004**, *14*, 2823–2827.

99. Wu, S., Janusz, J. Voltage-gated potassium channel inhibitors as class III antiarrhythmic agents (MEDI-008), *227th ACS National Meeting*, Anaheim, CA, March 28–April 1, **2004**.

100. Icagen, Eli Lilly, Potassium channel inhibitors, *WO 9962891*, **1999**.

101. Bristol-Myers Squibb, Heterocyclic dihydropyrimidines as potassium channel inhibitors, *WO 0140231*, **2001**.

102. Aventis Pharma Deutschland GmbH, Indanyl-substituted benzole carbonamide, method for the production of the same, use thereof as a medicament and pharmaceutical preparations containing the same, *WO 0100573*, **2001**.

103. Merck & Co., Isoquinolinone potassium channel inhibitors, *WO 0224655*, **2002**.

104. Cardiome, Imidazo[1,2-a]pyridine ether compounds useful as ion channel modulators, *WO 00196335*, **2000**.

105. Xention Discovery Ltd., Thionopyrimidine derivatives as potassium channel inhibitors, *WO 04111057*, **2004**.

106. Evans, B.E. et al. Methods for drug discovery: development of potent, selective, orally effective cholescystokinin antagonists. *J. Med. Chem.* **1988**, *31*, 2235–2246.

107. Chang, R.S. et al. Tifluadom, a κ-opiate agonist, acts as a peripheral cholecystokinin receptor antagonist. *Neurosci. Lett.* **1986**, *72*, 211–214.

108. For a recent overview of lead identification in the GPCR target class see: Klabunde, T., Hessler, G. Drug design strategies for targeting G-protein-coupled receptors. *ChemBioChem*, **2002**, *10*, 928–944.

109. Gerlach, U. IK$_s$ channel blockers: potential anti-arrhythmic agents. *Drugs Fut.* **2001**, *26*, 473.

110. Coghlan M. J., Carroll, W. A., Golapakrishnan, M. Recent developments in the biology and medicinal chemistry of potassium channel modulators: update from a decade of progress. *J. Med. Chem.*, **2001**, *44*, 1627.

111. UNITY Chemical Information Software, Version 4.1.1., Tripos, Inc. St. Louis, MO, **2000**.

112. Pirard. B., Brendel, J., Peukert, S. The discovery of Kv1.5 blockers as a case study for the application of virtual screening approaches, *J. Chem. Information Model.*, **2005**, *45*, 477–485.

113. Krapivinsky, G., Gordon, E. A., Wickman, K., Velimirovic, B., Krapivinsky, L., Clapham, D. E. The G-protein-gated atrial K$^+$ channel IK$_{ACh}$ is a heteromultimer of two inwardly rectifying K+-channel proteins. *Nature*, **1995**, *374*, 135.

114. McMorn, S. O., Harrison, S. M., Zang, W.-J., Yu, X.-J., Boyett, M. R. A direct negative inotropic effect of acetylcholine on rat ventricular myocytes. *Am. J. Physiol.* **1993**, *265*, H1393.

115. Dascal, N., Signalling via the G protein-activated K$^+$ channels. *Cell. Signal.* **1997**, *9*, 551.

116. Snyders, D. J., Yeola, S. W. Determinants of anti-arrhythmic drug action: electrostatic and hydrophobic components of block of the human cardiac hKv1.5 channel. *Circ. Res.* **1995**, *77*, 575–583.

117. Yeola, S. W., Rich, T. C., Uebele, V. N., Tamkun, M. M., Snyders, D. J. Molecular analysis of a binding site for quinidine in a human cardiac delayed rectifier K+ channel. Role of S6 in antiarrhythmic drug binding. *Circ. Res.* **1996**, *78*, 1105–1114.

118. Snyders, D. J., Knoth, K. M., Roberds, S. L., Tamkun, M. M. Time-, voltage-, and state-dependent block by quinidine of a cloned human cardiac potassium channel. *Mol. Pharmacol.* **1992**, *41*, 322–330.

119. Valenzuela, C., Delpon, E., Tamkun, M. M., Tamargo, J., Snyders, D. J. Stereoselective block of a human cardiac potassium channel (Kv1.5) by bupivacaine enantiomers. *Biophys. J.* **1995**, *69*, 418–427.

120. Caballero, R., Moreno, I., Gonzalez, T., Valenzuela, C., Tamargo, J., Delpon, E. Putative binding sites for benzocaine on a human cardiac cloned channel (Kv1.5). *Cardiovasc. Res.* **2002**, *56*, 104–117.

121. Luzhkov, V. B., Nilsson, J., Arhem, P., Aqvist, J. Computational modeling of the open-state Kv1.5 ion channel block by bupivacaine, *BBA – Proteins Proteom.* **2003**, *1652*, 35–51.

122. Decher, N., Pirard, B., Bundis, F., Peukert, S., Baringhaus, K.-H., Busch, A. E., Steinmeyer, K., Sanguinetti, M. C. Molecular basis for Kv1.5 channel block, *J. Biol. Chem.*, **2004**, *279*, 394–400.

7.4
Medicinal Chemistry of Ca²⁺-activated K⁺ Channel Modulators

Sean C. Turner and Char-Chang Shieh

7.4.1
Introduction

7.4.1.1 Calcium-activated K⁺ Channels

Ca^{2+}-activated K^+ channels (K_{Ca}) belong to a superfamily of potassium channels that link membrane excitability with intracellular Ca^{2+} signaling. On the basis of Ca^{2+} sensitivity, voltage-dependency and single channel conductance, there are three types of K_{Ca} channels – large-conductance (BK_{Ca}), intermediate-conductance (IK_{Ca}) and small-conductance (SK_{Ca}) Ca^{2+}-activated K^+ channels (Table 7.4.1). They play important roles in the regulation of muscle contraction, cellular proliferation, neuroendocrine secretion, neuronal firing and Ca^{2+} signaling in various neurons. These channels have been considered as viable therapeutic targets for the treatment of various human disorders (Table 7.4.2).

Tab. 7.4.1 Biophysical and molecular properties of Ca^{2+}-activated K^+ channels.

	Large-conductance Ca²⁺-activated K⁺ channel (BK_Ca)	Intermediate-conductance Ca²⁺-activated K⁺ channel (IK_Ca)	Small Ca²⁺-activated K⁺ channel (SK_Ca)
Single-channel conductance	100–300 pS	12–60 pS	1–20 pS
Voltage-dependence	Yes	No	No
Ca²⁺ sensitivity	$K_d = 13000$ nM at −50 mV $K_d = 500$ nM at +70 mV	$K_d = 270$ nM	$K_d = 300$ nM
Molecular components – α subunits	$K_{Ca}1.1$ (*Slo, KCNMA1*)	$K_{Ca}3.1$ (IK_Ca1, *KCNN4*)	$K_{Ca}2.1$ (SK_Ca1, *KCNN1*) $K_{Ca}2.2$ (SK_Ca2, *KCNN2*) $K_{Ca}2.3$ (SK_Ca3, *KCNN3*)
Molecular components – β subunits	β1 (*KCNMB1*) β2 (*KCNMB2*) β3 (*KCNMB3*) β4 (*KCNMB4*)		

Tab. 7.4.2 Therapeutic indications of Ca^{2+}-activated K^+ channels.

	Disorders	BK_{Ca}	IK_{Ca}	SK_{Ca}
CNS	Depression			B
	Epilepsy			O
	Memory and cognition	B		B
	Parkinson's disease			B
	Schizophrenia			O (???)
	Ataxia	O		O
	Anorexia nervosa			O
	Acute ischemic stroke	O		
	Acute subdural haematoma		B	
PNS/CNS	Pain (inflammatory/neuropathic)	O		O
Cardiovascular diseases	Hypertension	O		O
	Restenosis		O	
	Erectile dysfunction	O		
	Angioplasty restenosis		B	
Circulation	Sickle cell disease		B	
Respiratory	Asthma	O		
	Sleep apnea			B
	Sudden infant death			
	Cystic fibrosis		O	
Metabolism	Diabetes		B???	B???
Reproduction	Premature parturition	O		O (??)
Gastrointestine	Diarrhea		B	
Urological disorders	Incontinence	O	O	O
Immune system			B	
Cancer	Prostate		B	
	Breast		B	
	Pancreatic		B	

Note: O, opener; B, blocker; ?, not well defined yet.

7.4.1.2 **Therapeutic Implication**

7.4.1.2.1 **Large-conductance Ca^{2+}-activated K$^+$ Channels (K$_{Ca}$1.1)**

The K$_{Ca}$1.1 channel belongs to a heterogeneous family of proteins that consist of a pore-forming α subunit (*Slo, KCNMA1*) and auxiliary subunits β1–β4 (Fig. 7.4.1). Consistent with the general topology of voltage-gated K$^+$ channels, the BK$_{Ca}$ α subunit [1] is composed of six transmembrane spanning domains with the exception that there is the addition of a seventh membrane-spanning domain (S0), carrying the N-terminus of the protein to the extracellular side of membrane [2]. This extra transmembrane domain in the N-terminus might serve as a coupling site for β subunits. The long C-terminus is composed of four intracellular hydrophobic domains that contain highly conserved negatively charged amino acid residues that might form a highly selective Ca^{2+}-binding site [2]. Four auxiliary β subunits that modulate Ca^{2+} sensitivity, voltage-dependency and pharmacological properties of channels seem to have differential expression patterns, with β1 [3, 4] mainly expressed in smooth muscle cells, β2 in endocrine cells [5], β3 in testis and in intestinal tract, and β4 in the nervous system [6, 7]. The functional K$_{Ca}$1.1 channel can be formed by four α subunits as a tetramer that selectively conducts K$^+$ ions across the membrane with activation dependent on voltage and intracellular Ca^{2+}. The diversity of functional K$_{Ca}$1.1 channels can be revealed by the combination of α and different β subunits in a 1:1 stoichiometry. The splice variants of α subunit gene also contribute the functional diversity of K$_{Ca}$1.1 channels [8–10]. *KCNT2* (*Slo2*) and *KCMA3* (*Slo3*) genes encode channel proteins (K$_{Ca}$4.2 and K$_{Ca}$5.1, respectively) that share a high degree of homology with K$_{Ca}$1.1 channels. However, the channels encoded by *Slo2* and *Slo3* form Na$^+$-activated K$^+$ channels [11] and pH-sensitive K$^+$ channels [12], respectively, with functional and pharmacological properties incompatible with those expected for K$_{Ca}$1.1 channels.

The K$_{Ca}$1.1 is widely expressed in both excitable and non-excitable cells. The function of K$_{Ca}$1.1 channels that integrate membrane potential and intracellular Ca^{2+} via Ca^{2+} release from intracellular organelles, Ca^{2+} influx, Ca^{2+} buffering and diffusion, Ca^{2+} extrusion and uptake, and Ca^{2+} binding to the channel provides a fine tool to regulate intracellular Ca^{2+} and cellular excitability. Diffusible intracellular substances such as cyclic AMP (cAMP), cGMP, G-protein and metabolites of arachidonic acid can also modulate gating kinetics of BK$_{Ca}$ channels by direct phosphorylation of channel protein or by altering intracellular Ca^{2+} concentration [13–15]. The physiological functions of K$_{Ca}$1.1 involve regulation of muscle contraction, neuroendocrine release, neurotransmitter release, and synaptic plasticity.

The K$_{Ca}$1.1 channels are ideal drug targets for the treatment of neurological, vascular and urological diseases (Table 7.4.2). In central nervous system (CNS), the involvement of excessive excitatory amino acid release and pathologically high intracellular Ca^{2+} concentration resulting in the cell death by initiation of a cascade of neurotoxic events has been well established [16, 17]. As previously mentioned, the opening of K$_{Ca}$1.1 can provide a negative feedback to regulate Ca^{2+} entry and to prevent excessive intracellular Ca^{2+}. BMS-204352 and its enan-

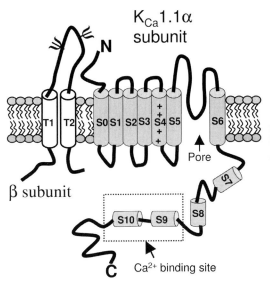

Modification of $K_{Ca}1.1$ properties by β subunits

β-subunits	Functional properties	Tissue expression
KCNMB1 (β1)	↑ Open probability ↑ Ca²⁺ sensitivity ↑ DHS-I sensitivity ↓ IbTx sensitivity ↓ Activation kinetic	Smooth muscle
KCNMB2 (β2)	↑Fast inactivation ↑ Ca²⁺ sensitivity ↓ Activation kinetic ↓ CTX sensitivity ↑ DHS-1 sensitivity	Endocrine tissue, ovary
KCNMB3 (β3)	Slight modification on $K_{Ca}1.1$	Testis
KCNMB4 (β4)	↑ Ca²⁺ sensitivity ↑17 β-estradiol sensitivity ↓ Activation kinetic ↓↓ CTX, IbTx sensitivity	Brain/nerve

Fig. 7.4.1 Schematic representation of the structure of the α and β subunits of BK$_{Ca}$ channels. The α subunit has seven transmembrane domains (S0 to S6) with the N-terminal (N) toward extracellular side and long C-terminal possessing four hydrophobic regions (S7–S10) in the cytosolic site. The linker between S5 and S6 transmembrane domains forms the major component of the conducting pore. The S4 domain containing positively charged amino acid residues serves as a voltage sensor. A stretch of negative amino acid residues (aspartic acid) between S9 and S10 constitutes a potential Ca²⁺ binding site. The S0 domain and exoplasmic N-terminus of the α subunit play an important role for β subunit modulation, which is composed of two transmembrane domains and multiple glycosylation sites (Ψ) in the extracellular loop. Inset: summary of the major tissue expression of four β subunits and their biophysical and pharmacological modulation of the BK$_{Ca}$ channel.

tiomer are two fluoro-oxindole $K_{Ca}1.1$ channel openers with fast brain penetration (Section 7.4.2.2.3) that provide significant reduction in cortical infarct volume when administered two hours after the onset of acute ischemic stroke in rats. Within the therapeutic concentration, these compounds did not affect blood pressure or cerebral blood flow in animals [18]. In addition to the activation of $K_{Ca}1.1$, BMS-204352 opens KCNQ channels [19]. BMS-204352 has also been demonstrated to be well tolerated in human patients after multiple iv dosing up to $2\,mg\,kg^{-1}$. In a Phase 3 study, however, BMS-204352 failed to show superior efficacy in acute stroke patients compared to placebo [20]. Nevertheless, the significant reduction in cortical infarct volume by BMS-204352 and its analogs demonstrate that $K_{Ca}1.1$ is a viable target for neuroprotection.

In many neurons, action potentials and associated Ca^{2+} influx are followed by repolarization and an afterhyperpolarization (AHP) that has three kinetic components – fast AHP (fAHP), medium AHP (mAHP) and slow AHP (sAHP) [21]. The fAHP is mediated though the opening of $K_{Ca}1.1$ that regulates the spike duration during repetitive firing in hippocampal pyramidal neurons [22]. Inhibition of $K_{Ca}1.1$ consequently contributes to spike broadening during the repetitive firing of hippocampal neurons and is likely to enhance neurotransmitter release at axon terminals. Thus, the brain penetrant $K_{Ca}1.1$ blockers that can modulate fAHP, leading to enhanced neurotransmitter release and influence synaptic plasticity, might have therapeutic indication to enhance cognition in conditions such as Alzheimer's disease.

The $K_{Ca}1.1$ knockout animal model uncovered important roles of $K_{Ca}1.1$ in cerebellar function, motor control and its involvement in cerebellar ataxia [23]. In control animals, cerebellar Purkinje cells fire spontaneously under basal conditions, thereby tonically inhibiting the deep cerebellar nuclei (DCN). However, the cerebellar Purkinje cells from $K_{Ca}1.1$ knockout mice significantly reduced spontaneous discharges, leading to a disinhibition of DCN through a depolarization block mechanism [23]. Therefore, the mice lacking $K_{Ca}1.1$ showed abnormal conditioned eye-blink reflex, abnormal locomotion and pronounced deficiency in motor coordination, which are likely consequences of cerebellar learning deficiency [23]. These results suggest $K_{Ca}1.1$ is involved in normal and pathological cerebellar function.

In peripheral nociception, the $K_{Ca}1.1$ shortens action potential duration, increases the speed of repolarization and contributes to fAHP in small diameter dorsal root ganglionic neurons (DRG) [24]. Activation of $K_{Ca}1.1$ with NS1619 in small diameter DRG neurons modulates action potential firing and suppresses 4-aminopyridine evoked hyperexcitability [25], suggesting the $K_{Ca}1.1$ channel opener may be utilized for neuropathic and/or inflammatory pain. Furthermore, the expression and gating of the $K_{Ca}1.1$ channel are responsible for electrical tuning of cochlear hair cells [26, 27]. Thus, modulation of $K_{Ca}1.1$ in cochlear hair cells might have therapeutic implication on hearing.

The role of $K_{Ca}1.1$ involved in the regulation of smooth muscle contractility has been well characterized in airway [28, 29], vascular [30], uterine [31], urinary [32] and corporal smooth muscle cells [33]. By coupling to Ca^{2+}-influx through voltage-dependent Ca^{2+} channels and Ca^{2+} sparks [34, 35], activation of $K_{Ca}1.1$ facilitates a negative feedback mechanism to oppose excitation and contraction of smooth muscle cells. Thus, $K_{Ca}1.1$ channels regulate resting membrane potential and initiate action potential repolarization, which consequently limits contraction frequency and amplitude. Indeed, deletion of β1 subunits leads to a decrease in Ca^{2+} sensitivity and a reduction in functional coupling of calcium sparks to $K_{Ca}1.1$ activation, which leads to increases in arterial tone and blood pressure [30, 36]. In β1 gene (*KCNMB1*) knock-out mouse, the urinary smooth muscle strip had elevated phasic contraction amplitude and decreased frequency when compared to control urinary smooth muscle strips [37]. Similarly, deletion of pore-forming α subunits in mouse enhanced spontaneous and electrically field stimulated bladder smooth muscle contraction as well as increases in urination frequency [32].

The therapeutic implication of targeting $K_{Ca}1.1$ openers for the treatment of over-active smooth muscle contractility has been further validated by reduction in bladder hyperactivity in rat after injection of $K_{Ca}1.1$ α subunit gene [38]. Similarly, intracoporal injection of *hSlo* cDNA in rats has been demonstrated to restore normal cavernous nerve-stimulated intracavernous pressure of age- and diabetes-related erectile dysfunction in rats [33, 39].

7.4.1.2.2 Intermediate-conductance Ca²⁺-activated K⁺ Channels (K$_{Ca}$3.1)

The intermediate-conductance Ca²⁺-activated K⁺ channel was first described as the Gardos channel in erythrocytes and has been shown to be a major pathway for sickle cell dehydration. Recent cloning revealed *KCNN4* gene encoding a Ca²⁺-activated K⁺ channel with intermediate-conductance that was expressed in placenta, prostate, colon, spleen, thymus, T lymphocytes, peripheral blood leukocytes [40–43] and smooth muscle cells [44, 45]. In addition, the interaction between $K_{Ca}4.1$ and $K_{Ca}1.1$ subunits forms a functional channel that might represent some types of $K_{Ca}3.1$ channels in the nervous system [46].

Unlike $K_{Ca}1.1$, the half maximal activation of $K_{Ca}3.1$ occurs at a significantly lower intracellular Ca²⁺ concentration than that required for $K_{Ca}1.1$ and activation of channel is not voltage-dependent (Table 7.4.1). The functional role of $K_{Ca}3.1$ by linking with intracellular Ca²⁺ is to set the membrane potential in the hyperpolarization value [47, 48] and thereby to establish large electrical gradients for the passive transport of ions such as Cl⁻ efflux that drives water and Na⁺ secretion in epithelial cells or Ca²⁺ influx that regulates cell proliferation in T-lymphocytes and smooth muscle cells. Indeed, in $K_{Ca}3.1$ knock-out mice, their parotid acinar cells expressed no $K_{Ca}3.1$, and their red blood cells lost K⁺ permeability. The volume regulation of T lymphocytes and erythrocytes was severely impaired in $K_{Ca}3.1$ knock-out mice but was normal in parotid acinar cells, although the animals were of normal appearance and fertility. Despite the loss of $K_{Ca}3.1$, activated fluid secretion from parotid glands was normal [49]. Thus, the $K_{Ca}3.1$ knock-out study confirmed, in part, the role of $K_{Ca}3.1$ in volume regulation and secretion. Activation of $K_{Ca}3.1$, in combination with SK_{Ca}, in the endothelial cells plays an important role in endothelium-derived hyperpolarizing factor (EDHF)-mediated vasodilation in carotid and mesenteric arteries [50, 51].

The $K_{Ca}3.1$ has been pursued as a therapeutic target for the treatment of sickle cell anemia, diarrhea, and rheumatoid arthritis [52]. ICA-17034, a recently identified selective and potent $K_{Ca}3.1$ blocker (Section 7.4.2.3.1), inhibits erythrocyte dehydration *in vitro* and significantly increases hematocrit and reduces mean corpuscular hemoglobin concentration in a transgenic mouse model of sickle cell disease [53]. It is currently in Phase 3 clinical trials for patients with sickle cell disease. In the immune system, $K_{Ca}3.1$ plays an important role in Ca²⁺ signaling, activation, adhesion and migration and is upregulated in T-lyphocytes following antigenic or mitogenic stimulation [54]. Inhibitors of $K_{Ca}3.1$ have long been regarded as attractive targets for immunotherapy. Specific and potent $K_{Ca}3.1$ blockers such as TRAM-34 and related compounds have been designed as immuno-suppressants [55]. IK_{Ca} channels are also involved in cell proliferation via Ca²⁺-activated mitogenic Ras/MAPK signaling pathway [56]. Inhibition of $K_{Ca}3.1$ suppresses cell proliferation of prostate cancer cells [56, 57], breast cancer cells [58] and human pancreatic cancer cells [59]. In proliferative vascular smooth muscle cells, $K_{Ca}3.1$ represents predominant Ca^{2+}-activated K^+ channels, which are greatly upregulated in proliferating cells [45]. In a rat model, upregulation of $K_{Ca}3.1$ was found two weeks after balloon catheter injury. Application of TRAM-34 suppresses epidermal growth factor-induced smooth muscle cell proliferation *in vitro* and significantly reduced intimal hyperplasia *in vivo* [60]. Thus, inhibition of $K_{Ca}3.1$ may represent a therapeutic avenue for the treatment of restenosis after angioplasty. In contrast, channel openers of $K_{Ca}3.1$ may be therapeutically beneficial in cystic fibrosis [61, 62] and urinary incontinence [44].

7.4.1.2.3 Small-conductance Ca^{2+}-activated K^+ Channels (SK_{Ca})

The small conductance Ca^{2+}-activated K^+ channel has a single channel conductance $< 20 \, pS$ and its activation by rise in intracellular Ca^{2+} with half maximal activation in the $\sim 400 \, nM$ range is voltage insensitive (Table 7.4.1). Three mammalian genes *KCNN1*, *KCNN2* and *KCNN3* encoding $K_{Ca}2.1$ ($SK_{Ca}1$), $K_{Ca}2.2$ ($SK_{Ca}2$) and $K_{Ca}2.3$ ($SK_{Ca}3$) channels were cloned that demonstrate a high degree of structural homology ($\sim 60\%$) and Ca^{2+} sensitivity [63]. The functional SK_{Ca} channels are heteromeric complexes composed of four pore-forming α subunits with six transmembrane spanning regions and Ca^{2+} binding protein calmodulin (Fig. 7.4.2). The Ca^{2+}-free calmodulin constitutively binds to the SK_{Ca} channel through the calmodulin-binding domain that is composed of a 91 amino acid spanning region in the C-terminus proximal to the pore of the α subunit [64]. Unlike voltage-gated K^+ channels, the opening of SK_{Ca} channels is mediated through a chemo-mechanical gating mechanism resulting from the binding of Ca^{2+} to the EF hands in the N-terminus of calmodulin, leading to a conformational change in the open state [65]. While $K_{Ca}2.1$, $K_{Ca}2.2$ and $K_{Ca}2.3$ share a high degree of primary sequence homology, they can be distinguished pharmacologically (Section 7.4.2.5). For example, the $K_{Ca}2.2$ is more sensitive to blockade by apamin and scyllatoxin than $K_{Ca}2.1$ and $K_{Ca}2.3$ [66], while dequalinium blocks $K_{Ca}2.3$ with potency 10-fold greater than those for $K_{Ca}2.1$ and $K_{Ca}2.2$ [67].

$K_{Ca}2.1$–$K_{Ca}2.3$ are not only widely expressed throughout the central nervous system but also in peripheral tissues [63, 68]. In brain, the expression of $K_{Ca}2.1$, $K_{Ca}2.2$ and $K_{Ca}2.3$ subunits partially overlaps with $K_{Ca}2.1$ and $K_{Ca}2.2$, showing extensive colocalization in the hippocampus and cortex, whereas $K_{Ca}2.3$ displays a complementary distribution in hypothalamus, thalamus and midbrain [69, 70]. In brain, the expression of $K_{Ca}2.1$, $K_{Ca}2.2$ and $K_{Ca}2.3$ subunits reveals distinct patterns with $K_{Ca}2.1$ and $K_{Ca}2.2$ colocalization in the hippocampus and cortex, whereas $K_{Ca}2.3$ displays a complementary distribution in hypothalamus, thalamus and midbrain [69, 70] Inhibition of SK_{Ca} channels with apamin increases the excitability of hippocampal neurons and facilitates the induction of synaptic plasticity [73, 74]. In mice or rats, treatment with apamin accelerates hippocampal-dependent spatial and non-spatial memory encoding and facilitates learning in several behavioral models [75, 74]. Indeed, age-dependent increases in $K_{Ca}2.3$

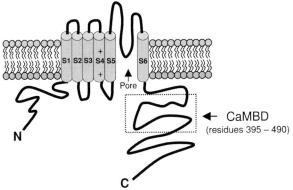

Fig. 7.4.2 Schematic representation of the structure of the α subunit of SK$_{Ca}$ channel, showing six transmembrane domains, similar to the topology of voltage-gated K$^+$ channels, a pore region, and a calmodulin-binding domain (CaMBD) in C-terminus (C). The S4 domain contains positively charged residues; however, this does not serve as a voltage sensor and does not gate channel opening. Instead, the charged residues have been proposed to form salt bridges with other residues of the channel for structural stability [63]. A stretch of amino acid residues in the C-terminus located adjacent to the S6 transmembrane domain (residues 395–490, rat K$_{Ca}$2.2) forms CaMBD that consists of two long α-helices, α1 and α2, connected by a loop. In the crystal structure, two CaMBDs align in an antiparallel fashion without being in direct contact. Two calmodulins are bound at the ends of this CaMBD dimer. The α1 is responsible for the constitutive interaction with the two EF hands in C lobe of calmodulin in a Ca^{2+}-independent manner. The binding of Ca^{2+} to the two EF hands in N lobe of calmodulins initiates the contact of calmodulin N lobe with α2 of CaMBD and dimerization of calmodulin-CaMBD occurs. This interaction results in a rotatory movement of the SK$_{Ca}$ channel gate, leading to channel opening in a chemo-mechanical gating mechanism [65].

expression contribute to reduced long-term potentiation and impaired trace fear conditioning, a hippocampus-dependent learning task in aged rats [76]. Infusion of K$_{Ca}$2.3 antisense into brain showed down-regulation of K$_{Ca}$2.3 transcript and protein production and reversed aged-related memory deficit [76]. Thus, inhibition of SK$_{Ca}$ channels can be a therapeutic approach to enhance cognition and memory in disorders such as Alzheimer's disease, as well as to prevent memory deficits associated with aging (Table 7.4.2).

K$_{Ca}$2.3 functions as a pacemaker that dynamically controls the frequency of spontaneous firing of dopaminergic neurons in hippocampus, the limbic system, midbrain regions and medulla oblongata involved in movement, cognition and reward behaviors, and processing of respiratory signals [77, 69]. Alteration in K$_{Ca}$2.3 function has been associated with schizophrenia, anorexia nervosa and ataxia [78, 79], although the evidence is not conclusive. Opening SK$_{Ca}$ channels with 1-ethyl-2-benzimidazolinone (1-EBIO, Section 7.4.2.4.1) increases both medium and slow AHP, strongly reducing electrical activity and the hyperexcitability induced by low Mg^{2+} in cultured cortical neurons [80]. Thus, SK$_{Ca}$ channel openers may have therapeutic implications for the management of CNS hyperexcitability disorders such as epilepsy and schizophrenia. Furthermore, overexpression of $K_{Ca}2.2$ prevents neuron death resulting from necrotic and apoptotic insults [81]. Thus, SK_{Ca} channel openers might serve as a therapeutic approach to prevent neuronal death due to acute ischemia, although the associated impairment of memory and learning can be a concern.

Transgenic mice with overexpression of $K_{Ca}2.3$ showed abnormal respiratory responses to hypoxia, in that they were unable to sustain a hyperpneic response [77], suggesting $K_{Ca}2.3$ may be a potential target for the treatment of sleep apnea or sudden infant death syndrome.

SK_{Ca} channels are also expressed in skeletal and many smooth muscles, including vascular, uterus and bladder. Injection of apamin to affected skeletal muscle suppresses myotonia [82]. Overexpression of $K_{Ca}2.3$ in mice results in compromised parturition [77]. Suppression of $K_{Ca}2.3$ led to a marked increase in non-voiding contractions although no changes in filling pressure, threshold pressure or bladder capacity were observed [83]. Suppression of endothelial $K_{Ca}2.3$ enhances pressure- and phenylephrine-induced muscle constrictions and leads to increases in blood pressure [84]. Thus, the SK_{Ca} channel may be a potential therapeutic target for the treatment of muscle disorders such as myotonic dystrophy, premature parturition, urinary incontinence and hypertension.

7.4.2
Medicinal Chemistry

As illustrated above, the great potential of Ca^{2+}-activated K^+ channels as therapeutic targets has resulted in a widespread effort by both academic and pharmaceutical research groups to identify small molecule channel modulators as pharmacological tools and as drug development candidates.

7.4.2.1 K_{Ca} 1.1 Channel Blockers
The availability of selective $K_{Ca}1.1$ blockers has been of great utility in the study of $K_{Ca}1.1$-related pathophysiologies. There is also potential for small molecule $K_{Ca}1.1$ blocking agents to be used in the treatment of neurological disorders such as depression and memory impairment [85].

7.4.2.1.1 Peptide Toxins
Charybdotoxin (ChTx), a 37-amino-acid peptide that is obtained from the venom of the scorpion *Leiuris quinquestriatus*, blocks $K_{Ca}1.1$ (IC_{50} 100 nM) [86] and has been widely used to study the role of $K_{Ca}1.1$. Due to its compact, well-organized, stabilized three-dimensional structure motif, ChTx has also been used as a molecular caliper to study the topology of both K^+ channels and drug–toxin binding sites [87–89]. Its effectiveness as a research tool, however, is limited by its action as a blocker of $K_{Ca}3.1$ and some delayed-rectifier K^+ (Kv) channels. Recently, using a structure-based design strategy, more selective ChTx analogs were reported that

showed ten-fold selectivity for $K_{Ca}3.1$ and $K_{Ca}1.1$ versus other voltage-gated K^+ channels [90]. Thus, the peptide toxins can serve as scaffolds to design selective and potent BK_{Ca} channel modulators.

Another example of a highly selective $K_{Ca}1.1$ blocker is the related 37-amino-acid peptide iberiotoxin (IbTx, IC_{50} 10 nM) [91], also isolated from scorpion venom (*Buthus tamulus*). This selectivity has led to the adoption of radiolabeled ^{125}I-IbTx as the standard reagent for determination of $K_{Ca}1.1$ opening activity in new synthetic series. Importantly, the different auxiliary β-subunits exert a strong influence on the sensitivity of the channel towards peptide blockers [e.g., the KCNMB4 (β4) shows greatly reduced sensitivity to IbTx, see Fig. 7.4.1].

7.4.2.1.2 Alkaloids

The indole alkaloids paxilline and penitrem A exhibit moderate $K_{Ca}1.1$-blocking activity [92], as does the quinoline alkaloid tetrandrine. In patch-clamp experiments, the $K_{Ca}1.1$-blocking activity of paxilline (K_i 30 nM) was greatly reduced on conversion into 13-desoxypaxilline (K_i 730 nM), indicating the key importance of the C-13 OH group [93]. Quinine and quinidine are also $K_{Ca}1.1$ blockers but also inhibit $K_{Ca}3.1$ and other K^+ channel subtypes.

7.4.2.1.3 Synthetic Blockers

There are no reports of selective synthetic $K_{Ca}1.1$ channel blockers to date. Tetra-ethylammonium chloride (TEA) has been widely employed as a $K_{Ca}1.1$-blocking research tool but exhibits poor selectivity against other K^+ channels. In patch-clamp experiments using both clonal GH3 cells and guinea-pig bladder myocytes, the Kv1.3 channel blocker WIN 17317-3 [1-benzyl-7-chloro-4-(n-pentylimino)-1,4-dihydroquinoline hydrochloride] was determined to be a reversible $K_{Ca}1.1$ blocker (GH3 K_i = 1.28 μM). The increased contractility of guinea-pig bladder observed following administration of WIN 17317-3 is suggested to result from an inhibition of $K_{Ca}1.1$ mechanism [94]. Clotrimazole (**15**, Fig. 7.4.4 below) (an imidazole antimycotic P-450 inhibitor used in clinical trials for the treatment of sickle cell anemia) inhibits BK currents in murine erythroleukemia cells and GH_3 lactotrophs at anterior pituitary GH3 cells. BK suppressing activity has also been reported for the related compounds ketoconazole and econazole [95]. These compounds are also inhibitors of other K^+ channels and of L-type Ca^{2+} channels [96].

7.4.2.2 $K_{Ca}1.1$ Channel Openers

Of the different categories of Ca^{2+}-activated K^+ channel modulators, it is the $K_{Ca}1.1$ openers that hold the most potential for therapeutic action on a wide range of disease states (Table 7.4.2). Although small molecule $K_{Ca}1.1$ openers have been reported from a growing number of chemical series, most have several structural features in common. These include two aromatic rings that are linked via a spacer group that is a heterocycle, a urea or a heterocycle fused to one of the aromatic rings. In addition, a key feature of many series is the 5-halo-2-hydroxy or 5-halo-2-methoxy substitution pattern present on one of the aromatic rings.

7.4.2.2.1 **Benzimidazolones**

The first $K_{Ca}1.1$ openers to have sufficient potency and selectivity to allow exploration of the therapeutic clinical potential were a series of substituted benzimidazolones, of which NS-4 (**1**) and NS-1619 (**2**) are key examples (Fig. 7.4.3), disclosed by Neurosearch. In a structure–activity relationship (SAR) study of the benzimidazolone series presented by Meanwell et al. [97], the importance of an electron-withdrawing substituent (e.g., the trifluoromethyl group in NS-4) on the benzimidazole core was apparent. In these structures, the carbonyl oxygen, together with either the amide hydrogen or the phenol hydroxyl group, may be viewed as the surrogate of a carboxylic acid, although the link between the heterocycle and the substituted phenol seems to be somewhat flexible. A structural search, based on a pharmacophore model of NS-4, led to the discovery of cyclic constrained flavonoids that are more efficacious in activating $K_{Ca}1.1$ than the parent compound, NS-4 [98].

NS-4 and NS-1619 have been shown to open $K_{Ca}1.1$ in nerve cells, bronchial cells, vascular smooth muscle cells and beta-cells. The limited solubility of these structures is a constraint [99]. NS-1619 activated $K_{Ca}1.1$ with EC_{50} 32 µM (single channel recording from rat motor neurons), but showed inhibition of delayed rectifiers and calcium currents in a similar concentration range [100].

Data relating to the *in vivo* effects of benzimidazolones are scarce. Administration of NS-4 increased the threshold for pentylenetetrazole-induced seizures in mice and reduced the degeneration of neurons during global or focal brain ischemia in gerbils and rats, respectively. In a study of nociception, intraplantar administration of NS-1619 did not significantly inhibit mechanical hyperalgesia induced by prostaglandin E_2 [101].

The insertion of a methylene spacer group between the aryl moiety and the benzimidazolone ring led to benzyl-benzimidazolones $K_{Ca}1.1$ openers such as BMS-189269 [102]. A recent patent has claimed benzimidazolones with the N-aryl group replaced by an N-alkyl (e.g., ethyl) as $K_{Ca}1.1$ modulators for pain relief [103]. Baragatti and coworkers [104] demonstrated that $K_{Ca}1.1$ opening activity was retained when the benzimidazolone core was replaced by a benzotriazole.

7.4.2.2.2 **Diphenylureas**

A ring-opening structural modification of the benzimidazolone NS-4 led to the identification of the potent $K_{Ca}1.1$ opener NS-1608 (**3**) – a phenyl urea analog reported by researchers at NeuroSearch, where it has been investigated as a potential pain-relieving agent [103].

In vitro $K_{Ca}1.1$ activity for NS-1608 has been shown for SMC from porcine coronary artery and guinea pig bladder [100]. Interestingly, the presence of the β subunit is not required for activity of NS-1608. In HEK 293 cells, NS-1608 opened $K_{Ca}1.1$ with an EC_{50} of 2.1 µM, and a maximum voltage shift of −74 mV. NS-1608 slows channel deactivation with the time constants for deactivation of tail currents being more than tripled.

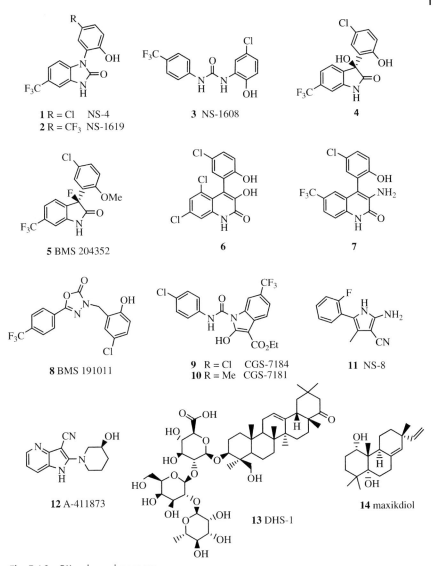

Fig. 7.4.3 BK$_{Ca}$ channel openers.

7.4.2.2.3 Aryloxindoles

Workers at Bristol Myers Squibb identified a series of aryloxindole BK$_{Ca}$ openers [105]. Structurally these bear a close resemblance to the benzimidazolones with the substitution of carbon for nitrogen to give an oxindole ring. Initial 3-hydroxy derivatives (e.g., **4**) were modified to give 3-fluoro analogs such as BMS-204352 (**5**, also known as Flindokalner or MaxiPost) with improved metabolic stability.

Replacement of the 6-CF$_3$ group with 6-iodo, 6-phenyl, or a fused phenyl was well tolerated.

In vitro, the (–)-enantiomer of **5** increased K$_{Ca}$1.1 currents in the *mSlo* channel expressed in *Xenopus* oocytes by 41% over control at 20 μM – greater than its (+)-enantiomer (24%) or NS-4 (32%). *In vivo*, **5** was reported to be neuroprotective and reduced infarct size in two preclinical rat stroke models (spontaneously hypertensive and Wistar rats) [106]. Efficacy was observed over a dose range of 10 μg kg^{-1} to 3 mg kg^{-1} (with a brain-to-plasma ratio of 7:11) [107]. Interestingly, at doses greater than 1 mg kg^{-1}, BMS-204352 displayed an inverted-U dose–response relationship. In healthy humans, multiple i.v. doses of BMS-204352 (0.2 mg kg^{-1}) were safe and well-tolerated but Phase 3 clinical trials for stroke were discontinued due to lack of efficacy [20].

7.4.2.2.4 Quinolinones

The replacement of the benzimidazolone ring of NS-4 with a quinoline led to the 3-hydroxy and 3-amino-4-aryl-quinolin-2-ones, which are potent BK openers active *in vitro* and *in vivo* [108, 109]. Within this series, the trend continues wherein the 5-halo-2-hydroxy substitution pattern is present on one of the aromatic rings.

Two related compounds (**6** and **7**) activated K$_{Ca}$1.1 currents by >50% over control at 20 μM in cloned *Slo* channels expressed in *Xenopus* oocytes. *In vivo*, **7** reduced infarct volume by 14% when dosed (0.001 mg kg^{-1}, i.v.) in a focal stroke model in rats [109]. Several compounds were found to be effective relaxants of precontracted rabbit corpus cavernosal strips *in vitro* and active in an *in vivo* rat model of erectile dysfunction (ED).

The analogous 3-thio structures have also been evaluated as potential ED therapeutics. In the rabbit corpus cavernosum strips, the compounds showed 22–59% inhibition at 10 μM (comparable to NS-4 at 31%). The low aqueous solubility of this series was improved to >1 mg mL^{-1} by introduction of a dialkylamino group [110].

7.4.2.2.5 Oxadiazolones

A series of 1,3,4-oxadiazolone compounds was also reported by Bristol-Myers Squibb [111]. A member of this series, BMS 191011 (**8**), was found to potentiate K$_{Ca}$1.1 currents by 26% over control values at a concentration of 1 mM. When administered i.v. in a focal stroke model using Wistar rats, **8** reduced infarct volume by 18% at a dose of 10 mg kg^{-1}. As a prelude to clinical evaluation, several prodrugs of BMS 191011 were investigated, with the deoxycarnitine ester giving the best balance of physiochemical properties and efficacy in a rat neocortical infarct model. Several structural analogs of this compound with variations in the attached heterocycle activate K$_{Ca}$1.1 currents *in vitro*. A related structure where the oxadiazole core was replaced by a triazole derivative has also been reported as a K$_{Ca}$1.1 opener [112].

7.4.2.2.6 Indole Ureas

A series of indole-3-carboxylic acid esters have been disclosed as $K_{Ca}1.1$ openers [113]. CGS-7184 (**9**) and CGS-7181 (**10**) are representative members of this class and bear some structural resemblance to the other BK openers discussed (a urea attached to substituted aromatic rings). Compounds from this series have been reported to potently open BK channels in both coronary artery and urinary bladder SMC, with a notable delay (6–8 min) between drug administration and maximal activation of BK_{Ca}. CGS-7184 and CGS-7181 appear to be more potent than the benzimidazolone analogs as the reported current enhancement occurs at an effective concentration around 20–50-fold lower than NS-1619 [113]. *In vivo* data for CGS-7184 have not been reported and development of the compound for the treatment of incontinence was discontinued following the merger of Ciba-Geigy with Sandoz to form Novartis.

7.4.2.2.7 Arylpyrroles

The discovery of the structurally novel arylpyrrole NS-8 (**11**), from Nippon Shinyaku, was of particular interest as it has little structural similarity to other previously reported $K_{Ca}1.1$ openers [114, 115]. NS-8 activates $K_{Ca}1.1$ currents in guinea pig bladder cells, relaxes precontracted guinea pig bladder strips (IC_{50} 0.54 µM) and suppresses the excitability of dorsal root ganglion neurons. It was efficacious *in vivo* in two different models of bladder overactivity. In rat cystometry, a dose-dependent 60–80% increase in bladder capacity was observed at 3 and 10 mg kg⁻¹ i.d. NS-8 also inhibited isovolumetric bladder contractions in rats when administered either intravenously at 0.03 to 1 mg kg⁻¹ or intravesically at concentrations from 30 to 300 mg mL⁻¹. No effect was observed when NS-8 was given by an intracerebroventricular route, suggesting that it exerts its bladder inhibitory effects via the afferent signaling pathway or at the bladder smooth muscle itself, rather than by a purely central mechanism. NS-8 has been advanced into Phase 2 clinical trials for the potential treatment of pollakiuria and incontinence.

7.4.2.2.8 Aminoazaindoles

Recently, a structurally novel class series of 2-amino-4-azaindoles has been reported by Turner and coworkers at Abbott as potent $K_{Ca}1.1$ openers [116]. The compounds originated from a program to design conformationally-restricted analogs of NS-8. SAR studies revealed that the 2-amino group was a key structural feature for potency, with the introduction of cyclic amines giving the most potent compounds. In an $^{86}Rb^+$ efflux assay, A-411873 (**12**) (a 3-hydroxypiperidyl analog) was shown to open $K_{Ca}1.1$ α channels in HEK293 cells with an EC_{50} of 5.07 µM. Whole-cell patch-clamp measurements showed a 108% increase in outward current in the presence of compound (10 mM) as % over control. A-411873 also suppressed electrically stimulated contractions of pig bladder smooth muscle strips with an EC_{50} of 9.26 µM (seven-fold greater than that of NS-8).

7.4.2.2.9 **Triterpenes**

Glycosylated triterpenes such as dehydrosoyasaponin-1 (**13**, DHS-1), maxikdiol (**14**) and L-735,334 were shown to be $K_{Ca}1.1$ channel openers in a $[^{125}I]ChTx$ binding assay. These compounds show BK activity only when applied intracellularly due to poor cell membrane permeability. DHS-1 shows $K_{Ca}1.1$ opening activity only when the α and β subunits are co-expressed [117] and acts as an allosteric inhibitor of charybdotoxin binding in smooth muscle preparations. Merck was investigating maxikdiol (**14**), an analog of the antifungal agent CAF-603 (8-daucene-3,4-diol), for the potential treatment of asthma [118]. A related terpenoid, pimaric acid, was reported to have higher potency than **14** (significant effect in patch-clamp assay when applied at 1 mM compared to 3 mM for maxikdiol) and better cell permeability [119, 120].

7.4.2.3 $K_{Ca}3.1$ **Channel Blockers**

$K_{Ca}3.1$ blockers have been reported in the last few years, with most structures falling into two distinct structural classes: triphenylmethyls and dihydropyridine-like compounds. TRAM-34 and TRAM-39 show selectivity for $K_{Ca}3.1$ versus SK_{Ca} channels.

7.4.2.3.1 **Triphenylmethyls**

Clotrimazole (**15**) and analogs such as TRAM-34 (**16**), TRAM-39 (**17**) and ICA-17043 (**18**) are highly potent IK_{Ca} inhibitors (Fig. 7.4.4). The $K_{Ca}3.1$ IC_{50} values of clotrimazole are in the 25–387 nM range, while TRAM-34 is more potent (IC_{50} 20 nM) [121]. In changing the imidazole in clotrimazole to pyrazole in TRAM-34 there is significant improvement in metabolic stability. In the presence of 10 μM clotrimazole or TRAM-34, proliferation of the BxPC-3 as well as the MiaPaCa-2 human pancreatic cancer cells was completely suppressed [59]. A more recent analog, ICA-17043, has an IC_{50} of 11 nM in human erythrocytes and inhibits red blood cell dehydration with a similar potency. $hK_{Ca}3.1$ suppressing activity has also been reported for the related compounds miconazole (IC_{50} 785 nM), econazole (IC_{50} 2.4–12 μM) [42, 54] and ketoconazole (IC_{50} 30 μM). *In vivo*, TRAM-34 and related structures do not inhibit CYP3A liver enzymes (which is observed with clotrimazole) and these compounds appear to have acceptable safety indices. ICA-17043 (10 mg kg^{-1} bid for 21 days) inhibited Gardos channel activity *in vivo* and decreased erythrocyte density in a transgenic SAD mouse model of sickle cell anemia (SCA) [106]. ICA-17043 has entered Phase 3 clinical trials for the potential treatment of SCA.

7.4.2.3.2 **Homopiperidines**

A novel class of $K_{Ca}3.1$ blockers are the homopiperidines reported by researchers at University College London [122]. The lead compound UCL-1608 (**20**) combines some features from the weak $K_{Ca}3.1$ blocker cetiedil (**19**) (IC_{50} 50 μM) with those of the triphenylmethyls discussed above. Further SAR efforts led to UCL-1606, with potent $K_{Ca}3.1$ blocking activity in rabbit blood cells (IC_{50} 1.5 μM). UCL-

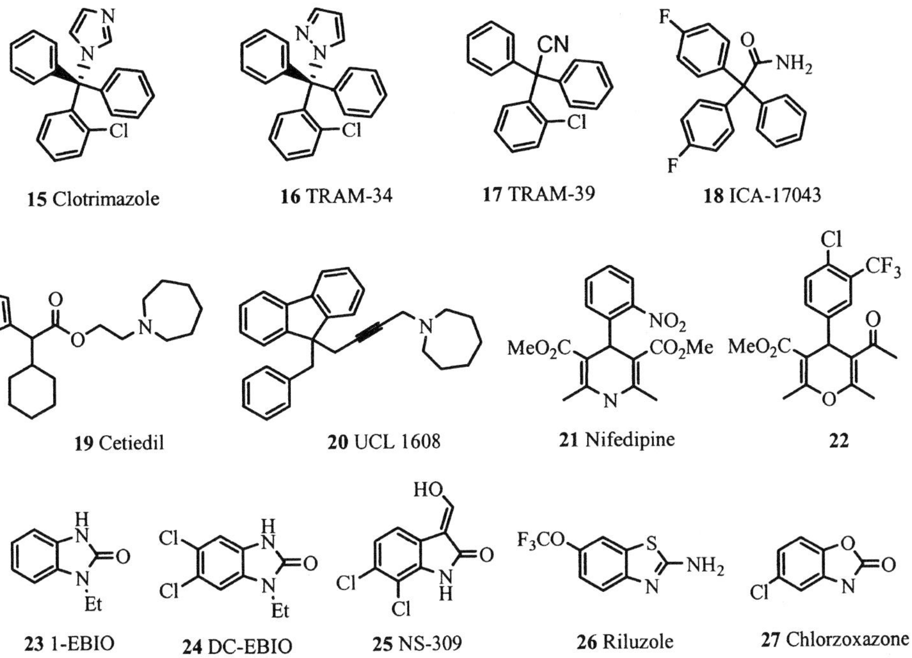

15 Clotrimazole   **16** TRAM-34   **17** TRAM-39   **18** ICA-17043

19 Cetiedil   **20** UCL 1608   **21** Nifedipine   **22**

23 1-EBIO   **24** DC-EBIO   **25** NS-309   **26** Riluzole   **27** Chlorzoxazone

Fig. 7.4.4 IK$_{Ca}$ channel modulators.

1269 (IC$_{50}$ 155 μM) and UCL-1274 (IC$_{50}$ 8.2 μM) are less active members of this series. An earlier lead, UCL-1495, has an IC$_{50}$ of 1.2 μM for blocking the K$_{Ca}$3.1 permeability of rabbit erythrocytes.

7.4.2.3.3 Dihydropyridines

The dihydropyridines nifedipine (**21**) and nitrendipine, which are potent L-type calcium channel blockers used in the treatment of hypertension, also block K$_{Ca}$3.1 with IC$_{50}$s of 1 μM and 0.027–0.9 μM, respectively [121, 54, 123]. However, the strong cardiovascular effects of these compounds preclude their use as investigative tools for K$_{Ca}$3.1.

7.4.2.3.4 4-Phenyl-4*H*-pyrans

An isoelectronic replacement of the central NH of nifedipine (known to be a key feature for its Ca²⁺-antagonistic activity) with oxygen, gave a series of pyrans recently reported as K$_{Ca}$3.1 blockers. Modification of the methyl ester of the nifedipine-like core to give a methyl ketone led to **22** with 100-fold improvement of K$_{Ca}$3.1 inhibition (IC$_{50}$ 8 nM) compared to nifedipine. This compound reduced the infarct volume, intracranial pressure and water content in a rat subdural hematoma model of traumatic brain injury after i.v. administration [123]. The same group has reported a related series of equipotent cyclohexadienes.

7.4.2.3.5 **Peptides**

Several peptide toxins (including some described in Section 7.4.2.1.1) are potent $K_{Ca}3.1$ blockers. In flux assays, apamin (an octadecapeptide component of honeybee venom) and ChTx have IC_{50}s of 2.5 and 5.4 nM, respectively [67]. Stichodactylatoxin, a peptide isolated from sea anemones, is a potent $K_{Ca}3.1$ blocker (IC_{50} 24.5 nM) but also displays high affinity for Kv1.1 and 1.3. Other peptide $K_{Ca}3.1$ blockers include maurotoxin (MTX, IC_{50} 1.4 nM) and margatoxin (MgTX, IC_{50} 459 nM), both isolated from scorpion venom [42, 124, 125].

7.4.2.4 $K_{Ca}3.1$ **Channel Openers**

Small molecule activators of the $K_{Ca}3.1$ known at present share a similar core motif: a substituted aryl group fused to a five-membered nitrogen heterocycle. None of these compounds are selective for $K_{Ca}3.1$ versus SK_{Ca} channels.

7.4.2.4.1 **Benzimidazolinones**

The first characterized pharmacological opener of $K_{Ca}3.1$ was the benzimidazolinone 1-EBIO (**23**) [126]. Although devoid of effect on $K_{Ca}1.1$, 1-EBIO also activates SK_{Ca} channels. SAR studies led to the identification of the 3,4-dichloronated-analog DC-EBIO (**24**), with around 30-fold improvement in $K_{Ca}3.1$ potency. 1-EBIO evoked concentration-dependent increases in $^{86}Rb^+$ efflux from LNCaP (EC_{50} 24.7 μM) and PC-3 (EC_{50} 32 μM) human prostate cancer cell lines and increased proliferation, indicating an important role for IK_{Ca} channels in the regulation of human prostate cancer cell proliferation [57].

Investigations by Neurosearch into modifications of the benzimidazolinone core led to the discovery of the oxime derivative NS-309 (**25**). NS-309 is a potent activator of human Ca^{2+}-activated K^+ channels of SK_{Ca} and $K_{Ca}3.1$ types, with $K_{Ca}3.1$ potency (IC_{50} 10 nM) of greater than 1000× 1-EBIO and 30× DC-EBIO respectively [127]. *In vivo*, **25** produced a dose-dependent relaxation of the urinary bladder in normal healthy rats. In a model of overactive bladder, (oxyhemoglobin-induced overactivity) NS-309 normalized both the micturition frequency and volume without affecting basal parameters [128].

7.4.2.4.2 **Riluzole**

Riluzole (**26**) is a neuroprotective agent with inhibitory effects on glutamatergic transmission and brain Na^+ channels that is used for the treatment of amyotrophic lateral sclerosis (ALS). Riluzole evoked concentration-dependent increases in $^{86}Rb^+$ efflux from LNCaP (EC_{50} 1.1 μM) and PC-3 (EC_{50} 1.5 μM) human prostate cancer cell lines [57].

7.4.2.4.3 **Benzooxazolones**

Chlorzoxazone (**27**) is employed as a centrally acting muscle relaxant and has been shown to activate IK_{Ca} channels in basolateral membranes and to increase apical membrane Cl^- conductance [129]. Chlorzoxazone reversibly increased $K_{Ca}3.1$ currents with an EC_{50} of 30 μM.

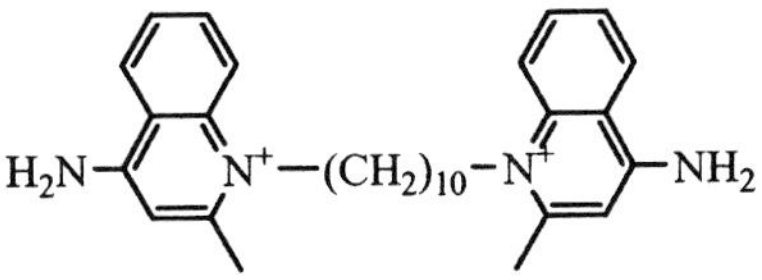

28 Dequalinium

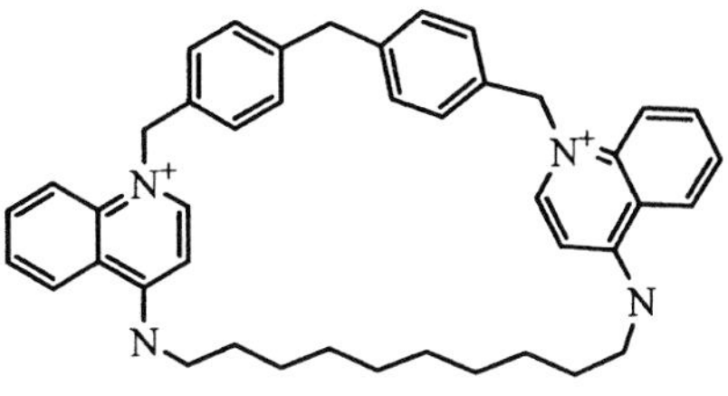

29 UCL-1530

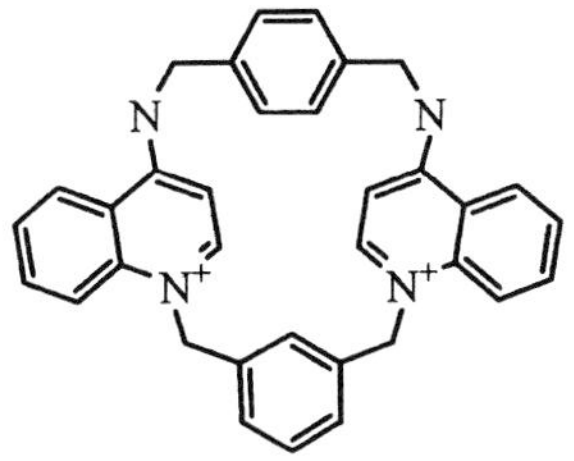

30 UCL 1684

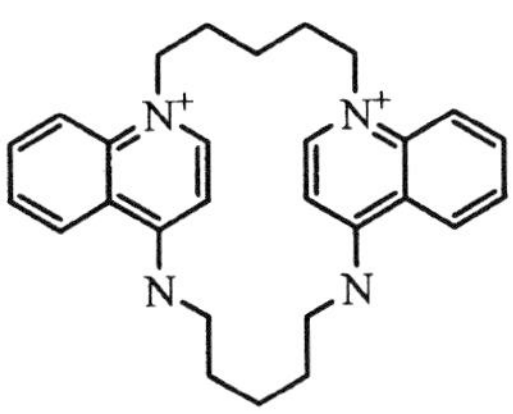

31 UCL-1848

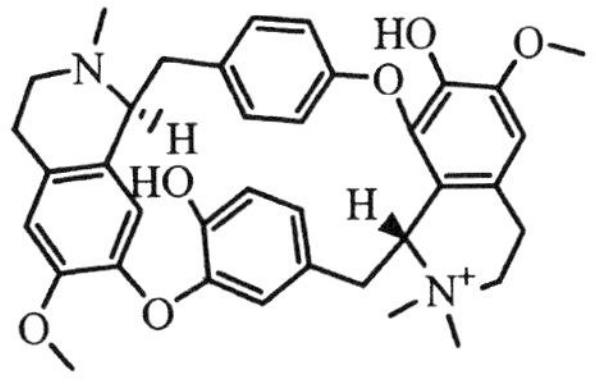

32 Tubocurarine

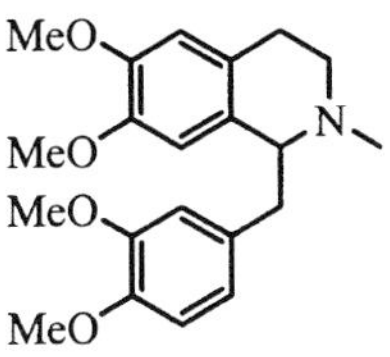

33 Laudanosine

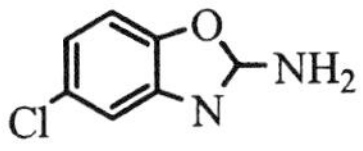

34 Zoxazolamine

Fig. 7.4.5 SK_{Ca} channel modulators.

7.4.2.5 SK_{Ca} Channel Blockers

The SK_{Ca} ($K_{Ca}2.1$–$K_{Ca}2.3$) blockers reported to date fall into three principal structural classes: (1) natural peptide toxins apamin and leiurotoxin I (scyllatoxin); (2) bis-quinoliniums such as UCL-1530 and UCL-1684; and (3) alkaloidal muscle blockers such as tubocurarine (Fig. 7.4.5).

7.4.2.5.1 Peptide Toxins

Apamin is a potent peptidic agent ($K_{Ca}2.2$ $IC_{50} = 6$–83 nM), whose blocking effect is only very slowly reversible. Despite limited selectivity, apamin has been invaluable in the study of SK_{Ca} channels, particularly in the elucidation of the different SK_{Ca} subtypes [130]. Scyllatoxin is a weaker blocker with similar selectivity to apamin. Recently, a novel 31-amino-acid peptide BmSKTx1 was reported with a K_d of 0.72 μM against SK_{Ca} channels in rat adrenal chromaffin cells [131].

7.4.2.5.2 **Quinoliniums**

A series of bis-quinolinium derivatives, discovered by UCL, have emerged as the most potent nonpeptidic SK_{Ca} channel blockers. The two charged sites (either protonizable or quaternized nitrogens) have been postulated to mimic the two arginine residues of apamin [130]. Dequalinium (**28**) was the first nonpeptidic SK_{Ca} channel blocker and is relatively potent and selective (Fig. 7.4.5) [132]. Restriction of the conformational flexibility of the quinolinium groups through their incorporation into a cyclophane structure has yielded bis-xylyl cyclophane blockers [133] such as UCL-1684 (**30**) with 100-fold greater potency than dequalinium (SK_{Ca} IC_{50} 3 nM in cultured rat sympathetic neurones) and selectivity for SK_{Ca} versus $K_{Ca}3.1$. Replacement of xylyl groups by alkane bridges led to UCL-1848 (**31**) (IC_{50} 2.7 nM). UCL-1848 is more selective than UCL-1684 for the $K_{Ca}2.2$ subtype of SK_{Ca} channels [134]. These blockers have proven to be useful tools for elucidating the physiological role of SK_{Ca} channels. Interestingly, UCL-1530 (**29**) selectively blocks neuronal SK_{Ca} channels relative to those expressed in either hepatocytes or jejunum (dequalinium shows no comparable expression discrimination) [135]. The same researchers also identified UCL-2027, a clotrimazole analog, which blocks the SK channel that underlies the sAHP that follows the action potential in hippocampal pyramidal neurons [136]. This is of particular interest because it has been suggested that age-related defects in memory may involve changes in this hippocampal AHP.

7.4.2.5.3 **Alkaloids**

The alkaloid laudanosine (**33**) (isolated from opium) and its quaternary derivatives are weak SK_{Ca} blockers. Intracellular recordings showed that laudanosine, methyl-laudanosine, and ethyl-laudanosine blocked the apamin-sensitive AHP of rat midbrain dopaminergic neurons with IC_{50}s of 152, 15, and 47 μM, respectively. In binding experiments on rat cerebral cortex membranes, methyl-laudanosine showed an IC_{50} of 4 μM [137]. Another alkaloid, d-tubocurarine (**32**) (the active component of the poison Curare), is a muscle relaxant that blocks apamin-sensitive SK_{Ca} channels [138].

7.4.2.6 **SK_{Ca} Channel Openers**

Many SK_{Ca} channel openers are also potent at $K_{Ca}3.1$ and come from the same bicyclic benzimidazole-like classes described in Section 7.4.2.4.

7.4.2.6.1 **NS-309, NS-4591**

Although NS-309 is a more potent opener of $K_{Ca}3.1$, it is significantly active at SK_{Ca} (IC_{50} 20–40 nM range). No selectivity is seen for any of the $K_{Ca}2.1$–$K_{Ca}2.3$ channels. NS-4591, an analog of NS-309, was reported to be orally active and to have improved pharmacokinetic properties [127].

7.4.2.6.2 Benzimidazolinones/Benzooxazolones/Riluzole

Cao and coworkers [139] have investigated the effects of 1-EBIO and related compounds on recombinant rat brain $K_{Ca}2.2$ channels expressed in HEK293 mammalian cells and determined the order of potency as 1-EBIO > chlorzoxazone > zoxazolamine (**34**). The benzoxazole chlorzoxazone activated r-$K_{Ca}2.2$ in *Xenopus* oocytes with a K_i of 87 µM [140]. Extracellular application of riluzole to r-$K_{Ca}2.2$-expressing HEK293 cells gave an EC_{50} for r-$K_{Ca}2.2$ channel activation of 43 µM [141].

7.4.3
Conclusions

As this chapter has demonstrated, Ca^{2+}-activated K^+ channels play an important role in a wide range of normal physiological and pathophysiological conditions. The key involvement of BK_{Ca} in stroke and bladder overactivity, IK_{Ca} in immunological response and cell proliferation, and SK_{Ca} in several neurological disorders are avenues of intensive ongoing research. Our understanding of the functional roles played by BK_{Ca} channels has been greatly assisted by the availability of knock-out mouse models. Much progress has been made in identifying selective small molecule modulators, in particular for $K_{Ca}1.1$ openers, where several development candidates have advanced into clinical trials. Although the identification of selective $K_{Ca}3.1$ and SK_{Ca} blockers has been a significant step forward, there is a clear need to identify selective activators of the IK_{Ca} and SK_{Ca} channels. With SK_{Ca} channels, the issue of subtype targeting will be important with regard to minimizing toxicity of future drug development candidates. The wide tissue distribution of Ca^{2+}-activated K^+ channels is illustrative of the great therapeutic potential for these agents in many disease states. However, there is also an element of risk associated with tissue selectivity and the potential for unwanted side effects. Doubtless, the exploration of the therapeutic potential of Ca^{2+}-activated K^+ channels modulators will continue to develop rapidly in the coming years.

References

1. L. Pallanck, B. Ganetzky. *Hum. Mol. Genet.* **1994**, *3*, 1239–1243.
2. L. Toro, M. Wallner, P. Meera, Y. Tanaka. *News Physiol. Sci.* **1998**, *13*, 112–117.
3. J. Tseng-Crank, N. Godinot, T. E. Johansen, P. K. Ahring, D. Strobaek, R. Mertz, C. D. Foster, S. P. Olesen, P. H. Reinhart. *Proc. Natl. Acad. Sci. U.S.A.* **1996**, *93*, 9200–9205.
4. P. Meera, M. Wallner, Z. Jiang, L. Toro. *FEBS Lett.* **1996**, *382*, 84–88.
5. M. Wallner, P. Meera, L. Toro. *Proc. Natl. Acad. Sci. U.S.A.* **1999**, *96*, 4137–4142.
6. R. Behrens, A. Nolting, F. Reimann, M. Schwarz, R. Waldschutz, O. Pongs. *FEBS Lett.* **2000**, *474*, 99–106.
7. R. Brenner, T. J. Jegla, A. Wickenden, Y. Liu, R. W. Aldrich. *J. Biol. Chem.* **2000**, *275*, 6453–6461.
8. M. Saito, C. Nelson, L. Salkoff, C. J. Lingle. *J. Biol. Chem.* **1997**, *272*, 11710–11717.
9. K. Ramanathan, T. H. Michael, P. A. Fuchs. *J. Neurosci.* **2000**, *20*, 1675–1684.
10. M. M. Zarei, N. Zhu, A. Alioua, M. Eghbali, E. Stefani, L. Toro. *J. Biol. Chem.* **2001**, *276*, 16232–16239.

11. A. Bhattacharjee, W. J. Joiner, M. Wu, Y. Yang, F. J. Sigworth, L. K. Kaczmarek. *J. Neurosci.* **2003**, *23*, 11 681–11 691.

12. M. Schreiber, A. Wei, A. Yuan, J. Gaut, M. Saito, L. Salkoff. *J. Biol. Chem.* **1998**, *273*, 3509–3516.

13. A. Alioua, Y. Tanaka, M. Wallner, F. Hofmann, P. Ruth, P. Meera, L. Toro. *J. Biol. Chem.* **1998**, *273*, 32 950–32 956.

14. S. Ling, J. Z. Sheng, A. P. Braun. *Am. J. Physiol. Cell Physiol.* **2004**, *287*, 5.

15. L. Tian, L. S. Coghill, H. McClafferty, S. H. MacDonald, F. A. Antoni, P. Ruth, H. G. Knaus, M. J. Shipston. *Proc. Natl. Acad. Sci. U.S.A.* **2004**, *101*, 11 897–11 902.

16. D. W. Choi. *Trends Neurosci.* **1995**, *18*, 58–60.

17. D. L. Small, P. Morley, A. M. Buchan. *Prog. Cardiovasc Dis.* **1999**, *42*, 185–207.

18. V. K. Gribkoff, J. E. Starrett, Jr., S. I. Dworetzky, P. Hewawasam, C. G. Boissard, D. A. Cook, S. W. Frantz, K. Heman, J. R. Hibbard, K. Huston, G. Johnson, B. S. Krishnan, G. G. Kinney, L. A. Lombardo, N. A. Meanwell, P. B. Molinoff, R. A. Myers, S. L. Moon, A. Ortiz, L. Pajor, R. L. Pieschl, D. J. Post-Munson, L. J. Signor, N. Srinivas, M. T. Taber, G. Thalody, J. T. Trojnacki, H. Wiener, K. Yeleswaram, S. W. Yeola. *Nat. Med.* **2001**, *7*, 471–477.

19. R. L. Schroder, T. Jespersen, P. Christophersen, D. Strobaek, B. S. Jensen, S. P. Olesen. *Neuropharmacology* **2001**, *40*, 888–898.

20. B. S. Jensen. *CNS Drug. Rev.* **2002**, *8*, 353–360.

21. P. Sah. *Trends Neurosci.* **1996**, *19*, 150–154.

22. L. R. Shao, R. Halvorsrud, L. Borg-Graham, J. F. Storm. *J. Physiol.* **1999**, *1*, 135–146.

23. M. Sausbier, H. Hu, C. Arntz, S. Feil, S. Kamm, H. Adelsberger, U. Sausbier, C. A. Sailer, R. Feil, F. Hofmann, M. Korth, M. J. Shipston, H. G. Knaus, D. P. Wolfer, C. M. Pedroarena, J. F. Storm, P. Ruth. *Proc. Natl. Acad. Sci. U.S.A.* **2004**, *101*, 9474–9478.

24. A. Scholz, M. Gruss, W. Vogel. *J. Physiol.* **1998**, *513*, 55–69.

25. X. F. Zhang, M. Gopalakrishnan, C. C. Shieh. *Neuroscience* **2003**, *122*, 1003–1011.

26. K. Ramanathan, T. H. Michael, G. J. Jiang, H. Hiel, P. A. Fuchs. *Science* **1999**, *283*, 215–217.

27. P. Langer, S. Grunder, A. Rusch. *J. Comp. Neurol.* **2003**, *455*, 198–209.

28. J. D. McCann, M. J. Welsh. *J. Physiol.* **1986**, *372*, 113–127.

29. H. M. Saunders, J. M. Farley. *J. Pharmacol. Exp. Ther.* **1991**, *257*, 1114–1120.

30. R. Brenner, G. J. Perez, A. D. Bonev, D. M. Eckman, J. C. Kosek, S. W. Wiler, A. J. Patterson, M. T. Nelson, R. W. Aldrich. *Nature* **2000**, *407*, 870–876.

31. K. Anwer, C. Oberti, G. J. Perez, N. Perez-Reyes, J. K. McDougall, M. Monga, B. M. Sanborn, E. Stefani, L. Toro. *Am. J. Physiol.* **1993**, *265*, C976–985.

32. A. L. Meredith, K. S. Thorneloe, M. E. Werner, M. T. Nelson, R. W. Aldrich. *J. Biol. Chem.* **2004**, *279*, 36 746–36 752.

33. A. Melman, W. Zhao, K. P. Davies, R. Bakal, G. J. Christ. *J. Urol.* **2003**, *170*, 285–290.

34. Y. Ohi, H. Yamamura, N. Nagano, S. Ohya, K. Muraki, M. Watanabe, Y. Imaizumi. *J. Physiol.* **2001**, *534*, 313–326.

35. G. J. Perez, A. D. Bonev, M. T. Nelson. *Am. J. Physiol. Cell Physiol.* **2001**, *281*, C1769–1775.

36. S. Pluger, J. Faulhaber, M. Furstenau, M. Lohn, R. Waldschutz, M. Gollasch, H. Haller, F. C. Luft, H. Ehmke, O. Pongs. *Circ. Res.* **2000**, *87*, E53–60.

37. G. V. Petkov, A. D. Bonev, T. J. Heppner, R. Brenner, R. W. Aldrich, M. T. Nelson. *J. Physiol.* **2001**, *537*, 443–452.

38. G. J. Christ, N. S. Day, M. Day, C. Santizo, W. Zhao, T. Sclafani, J. Zinman, K. Hsieh, K. Venkateswarlu, M. Valcic, A. Melman. *Am. J. Physiol. Regul. Integr. Comp. Physiol.* **2001**, *281*, R1699–1709.

39. G. J. Christ, N. Day, C. Santizo, Y. Sato, W. Zhao, T. Sclafani, R. Bakal, M. Salman, K. Davies, A. Melman. *Am. J. Physiol. Heart Circ. Physiol.* **2004**, *287*, H1544–553.

40. N. J. Logsdon, J. Kang, J. A. Togo, E. P. Christian, J. Aiyar. *J. Biol. Chem.* **1997**, *272*, 32 723–32 726.

41. T. M. Ishii, C. Silvia, B. Hirschberg, C. T. Bond, J. P. Adelman, J. Maylie. *Proc.*

Natl. Acad. Sci. U.S.A. **1997**, *94*, 11 651–11 656.

42. B. S. Jensen, D. Strobaek, P. Christophersen, T. D. Jorgensen, C. Hansen, A. Silahtaroglu, S. P. Olesen, P. K. Ahring. *Am. J. Physiol.* **1998**, *275*, C848–856.

43. R. Khanna, M. C. Chang, W. J. Joiner, L. K. Kaczmarek, L. C. Schlichter. *J. Biol. Chem.* **1999**, *274*, 14 838–14 849.

44. S. Ohya, S. Kimura, M. Kitsukawa, K. Muraki, M. Watanabe, Y. Imaizumi. *Jpn. J. Pharmacol.* **2000**, *84*, 97–100.

45. C. B. Neylon, R. J. Lang, Y. Fu, A. Bobik, P. H. Reinhart. *Circ. Res.* **1999**, *85*, e33–43.

46. W. J. Joiner, M. D. Tang, L. Y. Wang, S. I. Dworetzky, C. G. Boissard, L. Gan, V. K. Gribkoff, L. K. Kaczmarek. *Nat. Neurosci.* **1998**, *1*, 462–469.

47. R. Sullivan, S. K. Koliwad, D. L. Kunze. *Am. J. Physiol.* **1998**, *275*, C1342–1348.

48. X. Lu, A. Fein, M. B. Feinstein, F. A. O'Rourke. *J. Gen. Physiol.* **1999**, *113*, 81–96.

49. T. Begenisich, T. Nakamoto, C. E. Ovitt, K. Nehrke, C. Brugnara, S. L. Alper, J. E. Melvin. *J. Biol. Chem.* **2004**, *279*, 47 681–47 687.

50. J. M. Hinton, P. D. Langton. *Br. J. Pharmacol.* **2003**, *138*, 1031–1035.

51. I. Eichler, J. Wibawa, I. Grgic, A. Knorr, S. Brakemeier, A. R. Pries, J. Hoyer, R. Kohler. *Br. J. Pharmacol.* **2003**, *138*, 594–601.

52. C. Brugnara. *J. Pediatr. Hematol. Oncol.* **2003**, *25*, 927–933.

53. J. W. Stocker, L. De Franceschi, G. A. McNaughton-Smith, R. Corrocher, Y. Beuzard, C. Brugnara. *Blood* **2003**, *101*, 2412–2418.

54. S. Ghanshani, H. Wulff, M. J. Miller, H. Rohm, A. Neben, G. A. Gutman, M. D. Cahalan, K. G. Chandy. *J. Biol. Chem.* **2000**, *275*, 37 137–37 149.

55. H. Wulff, M. J. Miller, W. Hansel, S. Grissmer, M. D. Cahalan, K. G. Chandy. *Proc. Natl. Acad. Sci. U.S.A.* **2000**, *97*, 8151–8156.

56. S. G. Rane. *Biochem. Biophys. Res. Commun.* **2000**, *269*, 457–463.

57. A. S. Parihar, M. J. Coghlan, M. Gopalakrishnan, C. C. Shieh. *Eur. J. Pharmacol.* **2003**, *471*, 157–164.

58. H. Ouadid-Ahidouch, M. Roudbaraki, P. Delcourt, A. Ahidouch, N. Joury, N. Prevarskaya. *Am. J. Physiol. Cell Physiol.* **2004**, *287*, 25.

59. H. Jager, T. Dreker, A. Buck, K. Giehl, T. Gress, S. Grissmer. *Mol. Pharmacol.* **2004**, *65*, 630–638.

60. R. Kohler, H. Wulff, I. Eichler, M. Kneifel, D. Neumann, A. Knorr, I. Grgic, D. Kampfe, H. Si, J. Wibawa, R. Real, K. Borner, S. Brakemeier, H. D. Orzechowski, H. P. Reusch, M. Paul, K. G. Chandy, J. Hoyer. *Circulation* **2003**, *108*, 1119–1125.

61. D. P. Wallace, J. M. Tomich, J. W. Eppler, T. Iwamoto, J. J. Grantham, L. P. Sullivan. *Biochim. Biophys. Acta* **2000**, *15*, 69–82.

62. M. Mall, T. Gonska, J. Thomas, R. Schreiber, H. H. Seydewitz, J. Kuehr, M. Brandis, K. Kunzelmann. *Pediatr. Res.* **2003**, *53*, 608–618.

63. M. Kohler, B. Hirschberg, C. T. Bond, J. M. Kinzie, N. V. Marrion, J. Maylie, J. P. Adelman. *Science* **1996**, *273*, 1709–1714.

64. R. Wissmann, W. Bildl, H. Neumann, A. F. Rivard, N. Klocker, D. Weitz, U. Schulte, J. P. Adelman, D. Bentrop, B. Fakler. *J. Biol. Chem.* **2002**, *277*, 4558–4564.

65. M. A. Schumacher, A. F. Rivard, H. P. Bachinger, J. P. Adelman. *Nature* **2001**, *410*, 1120–1124.

66. M. Shah, D. G. Haylett. *Br. J. Pharmacol.* **2000**, *129*, 627–630.

67. D. Strobaek, T. D. Jorgensen, P. Christophersen, P. K. Ahring, S. P. Olesen. *Br. J. Pharmacol.* **2000**, *129*, 991–999.

68. M. Stocker, P. Pedarzani. *Mol. Cell Neurosci.* **2000**, *15*, 476–493.

69. J. Wolfart, H. Neuhoff, O. Franz, J. Roeper. *J. Neurosci.* **2001**, *21*, 3443–3456.

70. M. Stocker. *Nat. Rev. Neurosci.* **2004**, *5*, 758–770.

71. C. T. Bond, P. S. Herson, T. Strassmaier, R. Hammond, R. Stackman, J. Maylie, J. P. Adelman. *J. Neurosci.* **2004**, *24*, 5301–5306.

72. C. Villalobos, V. G. Shakkottai, K. G. Chandy, S. K. Michelhaugh, R. Andrade. *J. Neurosci.* **2004**, *24*, 3537–3542.

73. M. Stocker, M. Krause, P. Pedarzani. *Proc. Natl. Acad. Sci. U.S.A.* **1999**, *96*, 4662–4667.

74. R. W. Stackman, R. S. Hammond, E. Linardatos, A. Gerlach, J. Maylie, J. P. Adelman, T. Tzounopoulos. *J. Neurosci.* **2002**, *22*, 10 163–10 171.

75. F. J. van der Staay, R. J. Fanelli, A. Blokland, B. H. Schmidt. *Neurosci. Biobehav. Rev.* **1999**, *23*, 1087–1110.

76. T. Blank, I. Nijholt, M. J. Kye, J. Radulovic, J. Spiess. *Nat. Neurosci.* **2003**, *6*, 911–912.

77. C. T. Bond, R. Sprengel, J. M. Bissonnette, W. A. Kaufmann, D. Pribnow, T. Neelands, T. Storck, M. Baetscher, J. Jerecic, J. Maylie, H. G. Knaus, P. H. Seeburg, J. P. Adelman. *Science* **2000**, *289*, 1942–1946.

78. K. P. Figueroa, P. Chan, L. Schols, C. Tanner, O. Riess, S. L. Perlman, D. H. Geschwind, S. M. Pulst. *Arch. Neurol.* **2001**, *58*, 1649–1653.

79. M. Koronyo-Hamaoui, E. Gak, D. Stein, A. Frisch, Y. Danziger, S. Leor, E. Michaelovsky, N. Laufer, C. Carel, S. Fennig, M. Mimouni, A. Apter, B. Goldman, G. Barkai, A. Weizman. *Am. J. Med. Genet. B Neuropsychiatr. Genet.* **2004**, *131*, 76–80.

80. P. Pedarzani, J. Mosbacher, A. Rivard, L. A. Cingolani, D. Oliver, M. Stocker, J. P. Adelman, B. Fakler. *J. Biol. Chem.* **2001**, *276*, 9762–9769.

81. A. L. Lee, T. C. Dumas, P. E. Tarapore, B. R. Webster, D. Y. Ho, D. Kaufer, R. M. Sapolsky. *J. Neurochem.* **2003**, *86*, 1079–1088.

82. M. I. Behrens, P. Jalil, A. Serani, F. Vergara, O. Alvarez. *Muscle Nerve* **1994**, *17*, 1264–1270.

83. G. M. Herrera, M. J. Pozo, P. Zvara, G. V. Petkov, C. T. Bond, J. P. Adelman, M. T. Nelson. *J. Physiol.* **2003**, *551*, 893–903.

84. M. S. Taylor, A. D. Bonev, T. P. Gross, D. M. Eckman, J. E. Brayden, C. T. Bond, J. P. Adelman, M. T. Nelson. *Circ. Res.* **2003**, *93*, 124–131.

85. S. P. Olesen, E. Munch, P. Moldt, J. Drejer. *Eur. J. Pharmacol.* **1994**, *251*, 53–59.

86. J. Kang, J. R. Huguenard, D. A. Prince. *J. Neurophysiol.* **1996**, *76*, 4194–4197.

87. K. M. Giangiacomo, E. E. Sugg, M. Garcia-Calvo, R. J. Leonard, O. B. McManus, G. J. Kaczorowski, M. L. Garcia. *Biochemistry* **1993**, *32*, 2363–2370.

88. M. Stocker, C. Miller. *Proc. Natl. Acad. Sci. U.S.A.* **1994**, *91*, 9509–9513.

89. J. Aiyar, J. M. Withka, J. P. Rizzi, D. H. Singleton, G. C. Andrews, W. Lin, J. Boyd, D. C. Hanson, M. Simon, B. Dethlefs, et al. *Neuron* **1995**, *15*, 1169–1181.

90. H. Rauer, M. D. Lanigan, M. W. Pennington, J. Aiyar, S. Ghanshani, M. D. Cahalan, R. S. Norton, K. G. Chandy. *J. Biol. Chem.* **2000**, *275*, 1201–1208.

91. A. Nardi, V. Calderone, S. Chericoni, I. Morelli. *Planta Med.* **2003**, *69*, 885–892.

92. G. J. Kaczorowski, H. G. Knaus, R. J. Leonard, O. B. McManus, M. L. Garcia. *J. Bioenerg. Biomembr.* **1996**, *28*, 255–267.

93. L. K. McMillan, R. L. Carr, C. A. Young, J. W. Astin, R. G. Lowe, E. J. Parker, G. B. Jameson, S. C. Finch, C. O. Miles, O. B. McManus, W. A. Schmalhofer, M. L. Garcia, G. J. Kaczorowski, M. Goetz, J. S. Tkacz, B. Scott. *Mol. Genet. Genomics* **2003**, *270*, 9–23.

94. R. Vianna-Jorge, C. F. Oliveira, C. G. Ponte, G. Suarez-Kurtz. *Eur. J. Pharmacol.* **2001**, *428*, 45–49.

95. S. N. Wu, H. F. Li, C. R. Jan, A. Y. Shen. *Neuropharmacology* **1999**, *38*, 979–989.

96. S. N. Wu. *Curr. Med. Chem.* **2003**, *10*, 649–661.

97. N. A. Meanwell, S. Y. Sit, J. Gao, C. G. Boissard, J. Lum-Ragan, S. I. Dworetzky, V. K. Gribkoff. *Bioorg. Med. Chem. Lett.* **1996**, *6*, 1641–1646.

98. Y. Li, J. E. Starrett, N. A. Meanwell, G. Johnson, W. E. Harte, S. I. Dworetzky, C. G. Boissard, V. K. Gribkoff. *Bioorg. Med. Chem. Lett.* **1997**, 759–762.

99. V. K. Gribkoff, J. T. Lum-Ragan, C. G. Boissard, D. J. Post-Munson, N. A. Meanwell, J. E. Starrett, Jr., E. S. Kozlowski, J. L. Romine, J. T. Trojnacki, M. C. McKay, J. Zhong, S. I. Dworetzky. *Mol. Pharmacol.* **1996**, *50*, 206–217.

100. J. E. Starrett, S. I. Dworetzky, V. K. Gribkoff. *Curr. Pharm. Design* **1996**, *2*, 413–428.

101. D. P. Alves, A. C. Soares, J. N. Francischi, M. S. Castro, A. C. Perez, I. D. Duarte. *Eur. J. Pharmacol.* **2004**, *489*, 59–65.

102. V. Calderone. *Curr. Med. Chem.* **2002**, *9*, 1385–1395.

103. B. S. Jensen, G. J. Blackburn-Munro. *World Pat. 2004064835* **2004**, 1–34.

104. B. Baragatti, G. Biagi, V. Calderone, I. Giorgi, O. Livi, E. Martinotti, V. Scartoni. *Eur. J. Med. Chem.* **2000**, *35*, 949–955.

105. P. Hewawasam, M. Erway, S. L. Moon, J. Knipe, H. Weiner, C. G. Boissard, D. J. Post-Munson, Q. Gao, S. Huang, V. K. Gribkoff, N. A. Meanwell. *J. Med. Chem.* **2002**, *45*, 1487–1499.

106. V. K. Gribkoff. *Presentation* Princeton, New Jersey 2002 Second International Ion Channels in Drug Discovery and Development Symposium **2002**.

107. R. Krishna, H. Palme, J. Zeng, N. Srinivas. *Biopharm. Drug. Dispos.* **2002**, *23*, 227–231.

108. S. Y. Sit, N. A. Meanwell. *U.S. Pat.* 5,892,045 **1999**, 1–10.

109. P. Hewawasam, W. Fan, M. Ding, K. Flint, D. Cook, G. D. Goggins, R. A. Myers, V. K. Gribkoff, C. G. Boissard, S. I. Dworetzky, J. E. Starrett, Jr., N. J. Lodge. *J. Med. Chem.* **2003**, *46*, 2819–2822.

110. P. Hewawasam, W. Fan, D. A. Cook, K. S. Newberry, C. G. Boissard, V. K. Gribkoff, J. Starrett, N. J. Lodge. *Bioorg. Med. Chem. Lett.* **2004**, *14*, 4479–4482.

111. J. L. Romine, S. W. Martin, V. K. Gribkoff, C. G. Boissard, S. I. Dworetzky, J. Natale, Y. Li, Q. Gao, N. A. Meanwell, J. E. Starrett, Jr. *J. Med. Chem.* **2002**, *45*, 2942–2952.

112. P. Hewawasam, X. Chen, J. E. Starrett. *World Pat.* 2000044745 **2000**.

113. S. Hu, C. A. Fink, H. S. Kim, R. W. Lappe. *Drug Dev. Res.* **1997**, *41*, 10–21.

114. M. Tanaka, Y. Sasaki, Y. Kimura, T. Fukui, K. Hamada, Y. Ukai. *BJU Int.* **2003**, *92*, 1031–1036.

115. M. Tsuda, M. Tanaka, A. Nakamura. *World Pat.* 09640634 1–107.

116. S. C. Turner, W. A. Carroll, T. K. White, M. Gopalakrishnan, M. J. Coghlan, C. C. Shieh, X. F. Zhang, A. S. Parihar, S. A. Buckner, I. Milicic, J. P. Sullivan. *Bioorg. Med. Chem. Lett.* **2003**, *13*, 2003–2007.

117. O. B. McManus, G. H. Harris, K. M. Giangiacomo, P. Feigenbaum, J. P. Reuben, M. E. Addy, J. F. Burka, G. J. Kaczorowski, M. L. Garcia. *Biochemistry* **1993**, *32*, 6128–6133.

118. J. G. Ondeyka, R. G. Ball, M. L. Garcia, A. W. Dombrowski, G. Sabnis, G. J. Kaczorowski, D. L. Zink, G. F. Bills, M. A. Goetz, W. A. Schmalhofer, S. B. Singh. *Bioorg. Med. Chem. Lett.* **1995**, *5*, 733–734.

119. Y. Imaizumi, K. Sakamoto, A. Yamada, A. Hotta, S. Ohya, K. Muraki, M. Uchiyama, T. Ohwada. *Mol. Pharmacol.* **2002**, *62*, 836–846.

120. G. Edwards, A. H. Weston. *Expert Opin. Invest. Drugs* **1996**, *5*, 1453–1464.

121. B. S. Jensen, M. Hertz, P. Christophersen, L. S. Madsen. *Expert Opin. Ther. Targets* **2002**, *6*, 623–636.

122. C. J. Roxburgh, C. R. Ganellin, S. Athmani, A. Bisi, W. Quaglia, D. C. Benton, M. A. Shiner, M. Malik-Hall, D. G. Haylett, D. H. Jenkinson. *J. Med. Chem.* **2001**, *44*, 3244–3253.

123. K. Urbahns, E. Horvath, J. P. Stasch, F. Mauler. *Bioorg. Med. Chem. Lett.* **2003**, *13*, 2637–2639.

124. R. Kharrat, P. Mansuelle, F. Sampieri, M. Crest, R. Oughideni, J. Van Rietschoten, M. F. Martin-Eauclaire, H. Rochat, M. El Ayeb. *FEBS Lett.* **1997**, *406*, 284–290.

125. N. A. Castle, D. O. London, C. Creech, Z. Fajloun, J. W. Stocker, J. M. Sabatier. *Mol. Pharmacol.* **2003**, *63*, 409–418.

126. D. C. Devor, A. K. Singh, R. A. Frizzell, R. J. Bridges. *Am. J. Physiol.* **1996**, *271*, L775–784.

127. D. Stroaek, L. Teuber, T. D. Jorgensen, P. K. Ahring, K. Kaer, R. S. Hansen, S. P. Olesen, P. Christophersen, B. Skaaning-Jensen. *Biochim. Biophys. Acta* **2004**, *11*, 1–2.

128. B. S. Jensen. *Presentation* Princeton, New Jersey 2002 Second International Ion Channels in Drug Discovery and Development Symposium **2002**.

129. A. K. Singh, D. C. Devor, A. C. Gerlach, M. Gondor, J. M. Pilewski, R. J. Bridges. *J. Pharmacol. Exp. Ther.* **2000**, *292*, 778–787.

130. J. F. Liegeois, F. Mercier, A. Graulich, F. Graulich-Lorge, J. Scuvee-Moreau, V. Seutin. *Curr. Med. Chem.* **2003**, *10*, 625–647.

131. C. Q. Xu, B. Brone, D. Wicher, O. Bozkurt, W. Y. Lu, I. Huys, Y. H. Han, J. Tytgat, E. Van Kerkhove, C. W. Chi. *J. Biol. Chem.* **2004**, *279*, 34562–34569.

132. N. A. Castle, D. G. Haylett, J. M. Morgan, D. H. Jenkinson. *Eur. J. Pharmacol.* **1993**, *236*, 201–207.

133. C. J. Rosa, D. Galanakis, C. R. Ganellin, P. M. Dunn, D. H. Jenkinson. *J. Med. Chem.* **1998**, *41*, 2–5.

134. J. Q. Chen, D. Galanakis, C. R. Ganellin, P. M. Dunn, D. H. Jenkinson. *J. Med. Chem.* **2000**, *43*, 3478–3481.

135. C. J. Rosa, B. M. Beckwith-Hall, D. Galanakis, C. R. Ganellin, P. M. Dunn, D. H. Jenkinson. *Bioorg. Med. Chem. Lett.* **1997**, *7*, 7–10.

136. M. M. Shah, Z. Miscony, M. Javadzadeh-Tabatabaie, C. R. Ganellin, D. G. Haylett. *Br. J. Pharmacol.* **2001**, *132*, 889–898.

137. J. Scuvee-Moreau, J. F. Liegeois, L. Massotte, V. Seutin. *J. Pharmacol. Exp. Ther.* **2002**, *302*, 1176–1183.

138. T. M. Ishii, J. Maylie, J. P. Adelman. *J. Biol. Chem.* **1997**, *272*, 23 195–23 200.

139. Y. Cao, J. C. Dreixler, J. D. Roizen, M. T. Roberts, K. M. Houamed. *J. Pharmacol. Exp. Ther.* **2001**, *296*, 683–689.

140. C. A. Syme, A. C. Gerlach, A. K. Singh, D. C. Devor. *Am. J. Physiol. Cell Physiol.* **2000**, *278*, C570–581.

141. Y. J. Cao, J. C. Dreixler, J. J. Couey, K. M. Houamed. *Eur. J. Pharmacol.* **2002**, *449*, 47–54.

7.5
Drugs Active at ATP-sensitive K$^+$ Channels

William A. Carroll

7.5.1
Introduction

ATP-sensitive potassium (K$_{ATP}$) channels are present in a wide variety of tissue types such as the pancreas, central nervous system, heart, skeletal muscle and in diverse smooth muscle tissues such as the bladder and peripheral vasculature. Being inhibited by intracellular ATP, K$_{ATP}$ channels play an important role in linking cellular energy metabolism to membrane potential and cellular excitability [1, 2]. Opening of K$_{ATP}$ channels gives rise to an efflux of potassium ions from the cell and an accompanying reduction in membrane potential (hyperpolarization) that inhibits calcium entry through L-type calcium channels, thereby dampening cellular excitability.

K$_{ATP}$ channels are heterooctameric complexes composed of four inwardly-rectifying potassium channel (Kir) subunits that form the pore, and four regulatory sulfonylurea receptor (SUR) subunits [3, 4]. The Kir subunit consists of two membrane-spanning segments with cytoplasmic C- and N-termini, whereas the SUR subunit is composed of 17 transmembrane segments that can be broken down into one group of five transmembrane helices and two groups of six transmembrane helices [5, 6]. The SUR subunit also possesses two long intracellular loops that are important for nucleotide recognition and channel function. K$_{ATP}$ openers are thought to bind at specific sites on the SUR subunit whereas blockers have been found that exert their effects by interacting with either the SUR or Kir subunit [7, 8].

Molecular biological advances have led to the identification of distinct Kir (Kir6.1 and Kir6.2) and SUR (SUR1, SUR2A and SUR2B) subtypes that are assembled in tissue-specific combinations to form functioning channels with unique pharmacological sensitivities [5, 8]. Additional splice variants of SUR1 and SUR2 have also been described [7, 9]. The pancreatic β-cell K$_{ATP}$ that regulates insulin release is composed of the combination of SUR1 with Kir6.2. The SUR2A/Kir6.2 combination has been identified as the sarcolemmal cardiac K$_{ATP}$ whereas the smooth muscle K$_{ATP}$ consists of SUR2B with either Kir6.1 or Kir6.2. Additionally, a mitochondrial K$_{ATP}$ (mito-K$_{ATP}$) channel has been characterized pharmacologically; however, the molecular structure of this channel has yet to be revealed.

K$_{ATP}$ openers and blockers have been investigated for numerous therapeutic applications, prominent among which are diabetes, hyperinsulinemia, hypertension, angina, cardioprotection, ventricular arrhythmia, asthma, alopecia and bladder overactivity. Table 7.5.1 shows currently marketed drugs that exert their therapeutic effect predominantly via modulation of K$_{ATP}$ channel activity. The K$_{ATP}$ channel openers diazoxide (**3**), pinacidil (**2**) and cromakalim (**1**) and the K$_{ATP}$ blocker glyburide (**4**) have been the most widely used pharmacological tools employed in preclinical studies (Fig. 7.5.1). Much of the early focus in the area of

Tab. 7.5.1 Marketed K$_{ATP}$ modulating drugs.

Generic name	Brand name	Indications	K$_{ATP}$ activity
Diazoxide (3)	Hyperstat®	Hypertension	SUR2B/Kir6.2 opener
	Proglycem®	Hyperinsulinemia	SUR1/Kir6.2 opener
Pinacidil (2)	Pindac®	Hypertension	SUR2B/Kir6.2 opener
Glyburide (4)	Diabeta®	Type 2 diabetes	SUR1/Kir6.2 blocker
Minoxidil (29)	Rogaine®	Alopecia	SUR2B/Kir6.2 opener
Nicorandil	Ikorel®	Angina	SUR2B/Kir6.2 opener

K$_{ATP}$ channel openers was directed towards exploring their potential as antihypertensive agents. Although generally effective blood pressure lowering drugs, K$_{ATP}$ openers failed to show clear advantages over existing antihypertensive drug classes such as calcium channel blockers and ACE inhibitors [10]. Diazoxide (Hyperstat®, I.V.) has advanced to the stage of a commercial antihypertensive drug, although its use for this indication is primarily limited to acute hypertensive crises in the hospital setting [11]. Likewise, pinacidil (Pindac®) has been approved for use as an antihypertensive; however, its use has been rather restricted. Diazoxide (Proglycem®) is also available in oral dosages for managing hypoglycemia due to hyperinsulinemia [12], whereas the K$_{ATP}$ channel blocker glyburide (Diabeta®) is a widely used antidiabetic agent in the clinic.

In the past 10–15 years, research in the K$_{ATP}$ opener field has shifted away from the emphasis on hypertension towards more poorly treated medical conditions such as those listed in the previous paragraph. Unfortunately, divorcing the blood pressure lowering effects of K$_{ATP}$ openers from their desired therapeutic effects elsewhere has proven to be a considerable challenge in every field of investigation. This can largely be attributed to the difficulty in effectively discriminat-

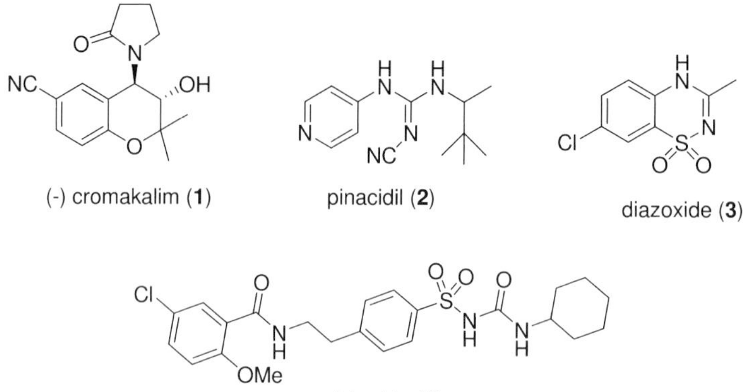

(-) cromakalim (**1**)   pinacidil (**2**)   diazoxide (**3**)

glyburide (**4**)

Fig. 7.5.1 Compounds 1–4.

ing between the K_{ATP} channel subtypes present in the target organs versus the vasculature. Furthermore, K_{ATP} channel subunit composition between tissue types may be highly homologous, thus increasing the challenge of identifying selective agents. For K_{ATP} openers, there are few examples of molecular selectivity for a particular SUR/Kir combination that can be correlated to an improved therapeutic index *in vivo*. The most notable among these are the SUR1/Kir6.2 selective openers being investigated for insulin-releasing disorders (*vide infra*). In the area of K_{ATP} blockers, advances have also been made toward identifying SUR2A/Kir6.2 selective agents for use in the treatment of ventricular fibrillation (*vide infra*). Tissue and/or organ selectivity has certainly been observed for a wider class of agents than those just mentioned, albeit without a solid understanding of the molecular underpinnings, if any, that are responsible. Since many of the specific therapeutic applications of current interest for K_{ATP} openers and blockers have been reviewed in detail in recent years, it is not the purpose of this chapter to extensively recapitulate that information. Rather, the intent is to provide a general overview for the non-expert in the field of the current state of development, directions and potential for additional therapeutic applications of K_{ATP} openers and blockers.

7.5.2
Mitochondrial K_{ATP} Channel Openers for Myocardial Ischemia

A significant body of evidence has accumulated that implicates the involvement of a mitochondrial K_{ATP} channel (mito-K_{ATP}), located in cardiomyocytes, in the cardioprotection conferred by certain K_{ATP} openers, or by brief preconditioning periods of ischemia. There has been considerable progress in recent years toward elucidating the potential mechanisms that link mito-K_{ATP} channel activation, whether pharmacological or due to ischemic preconditioning, with cardioprotective effects [10, 13–16]. Studies have focused on the role of the mito-K_{ATP} channel as trigger or effector of cardioprotection, as well as the roles of various mediators [kinases, reactive oxygen species (ROS), NO, Ca^{2+}, K^+] in the signal transduction pathway. Although the mito-K_{ATP} channel has garnered much attention in recent years as the major player in cardioprotection, there is still believed to be a role for the sarcolemmal K_{ATP} channel in this process. In fact, it has been suggested that the mito-K_{ATP} and sarc-K_{ATP} may act in complementary fashion to regulate distinct steps in the overall process of cardioprotection [17].

Despite advances in understanding of the mechanism of mito-K_{ATP}-mediated cardioprotection, the molecular structure of this channel has yet to be determined. The pharmacology of the mito-K_{ATP} channel is reported to closely resemble SUR1/Kir6.1 [18] but other evidence suggests the participation of a SUR2-like protein in either heart [10, 19] or brain [20]. A more satisfactory understanding of the relationship between the proposed steps in mito-K_{ATP} mediated cardioprotection would be enabled by the resolution of its protein structure.

Exploration into the therapeutic potential of modulating the mito-K_{ATP} channel continues to be an area of active scientific exploration. Although early generation

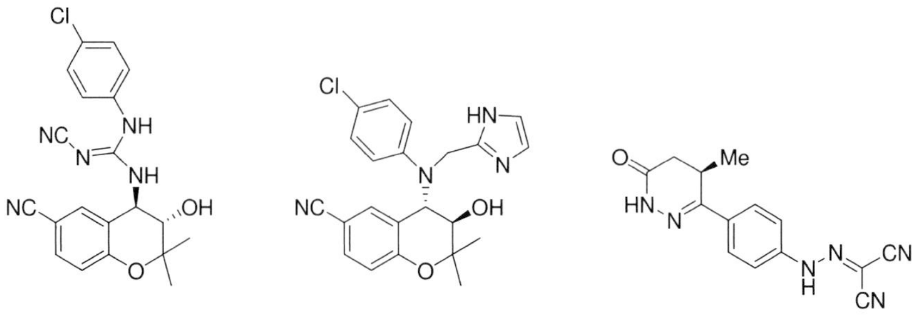

BMS-180448 (**5**)   BMS-191095 (**6**)   levosimendan (**7**)

Fig. 7.5.2 Compounds **5**–**7**.

K_{ATP} openers such as cromakalim, pinacidil and diazoxide were found to be cardioprotective [14, 17, 21], the clinical utility of these agents is limited by their potent vasodilator activities due to activity at smooth muscle K_{ATP} channels. Researchers at Bristol-Meyers Squibb have been at the forefront of efforts to identify novel agents with improved selectivity for the mito-K_{ATP} channel over the smooth muscle and cardiac sarcolemmal channels. Several benzopyran-derived mito-K_{ATP} openers have been identified that possess improved selectivity for cardioprotective effects compared to first generation agents such as cromakalim. Two representatives of the benzopyran class, BMS-180448 (**5**) and BMS-191095 (**6**) (Fig. 7.5.2), have been well characterized in the literature with regard to their *in vitro* and *in vivo* cardioprotective activities [5].

BMS-191095 has thus far demonstrated the most attractive preclinical profile for a mito-K_{ATP} opener, superior to BMS-180488. BMS-191095 has no effect *in vitro* either on cardiac sarcolemmal K_{ATP} channels (electrophysiology) or smooth muscle K_{ATP} channels (tissue bath) while demonstrating mito-K_{ATP} activity in two different *in vitro* systems and also exhibiting cardioprotective effects in isolated rat hearts [22]. *In vivo*, BMS-191095 reduced infarct size up to 70% upon intravenous administration in the dog at doses (0.6–$3.5\,\mathrm{mg\,kg}^{-1}$) that had no significant effect on hemodynamic parameters [23]. In a separate study, the dose required to reduce systemic blood pressure by 20% was at least ten-fold higher than the aforementioned cardioprotective doses, with compound solubility apparently preventing the attainment of an actual ED_{20} [22]. Additionally, BMS-191095 displayed no effect on plasma insulin levels, no negative inotropic activity, nor any hemodynamic or electrophysiological effects that could lead to proarrhythmic activity [24]. Although BMS-191095 did advance to Phase I human clinical trials for myocardial ischemia it has, unfortunately, been dropped from clinical development due to non-mechanism related neuronal toxicity [24]. Meanwhile, BMS-180448 did progress as far as Phase II trials but development has also been discontinued, although the reasons have not been disclosed [25].

With the increased appreciation of the cardioprotective potential of a mito-K_{ATP} channel opener, older drugs are being reinvestigated for activity at this channel. Minoxidil (**29**), an early K_{ATP} opener and topical treatment for alopecia [26] (*vide* *infra*), and levosimendan (**7**, Fig. 7.5.2), a calcium sensitizing positive inotropic agent for use in heart failure [27, 28], have recently been found to be mito-K$_{ATP}$ activators [29, 30]. The finding of mito-K$_{ATP}$ activity with levosimendan was particularly intriguing in light of the earlier report that it reduced myocardial infarct size in dogs in a glyburide-sensitive manner [31]. Although further study is required to ascertain the precise cardioprotective mechanism(s) of levosimendan, current evidence suggests this may be due at least partly to mito-K$_{ATP}$ activity [32]. Levosimendan has already been launched in several countries for heart failure and is currently undergoing clinical evaluation for myocardial infarction [25].

With the discontinuation of development of BMS-180448 and BMS-191095, additional selective agents are needed [33]. Unfortunately, the continued absence of a recombinant cell line expressing the mito-K$_{ATP}$ channel limits the ability of researchers to quickly screen for novel leads. Evaluating compounds for mito-K$_{ATP}$ activity and selectivity continues to involve relatively low throughput tissue bath assays that act as a bottleneck at the first step in the process of identifying promising leads. The present situation has changed little since this issue was spelled out by Atwal, nearly 10 years ago [34]. Clearly, the establishment of a cell-based assay allowing for the use of high-throughput screening tools amenable to ion channels is a requirement for further significant medicinal chemistry advances in this area.

7.5.3
Sarc-K$_{ATP}$ Blockers for Ventricular Arrhythmia

The reduction in ATP concentration during periods of cardiac ischemia opens sarc-K$_{ATP}$ channels, causing a shortening of the cardiac action potential, increasing extracellular K$^+$ concentrations that in turn can partially depolarize surrounding non-ischemic tissue and set the stage for potential ventricular tachycardias and fibrillation that can lead to sudden cardiac death [35]. Since the sarc-K$_{ATP}$ channel is closed under normal physiological conditions and does not participate in cardiac repolarization, a specific sarc-K$_{ATP}$ blocker may be able to antagonize the action potential shortening seen during ischemia while having no effect on normoxic tissue. Indeed, antidiabetic sulfonylureas like glyburide have demonstrated anti-arrhythmic activity in both clinical and preclinical studies. Unfortunately, utilization of such agents is limited by their insulin secreting activity that is seen at anti-arrhythmic doses [36]. On a molecular level this can be understood by considering that glyburide displays significantly greater affinity for the pancreatic K$_{ATP}$ subtype (SUR1/Kir6.2) over the cardiac K$_{ATP}$ subtype (SUR2A/Kir6.2) [8, 10].

The potential application of SUR2A/Kir6.2 blockade as a mechanism for the prevention of sudden cardiac death has spurred interest into identifying selective blockers of this channel. Although glyburide possessed greater potency to inhibit SUR1/Kir6.2, it nonetheless represented an attractive starting point from which to further explore the opportunities to synthesize novel agents with SUR2A/Kir6.2 selectivity. Workers at Aventis made the key discovery that replacing the sulfonylurea with sulfonylthiourea and shifting it to the meta position relative

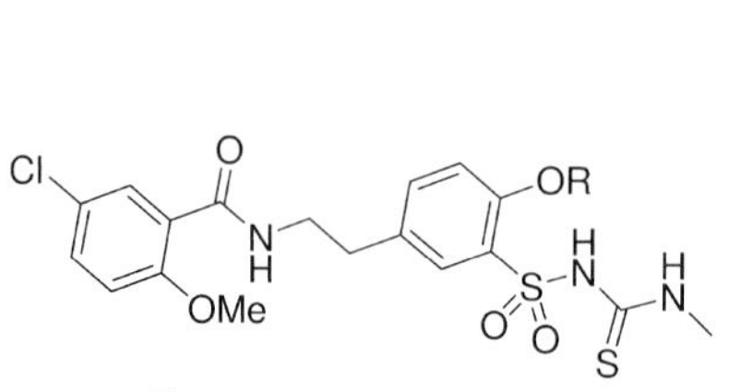

Fig. 7.5.3 Compounds **8**–**10**.

R		
Me	HMR1883 (**8**)	
(CH$_2$)$_2$OMe	HMR1402 (**9**)	

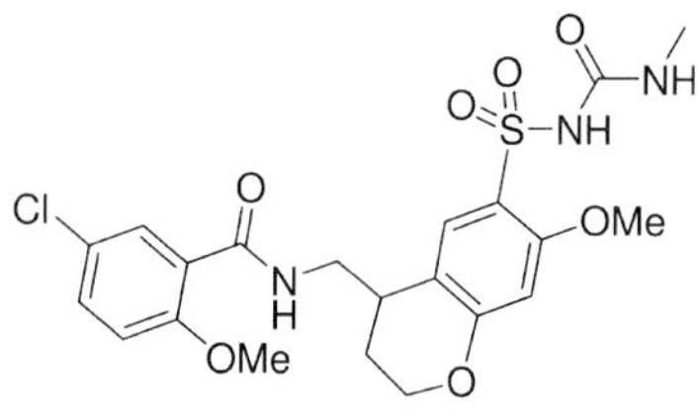

to the benzamide could yield SUR2A/Kir6.2 selective compounds as long as appropriate substituents were present on the central aromatic ring, as in the case of HMR1883 (**8**, Fig. 7.5.3) [37]. The presence of an electron-donating moiety adjacent to the sulfonylthiourea was a requirement for activity, but groups other than methoxy are tolerated (HMR1402, **9**). Although additional aromatic substituents generally reduced activity, the chromane derivative (–)-**10** did show potency and selectivity for SUR2A. For a discussion of the SAR of this series in greater detail, the reader is referred to a recent review [35].

In addition to selectivity for SUR2A within the K$_{ATP}$ family, HMR1883 is also selective versus the cardiac potassium channels that control the I$_{kr}$ and I$_{ks}$ currents [37]. *In vivo* studies in rats and dogs with HMR1883 revealed that the compound possessed antifibrillatory activity and prevented sudden cardiac death. Over the active dose range, there were no untoward effects observed on blood glucose levels or coronary blood flow. Additionally, HMR1883 had no effect on myocardial preconditioning due to the lack of blocking activity at the mito-K$_{ATP}$ channel [37]. HMR1883 did advance as far as Phase II clinical trials as the sodium salt (HMR1098) but further development has been discontinued, although the reason has not been disclosed [38].

7.5.4
SUR1/Kir6.2 Openers for Diabetes and Hyperinsulinemia

As mentioned above, the pancreatic β-cell K$_{ATP}$ (SUR1/Kir6.2) is an important mediator of insulin release. Increased ATP concentrations lead to channel block and promotion of insulin release whereas lower ATP concentrations produce channel opening and a reduction of insulin secretion [6]. Several SUR1 blockers such as glyburide have been used clinically in the treatment of type 2 diabetes due to their insulin-releasing properties [35]. Conversely, conditions associated with excessive insulin release (hyperinsulinemia) can be ameliorated by administration of SUR1 openers. Diazoxide, for example, has shown clinical efficacy in patients with nesidioblastosis (Persistent Hyperinsulinemic Hypoglycemia of Infancy, PHHI), Polycystic Ovary Syndrome, and insulin producing tumors (insulinomas) [6]. It has also been hypothesized that insulin hypersecretion due to pancreatic

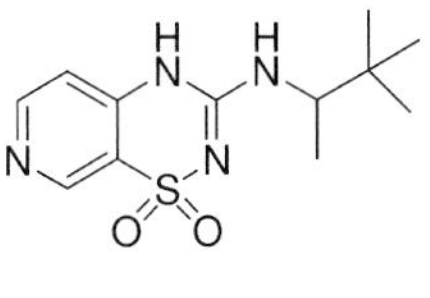

BPDZ 44 (**11**) 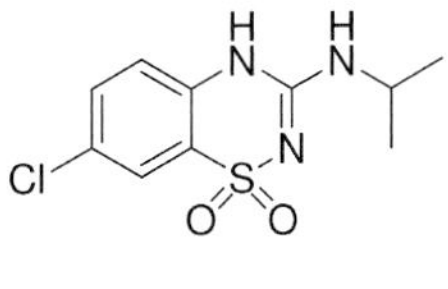   BPDZ 73 (**12**) 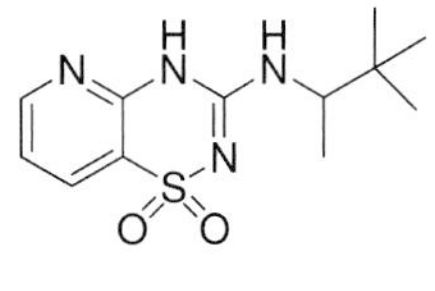   BPDZ 79 (**13**)

Fig. 7.5.4 Compounds **11**–**13**.

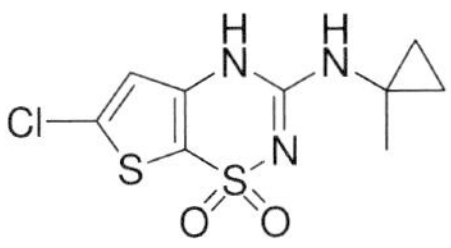

NN 414 (**14**) 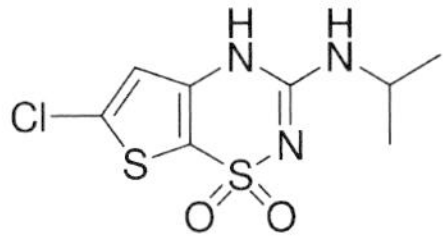   NNC 55-0118 (**15**)   **Fig. 7.5.5** Compounds **14** and **15**.

β-cell overwork may be responsible for the development of glucose intolerance and β cell degeneration in type 2 diabetes [6]. In obese, hyperinsulinemic rats, diazoxide was effective in reducing insulin secretion, improving glucose tolerance and preventing the development of insulin resistance and type 2 diabetes [39, 40]. Similar results were found in a small placebo controlled clinical trial where diazoxide provided a significant attenuation of hyperinsulinemia in obese adults [41]. Furthermore, a combination of SUR1 activation and exogenous insulin has been suggested to serve as a treatment modality to improve β cell function in type 1 or type 2 diabetes through β cell rest. Diazoxide treatment has in fact been shown to improve β cell function in type 2 diabetic patients either alone or in combination with insulin [42, 43] and also to delay the progression of type 1 diabetes in recently diagnosed patients [44]. More widespread use of this approach has, however, been severely limited by the lack of selectivity of diazoxide for SUR1, leading to SUR2-mediated side effects (edema, headache) related to its vasodilatory properties.

Drug discovery efforts have focused significant attention on modification of the diazoxide structure to enhance selectivity for SUR1 over SUR2A and SUR2B. Hybrid structures between pinacidil and diazoxide led to several interesting agents that displayed selectivity for pancreatic or cardiovascular tissue that was dependent on relatively minor structural variations. For example, BPDZ 44 (**11**) and BPDZ 73 (**12**) were early compounds selective for β cell effects whereas BPDZ 79 (**13**) showed greater selectivity towards cardiovascular tissue (Fig. 7.5.4). More recently, the thiophene-fused derivatives NN414 (**14**) and NNC 55-0118 (**15**) have been discovered to be potent and selective openers of SUR1/Kir6.2 (Fig. 7.5.5).

Detailed SAR studies in both the thieno- and benzo-fused systems have shed further light on the factors governing pancreatic selectivity [45, 46]. The alkylamino side chain appears to play a particularly important role in modulating tissue selectivity. In the thieno series, α-branched alkyl increased potency in β cells whereas a six-carbon chain imparted greater vasorelaxant potency. Incorporation of α-cycloalkyl moieties on the nitrogen led to further increases in potency such as in NN414. Interestingly, in the benzo-fused series (i.e., BPDZ 73), replacement of the 7-chloro with 7-fluoro was well tolerated but significantly attenuated the sensitivity towards variation of the alkyl substituent. Thus, whereas the N-isopropyl and N-t-butyl analogs displayed the opposite tissue selectivity in the 7-chloro variant, these same two moieties were nearly identical in terms of potency and selectivity with 7-fluoro. Much of the structure–activity relationships of this series of compounds have been recently reviewed [6].

Preclinical *in vivo* studies with NN414 have demonstrated significant potential for the treatment of disorders resulting from excessive insulin release. In a three-week study, NN414 improved first phase insulin release *ex vivo* and glucose tolerance *in vivo* in a mildly diabetic VDF rat strain [47]. Importantly, the beneficial effects of NN414 were maintained with continuous dosing. In other words, drug discontinuation prior to oral glucose tolerance test (OGTT) was not required. The long-term improvements in insulin responsiveness have been speculated to be due to β cell rest. Also, in a longer six-week study, NN414 reduced hyperinsulinemia in a dose-dependent manner in obese Zucker rats and improved glucose responsiveness [48]. NN414 (tifenazoxide) has entered human clinical trials where it was found to inhibit insulin release 1 hour post dosing in both healthy volunteers and type 2 diabetic patients. In the diabetic group, there was a trend towards improved β-cell secretory function [49]. Unfortunately, development of NN414 was recently suspended, apparently because of elevated liver enzymes that were observed in the clinic [25].

Although molecular selectivity for SUR1/Kir6.2 over either SUR2A/Kir6.2 or SUR2B/Kir6.2 expressed in oocytes has been reported for selected analogs, most of the selectivity data available in the literature was generated from studies using isolated tissue sources. Assessment of potential cardiovascular effects has been reported using either rat aorta rings or a mesenteric artery preparation. Although the latter two assays are widely used, they suffer from being relatively low throughput, and in the case of NN414 analogs, the potencies in rat aorta and mesenteric artery showed a weak correlation with each other. Given the advances in technology to measure ion channel function [50, 51] and the existence of clearly defined molecular targets that can be cloned and stably expressed, efficient means for identifying SUR1 selective compounds are now available. This situation differs considerably from that encountered in the field of mitochondrial K_{ATP} channels where the molecular identity of the channel remains in question, leaving researchers with only relatively low throughput assays to search for improved analogs.

7.5.5
SUR2B/Kir6.2 Openers for Overactive Bladder (OAB)

The presence of SUR2B/Kir6.2 K_{ATP} channels in bladder smooth muscle coupled with the ability of K_{ATP} openers to relax smooth muscle has sparked considerable interest in examining this pharmacological class as a source of potential agents to treat overactive bladder [5, 8]. The condition of OAB can be broadly classified as being of either neurogenic [52] or myogenic [53] etiology. In the former case, involuntary bladder contractions are attributed to hyperactivity of the afferent/efferent nerve pathway between the bladder and spinal cord whereas in the latter case spontaneous contractions of the bladder smooth muscle itself, in the absence of nerve input, are hypothesized to initiate an undesired micturition event. Given the ability of K_{ATP} openers to dampen cellular excitability of smooth muscle cells through their actions on lowering membrane potential, this class of agents might be expected to produce particularly beneficial effects on inhibiting bladder contractions of a myogenic origin. Numerous studies have shown that K_{ATP} openers hyperpolarize bladder cells and relax bladder strips *in vitro* [54–57].

Despite the compelling scientific rationale for the use of K_{ATP} openers in the treatment of OAB and the plethora of compounds available, very little convincing clinical evidence has been gathered to date. The apparent reason is that cardiovascular effects related to the vasorelaxant properties of all first generation K_{ATP} openers has prevented the administration of doses that might be expected to produce robust efficacy. A widely cited pilot study with cromakalim [58] has often been put forward as evidence for the utility of K_{ATP} openers in OAB, yet the small size of the trial and lack of placebo control makes it difficult to ascribe the observed effects to the K_{ATP} opening properties of cromakalim. In terms of recent clinical development of K_{ATP} openers, two compounds from AstraZeneca, ZD6169 (**16**) and ZD0947 (**17**) (Fig. 7.5.6), entered clinical trials for OAB, and advanced as far as phase 2 before discontinuation [38]. The reasons for terminating development of these compounds have not been disclosed.

When examined on a molecular level, the difficulties associated with identifying agents that are selective for bladder over vascular muscle become readily apparent. The predominant bladder K_{ATP} of SUR2B/Kir6.2 is also recognized to be widely expressed in the vasculature where its modulation can affect systemic

Tertiary carbinol   Dihydropyridine   Arylsquarate

ZD6169 (**16**)   ZD0947 (**17**)   WAY-133537 (**18**)

Fig. 7.5.6 Compounds **16–18**.

blood pressure. This likely explains why various labs have consistently failed to observe selectivity *in vitro* for bladder tissue from various species relative to a cardiovascular tissue like rat aorta or portal vein. Recently, a novel splice variant of SUR2B lacking exon 17 has been identified that predominates in the bladder, whereas the SUR2B exon 17+ has been found to be the main SUR2B isoform elsewhere [9]. Pharmacological evaluation of the two isoforms, however, failed to reveal any noteworthy selectivity of various known K_{ATP} openers, including ZD6169, a compound reported to possess bladder selective actions *in vivo* [59]. More extensive screening of structurally diverse compound libraries against these two splice variants is clearly required to more definitively establish the potential for identifying splice variant selective molecules. In this regard selectivity is clearly a more important determinant than potency and hence it may be necessary to look beyond the well-known chemotypes toward more novel structural variations. Presently, however, reported bladder selective actions of certain K_{ATP} openers (*vide infra*) are likely not attributable to selectivity on a molecular level.

The existence of a clear clinical need for safe and effective agents to treat OAB has maintained interest in pursuing K_{ATP} openers despite the uncertainties and challenges associated with this therapeutic application. Several second generation agents spanning a range of chemotypes have been identified over the past decade that reportedly possess selectivity *in vivo* to inhibit bladder overactivity with reduced effects on blood pressure relative to first generation compounds like cromakalim [5, 8]. Prominent among this group are ZD6169 (**16**), WAY-133537 (**18**), ZM244085 (**19**) and A-278637 (**20**) (Figs. 7.5.6 and 7.5.7). The most compelling preclinical data to support bladder selective actions of K_{ATP} openers has been obtained in studies that measured bladder overactivity of predominantly myogenic origin. The conclusions from these studies have added significance since similar data was generated in independent labs. This is not to say that K_{ATP} openers are ineffective at relaxing neurogenically mediated bladder contractions, just that higher doses are required. Cromakalim, ZD6169 and WAY-133537 were all more potent to inhibit myogenic bladder contractions than to inhibit neurogenic contractions when dosed intravenously in rats [60]. Similar myogenic selectivity was demonstrated with pig bladder strips *in vitro* from a range of chemical classes [57, 61, 62]. With regard to selectivity versus cardiovascular effects, WAY-133537, ZD6169 and A-278637 (**20**, Fig. 7.5.7) have all been reported to show some measure of selectivity towards the reduction of spontaneous myogenic bladder contractions on oral dosing in rats [63, 64]. Additionally, A-278637 showed selectivity on a plasma level basis when dosed intravenously in pigs and was incrementally improved over cromakalim in this regard [61]. Reported selectivities in all these studies range from 3- to approximately 20-fold. Selectivity has commonly been defined as the dose that inhibits bladder activity by 50% compared to the dose that lowers mean arterial pressure by 20%. The observed selectivities may be explained in terms of a greater sensitivity of spontaneous myogenic activity towards inhibition by K_{ATP} openers. In other words, inhibition of spontaneous myogenic activity may occur with a smaller degree of membrane hyperpolarization (hence lower dose) than that required for more global smooth muscle relaxation that

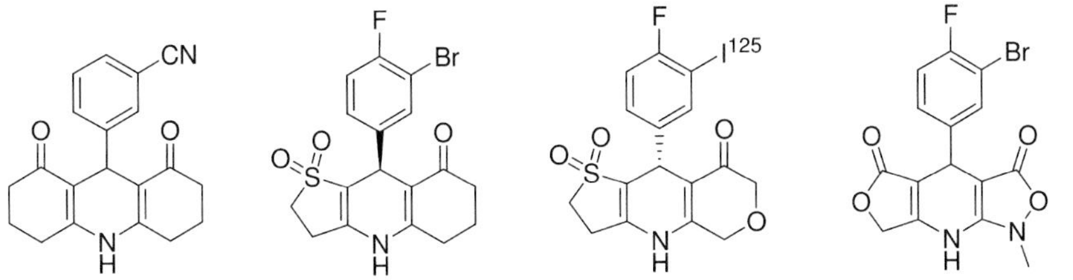

ZM244085 (**19**)   A-278637 (**20**)   A-312110 (**21**)   **22**

Fig. 7.5.7 Compounds **19–22**.

would lead to vasodilation. The relevant selectivity in the context of a clinically meaningful response remains an open question since none of the preclinical models of efficacy currently in practice have been validated with a corresponding clinical correlate.

Recent SAR studies around the two dihydropyridine-based K_{ATP} openers ZM244085 and A-278637 (Fig. 7.5.7) have significantly expanded our appreciation for the structural features affecting potency and functional efficacy at the K_{ATP} channel within this chemotype [61, 65, 66]. While only a fairly narrow range of substituents are tolerated on the aromatic ring (preferably 3,4-substituted with electron-withdrawing groups or halogens), numerous variations of the dihydro-pyridine core involving oxygen or nitrogen incorporation into the flanking rings produce highly potent K_{ATP} openers like the novel radioligand A-312110 (**21**) (K_D = 5 nM) [67] and compound **22** (EC_{50} = 13 nM) (Fig. 7.5.7). Stereochemistry of the aryl substituent was also an important determinant of potency in certain core variations [61, 65]. Although several new analogs were found from this diverse compound collection with selectivity *in vitro* for spontaneous bladder contractions, agents with substantially improved selectivity versus cardiovascular effects relative to A-278367 were not identified. A verifiable selectivity structure–activity relationship (SAR) also failed to emerge from these efforts, in large part because the assays employed were not sensitive enough to distinguish any subtle differences that may exist between compounds [61, 65].

Substitution on the tricyclic dihydropyridine core also plays an important role in modulating functional efficacy at the K_{ATP} channel. Thus, gem-dimethyl substitution on the acridinedione structure converted the potent K_{ATP} opener A-184208 (**23**) (rat aorta pEC_{50} = 7.02) into the blocker A-184209 (**24**) (rat aorta pA_2 = 6.34) (Fig. 7.5.8) [68]. Unlike glyburide, A-184209 did not discriminate substantially between pancreatic, cardiac and smooth muscle K_{ATP} channels, showing similar potency in all three tissues. A-184209 was competitively displaced by both [³H]glyburide and [¹²⁵I]-A-312110, indicating that it interacts with both opener and blocker binding sites on the SUR protein. This ability to modulate functional efficacy within a pharmacophore is reminiscent of effects previously observed in the cyanoguanidine class of K_{ATP} openers where the opener P1075 differs from the blocker PNU99963 by the hydrophobic group attached to one side of the molecule [8]. As further evidence for the ability of alkyl substitution to affect efficacy,

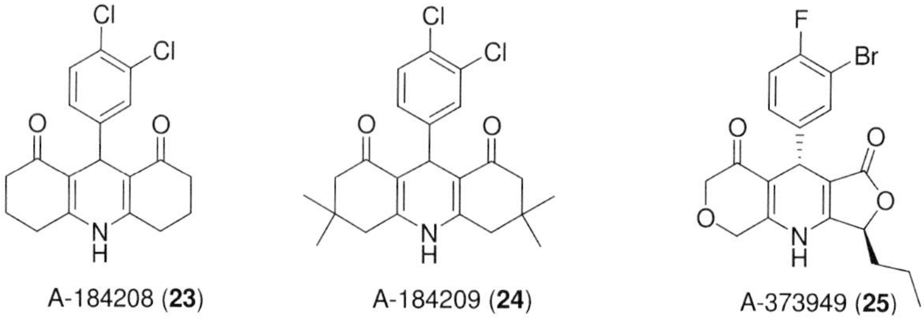

A-184208 (**23**)   A-184209 (**24**)   A-373949 (**25**)

Fig. 7.5.8 Compounds **23–25**.

certain specific combinations of core and alkyl group were found to produce apparent partial agonists at the K_{ATP} channel. A-373949 (**25**, Fig. 7.5.8) was found to possess the intriguing property of enhanced functional selectivity for myogenic over neurogenic contractions *in vitro*, potently and completely inhibiting spontaneously contracting pig bladder strips but only partially inhibiting (34%) electrically driven contractions in pig bladder strips [69].

Importantly, the myogenic functional selectivity described above *in vitro* and *in vivo* differs in principle from molecular selectivity as the channel in question is presumably the same in either the myogenic, neurogenic or cardiovascular assay. Since potency to inhibit neurogenically mediated bladder contractions has tended to track with potency to reduce systemic blood pressure [60], enhanced selectivity for spontaneous contractions through modulation of functional efficacy may provide a path forward towards agents with a greater separation between effects on diseased bladder contractions and cardiovascular effects. Additionally, a preferential reduction in functional efficacy for non-myogenic activity (partial agonism) might be expected to produce a blunted maximal hypotensive response, thus limiting the potential adverse consequences of a precipitous drop in blood pressure. Although another potential consequence of partial agonism is an eventual ceiling on the desired bladder relaxant effect, this may prove indirectly beneficial as complete bladder stasis is also undesirable due to the likelihood of producing urinary retention. One practical impediment to the identification of partial agonist K_{ATP} openers is the limitation of existing assays to distinguish potent, low efficacy compounds from assay noise. For example, the efficacious plasma concentration (EC_{50}) of A-278637 in a pig model of myogenic instability was only 8 nM [61], a concentration at which this compound produced < 20% response in a fluorescence-based *in vitro* functional assay [62]. Although this may be additional evidence that full efficacy *in vitro* is not a requirement for robust efficacy *in vivo*, it certainly presents a screening challenge from a drug discovery perspective. In this regard, a binding assay with the radioligand A-312110 may prove useful as a screen to look for potent binders that exhibit low efficacy or appear otherwise inactive as K_{ATP} openers.

As an outgrowth of efforts to expand the diversity of chemotypes available from which to explore for bladder selective agents, several novel pharmacophores or structural variations have been reported recently (Fig. 7.5.9). The dihydropyrimi-

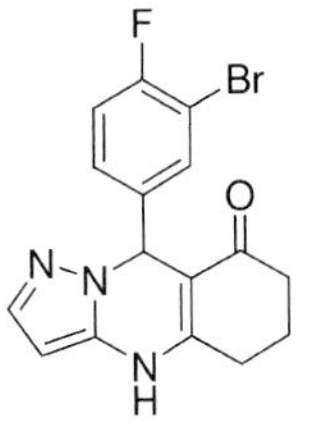

Fig. 7.5.9 Compounds **26–28**.

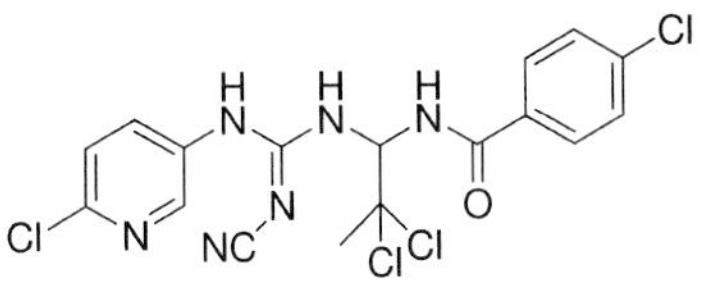

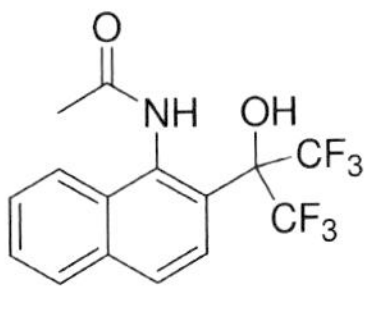

dine **26** [70] represents a novel hybrid structure between ZM244085 and a pyrazo-lopyrimidine high-throughput screening lead. Much of the SAR with this novel structure paralleled that seen with the dihydropyridine-based cores and a number of highly potent (EC$_{50}$ < 50 nM) analogs were identified.

The aminal-containing cyanoguanidine **27** has some similarities to pinacidil yet possesses a very different amide appendage on the right-hand side that is clearly distinct from the hydrophobic group of the predecessor compounds [71, 72]. Compound **27** demonstrated similar potency to cromakalim *in vitro* (pig bladder strip pEC$_{50}$ = 6.6) and, contrary to the expectation of low stability for the aminal moiety, had excellent pharmacokinetic properties in the dog (oral bioavailability = 100%, $t_{1/2}$ = 15 h). Although various substituents on the two aromatic rings could be tolerated, 4-substitution on the right-hand side and 5- or 6-substitution on the left-hand pyridine ring provided the most potent compounds. Replacement of the cyanoguanidine with several isosteres led to a reduction in potency whereas dichloroethyl, trichloromethyl and *tert*-butyl were optimal at the aminal carbon. This series of compounds further illuminates the possibilities to expand structural diversity of K$_{ATP}$ openers in unforeseen directions.

Finally, the tertiary carbinol A-151892 (**28**, Fig. 7.5.9) bears only minimal resemblance to ZD6169 among known K$_{ATP}$ openers and was found to be considerably more potent [73]. In this series various alkyl, aryl and heteroaryl amides provided potent (EC$_{50}$ < 50 nM) K$_{ATP}$ openers at the recombinant SUR2B/Kir6.2 splice variant present in bladder.

Despite significant efforts over the past decade by several pharmaceutical companies to develop a K$_{ATP}$ opener for OAB, none of the second generation agents have yet advanced towards late stage clinical evaluation. Identifying agents with adequate selectivity versus cardiovascular effects continues to be the primary challenge in this regard and those compounds that have emerged to date were found to be selective only using laborious *in vivo* assays. *In vivo* efficacy to inhibit myogenic bladder overactivity has been described for both the novel chemotypes **27** and A-151892 but significant improvements in selectivity versus cardiovascular effects from these structures have yet to be reported. A critical limitation confronting researchers in this field is the highly homologous nature of the K$_{ATP}$ channel subtype in both bladder and vasculature that may render true molecular selectivity unattainable. Although functional selectivity for spontaneous contractions may re-

minoxidil (**29**) 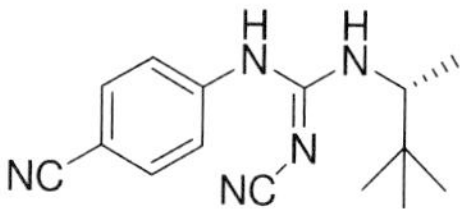   naminidil (**30**)

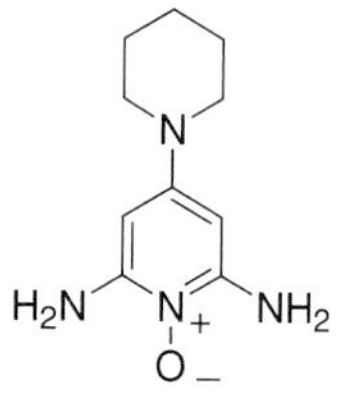

Fig. 7.5.10 Compounds **29** and **30**.

present a path forward, assay methodology to identify such agents either involves the use of low-throughput *in vivo* or tissue bath experiments or awaits the implementation of more sensitive cell-based ion channel screening assays that can detect subtle channel modulating effects of drugs.

7.5.6
K$_{ATP}$ Openers for Alopecia

It has been known since the early 1970s that the K$_{ATP}$ opener minoxidil (**29**, Fig. 7.5.10) causes hypertrichosis, including the promotion of hair regrowth in balding men [26]. These initial observations were made in the context of side-effects noted in patients taking minoxidil for hypertension. In the K$_{ATP}$ opener class, hypertrichosis is not unique to minoxidil, having also been observed in patients treated with diazoxide or pinacidil. Unlike these latter two agents, however, minoxidil is currently available as a topical treatment for alopecia in both men and women (Rogaine®).

Although minoxidil is a clinically proven agent that can stimulate hair growth, its precise mechanism of action is not well understood. There is certainly some evidence that points to the involvement of K$_{ATP}$ channels; however, to date, K$_{ATP}$ channels have not been shown to be present in hair follicles. The recent review by Messenger and Rundegren nicely summarizes the current evidence related to a potential K$_{ATP}$-mediated mechanism as well as other possible mechanisms [26]. Naminidil is a cyanoguanidine K$_{ATP}$ opener related to pinacidil that has recently surfaced as a potential drug candidate for alopecia [38].

7.5.7
Conclusions

The K$_{ATP}$ channel in cardiomyocytes was first described over 20 years ago [74]. In the intervening years, tremendous advances have been made in terms of our understanding of the diversity, structure and function of this ubiquitous K$^+$ channel. These biological advancements have been paralleled by and in many cases preceded by an enhanced appreciation on the chemistry side for the structural types that modulate K$_{ATP}$ channel subtype activity. Until very recently, K$_{ATP}$ subtype selectivity was assessed only in terms of effects on isolated cells or tissues,

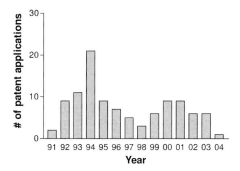

Fig. 7.5.11 K_{ATP} drug-related patent applications by year. (Data obtained from analysis of search results in SciFinder® for K_{ATP} openers and blockers.)

or in whole animal experiments. The use of such methods in drug discovery, although critical at the latter stages of a program, are no longer adequate in the early stages of lead identification/optimization due to their limited capacity for data turnaround.

Although more efficient methods of assessing function and potency at most K_{ATP} channel subtypes now exist, the availability of such techniques arrived after the apparent peaks of interest in exploring the therapeutic potential of K_{ATP} channel modulation. As shown in Fig. 7.5.11, patent activity in the K_{ATP} area since 1991 has been cyclical in nature with peaks around 1993–1994 and 2000–2001 as different players have entered and left the field. So although better tools now provide an opportunity to fully exploit the potential of the K_{ATP} channel, pharmaceutical interest appears to be presently experiencing a lull. Whether this is only a temporary lull resulting from a refocusing of attention to the most promising therapeutic applications from the standpoint of current medical need and potential for clinical success remains to be seen.

The recent downturn in interest in K_{ATP} channel modulation in a broad sense (i.e., widespread interest in multiple therapeutic applications) can be best understood in the context of the cost and difficulty associated with obtaining truly selective agents. Few selective 2nd generation K_{ATP} modulating drugs have entered clinical trials (Table 7.5.2) and thus far none have made it to the market despite 20 years of advancement of the science in both industrial and academic labs. The only currently marketed K_{ATP} openers and blockers (Table 7.5.1) continue to be those that were discovered before the recognition of their mechanisms of action, and although levosimendan has recently been approved in some markets for heart failure this is not believed to be related to its K_{ATP} activity. The mito-K_{ATP} activity of levosimendan is still being evaluated in clinical trials for myocardial ischemia.

Selectivity versus hypotensive effects mediated by activity at SUR2B has been and continues to be the primary challenge to be overcome for a K_{ATP} opener. In this regard, the most promising area poised to take advantage of the biological advances in the K_{ATP} field are SUR1/Kir6.2 openers for insulin-releasing disorders since discrimination between SUR1 and SUR2B by K_{ATP} openers has already been demonstrated and the molecular tools are available. Likewise SUR2A/Kir6.2 blockers for ventricular arrhythmia also benefit from the availability of reagents

Tab. 7.5.2 Selective 2nd generation K$_{ATP}$ modulating drugs.

Drug	Indications	K$_{ATP}$ activity	Clinical result
BMS-180448 (5)	Myocardial ischemia	Mito K$_{ATP}$ opener	Discontinued phase 2
BMS-191095 (6)	Myocardial ischemia	Mito K$_{ATP}$ opener	Discontinued phase 1
Levosimendan (7)	Myocardial ischemia	Mito K$_{ATP}$ opener	Phase 2 trials
HMR1883	Ventricular fibrillation	SUR2A/Kir6.2 blocker	Discontinued phase 2
NN 414 (14)	Hyperinsulinemia	SUR1/Kir6.2 opener	Suspended phase 2
ZD6169 (16)	Overactive bladder	SUR2B/Kir6.2 opener	Discontinued phase 2
ZD0947 (17)	Overactive bladder	SUR2B/Kir6.2 opener	Discontinued phase 2

and techniques to continue progress in this field. Conversely, substantial progress in the mito-K$_{ATP}$ field would be greatly facilitated by the cloning and expression of this intriguing channel. Future opportunities for K$_{ATP}$ openers in OAB depend upon the ability to translate the selectivity observed preclinically for inhibiting myogenic bladder overactivity to the clinical setting. Beyond that, the field awaits novel structures that can specifically discern the SUR2B splice variant that predominates in the bladder, or other agents capable of more subtly modulating channel function without producing hypotensive responses.

References

1. Quayle, J. M., Nelson, M. T., Standen, N. B. ATP-sensitive and inwardly rectifying potassium channels in smooth muscle. *Physiol. Rev.* **1997**, *77*, 1165–1232.
2. Aguilar-Bryan, L., Clement, J. P. I. V., Gonzalez, G., Kunjilwar, K., Babenko, A., Bryan, J. Toward understanding the assembly and structure of K$_{ATP}$ channels. *Physiol. Rev.* **1998**, *78*, 227–245.
3. Ashcroft, F. M., Gribble, F. M. Correlating structure and function in ATP-sensitive K$^+$ channels. *Trends Neurosci.* **1998**, *21*, 288–294.
4. Miki, T., Inagaki, N., Nagashima, K., Gonoi, T., Seino, S. Structure and function of ATP-sensitive potassium channels. *Curr. Top. Membr.* **1999**, *46*, 373–385.
5. Mannhold, R. K$_{ATP}$ channel openers: structure-activity relationships and therapeutic potential. *Med. Res. Rev.* **2004**, *24*, 213–266.
6. Bondo Hansen, J., Arkhammar, P. O. G., Bodvarsdottir, T. B., Wahl, P. Inhibition of insulin secretion as a new drug target in the treatment of metabolic disorders. *Curr. Med. Chem.* **2004**, *11*, 1595–1615.
7. Hambrock, A., Preisig-Müller, R., Russ, U., Piehl, A., Hanley, P. J., Ray, J., Daut, J., Quast, U., Derst, C. Four novel splice variants of sulfonylurea receptor 1. *Am. J. Physiol. Cell Physiol.* **2002**, *283*, C587–C598.
8. Coghlan, M. J., Carroll, W. A., Gopalakrishnan, M. Recent developments in the biology and medicinal chemistry of potassium channel modulators: update from a decade of progress. *J. Med. Chem.* **2001**, *44*, 1627–1653.
9. Scott, V. E., Davis-Taber, R. A., Silvia, C., Hoogenboom, L., Choi, W., Kroeger, P., Whiteaker, K. L., Gopalakrishnan, M. Characterization of human urinary bladder K$_{ATP}$ channels containing SUR2B

splice variants expressed in L-cells. *Eur. J. Pharmacol.* **2004**, *483*, 195–205.

10. Miura, T., Miki, T. ATP-sensitive K$^+$ channel openers: old drugs with new clinical benefits for the heart. *Curr. Vasc. Pharmacol.* **2003**, *1*, 251–258.

11. http://www.healthdigest.org/drugs/diazoxideiv.html

12. http://www.healthdigest.org/drugs/diazoxideoral.html

13. O'Rourke, B. Evidence for mitochondrial K$^+$ channels and their role in cardioprotection. *Circ. Res.* **2004**, *94*, 420–432.

14. Patel, H. H., Gross, G. J. Mitochondrial K$_{ATP}$ channels and cardioprotection. *Drug Dev. Res.* **2002**, *55*, 17–21.

15. Grover, G. J., Garlid, K. D. ATP-sensitive potassium channels: A review of their cardioprotective pharmacology. *J. Mol. Cell. Cardiol.* **2000**, *32*, 677–695.

16. Oldenburg, O., Cohen, M. V., Yellon, D. M., Downey, J. M. Mitochondrial K$_{ATP}$ channels: role in cardioprotection. *Cardiovasc. Res.* **2002**, *55*, 429–437.

17. Peart, J. N. Gross, G. J. Sarcolemmal and mitochondrial K$_{ATP}$ channels and myocardial ischemic preconditioning. *J. Cell. Mol. Med.* **2002**, *6*, 453–464.

18. Liu, Y., Ren, G., O'Rourke, B., Marbán, E., Seharaseyon, J. Pharmacological comparison of native mitochondrial K$_{ATP}$ channels with molecularly defined surface K$_{ATP}$ channels. *Mol. Pharmacol.* **2001**, *59*, 225–230.

19. Lacza, Z. Snipes, J. A., Busija, D. W. Characterization of the mitochondrial ATP-dependent K$^+$ channel in the mouse heart [Abstract]. *Circulation* **2002**, *106* (Suppl II), II-194.

20. Lacza, Z., Snipes, J. A., Kis, B., Csaba, S., Grover, G. Busija, D. Investigation of the subunit composition and pharmacology of the mitochondrial ATP-dependent K$^+$ channel in the brain. *Brain Res.* **2003**, *994*, 27–36.

21. Garlid, K. D., Paucek, P., Yarov-Yarovoy, V., Murray, H. N., Darbenzio, R. B., D'Alonzo, A. J., Lodge, N. J., Smith, M. A., Grover, G. J. Cardioprotective effect of diazoxide and its interaction with mitochondrial ATP-sensitive K$^+$ channels: possible mechanism of cardioprotection. *Circ. Res.* **1997**, *81*, 1072–1082.

22. Grover, G. J., D'Alonzo, A. J., Garlid, K. D., Bajgar, R., Lodge, N. J., Sleph, P. G., Darbenzio, R. B., Hess, T. A., Smith, M. A., Paucek, P., Atwal, K. S. Pharmacologic characterization of BMS-191095, a mitochondrial K$_{ATP}$ opener with no peripheral vasodilator or cardiac action potential shortening activity. *J. Pharmacol. Exp. Ther.* **2001**, *297*, 1184–1192.

23. Grover, G. J., D'Alonzo, A. J., Darbenzio, R. B., Parham, C. S., Hess, T. A., Bathala, M. S. In vivo characterization of the mitochondrial selective KATP opener (3*R*)-*trans*-4-(4-chlorophenyl)-N-(1*H*-imidazol-2-ylmethyl)dimethyl-2*H*-1-benzopyran-6-carbonitrile monohydrochloride (BMS-191095): cardioprotective, hemodynamic and electrophysiological effects. *J. Pharmacol. Exp. Ther.* **2002**, *303*, 132–140.

24. Grover, G. J., Atwal, K. S. Pharmacologic profile of the selective mitochondrial-K$_{ATP}$ opener BMS-191095 for treatment of acute myocardial ischemia. *Cardiovasc. Drug Rev.* **2002**, *20*, 121–136.

25. Adis R&D Insight Database (accessed January, 2005).

26. Messenger, A. G., Rundegren, J. Minoxidil: mechanisms of action on hair growth. *Br. J. Dermatol.* **2004**, *150*, 186–194.

27. Toller, W. G., Pagel, P. S., Kersten, J. R., Warltier, D. C. Levosimendan. *Drugs Future* **2000**, *25*, 563–568.

28. Nawarskas, J. J., Anderson, J. R. Levosimendan: A unique approach to the treatment of heart failure. *Heart Disease* **2002**, *4*, 265–271.

29. Sato, T., Li, Y., Saito, T., Nakaya, H. Minoxidil opens mitochondrial K$_{ATP}$ channels and confers cardioprotection. *Br. J. Pharmacol.* **2004**, *141*, 360–366.

30. Kopustinskiene, D. M., Pollosello, P., Saris, N.-E. L. Levosimendan is a mitochondrial K$_{ATP}$ channel opener. *Eur. J. Pharmacol.* **2001**, *428*, 311–314.

31. Kersten, J. R., Montgomery, M. W., Pagel, P. S., Warltier, D. C. Levosimendan, a new positive inotropic drug, decreases myocardial infarct size via activation of K$_{ATP}$ channels. *Anesth. Analg.* **2000**, *90*, 5–11.

32. Kopustinskiene, D. M., Pollosello, P., Saris, N.-E. L. Potassium-specific effects of levosimendan on heart mitochondria. *Biochem. Pharmacol.* **2004**, *68*, 807–812.

33. Lee, B. H., Seo, H. W., Yoo, S.-E. Cardioprotective effects of (2S, 3R, 4S)-N′-benzyl-N′′-cyano-N-(3,4-dihydro-2-dimethoxymethyl-3-hydroxy-2-methyl-6-nitro-2H-benzopyran-4-yl)-guanidine (KR-31372) in rats and dogs. *Pharmacology* **2004**, *70*, 74–82.

34. Atwal, K. S. Myocardial protection with the ATP-sensitive potassium channel openers. *Curr. Med. Chem.* **1996**, *3*, 227–238.

35. Englert, H. C., Heitsch, H., Gerlach, U., Knieps, S. Blockers of the ATP-sensitive potassium channel SUR2A/Kir6.2: A new approach to prevent sudden cardiac death. *Curr. Med. Chem.* **2003**, *1*, 253–271.

36. Billman, G. E. Animal models of lethal arrhythmias *Drug Dev. Res.* **2002**, *55*, 59–72.

37. Englert, H. C., Gerlach, U., Goegelein, H., Hartung, J., Heitsch, H., Mania, D., Scheidler, S. Cardioselective K_{ATP} channel blockers derived from a new series of m-anisamidoethylbenzenesulfonylthioureas. *J. Med. Chem.* **2001**, *44*, 1085–1098.

38. Pharmaprojects database access January 2005.

39. Aizawa, T., Taguchi, N., Sato, Y., Nakabayashi, T., Kobuchi, H., Hidaka, H., Nagasawa, T., Ishihara, F., Itoh, N., Hashizume, K. Prophylaxis of genetically determined diabetes by diazoxide: A study in a rat model of naturally occurring obese diabetes. *J. Pharmacol. Exp. Ther.* **1995**, *275*, 194–199.

40. Alemzadeh, R., Slonim, A. E., Zdanowicz, M. M., Maturo, J. Modification of insulin resistance by diazoxide in obese Zucker rats. *Endocrinology* **1993**, *133*, 705–712.

41. Alemzadeh, R., Langley, G., Upchurch, L., Smith, P., Slonim, A. E. Beneficial effect of diazoxide in obese hyperinsulinemic adults. *J. Clin. Endocrinol. Metab.* **1998**, *83*, 1911–1915.

42. Greenwood, R. H., Mahler, R. F., Hales, C. N. Improvement in insulin secretion in diabetes after diazoxide. *The Lancet* **1976**, 444–446.

43. Guldstrand, M., Grill, V., Björklund, A., Lins, P. E., Adamson, U. Improved beta cell function after short-term treatment with diazoxide in obese subjects with type 2 diabetes. *Diabetes Metab. (Paris)* **2002**, *28*, 448–456.

44. Bjork, E., Berne, C., Kampe, O., Wibell, L., Oskarsson, P., Karlsson, F. A. Diazoxide treatment at onset preserves residual insulin secretion in adults with autoimmune diabetes. *Diabetes* **1996**, *45*, 1427–1430.

45. Nielsen, F. E., Bodvarsdottir, T. B., Worsaai, A., MacKay, P., Stidsen, C. E., Boonen, H. C. M., Pridal, L., Arkhammar, P. O. G., Wahl, P., Ynddal, L., Junager, F., Dragsted, N., Tagmose, T. M., Mogensen, J. P., Koch, A., Treppendahl, S. P., Bondo Hansen, J. 6-Chloro-3-alkylamino-4H-thieno[3,2-e]-1,2,4-thiadiazine 1,1-dioxide derivatives potently and selectively activate ATP sensitive potassium channels of pancreatic β-cells. *J. Med. Chem.* **2002**, *45*, 4171–4187.

46. de Tullio, P., Becker, B., Boverie, S., Dabrowski, M., Wahl, P., Antoine, M.-H., Somers, F., Sebille, S., Ouedraogo, R., Bondo Hansen, J., Lebrun, P., Pirotte, B. Toward tissue-selective pancreatic B-cells K_{ATP} channel openers belonging to 3-alkylamino-7-hal-4H-1,2,4-benzothiadiazine 1,1-dioxides. *J. Med. Chem.* **2003**, *46*, 3342–3353.

47. Carr, R. D., Brand, C. L., Bodvarsdottir, T. B., Bondo Hansen, J., Sturis, J. NN414, a SUR1/Kir6.2-selective potassium channel opener, reduces blood glucose and improves glucose tolerance in the VDF Zucker rat. *Diabetes*, **2003**, *52*, 2513–2518.

48. Alemzadeh, R., Fledelius, C., Bodvarsdottir, T., Sturis, J. Attenuation of hyperinsulinemia by NN414, a SUR1/Kir6.2 selective K+-adenosine triphosphate channel opener, improves glucose tolerance and lipid profile in obese Zucker rats. *Metabolism* **2004**, *53*, 441–447.

49. Choi, J.-K. NN-414. *Curr. Opin. Invest. Drugs* **2003**, *4*, 455–458.

50. Bennett, P. B., Guthrie, H. R. E. Trends in ion channel drug discovery: Advances in screening technologies. *Trends Biotechnol.* **2003**, *21*, 563–569.

51. Worley, J. F., Main, M. J. An industrial perspective on utilizing functional ion channel assays for high throughput

screening. *Receptors Channels* 2002, *8*, 269–282.

52. De Groat, W. C. A neurologic basis for the overactive bladder. *Urology* 1997, *50* (Suppl A), 36–52.

53. Brading, A. F. A myogenic basis for the overactive bladder. *Urology* 1997, *50* (Suppl A), 57–67.

54. Turner, W. H., Brading, A. F. Smooth muscle of the bladder in the normal and the diseased state: Pathophysiology, diagnosis and treatment. *Pharmacol. Ther.* 1997, *75*, 77–110.

55. Gopalakrishnan, M., Whiteaker, K. L., Molinari, E. J., Davis-Taber, R., Scott, V. E. S., Shieh, C.-C., Buckner, S. A., Milicic, I., Cain, J. C., Postl, S., Sullivan, J. P., Brioni, J. D. Characterization of the ATP-sensitive potassium channels (K_{ATP}) expressed in guinea pig bladder smooth muscle cells. *J. Pharmacol. Exp. Ther.* 1999, *289*, 551–558.

56. Buckner, S. A., Milicic, I., Daza, A., Davis-Taber, R., Scott, V. E. S., Sullivan, J. P., Brioni, J. D. Pharmacological and molecular analysis of ATP-sensitive K^+ channels in the pig and human detrusor. *Eur. J. Pharmacol.* 2000, *400*, 287–295.

57. Buckner, S. A., Milicic, I., Daza, A. V., Coghlan, M. J., Gopalakrishnan, M. Spontaneous phasic activity of the pig urinary bladder smooth muscle: Characteristics and sensitivity to potassium channel modulators. *Br. J. Pharmacol.* 2002, *135*, 639–648.

58. Nurse, D. E., Restorick, J. M., Mundy, A. R. The effect of cromakalim on the normal and hyperreflexic human detrusor muscle. *Br. J. Urol.* 1991, *68*, 27–31.

59. Howe, B. B., Halterman, T. J., Yochim, C. L., Do, M. L., Pettinger, S. J., Stow, R. B., Ohnmacht, C. J., Russell, K., Empfield, J. R., Trainor, D. A., Brown, F. J., Kau, S. T. ZENECA ZD6169: A novel K_{ATP} channel opener with *in vivo* selectivity for urinary bladder. *J. Pharmacol. Exp. Ther.* 1995, *274*, 884–890.

60. Fabiyi, A. C., Gopalakrishnan, M., Lynch, J. J., Brioni, J. D., Coghlan, M. J., Brune, M. E. *In vivo* evaluation of the potency and bladder-vascular selectivity of the ATP-sensitive potassium channel openers (-) cromakalim, ZD6169 and WAY-133537 in rats. *BJU Int.* 2003, *91*, 284–290.

61. Carroll, W. A., Altenbach, R. A., Bai, H., Brioni, J. D., Brune, M. E., Buckner, S. A., Cassiday, C., Chen, Y., Coghlan, M. J., Daza, A. V., Drizin, I., Fey, T. A., Fitzgerald, M., Gopalakrishnan, M., Gregg, R. J., Henry, R. F., Holladay, M. W., King, L. L., Kort, M. E., Kym, P. R., Milicic, I., Tang, R., Turner, S. C., Whiteaker, K. L., Yi, L., Zhang, H., Sullivan, J. P. Synthesis and structure-activity relationships of a novel series of 2,3,5,6,7,9-hexahydrothieno[3,2-*b*]quinolin-8(4*h*)-one 1,1-dioxide K_{ATP} channel openers: discovery of (-)-(9*S*)-9-(3-bromo-4-fluorophenyl)-2,3,5,6,7,9-hexahydrothieno[3,2-*b*]quinolin-8(4*H*)-one 1,1-dioxide (A-278637), a potent K_{ATP} opener that selectively inhibits spontaneous bladder contractions. *J. Med. Chem.* 2004, *47*, 3163–3179.

62. Gopalakrishnan, M., Buckner, S. A., Whiteaker, K. L., Shieh, C.-C., Molinari, E. J., Milicic, I., Daza, A. V., Davis-Taber, R., Scott, V. E., Sellers, D., Chess-Williams, R., Chapple, C. R., Liu, Y., Liu, D., Brioni, J. D., Sullivan, J. P., Williams, M., Carroll, W. A., Coghlan, M. J. (-)-(9*S*)-9-(3-Bromo-4-fluorophenyl)-2,3,5,6,7,9-hexahydrothieno[3,2-*b*]quinolin-8(4*H*)-one 1,1-dioxide (A-278637): A novel ATP-sensitive potassium channel opener efficacious in suppressing urinary bladder contractions. I. In vitro characterization. *J. Pharmacol. Exp. Ther.* 2002, *303*, 379–386.

63. Butera, J. A., Antane, M. M., Antane, S. A., Argentieri, T. M., Freeden, C., Graceffa, R. F., Hirth, B. H., Jenkins, D., Lennox, J. R., Matelan, E., Norton, N. W., Quagliato, D., Sheldon, J. H., Spinelli, W., Warga, D., Wojdan, A., Woods, M. Design and SAR of novel potassium channel openers targeted for urge urinary incontinence I. N-cyanoguanidine bioisosteres possessing in vivo bladder selectivity. *J. Med. Chem.* 2000, *43*, 1187–1202.

64. Lynch, J. J., Brune, M. E., Lubbers, N. L., Coghlan, M. J., Cox, B. F., Polakowski, J. S., King, L. L., Sullivan, J. P., Brioni, J. D. K-ATP opener-mediated attenuation of spontaneous bladder contractions in ligature-intact, partial bladder outlet obstructed rats. *Life Sci.* 2003, *72*, 1931–1941.

65. Carroll, W. A., Agrios, K. A., Altenbach, R. A., Buckner, S. A., Chen, Y., Coghlan, M. J., Daza, A. V., Drizin, I., Gopalakrishnan, M., Henry, R. F., Kort, M. E., Kym, P. R., Milicic, I., Smith, J. C., Tang, R., Turner, S. C., Whiteaker, K. L., Zhang,H., Sullivan, J. P. Synthesis and structure-activity relationships of a novel series of tricyclic dihydropyridine-based K_{ATP} openers that potently inhibit bladder contractions in vitro. *J. Med. Chem.* **2004**, *47*, 3180–3192.

66. Drizin, I., Altenbach, R. A., Buckner, S. A., Whiteaker, K. L., Scott, V. E., Darbyshire, J. F., Jayanti, V., Henry, R. F., Coghlan, M. J., Gopalakrishnan, M., Carroll, W. A. Structure-activity studies for a novel series of tricyclic dihydropyridopyrazolones and dihydropyridoisoxazolones as K_{ATP} channel openers. *Bioorg. Med. Chem.* **2004**, *12*, 1895–1904.

67. Davis-Taber, R., Molinari, E. J., Altenbach, R. J., Whiteaker, K. L., Shieh, C.-C., Rotert, G., Buckner, S. A., Malysz, J., Milicic, I., McDermott, J. S., Gintant, G. A., Coghlan, M. J., Carroll, W. A., Scott, V. E., Gopalakrishnan, M. [125I]A-312110, a novel high-affinity 1,4-dihydropyridine ATP-sensitive K+ channel opener: Characterization and pharmacology of binding. *Mol. Pharmacol.* **2003**, *64*, 143–153.

68. Gopalakrishnan, M., Miller, T. R., Buckner, S. A., Milicic, I., Molinari, E. J., Whiteaker, K. L., Davis-Taber, R., Scott, V. E., Cassidy, C., Sullivan, J. P., Carroll, W. A. Pharmacological characterization of a 1,4-dihydropyridine analogue, 9-(3,4-dichlorophenyl)-3,3,6,6-tetramethyl-3,4,6,7,9,10-hexahydro-1,8-(2H,5H)-acridinedione (A-184209) as a novel KATP channel inhibitor. *Br. J. Pharmacol.* **2003**, *138*, 393–399.

69. Carroll, W. A., Brune, M. E., Buckner, S. A., Gopalakrishnan, M., Coghlan, M. J., Scott, V. E. S., Shieh, C.-C., Sullivan, J. P. Targeting the inhibition of spontaneous myogenic contractions as a basis for KATP openers with selectivity for bladder. *227th American Chemical Society National Meeting*, Anaheim, CA, March **2004**, Division of Medicinal Chemistry Abstract No. 007.

70. Drizin, I., Holladay, M. W., Yi, L., Zhang, H. Q., Gopalakrishnan, S., Gopalakrishnan, M., Whiteaker, K. L., Buckner, S. A., Sullivan, J. P., Carroll. W. A. Structure-activity studies for a novel series of tricyclic dihydropyrimidines as K_{ATP} openers (KCOs). *Bioorg. Med. Chem. Lett.* **2002**, *12*, 1481–1484.

71. Perez-Medrano, A., Buckner, S. A., Coghlan, M. J., Gregg, R. J., Gopalakrishnan, M., Kort, M. E., Lynch, J. K., Scott, V. E., Sullivan, J. P., Whiteaker, K. L., Carroll, W. A. Design and synthesis of novel cyanoguanidine ATP-sensitive potassium channel openers for the treatment of overactive bladder. *Bioorg. Med. Chem. Lett.* **2004**, *14*, 397–400.

72. Perez-Medrano, A., Kort, M. E., Gregg, R. J., Altenbach, R. J., Carroll, W. A., Gopalakrishnan, M., Scott, V. E., Whiteaker, K. L., Buckner, S. A., Brune, M. E., Coghlan, M. J. Design and synthesis of aminal-containing ATP-sensitive potassium channel openers for the treatment of bladder overactivity. *224th American Chemical Society National Meeting*, Boston, MA, August **2002**, Division of Medicinal Chemistry Abstract No. 352.

73. Turner, S. C., Carroll, W. A., White, T. K., Brune, M. E., Buckner, S. A., Gopalakrishnan, M., Fabiyi, A., Coghlan, M. J., Scott, V. E., Castle, N. A., Daza, A. V., Milicic, I., Sullivan, J. P. Structure-activity relationship of a novel class of naphthyl amide K_{ATP} channel openers. *Bioorg. Med. Chem. Lett.* **2003**, *13*, 1741–1744.

74. Noma, A. ATP-regulated K+ channels in cardiac muscle. *Nature* **1983**, *305*, 147–148.

7.6
Compounds that Activate KCNQ(2–5) Family of Potassium Ion Channels

Grant McNaughton-Smith and Alan D. Wickenden

7.6.1
Introduction

KCNQ channels are a family of six transmembrane domain, single pore-loop, voltage-gated K^+ channels. Five members of the family have been identified to date, including the cardiac channel KCNQ1 (formerly known as KvLQT1) and four neuronal KCNQ channels, KCNQ(2–5). In the most recently agreed nomenclature, KCNQ(2–5) channels have been designated as Kv7.2–7.5, respectively [1]. The biology of KCNQ channels is described in detailed elsewhere in Chapter 7. For additional background information, the reader is referred to several excellent recent reviews on the KCNQ family of K^+ channels [2, 3]. Briefly, KCNQ2 and KCNQ3 are voltage-gated K^+ channels expressed predominantly in the central nervous system. They can be found both pre- and post-synaptically in brain regions that are important for the control of neuronal network oscillations and synchronization [4]. KCNQ2 and KCNQ3 channels are co-expressed in many areas of the brain, where they probably form heterotetramers. Heterotetrameric KCNQ2/Q3 channels are thought to underlie the neuronal M-current, a non-inactivating, slowly deactivating, sub-threshold current [5, 6] $I_K(M)$, has long been known to exert a powerful influence on neuronal excitability. As such, modulators of these channels have the potential to influence neuronal activity in various tissues and are of much interest as therapeutic drug targets. In particular, there is a significant (and growing) body of genetic, molecular, physiological and pharmacological evidence that now exists to support the premise that neuronal KCNQ-based currents represent particularly interesting targets for the treatment of diseases such as epilepsy and neuropathic pain [7].

The present chapter reviews compounds reported to activate KCNQ(2–5) channels. The compounds were grouped based upon their chemical similarities and chemotypes. Direct comparison between most compounds is, unfortunately, difficult due to the use of various KCNQ(2–5) containing cell lines, species and methodological differences reported in each of the citations. Wherever possible, *in vivo* pharmacology data have been supplied along with a status report.

7.6.2
Flupirtine, Retigabine and Related Compounds

7.6.2.1 **Structures and *In Vitro* Activities**
The novel, orally active anticonvulsant flupirtine (**1**) (Fig. 7.6.1) was identified via random screening of a collection of compounds from Asta Medica, through the National Institute of Health's sponsored Antiepileptic Drug Development program (ADD) [8]. Further investigations indicated flupirtine to have a better profile as a centrally acting analgesic and it is marketed in Europe under the trade name

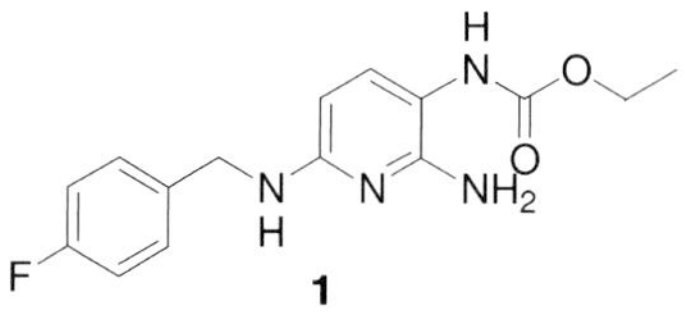

Fig. 7.6.1 Structures of flupirtine (**1**) and retigabine (**2**).

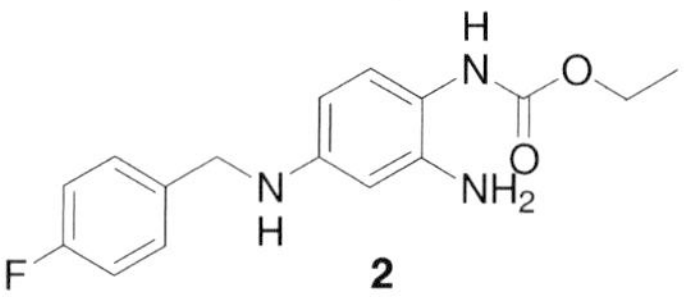

Katadolon® (Flupirtine maleate, Asta Medica) for the treatment of tumor and cancer pain. Additional chemistry and molecular modeling studies by Asta Medica focused on improving the anticonvulsant activity of flupirtine while lowering the analgesic properties [9]. A change of the pyridine core to phenyl led to the generation of retigabine (also known as D-23129) (**2**) (Fig. 7.6.1), which possessed the desired anticonvulsant properties [10, 11]. The anticonvulsant efficacy illustrated by retigabine was initially believed to be due to its ability to enhance GABAergic transmission in the central nervous system. Thus, retigabine has been shown to increase the synthesis of GABA in rat hippocampal slices and to enhance GABA-induced chloride currents in cultured rat cortical neurons [12, 13]. Retigabine may also possess weak sodium and calcium channel blocking activity [12]. More recently, it was shown to exert potassium channel opening activity in neuronal cells [14, 15]. For example, in differentiated NG108-15, human NT and PC12 cells, retigabine increased a potassium conductance that was Ba^{2+} and tetraethylammonium sensitive but 4-aminopyridine insensitive. Similar results were obtained in isolated mouse cortical neurons. Further investigations revealed that retigabine was able to open the potassium current $I_K(M)$, a channel that governs the resting membrane potential in certain central and peripheral neurons, at sub-micromolar concentrations [16, 17]. The ability of retigabine to open potassium channels in neuronal cells differentiates from presently available anticonvulsant agents such as phenytoin, carbamazepine and valproate [14].

As mentioned earlier, heteromeric KCNQ2/Q3 or KCNQ3/Q5 channels represent the molecular correlate(s) of $I_K(M)$ depending on the cell type and location [5, 18]. Retigabine (**2**) and flupirtine (**1**) enhance the activation of these channels at concentrations that are devoid of activity on other ion channels. Within the KCNQ family, retigabine appears highly selective for the neuronal KCNQ(2–5) channels over the more peripherally distributed KCNQ1 channel, but it possesses little selectivity between the neuronal family members (Table 7.6.1) [16, 17, 19, 20]. In addition to retigabine, recent data suggests that flupirtine (**1**) also activates KCNQ2/Q3 channels and native M-currents [21]. To date, there have been no reports of KCNQ sub-type selectivity for flupirtine and surprisingly no KCNQ associated structure–activity relationships have been published for either retigabine (**2**) or flupirtine (**1**). Lundbeck, however, has recently published two patent applications that have utilized the retigabine scaffold as a template, on which they have built in new heterocyclic substituents and/or novel ring constraints (**3–7**, Fig. 7.6.2) [22, 23]. No specific KCNQ activities were discussed for any of the com-

Tab. 7.6.1 Electrophysiology determined EC_{50} data for retigabine (**2**) against several KCNQ channels

Channel	EC_{50} (µM)	Ref.
KCNQ1	1001.1 ± 6.5	17
KCNQ2	2.5 ± 0.6	17
KCNQ2/3	1.9 ± 0.2	17
KCNQ4	5.2 ± 0.9	17
KCNQ3/5	1.4 ± 0.2	20

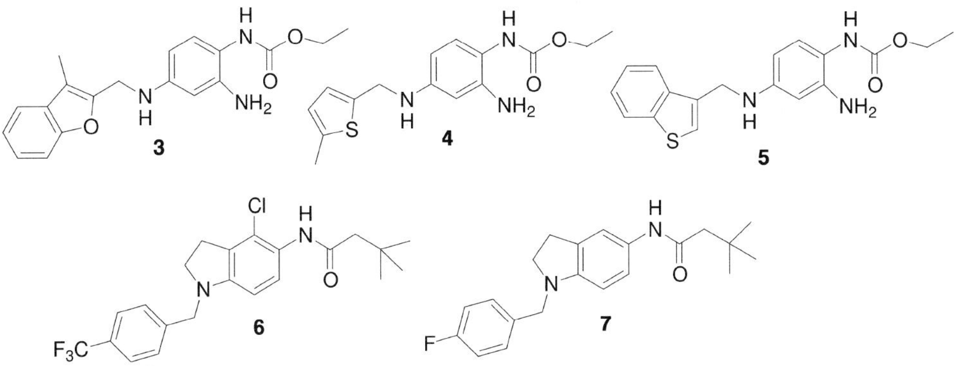

Fig. 7.6.2 Example structures from Lundbeck patent applications WO 04058739 and WO 04096767.

pounds, but the applications stated that the vast majority of exemplified compounds had EC_{50}

7.6.2.2 *In Vivo* Pharmacology (Animal)

Retigabine (**2**) is now considered the prototypical KCNQ agonist and has therefore been assessed in several animal models of epilepsy and pain. For example, across several rodent anticonvulsant models it prevents seizures induced by pentylenetetrazol, maximal electroshock, kainite and picrotoxin [10, 24]. In addition, **2** was also protective against audiogenic seizures in DBA/2J mice, seizures in epilepsy prone rats, and seizures in an amygdala-kindling model (Table 7.6.2) [10, 25, 26]. Flupirtine (**1**) was effective against pentylenetetrazol-induced seizures in mice and possessed a reasonable therapeutic index [10, 27].

In addition to their activity as antiepileptics, retigabine (**2**) and flupirtine (**1**) are efficacious in a wide range of pain models. Blackburn-Monro and Jensen tested **2** in several rat models of nociceptive, persistent and chronic pain [28]. In the chronic constriction injury model of neuropathic pain, retigabine attenuated mechanical hypersensitivity to pin prick stimulation and cold allodynia, but had no

Tab. 7.6.2 *In vivo* pharmacology of retigabine (**2**) in various animal models related to epilepsy and pain.

Assay	Species	Route	Dose(s) mpk	Result	Ref
MES (anticonvulsant)	Mice	PO	Several	ED50 = 27 mpk	10
MES (anticonvulsant)	Rat	PO	Several	ED50 = 3 mpk	10
PTZ (anticonvulsant)	Mice	PO	Several	ED50 = 14 mpk	10
Pic (anticonvulsant)	Mice	PO	Several	ED50 = 19 mpk	10
Tail flick test (analgesia)	Mice	IP	1, 3, 10	Small but significant increase in withdraw latency at 10 mpk	29
Tail flick test (analgesia)	Rat	PO	5, 10, 20	Significant increase at all doses.	29
Tail flick test (analgesia)	Rat	PO	20	No effect.	28
Formalin (persistant pain)	Rat	PO	20	Significant reduction in flinching. Nulified with XE-991.	28
Chung (mechanical allodynia)	Rat	PO	5, 10, 20	Efficacious at 10 & 20 mpk. Effect nulified by co-administration of linopiridine.	29
Chung (thermal hyperalgesia)	Rat	PO	3, 5, 10, 20	Increased withdrawal latency at 5, 10 & 20 mpk	29
CCI (mechanical allodynia)	Rat	PO	5, 20	No effect at either dose	28
CCI (mechanical hyperalgesia)	Rat	PO	5, 20	Significant reduction in withdrawal duration at both doses.	28
CCI (cold allodynia)	Rat	PO	5, 20	Significant reversal of allodynia at 20 mpk.	28
Spared nerve injury	Rat	PO	20	No effect.	28
Spared nerve injury	Rat	PO	20	Significant reduction in withdrawal at 60 min.	28
Spared nerve injury	Rat	PO	20	No effect.	28
Carrageenan (inflammatory pain)	Rat	PO	5	Significant redistribution of weight bearing.	30

effect on tactile allodynia (von Frey hairs). Attenuation of hyperalgesia to a pin prick response was also observed with retigabine in the spared nerve model of neuropathic pain, but again this compound produced no anti-allodynic effect when measured using von Frey hairs. In the formalin model of persistent pain, retigabine attenuated phase 2 flinching (a response thought to result from secondary spinal sensitization). Interestingly, this effect could be completely reversed by the selective KCNQ channel blocker XE-991. Acute pain, as measured by phase 1 flinching in the formalin test and in the tail flick model, was unaffected by retigabine or XE-991. Importantly, retigabine did not appear to impair motor coordination at the doses tested in the pain models described above. The attenuation of behavioral responses to painful stimuli therefore appears to represent a genuine anti-nociceptive property of **2**.

Dost and colleagues also tested retigabine against the thermal hyperalgesia produced by either ligation or transection of the L5 spinal nerve [29]. In both models, retigabine (**2**) displayed anti-hyperalgesic activity, as evidenced by increased paw withdrawal latencies to thermal stimuli. Retigabine was also tested in the formalin model, and like Blackburn-Munro and Jensen [28], they found no activity against phase 1 flinching but a significant reduction in phase 2 flinching.

In a third study, Passmore and colleagues examined retigabine in the rat carrageenan model of inflammatory pain [30]. Weight distribution was used as a measure of nociception following intraplantar administration of 2% carrageenan. In this model, carrageenan injection produced a significant redistribution of weight bearing such that the inflamed paw bore only 21% of the hind paw load (normal weight bearing would be 50% on each hind paw). Administration of retigabine (**2**) increased the weight born by the inflamed paw up to 41%, an effect that was blocked by co-administration of XE-991.

Flupirtine (**1**) was also shown to be active in the rat-tail flick assay and in the spinal nerve ligation model of neuropathic pain [21, 31]. These data provide further support for the role of KCNQ2/Q3 channels in pain processing and perception.

7.6.2.3 *In Vivo* Pharmacology (Man)

Retigabine is in clinical development for the treatment of epilepsy and has been evaluated in five Phase 2a (efficacy and dose-range-finding) trials as well as a long-term extension study [32]. Data from two add-on, open label, studies in patients with treatment-resistant partial seizures (>4 seizures a month), demonstrated that 12 out of 35 patients that completed the study had greater than 50% reduction in seizure frequency [33]. In a larger, randomized, double blind, placebo-controlled dose-ranging add-on study in 399 patients, retigabine at 900 and 1200 mg resulted in statistically significant reductions in seizures. Rates of patients with a 50% reduction in seizures (known as the responder population) were 16, 23, 32 and 33%, for placebo and daily doses of 600, 900 and 1200 mg respectively [32]. Retigabine also reduced seizure frequency in a set of patients with treatment-resistant partial seizures. In addition to being marketed in Germany as a pain reliever, flupirtine has been evaluated in a small-scale clinical trial involving four patients with refractory epilepsy. Flupirtine ($400\,mg\,day^{-1}$) was administered in conjunction with existing anti-epileptic therapy and all four patients showed a decrease in seizure frequency [34]. In a second trial, 400–$800\,mg\,day^{-1}$ flupirtine, administered with existing therapy, reduced seizure frequency in eight out of nine patients [35].

7.6.2.4 Current Status
Flupirtine (Katadolon®): marketed in Europe as a centrally acting analgesic.
   Retigabine: Currently beginning phase III trials for drug-resistant epilepsy.
   Lundbeck compounds: Unknown.

7.6.3
Benzanilide, Benzisoxazole and Indazole Derivatives

7.6.3.1 Structures and In Vitro Activities
One of the earliest collections of compounds to be disclosed as novel KCNQ agonists were a series of 6-substitued-pyridin-3-yl based benzamides (8–10) (Fig. 7.6.3) [36]. Several of the compounds were reported to have EC_{50} values against cloned KCNQ2/Q3 channels of less than 500 nM. Further modifications around the difluorophenyl moiety led to the chain extended analogues 11 and 12, which also possessed sub-micromolar activity against Q2/Q3 channels [37]. A subsequent patent disclosed that the pyridine could be effectively replaced with the corresponding pyrimidine (e.g., 13) without apparent loss of in vitro activity [38].

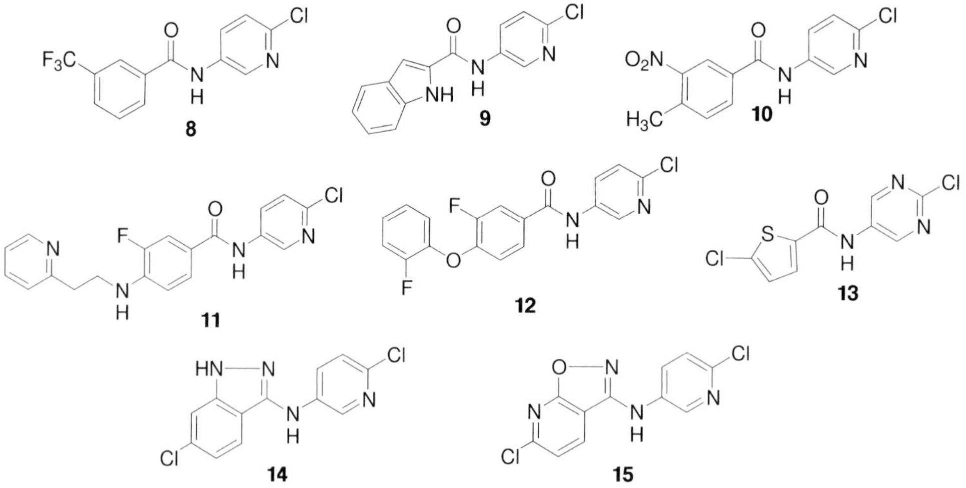

Fig. 7.6.3 Selected exemplified benzanilide, benzisoxazole and indazole derivatives from Refs. [36–39].

Tab. 7.6.3 Rubidium efflux and whole-cell patch clamp electrophysiology (EP) data for ICA-D1.

Channel	^{86}Rb$^+$ efflux EC$_{50}$ or (IC$_{50}$) (μM)	EP, EC$_{50}$ or (IC$_{50}$) (μM)
KCNQ1/minK	(>100)	814)
KCNQ2/Q3	0.36	1.2
KCNQ4	9.8	~20
KCNQ3/Q5	30	>30
SCN2A	–	(>10)
SCN5A	–	(>100)
GABA$_A$ (rα1γ2)	–	>30

The utilization of conformational constraints and amide bond bioisosteres led to the identification of the corresponding indazoles **14** and benzisoxazoles **15**, which were reported to possess good *in vitro* KCNQ2/Q3 activity [39]. The patent literature provides little information regarding the potency or selectivity of these particular compounds. However, information about a structurally undisclosed compound (ICA-D1), from the 6-substitued-pyridin-3-yl based benzamides collection, was recently presented [40]. In a rubidium efflux assay and whole-cell patch clamp experiments, ICA-D1 was found to activate M-currents with EC$_{50}$s of 0.36 and 1.2 μM, respectively (Table 7.6.3). Furthermore, unlike retigabine, ICA-D1 appeared to exhibit selectivity for KCNQ2/Q3 channels over other neuronal KCNQ family members and had no activity against a series of recombinant potassium, sodium, calcium or GABA-activated currents.

7.6.3.2 *In Vivo* Pharmacology (Animal)

From the limited published information on ICA-D1 [40], this compound appears to be an effective anticonvulsant agent with broad-spectrum activity across a range of animal epilepsy models (Table 7.6.4). For example, ICA-D1 prevented seizures induced by maximal electroshock and pentylenetretrazol in both mice and rats (ED$_{50}$ < 10 mg kg^{-1}) and was efficacious in both the mouse 6 Hz "psychomotor" test and the rat amygdala-kindling model. Like retigabine (**2**), ICA-D1 was also reported to be effective in several rat models of pain, including the carrageenan-induced inflammatory pain model and phase 2 of the formalin model. Interestingly, ICA-D1 was active in the spinal nerve ligation (SNL) (Chung) model (Table 7.6.4). Notably, the anticonvulsant, analgesic and anti-allodynic activities were observed at doses that did not exert motor impairment.

Tab. 7.6.4 *In vivo* efficacy of ICA-D1 in anticonvulsant and pain animal models.

Model	Active dose (range) (mg kg^{-1}, PO)
MES (mouse/rat)	3–9
PTZ (mouse/rat)	2–5
6 Hz test (mouse)	10–25
Carrageenan (rat)	30
Formalin (rat)	30
SNL (rat)	5–25

7.6.3.3 Current Status

Unknown.

7.6.4
Oxindoles and Quinolinones

7.6.4.1 Structures and *In Vitro* Activities

In 2001, researchers at Neurosearch reported that racemic BMS-204352 (**16**) activated several of the neuronal homomeric and heteromeric KCNQ2-Q4 channels (Table 7.6.5) [41, 42]. The pure (*S*)-enantiomer of BMS-204352, formerly considered a selective and potent agonist of the large conductance calcium activated potassium channel, had previously entered phase III clinical trials for the treatment of stroke under the trade name MaxiPost™ [43]. When applied to human KCNQ5 cloned channels MaxiPost™ significantly increases the size of the current (Table 7.6.5) [44]. Further investigation, this time by Bristol-Myers Squibb scientists, revealed that closely related analogues **17–23** were also able to activate murine KCNQ2 channels to varying degrees (Table 7.6.6) [45].

Extrapolation from the oxindole structure of BMS-204352 led to a collection of quinolinones (**24–30**, Table 7.6.7) that were potent KCNQ2 agonists [46]. Similar to the oxindoles, the KCNQ2 efficacy of the quinolinones was enhanced by the presence of electron-withdrawing groups on the quinolinone core, while the presence of either a hydroxyl or an ether functionality at R^4 also proved beneficial. Also notable was the positive impact on activity of the triflate group on the amine functionality. Compound **30** was the most potent KCNQ2 agonist in this series with an EC$_{50}$ of 663 nM [46].

Tab. 7.6.5 Whole-cell electrophysiology (EP) activity
of (±)-BMS-204352 (**16**) against several KCNQ channels [41].

KCNQ channel	Fold increase in current caused by 10 μM (**16**) at −30 mV
2	2.3 ± 0.3
2/3	1.3 ± 0.03
4	2.1 ± 0.3
3/4	1.6 ± 0.1
5	12.4 ± 1.1[a]

[a] (S)-Enantiomer used [43].

Tab. 7.6.6 Activation of mouse KCNQ2 currents expressed in
Xenopus oocytes, represented as a percent increase over
KCNQ2 current in control, by a collection of substituted oxindoles [42].

ID	R^1	R^2	% Increase over control current after application of 20 μM of compound
(±)-**17**	6-CF$_3$	CH$_2$CF$_3$	>150[a]
(+)-**18**	6-CF$_3$	CH$_2$CF$_3$	>150[a]
(±)-**19**	5,6-di-Cl	CH$_2$CF$_3$	120–150
(±)-**20**	5,6-di-F	CH$_2$CF$_3$	120–150
(+)-**21**	3-Cl	CH$_2$CF$_3$	>150
(±)-**22**	3-Cl	CH$_2$CF$_3$	>150
(±)-**23**	6-CF$_3$	CH$_2$CF$_2$F	120–150[a]

[a] % over control at 10 μM.

Tab. 7.6.7 Percent increase in KCNQ2 current, expressed in *Xenopus* oocytes, by a selection of quinolinones [46].

ID	R^1	R^2	R^3	R^4	R	% Increase over control at 20 μM (−40 mV)
24	H	H	H	H	H	103 ± 4
25	H	H	H	H	SO_2CF_3	131 ± 3
26	H	CF_3	OH	Cl	H	122 ± 11
27	H	CF_3	OH	Cl	SO_2CF_3	284 ± 14
28	CF_3	H	H	H	SO_2CF_3	223 ± 12
29	CF_3	H	OMe	H	SO_2CF_3	254 ± 17
30	H	CF_3	OH	Cl	SO_2CF_3	231 ± 24
Retigabine (2)						269 ± 14

7.6.4.2 *In vivo* Pharmacology (Animal)

The oxindole **17** was tested intravenously in an animal model of acute migraine. A dose-dependent reduction in the superior sagital, sinus-stimulated, trigeminal field response in rats was observed following administration of 0.1, 1, 10, 30 and 50 mg kg^{-1}, with the highest dose producing almost complete blockade of the superior sagital, sinus-stimulated response [43].

7.6.4.3 Current Status

Unknown.

7.6.5
2,4-Disubstituted Pyrimidine-5-carboxamides Derivatives

7.6.5.1 Structures and *In Vitro* Activities

A well-defined collection of 2,4-disubstituted pyrimidine-5-carboxamides (**31–35**, Table 7.6.8) discovered by Bristol-Myers Squibb scientists were reported to be activators of the KCNQ2 channel [47]. Treatment of murine KCNQ2 channels with 20 μM concentrations of these pyrimidines resulted in a significant increase in the size of the current compared to the vehicle control. The most potent compound

Tab. 7.6.8 Percent increase in mKCNQ2 current, expressed in *Xenopus* oocytes, upon application of compounds (**31–35**) [47].

ID	R^1	R^2	R^3	% Increase over control at 20 μM
31	C$_6$H$_{11}$	Morpholino	F	>200
32	C$_6$H$_{11}$	Pyrollidine	F	>200
33	CF$_3$	Pyrollidine	CF$_3$	151–200[a]
34	CF$_3$	Benzylamine	CF$_3$	151–200
35	Pyrollidine	4-Chlorobentylamine	CF$_3$	125–150

[a] 5 μM concentration used.

exemplified (**33**) was able to significantly increase the size of the current at concentrations as low as 5 μM.

7.6.5.2 *In vivo* Pharmacology (Animal)
The patent application in Ref. [47] indicated that **33** produced a significant reduction in the superior sagital, sinus-stimulated trigeminal field response in rats (migraine model) after intravenous administration of a 1 mg kg^{-1} dose, when compared with vehicle [47].

7.6.5.3 Current Status
Unknown.

7.6.6
Cinnamide Derivatives and Analogues

7.6.6.1 Structures and *In Vitro* Activities
A third chemotype identified by Bristol-Myers Squibb was the cinnamides (Tables 7.6.9–7.6.14). This chemotype was extensively investigated and resulted in the identification of many chemically related, potent KCNQ agonists. The initial patent application outlined a series of α-methylbenzylamine based cinnamides that increased current flow through murine KCNQ2 channels heterologously expressed in either oocytes or HEK 293 cells [48]. Several of the compounds exemplified in the application are illustrated in Table 7.6.9, along with their electrophysiologically determined mKCNQ2 EC$_{50}$ values. Several fused bicyclic heterocycles,

including dihydro-2*H*-benzo[1,4]oxazines, benzo[1,3]dioxoles and tetrahydroqui-nolines, were tolerated on the right-hand side and resulted in compounds with a wide range of *in vitro* activity. Compounds **42** and **43** were very potent agonists of murine KCNQ2 channels, with EC_{50}s of 6 and 9 nM, respectively. Addition of halogens, trifluoromethoxy, methoxy or nitro groups to the cinnamoyl phenyl ring was also tolerated. Further chemistry indicated that a hydroxyl functionality could be added to the α-methyl group without any detrimental effect to KCNQ2 activity (**44–46**, Table 7.6.10) [49].

A subsequent patent application and publication focused on the activity of the 3'-substituted nitrogen-containing heteroaryls, most notably 1,2,4-triazoles, 1,2-pyrazoles and 3-pyridyls **47–49**, Table 7.6.10) [50, 51]. Electrophysiology data indicated that EC_{50}s below 500 nM could be attained for these modifications. Another recent application disclosed that a piperazine spacer between the right-hand side phenyl and the 3'-pyridyl group also yielded extremely potent murine KCNQ2 agonists (**50 and 51**, Table 7.6.10) [52]. Replacement of the alkene group for a trans-cyclopropyl moiety (**52 and 53**, Table 7.6.10) also generated potent KCNQ2 agonists [53].

The 3'-morpholino-substituted cinnamide analogues have been featured in several publications [48–50]. Compound **54** (Table 7.6.11) was revealed to be a potent activator of mKCNQ2 in *Xenopus* oocytes and HEK293 cells (EC_{50}s of 0.6 and 3.3 µM, respectively) and also to activate native M-currents in SH-SY-5Y neuroblastoma cells with an EC_{50} of 0.69 µM. Table 7.6.12 gives a detailed summary

Tab. 7.6.9 Whole-cell electrophysiology (EP) patch clamp EC_{50} recordings from HEK 293 cells expressing mKCNQ2 channels, for a collection of cinnamide derivatives [48].

(37-44)

ID	R	R^1	R^2	R^3	mKCNQ2 EP HEK 293 (−40 mV) EC_{50} (µM)
36	(S)-Me	4F	–Ph–		9.2
37	Me	2F, 5F	–OCH$_2$O–		0.42
38	(S)-Me	2F, 5F	Morpholino	H	1.0
39	(S)-Me	4F	Morpholino	F	1.2
40	Me	2Cl	–OCH$_2$O–		0.0006
41	Me	2F	–N(Et)CH$_2$CH$_2$O–		0.0009
42	Me	3F, 5F	–(CH$_2$)$_3$NH–		5.0
43	Me	2F, 5F	–N(Et)(CH$_2$)$_3$–		1.4

Tab. 7.6.10 Structural modifications to the original cinnamides and their thallium flux based mouse KCNQ2 EC_{50} values [49–53].

ID	R	R^1	R^2	Thallium flux mKCNQ2 EC_{50} (μM)
44	CH_2OH	2Cl	–	0.22
45	CH_2OH	2F	–	1.44
46	CH_2OH	4F	–	0.012
47	CH_3	2F	Triazole	$0.33^{[a]}$
48	CH_3	H	Pyrazole	$0.12^{[a]}$
49	CH_3	4F	Pyrazole	$1.24^{[a]}$
50	CH_3	2F	3-Pyridyl	0.48
51	CH_3	2,5-diF	3-Pyridyl	0.004
52	CH_3	H	H	0.001
53	CH_3	3F	Cl	2.21

[a] EC_{50} values determined by whole-cell patch clamp recortdings from HEK 293 cells expressing mKCNQ2.

of the structure–activity relationship of these leads [56]. In general only the (S)-enantiomers increased current flow through the KCNQ2 channels, as shown by the difference in the ability of compounds (S)-(**54**) and (R)-(**54**) to increase the current flow compared to vehicle control (163 and 96% respectively). Comparison of the activity levels for **55** and (S)-(**54**) indicated that incorporation of the methyl group at the benzylic site was very important for activity. As with other KCNQ agonists, the secondary amide functionality appeared to be important for activity. Removal of the carbonyl to provide the corresponding amine **56** or replacement of the amide hydrogen for a methyl group **57** resulted in a significant decrease in the ability of these compounds to activate the KCNQ2 channel.

In addition, the ability to increase the amplitude of the KCNQ2 current was also affected by the substituents located at the X or Y positions in these molecules. While the incorporation of a fluoro-group at X was tolerated (**60**), the addition of the larger electron-donating methoxy-group in either the X or Y positions (**58**

Tab. 7.6.11 *In vitro* potency of (**54**) in *Xenopus* oocytes and HEK 293 cells expressing mKCNQ2, and against native M-currents in SH-SY-5Y human neuroblastoma cells.

ID	Oocytes mKCNQ2 EP(−40 mV) EC_{50}	HEK mKCNQ2 EP (−40 mV) EC_{50}	SH-SY-5Y EC_{50}
54	Estimated = 0.6 μM	3.3 ± 0.6 μM	0.69 ± 0.08 μM

Tab. 7.6.12 Electrophysiologically determined relative activation of mKCNQ2 channels, expressed in *Xenopus* oocytes, by the presence of 10 μM of a series of morpholino based cinnamides [56].

ID	Z	R^1	R^2	X	Y	% Increase in mKCNQ2 current compared to control (EP −40 mV)
(S)-55	O	H	H	H	H	110 ± 3
(S)-54	O	Me	H	H	H	163 ± 9
(S)-54	O	Me	H	H	H	96 ± 2
(S)-56	H,H	Me	H	H	H	98 ± 2
(S)-57	O	Me	Me	H	H	72 ± 3
(S)-58	O	Me	H	H	OMe	98 ± 2
(S)-59	O	Me	H	OMe	H	96 ± 1
(S)-60	O	Me	H	F	H	129 ± 9

or **59**) was detrimental to *in vitro* potency. Replacement of the morpholino moiety with other tertiary-amino groups was also investigated (Table 7.6.13). The smaller dimethylamino and the slightly larger seven-membered homomorpholino-groups both produced only modest increases in current while the bulkier 2,6-*cis*-dimethylmorpholinyl group produced an increase in current comparable to that

Tab. 7.6.13 Comparison of the activation of mKCNQ2 channels, expressed in *Xenopus* oocytes, for various 3′-amino appendages [56].

ID	R^{1}R^{2}N	% Increase in mKCNQ2 current compared to control 10 µM (EP –40 mV)
(S)-54	Morpholinyl	163 ± 9
61	2,6-*cis*-Dimethylmorpholinyl	138 ± 10
62	Homomorpholinyl	110 ± 4
63	Dimethylamino	110 ± 2

seen with **54**. To date, no KCNQ subtype selectivity data has been reported for any of the cinnamide-based compounds mentioned above.

7.6.6.2 *In Vivo* Pharmacology (Animals)

Several of the cinnamide compounds have been tested in animal models of migraine, neuropathic pain and anxiety (Table 7.6.14). Although compounds **39** and **54** were reported to have good oral pharmacokinetic properties, all of the reported *in vivo* studies used the intravenous route of administration [54, 55]. In general, compounds from this class reduced the number of depolarizations in a cortical spreading depression model of migraine. For example, compound **38** reduced the number of depolarizations by 37% after administration of a 1 mg kg^{-1} dose. All of the compounds tested in the migraine model were as efficacious as the reference drug valproic acid (100 mg kg^{-1}). Compound **38** (3 mg kg^{-1}) also reversed the tactile allodynia in two different models of neuropathic pain (Chung and streptozoticin). In the streptozotocin model, the reversal was 56% and in the Chung model the reversal was 13%. In the canopy test for anxiety, compound **36** reduced stretch attended postures with an approximate ED$_{50}$ of 1.5 mg kg^{-1} [48] Although none of the compounds tested completely reversed the allodynia produced in either the Chung or streptozoticin models, levels were roughly comparable to those produced by the reference compound gabapentin (reduced allodynia by 50% after a 100 mg kg^{-1} intravenous dose) [48].

7.6.6.3 Current Status

Unknown.

Tab. 7.6.14 *In vivo* studies and results conducted with selected cinnamide KCNQ agonists [48].

ID	Cortical spreading depression (migraine model), % reduction in depolarizations at 1 mg kg⁻¹	Tactile allodynia (Chung neuropathic pain model), % reversal, dose	Tactile allodynia (streptozoticin diabetic pain model), % reversal, dose	Canopy test (anxiety model), % reduction in stretched attended postures, dose
54	27	25% at 10 mg kg⁻¹	23% at 10 mg kg⁻¹	28% at 3 mg kg⁻¹
36	25	8% at 3 mg kg⁻¹	Not tested	44% at 1.25mg kg⁻¹
39	25	31% at 10 mg kg⁻¹	12%at 10 mg kg⁻¹	18% at 3 mg kg⁻¹
38	37	13 at 3 mg kg⁻¹	56% at 3 mg kg⁻¹	Not tested

7.6.7
5-Carboxamide-thiazole Derivatives

7.6.7.1 Structures and *In Vitro* Activities

The most recent group of KCNQ channel agonists described by Bristol-Myers Squibb were the 5-carboxamide-thiazole derivatives **64–75** (Table 7.6.15) [57] Structurally, these compounds resemble the cinnamide derivatives mentioned above, where the trifluoromethyl-substituted thiazole group acts as spatial replacement for the alkene/cyclopropyl linker found in the cinnamide compounds. The α-methyl substituted benzylic compounds were once again a prominent feature found in the most potent subset of compounds, although it was not critical for activity. Several compounds lacking this group such as **69** possessed moderate *in vitro* activity. Interestingly, many of the active compounds contained tertiary amino groups tethered either in the R^1 position **70–74** or both R^1 and R^2 positions **75**.

Compound **64** was found to be a very potent agonist of murine KCNQ2 channels with an EC_{50} of 3 nM in whole-cell patch clamp studies using stably expressed in HEK293 cells. A subsequent patent application established that the left-hand side aryl moiety could be replaced with an alkyl spacer group [58]. Once again a basic tertiary amine was tolerated and may in fact be an important feature for efficacy. The representative compounds **76** and **77** (Fig. 7.6.4) possessed mKCNQ2 EC_{50}s of 3.6 and 1.2 μM, respectively, in patch-clamp experiments.

7.6.7.2 *In Vivo* Pharmacology (Animal)

In vivo activity was only described for compound **64**. Following intravenous administration of a 10 mg kg⁻¹ dose, **64** produced a 28% reversal of tactile allodynia in the Chung model. By comparison, the positive control, gabapentin (100 mg kg⁻¹,

Tab. 7.6.15 Structure–activity relationship for a set of 5-carboxamide-thiazoles [57].

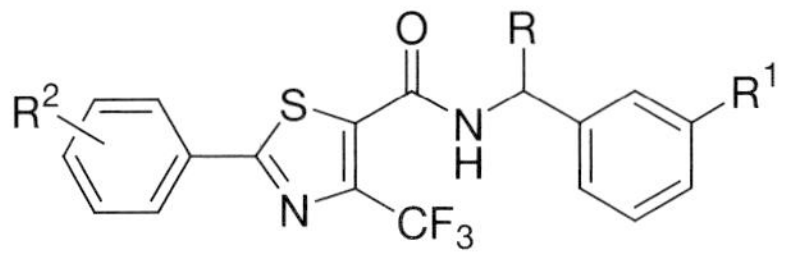

ID	R	R¹	R²	EC$_{50}$[a]
64	CH$_3$	OCF$_3$	2-F	+++
65	CH$_3$	OCF$_3$	4-F	+++
66	CH$_3$	OCF$_3$	2-OCH$_3$	+++
67	CH$_3$	OCF$_3$	3-OCH$_3$	++
68	CH$_3$	OCF$_3$	4-OCH$_3$	+
69	H	OCF$_3$	2-CH$_2$N(Et)$_2$	++
70	CH$_3$	Morpholino	2-OCH$_3$	++
71	CH$_3$	N-Piperazine	2-OCH$_3$	+++
72	CH$_3$	N-Piperidine	2-OCH$_3$	++
73	CH$_3$	NH$_2$	2-OCH$_3$	+
74	CH$_3$	NHC(=NH)NH$_2$	2-OCH$_3$	+
75	CH$_3$	Morpholino	CH$_2$N(Et)pyridyl	++

[a] mKCNQ EC$_{50}$ values (thallium flux assay): +++ = < 50 nM;
++ = 50–1000 nM; + = 1000–2000 nM.

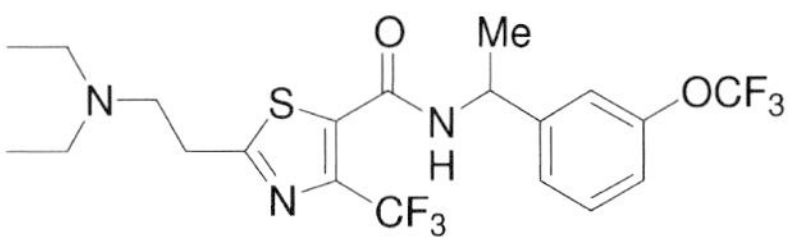

(76) 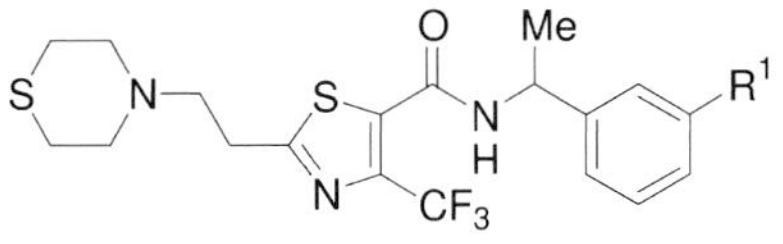   (77)

Fig. 7.6.4 Two exemplified compounds from patent application US 2004/0138268 [58].

intravenous), produced 50% reversal of tactile allodynia [57]. A similar result was seen in the streptozotocin model, where 10 mg kg^{-1} (intravenous) produced a 28% reversal of tactile allodynia.

7.6.7.3 Current Status
Unknown.

7.6.8
Benzothiazoles as KCNQ(2–5) Agonists

7.6.8.1 Structures and *In Vitro* Activities

A series of benzothiazoles were reported by Icagen to be KCNQ agonists (Table 7.6.16) with efficacy in both rodent anticonvulsant and pain models [59]. Initial structure–activity investigations revealed that the conversion of the secondary amide into the methyl substituted tertiary amide, compound **79**, was detrimental to KCNQ activity. Likewise, replacement of the benzothiazole core for the benzoxazole system, **80**, resulted in a loss of KCNQ2/Q3 activity. Various electron-withdrawing groups in various positions on the benzothiazole core were generally tolerated (illustrated by **81–86**). Substitution of the large adamantyl group found in the initial hits with smaller branched or cyclic alkyl groups also resulted in potent KCNQ2/Q3 activators, e.g., **87** and **89**. Replacement of the amide with either a urea or sulfonamides functionality was not beneficial, while carbamates such as **90** afforded compounds that possessed both *in vitro* and *in vivo* activity [59].

Tab. 7.6.16 *In vitro* and *in vivo* structure–activity relationship of 3*H*-benzothiazoles based KCNQ modulators [59].

ID	R	X	R^1	R^2	KCNQ2/Q3 Rb^{86+} efflux assay EC$_{50}$ (μM)	Mouse MES (10 mg kg^{-1}, PO)
78	H	S	H	Adamantyl	0.07	0/8
79	CH$_3$	S	H	Adamantyl	>10	nt
80	H	O	H	Adamantyl	>10	nt
81	H	S	4F	Adamantyl	0.38	0/8
82	H	S	6F	Adamantyl	0.05	1/8
83	H	S	6CF$_3$	Adamantyl	0.2	0/8
84	H	S	6OCF$_3$	Adamantyl	0.03	5/8
85	H	S	4CF$_3$	Adamantyl	0.09	0/8
86	H	S	7CH$_3$	Adamantyl	3.1	0/8
87	H	S	6OCF$_3$	CH$_2$tBu	0.05	7/8
88	H	S	6OCF$_3$	CH$_2$iPr	0.27	7/8
89	H	S	6OCF$_3$	CH$_2$(C$_5$H$_9$)	0.04	5/8
90	H	S	6OCF$_3$	OCH$_2$iPr	0.16	5/8

Tab. 7.6.17 Selected oral *in vivo* data **90**.

(90)

ID	Mouse MES ED$_{50}$	Rat PTZ ED$_{50}$	% Reduction in flinches (phase II formalin)	% Reduction in allodynia score (Chung)
90	9 mg kg^{-1}	3 mg kg^{-1}	45% at 17 mg kg^{-1}	100% at 10 mg kg^{-1}

7.6.8.2 *In Vivo* Pharmacology (Animal)

Many of the initial hits possessed sub-micromolar activity against KCNQ2/Q3 and protected mice in the maximal electroshock model after intraperitoneal administration. With the exception of the 6-trifluoromethoxy analogues most of the initial compounds were not active when administered orally. *In vivo* testing of carbamate **90** revealed that it was a broad-spectrum anticonvulsant, with efficacy in both the maximal electroshock and pentylenetetrazol induced epilepsy models (ED$_{50}$ of 9 and 3 mg kg^{-1}, respectively) [59] This compound was also assessed in the formalin and Chung pain models (Table 7.6.17). In the formalin model, the numbers of flinches recorded in phase 2 were significantly reduced following a 17 mg kg^{-1} oral dose of **90** while a 10 mg kg^{-1} oral dose fully reversed the tactile allodynia in the Chung neuropathic pain model [59].

7.6.8.3 Current Status

No further information has been published.

7.6.9
Quinazolinones Derivatives

7.6.9.1 Structures and *In Vitro* Activities

A series of 2-thio-quinazolinones were disclosed as KCNQ agonists in a recent patent application [60]. Structurally, they resemble the benzothiazoles discussed above, having a similar connectivity between the left-hand phenyl and the amide functionality. The compounds disclosed included several with modifications on the right-hand side of the molecule through an amide or urea linkage. Substituents on the 2-thioether were limited in the preferred claims to a subset of small alkyl groups. Relative compound activity was assessed against native M-currents in NG108-15 cells using a fluorescence-based assay. Table 7.6.18 summarizes the results for a series of fluoro, or trifluoromethyl-substituted quinazo-

Tab. 7.6.18 Activation of M-currents in the NG-108-15 cell line; determination of EC_{50} values for a collection of quinazolinones [60].

ID	R^1	R^2	R^3	NG-108-15 EC_{50} (μM)
91	H	Et	4F	0.85
92	4F	Et	4F	1.00
93	4F	Me	H	0.46
94	4F	Me	4F	0.48
95	6F, 7F	Me	4F	0.81
96	7F	Me	H	0.24
97	7F	Me	4F	0.23
98	H	iPr	H	0.17
99	7F	iPr	H	0.09
100	7CF$_3$	Me	H	0.19
101	7CF$_3$	iPr	H	0.10

linones. These values indicated that while the incorporation of a 4-fluoro to the right-hand side phenyl did not appear to significantly alter the *in vitro* potency (comparison of **93** with **94**, and **96** with **97**), the position, number and choice of electron-withdrawing groups on the quinazolinone core dramatically affected their ability to increase the opening of the M-channels. Both the methyl and ethyl thioether derivatives were effective KCNQ agonists; however, the isopropyl moiety appeared to generate the most potent agonists, e.g., **99** and **101**, exhibiting EC_{50} values of 90 and 100 nM, respectively [60].

7.6.9.2 *In Vivo* Pharmacology
No data has been reported to date.

7.6.9.3 Current Status
Unknown.

Tab. 7.6.19 Exemplified compounds with associated hKCNQ4 (EP) data [61].

ID	% Increase in control hKCNQ4 current
102	115
103	115

7.6.10
Salicylic Acid Derivatives

7.6.10.1   Structures and *In Vitro* Activities
A novel collection of KCNQ agonists has appeared in a patent application from Neurosearch [61]. These new compounds contain strongly acidic moieties, a feature not shared by any of the previously described KCNQ2–5 agonists. Although very little biological data was provided in the patent application two exemplified compounds, **102** and **103**, modestly increased the current through human KCNQ4 channels stably expressed in HEK293 cells (Table 7.6.19). No sub-type selectivity data were reported.

7.6.10.2   *In Vivo* Pharmacology
None reported.

7.6.10.3   Current Status
Unknown.

7.6.11
Melcofenamic Acid and Diclofenac-based KCNQ(2–5) Agonists

7.6.11.1   Structures and *In Vitro* Activities
Professor Attali and his research team at the University of Tel Aviv discovered that meclofenamic acid (**104**) and diclofenac (**105**), two well-known anti-inflammatory agents, as well as several closely related analogues **106–110**, possessed

Tab. 7.6.20 Efficacy of new KCNQ agonists (**105–111**), against spontaneous firing in rat cortical neurons (*in vitro*) [62].

Compound	R^1	R^2	R^3	R^4	R^5	X	Concentration that significantly reduced spontaneous neuronal firing (μM)
104	–	–	–	–	–	–	5
105	–	–	–	–	–	–	25
106	H	CF_3	H	H	H	–	20
107	CH_3	CH_3	H	H	H	–	10
108	Cl	H	H	Cl	H	NH	20
109	Cl	H	H	Cl	H	O	20
110	Cl	H	H	Cl	CH_3	O	5

KCNQ2/Q3 opening capabilities (Table 7.6.20) [62]. Meclofenamic acid and diclofenac were devoid of activity against several other voltage-gated potassium ion channels such as Kv1.2, Kv1.5 and Kv2.1 at concentrations that significantly opened KCNQ2/Q3. Of particular note was their lack of activity against the closely related family member KCNQ1 in either its homomeric form or when co-expressed with KCNE1. All of the compounds in Table 7.6.20 increase current flow through KCNQ2/Q3 channels by causing a leftward shift in the voltage-dependence of channel activation. Potency was determined as the concentration to produce a half-maximal shift in the KCNQ2/Q3 activation curve. The EC_{50} for meclofenamic acid (**104**) and diclofenac (**105**) determined in this way were 25 and 2.6 μM, respectively. These two compounds, as well as the related compounds

106–110, also increased the amplitude of native M-currents and reduced both electrically or spontaneously evoked action potential firing in rat cortical neurons, albeit at relatively high micromolar concentrations.

7.6.11.2 *In Vivo* Pharmacology (Animal)

Pretreatment of mice (intraperitoneal) with either meclofenamic acid (**104**) or diclofenac (**105**) protected animals from tonic convulsions produced by electrical stimuli (maximal electroshock model). The ED_{50} for diclofenac (**105**) was 43 mg kg^{-1} while meclofenanmic acid (**104**) was less efficacious, with 17% of the animals fully protected at 50 mg kg^{-1}. Higher doses of meclofenamic acid (200 mg kg^{-1}) caused proconvulsive toxic effects and hyperactivity [62].

7.6.11.3 Current Status

Unknown.

7.6.12 Summary

Since their discovery and correlation with the potassium current $I_K(M)$, compounds that activate KCNQ2/Q3 and/or KCNQ3/Q5 heteromultimers have become targets for several companies and research groups throughout the world. The discovery that retigabine (**2**) selectively activates KCNQ(2–5) channels and exhibits interesting pharmacology in both animals and humans has increased the interest in these channels as therapeutic targets. The discovery of additional KCNQ sub-type selective compounds should promote further evidence for neuronal KCNQ channels as therapeutic targets and possibly generate better anticonvulsants and pain relievers.

Disclaimer

The opinions expressed in this paper are solely those of the authors and are not to be attributed to Icagen, Inc.

References

1. Gutman, G.A, et al. International Union of Pharmacology. XLI. Compendium of voltage-gated ion channels: potassium channels. *Pharmacol. Rev.* **2003**, 55(4), 583–586.
2. Jentsch, T.J. Neuronal KCNQ potassium channels: physiology and role in disease. *Nat. Rev. Neurosci.* **2000**, 1, 20–30.
3. Gribkoff, V.K. The therapeutic potential of neuronal KCNQ channel modulators. *Expert Opin. Ther. Targets* **2003**, 7(6), 737–748.
4. Cooper, E.C. Harrington, E., Jan, Y.N., Jan, L.Y. M-channel KCNQ2 subunits are localized to key sites for control of neuronal network oscillations and synchronization in mouse brain. *J. Neurosci.* **2001**, 21, 9529–9540.
5. Wang, H-S., Pan, Z., Shi, W., Brown, B.S., Wymore, R.S., Cohen, U.S., Dixon, J.E., McKinnon, D. KCNQ2 and KCNQ3 potassium channel subunits: molecular correlates of the M-channel. *Science,* **1998**, 282, 1890–1893.
6. Brown, D.A., Adams, P.R. Muscarinic suppression of a novel voltage-sensitive K$^+$ current in a vertebrate neuron. *Nature,* **1980**, 283, 673–676.
7. Wickenden, A.D, Roeloffs, R., McNaughton-Smith, G.A., Rigdon, G.C. *Expert Opin. Ther. Patents* **2004**, 14(4), 457–469.
8. Seaman, C.A., Sheridan, P.H., Engel, J., Molliere, M., Narang, P.K., Nice, F.J. Flupirtine. pp. 135–146. In *New Anticonvulsant Drugs*, Meldrum, B.S., Porter R.J. (eds), Libbey, London, **1985**.
9. Seydel, J.K., Schaper, K-J., Coates, E.A., Cordes, H.P., Emig, P., Engel, J., Kutscher, B., Polymeropoulos, E.E. Synthesis and quantitative structure-activity relationship of anticonvulsant 2,3,6-triaminopyridines. *J. Med. Chem.* **1994**, 37, 3016–3022.
10. Rocstock, A. Tober, C., Rundfeldt, C., Bartsch, R., Engel, J., Polymeropoulos, E.E, Kutscher, B., Löscher, W., Hönack, D., White, H.S., Wolf, H.H. D-23129: a new anticonvulsant with broad-spectrum activity in animal models of epileptic seizures. *Epilepsy Res.* **1996**, 23, 211–223.
11. Kapetanovic, I.M., Rundfeldt, C. D-23129: a new anticonvulsant compound. *CNS Drug Rev.* **1996**, 2, 308–321.
12. Rundfeldt, C. Armand, V., Heinemann, U. Multiple actions of the new anticonvulsant D-23129 on voltage-gated inward currents and GABA-induced currents in cultured neuronal cells (abstract). *Naunyn-Schmiedeberg's Arch. Pharmacol.* **1995**, 351(suppl), R160.
13. Kapetanovic, I.M. Yonekawa, W.D. Kupferberg, H.J. The effects of D-23129, a new experimental anticonvulsant drug, on neurotransmitter amino acids in the rat hippocampus *in vivo. Epilepsy Res.* **1995**, 22, 167–173.
14. Rundfeldt, C. The new anticonvulsant retigabine (D-23129) acts as an opener of K$^+$ channels in neuronal cells. *Eur. J. Pharmacol.* **1997**, 336, 243–249.
15. Rundfeldt, C. Characterization of the K$^+$ channel opening effect of the anticonvulsant retigabine in PC12 cells. *Epilepsy Res.* **1999**, 35, 99–107.
16. Wickenden, A.D., Yu, W., Zou, A., Jegla, T., Wagoner, P.K. Retigabine, a novel anticonvulsant, enhances activation of KCNQ2/Q3 potassium channels. *Mol. Pharmacol.* **2000**, 58, 591–600.
17. Tatulian, L., Delmas, P., Abogadie, F.C., Brown, D.A. Activation of expressed KCNQ potassium currents and native neuronal M-type potassium currents by the anti-convulsant drug retigabine. *J. Neurosci.* **2001**, 21(15), 5535–5545.
18. Jentsch, T. J. Neuronal KCNQ potassium channels: Physiology and role in disease. *Nat. Rev. Neurosci.* **2000**, 1, 21–30.
19. Rundfeldt, C., Netzer, R. The novel anticonvulsant retigabine activates M-currents in Chinese hamster ovary-cells transfected with human KCNQ2/3 subunits. *Neurosci. Lett.* **2000**, 282, 73–76.
20. Wickenden, A.D., Zou, A., Wagoner, P.K, Jegla, T. Characterization of KCNQ5/Q3 potassium channels expressed in mammalian cells. *Br. J. Pharmacol.* **2001**, 132, 381–384.
21. Ilyin, V.I., Carlin, K.P., Hodges, D.D., Robledo, S, Woodward, R.M. Flupritine a positive modulator of heteromeric

KCNQ2/Q3 channels. *Soc. Neurosci. Abstr.* **2002**, 758.10.

22. H. Lundbeck A/S. 1,2,4-Triaminoben-zene derivatives useful for the treating disorders of the central nervous system. *WO 04058739.*

23. H. Lundbeck A/S. Substituted indoline and indole derivatives. *WO04096767.*

24. Nikel, B., Shandara, A.A., Godlevsky, L.S., Mazarati, A.M., Kupferberg, H.J, Szelenyi, I. Anticonvulsant activity of a new drug D-20443 (abstract). *Epilepsia,* **1993b**, 34(suppl), S95.

25. Dailey, J.W., Cheong, J.H., Ko, K.H., Adams-Curtis, L.E, Jobe, P.C. Anticon-vulsant properties of D-20443 in geneti-cally epilepsy-prone rats: prediction of clinical response. *Neurosci. Lett.* **1995**, 195, 77–80.

26. Tober, C., Rocstock, A., Rundfeldt, C., Bartsch, R. D-23129: a potent anticon-vulsant in the amygdala kindling model of complex partial seizures. *Eur. J. Phar-macol.* **1996**, 303, 163–169.

27. Porter A.J. et al. Effect of flupirtine on uncontrolled partial or absence seizures. *Epilepsia,* **1983**, 24, 253–254.

28. Blackburn-Munro, G., Jensen, B.S. The anticonvulsant retigabine attenuates no-ciceptive behaviours in rat models of persistent and neuropathic pain. *Eur. J. Pharamcol.* **2003**, 460, 109–116.

29. Dost, R., Rocstock, A., Rundfeldt, C. The anti-hyperalgesic activity of retigabine is mediated by KCNQ channel activation. *Naunyn-Schmiedeberg's Arch. Pharmacol.* **2004**, 369, 382–390.

30. Passmore, G.M., Selyanko, A.A., Mistry, M. et al. KCNQ/M currents in sensory neurons: significance for pain therapy. *J. Neurosci.,* **2003**, 23, 7227–7236.

31. Carlsson, K.H., Jurna, I. Depression by flupirtine, a novel analgesic, of motor and sensory responses of the nociceptive system in rat spinal cord. *Eur. J. Pharamcol.* **1987**, 143, 89–99.

32. Bialer, M., Johannessen, S.I. Kupferberg, H.J., Levy, R.H., Loiseau, P., Perucca, E. Progress report on new antiepileptic drugs: a summary of the Sixth Eilat Conference (EILAT VI). *Epilepsy Res.* **2002**, 51, 31–71.

33. Bialer, M., Johannessen, S.I. Kupferberg, H.J., Levy, R.H., Loiseau, P., Perucca, E.

Progress report on new antiepileptic drugs: a summary of the Fifth Eilat Conference (EILAT V). *Epilepsy Res.* **2001**, 43, 11–58.

34. Porter, A.J., Gratz, E., Narang, P.K. et al. Effect of flupirtine on uncontrolled par-tial or absence seizures. *Epilepsia,* **1983**, 24, 253–254.

35. Sheridan, P.H., Seaman, C.A., Narang, P.K. et al. Pilot study of flupirtine in re-fractory seizures. *Neurology,* (abstract), **1986**, 36(suppl), 85.

36. Benzanilides as potassium channel openers. *US 6372767* (filed August 4, 1999).

37. Pyridine substituted benzanilides as po-tassium channel openers. *US 6495550* (filed August 4, 1999).

38. Pyrimidines as novel openers of potas-sium ion channels. *WO 03068767* (filed February 14, 2002).

39. Bisarylamines as potassium channel openers. *US 6593349* (filed March 19, 2001).

40. Wickenden, A.D. Identification of novel, orally active, KCNQ2/Q3 activators for the treatment of epilepsy and pain. IBC 2nd International Ion Channel Meeting, Boston, October 20–22nd, **2003**.

41. Schrøder, R. L., Jespersen, T., Christo-phersen, P., Strobaek, D., Jensen, B. S., Olesen, S. P., KCNQ4 channel activation by BMS-204352 and retigabine. *Neuro-pharmacology,* **2001**, 40, 888–898.

42. Neurosearch A/S. Use of 3-oxindole de-rivatives as KCNQ potassium channel modulators. *WO 02000217.*

43. Gribkoff, V.K. et al. *Nat. Med.,* **2001**, 7 (4), 471–477.

44. Dupuis, D.S., Schrøder, R.L., Jespersen, T., Christensen, J.K., Christophersen, P., Jensen, B.S., Olesen, S.P. Activation of KCNQ5 channels stably expressed in HEK293 cells by BMS-204352. *Eur. J. Pharamcol.* **2002**, 437, 129–137.

45. Bristol-Myers Squibb Company. Fluoro-oxindole derivatives as modulators of KCNQ potassium channels. *US 6469042.*

46. Hewawasam, P. et al. The synthesis and structure-activity relationship of 3-amino-4-benzylquinolin-2-ones: discovery of novel KCNQ2 channel openers. *Bioorg. Med. Chem. Lett.* **2004**, 14, 1615–1618.

47. Bristol-Myers Squibb Company. 2,4-Dis-ubstituted pyrimidine-5-carboxamides as KCNQ potassium channel modulators. *WO 02066036.*

48. Bristol-Myers Squibb Company. Cinna-mides as KCNQ potassium channel modulators. *WO 02096858.*

49. Bristol-Myers Squibb Company. 1-Aryl-2-hydroxethyl amides as potassium chan-nel openers. *WO 04047743.*

50. Bristol-Myers Squibb Company. 3-Het-erocyclic benzylamide derivatives as po-tassium channel openers. *WO 04047744.*

51. L'Heureux, A. et al., *(S,E)*-*N*-[1-(3-Het-eroarylphenyl)ethyl-3-(2-fluorophenyl)-acrylamides: synthesis and KCNQ2 po-tassium channel opener activity. *Bioorg. Med. Chem. Lett.* **2005**, 15, 363–366.

52. Bristol-Myers Squibb Company. 3-(Pyri-dyl-piperazin-1-yl)-phenethyl amides as potassium channel openers. *WO 04047745.*

53. Bristol-Myers Squibb Company. Arylcy-clopropylcarboxylic amides as potassium channel openers. *WO 04047738.*

54. Wu, Y.-J. et al. (S)-N-[1-(3-Morpholin-4-ylphenyl)ethyl]-3-phenylacrylamide: An orally bioavailable KCNQ2 opener with significant activity in a cortical spreading depression model of migraine. *J. Med. Chem.* **2003**, 46, 3197–3200.

55. Wu, Y.-J. et al. Fluorine substitution can block CYP3A4 metabolism-dependent inhibition: Identification of (S)-N-[1-(4-fluoro-3-morpholin-4-ylphenyl)ethyl]-3-(4-fluorophenyl)acrylamide as an orally bioavailable KCNQ2 opener devoid of CYP3A4 metabolism-dependent inhibi-tion. *J. Med. Chem.* **2003**, 46, 3778–3781.

56. Wu, Y.-J., He, H., Sun, L.Q., L'Heureux, A., Chen, J., Dextraze, P., Starrett, J.E., Boissard, C.G., Gribkoff, V.K., Natale, J., Dworetzky, S.I. Synthesis and structure-activity relationship of acrylamides as KCNQ2 potassium channel openers. *J. Med. Chem.* **2004**, 47, 2887–2896.

57. Bristol-Myers Squibb Company. 2-Aryl thiazole derivatives KCNQ modulators. *WO 04060281.*

58. Bristol-Myers Squibb Company. Ami-noalkyl thiazole derivatives KCNQ mod-ulators. *US 2004/0138268.*

59. McNaughton-Smith, G.A., et al. Novel KCNQ agonists as potential new thera-peutics. *SERMACS*, Atlanta, November 16–19th, **2003**.

60. ICAGEN, Inc. Quinazolinones as potas-sium channel modulators. *WO 04058704.*

61. Neurosearch A/S. Novel KCNQ channel modulating compounds and their use. *WO 04080377.*

62. Ramot at Tel Aviv University Ltd. Deri-vatives of N-phenylanthranilic acid and 2-benzimidazolon as potassium channel and/or cortical neuron activity modula-tors. *WO 04035037.*

8
Genetic and Acquired Channelopathies

8.1
Inherited Disorders of Ion Channels
Kate Bracey and Dennis Wray

8.1.1
Introduction

One of the most fascinating areas of current research on ion channels is the field of channelopathies – human inherited disorders of ion channels. Recently, there has been a marked increase in the knowledge available on these types of disorders. The field is truly interdisciplinary, employing as it does the techniques of genetics, molecular biology, structural biophysics, physiology, electrophysiology and pathology. Together with the detailed knowledge about the plethora of ion channels, great advances have been made in identifying the many mutations that underlie these inherited disorders of ion channels, and these have been widely reviewed [1–27].

Nowadays, all the human ion channels have been cloned, and they have been assigned to families on the basis of their homology, their membrane disposition, their selectivity for ions that they pass, and their mechanism of activation. Determination of chromosomal localization and exon structure of the corresponding genes is straightforward by comparison with the wonderful known sequence of the human genomic DNA. Identification of genes underlying channelopathies has continued starting from either educated guesses about the likely gene that involves a disorder (candidate gene approach), or by genetic mapping using linkage studies of families with inherited disorders (positional cloning). For the latter approach, localization of a gene that is involved in an inherited disorder is made much easier by the availability of many close markers on the genomic DNA (sequence tagged sites, STSs), and final identification of the gene involved in a disorder can readily be made by comparison with the vast genomic databases. However, to really show that a gene is involved in an inherited disorder, it is necessary to establish and identify inherited mutations in the gene. This is also nowadays straightforward; exons of the appropriate gene can be amplified from patients by PCR (polymerase chain reaction), and the different mobilities of mutant and wild-type exons can be exploited to identify a mutant-containing exon,

Voltage-Gated Ion Channels as Drug Targets. Edited by D. Triggle
Copyright © 2006 WILEY-VCH Verlag GmbH & Co. KGaA, Weinheim
ISBN 3-527-31258-7

which can then be sequenced completely by modern automated DNA sequencing techniques.

It is not necessary to review here the basic science of the different families of ion channels because they are covered in great detail in the rest of this book. Our aim in this chapter is to overview the main features of all the inherited disorders of ion channels that are currently known. Thus we have covered potassium channels (both voltage-dependent and inwardly rectifying families), non-selective cation channels, sodium channels (voltage-dependent and amiloride sensitive families), voltage-dependent calcium channels, chloride channel families, and ligand operated channels including those activated by acetylcholine, glycine and GABA. Because this is such a vast field, we have simply tried to give the basic features of the disorders and their aetiology via the channel mutations. Considerable care was taken in constructing figures showing the locations of the mutations on the ion channel structures as they are known to date, taking into account as many of the recent papers as we could find. We would like to think that these figures are reasonably complete, although new mutations are being reported all the time.

One of the most interesting areas of current research is how the mutations (having been identified) actually cause the disorder in question. In some cases this is well understood, for instance in Liddle's syndrome it is clear how the mutations generally prevent the down-regulation of the epithelial sodium channel, leading to excess sodium reabsorption in the kidney. But, for many inherited disorders, the way mutations cause the disorders is not really understood, for instance for the epileptic disorders the detailed mechanisms are rather unclear. However, much progress is being made by heterologous expression of mutant channels in cell lines and oocytes, followed by electrophysiological analysis of the ion channel currents. Unfortunately, in this overview space does not allow a detailed discussion of all the experimental results concerning the precise effects of the mutations on ion channel function. However, there are several recurrent effects that mutations can have on ion channels. In many cases there is a loss of function of the ion channel caused by the mutation, and this can occur, for instance, by defects in trafficking within the cell and failure to express in the cell membrane (perhaps by protein folding defects), or the ion channel may reach the membrane but be non-functional, or the channel may have altered function and gating. In other cases, mutations may cause increased channel currents (i.e., gain of function); for instance, this might occur if inactivation is removed from a channel by the mutation. Often loss of function mutations underlie recessive disorders (because a wild-type allele as exists in the heterozygote would be able to provide sufficient ionic current), and gain of function mutations underlie dominant disorders (because a single mutant allele then displays the aberrant channel current). However, this is by no means always the case; for instance, in a heterozygous case where an ion channel is made up from several subunits, a mutant subunit with loss of function can combine with a wild-type subunit to give a channel with overall loss of function – a dominant effect. Another complication is that in some cases mutations lead to susceptibility to a disorder rather than being an absolute cause, and mutations may also involve other interacting mutant genes or environmental factors.

The authors hope the reader will find this wide-ranging overview of great value in understanding and indeed in simplifying the initially bewildering nature of this important class of inherited disorders: the channelopathies.

8.1.2
Potassium Channels

8.1.2.1 Inward Rectifier Families

8.1.2.1.1 Kir1.1
The Kir1.1 channel belongs to the inward rectifier family of potassium channels, with two membrane-spanning regions in each subunit of this tetrameric channel. This channel was the first member of this family to be cloned, and the corresponding gene (KCNJ1) is located on chromosome 11. This channel is predominantly expressed in the kidney where it is located mainly on the apical membrane of the ascending limb of the loop of Henle. In this location the Kir1.1 channel plays a major role in potassium and salt excretion, and therefore the control of blood pressure and potassium homeostasis. The channelopathy associated with mutations in the KCNJ1 gene is Bartter's syndrome type 2 [21, 22, 28], which is an autosomal recessive disease. This condition is characterized by hypokalaemia, alkalosis, salt wasting and several other conditions associated with interference in salt and water reabsorption. The mutations causing Bartter's syndrome result in either a partial or complete loss of function of the Kir1.1 channel, and are distributed throughout the channel protein (Fig. 8.1.1), including the intracellular regions. The result of the lack of function of the channel is an inability to recycle potassium from cells of the thick ascending loop of Henle back into the renal tubule, which in turn impairs reabsorption of sodium because of interference with the Na-K-2Cl co-transporter. This leads to increased sodium load at the distal tubule, which stimulates potassium and hydrogen ion secretion, leading to the hypokalaemic alkalosis and salt wasting characteristic of the disorder.

8.1.2.1.2 Kir2.1
A further member of the Kir family associated with ion channelopathies is Kir2.1, with corresponding gene KCNJ2 located on chromosome 17. This channel is expressed in the brain, heart and skeletal muscle, and it functions to maintain the resting membrane potential and to modulate action potentials in these tissues. Mutations in the KCNJ2 gene give rise to the autosomal dominant Andersen-Tawil syndrome (ATS) [22, 29–39], a disorder with periodic paralysis and/or cardiac arrhythmias and/or developmental abnormalities. Mutants show loss of function; co-expression of mutant channel with wild-type channels almost always reduces current amplitudes, underlying the dominant nature of the disorder. Mutations are located mainly in the N and C terminal intracellular regions, but

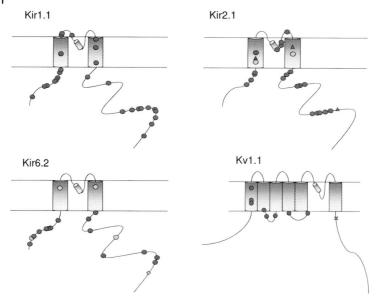

Mutations: ○ missense ◇ frameshift ✕ nonsense △ in frame deletion

Fig. 8.1.1 Mutations in Kir1.1 (Bartter's syndrome type 2), Kir2.1 (red symbols, Andersen-Tawil syndrome; green symbols, short QT), Kir6.2 (red symbols, diabetes; green symbols, persistent hyperinsulinaemic hypoglycaemia of infancy), and Kv1.1 (episodic ataxia type 1). The key shows the type of mutation (missense, frameshift and nonsense).

also in the pore selectivity filter (Fig. 8.1.1). For mutations in the N terminal region and the pore, trafficking to the membrane is usually normal, but the channels are non-functional, apparently by interference with channel gating or the pore itself. Some mutations in the C terminal region also traffic normally to the membrane but interfere with PIP_2 binding that is essential for normal Kir function; also a few mutants show defective trafficking. The symptoms of the condition correlate with the location of the expression of the KCNJ2 gene. In the heart, the potassium current involved in repolarization, I_{K1}, is carried by Kir2 channels, and the reduction in Kir2.1 currents prolongs the action potential, leading to a modest lengthening of the QT interval of the electrocardiogram (ECG), and indeed this disorder was initially classified as long QT type 7. This lengthening of the QT interval confers a susceptibility to dangerous arrhythmias, fainting or sudden death. However, the phenotype of this disorder is variable, and lengthening is not seen in all cases, so this channelopathy is perhaps best labeled ATS type 1 instead of LQT7. In skeletal muscle, the depolarization induced by the nonfunctional mutant Kir2.1 channels may lead to depolarized resting potentials and hence inactivation of sodium channels and underlying action potentials, so leading to paralysis. The way mutant Kir2.1 channels cause developmental abnormalities is still under investigation.

Recently, gain of function mutations have been reported in KCNJ2 that lead to an "opposite" disorder with short QT interval [40–42], characterized by increased susceptibility to arrhythmias such as atrial fibrillation and sudden cardiac death. The gain of function mutations are located in the M1 and M2 regions (Fig. 8.1.1), and probably affect channel conductance or rectification properties. The increased potassium current via the mutated channels leads to shortening of the action potential duration by accelerated repolarization (and hence short QT interval), and underlies the symptoms of this autosomal dominant disorder.

8.1.2.1.3   Kir6.2
Another inward rectifier channel underlying inherited disorders is Kir6.2, which has the corresponding gene KCNJ11 and is located on chromosome 11. Kir6.2 co-assembles with the structurally unrelated ATP binding cassette protein, the sulfonylurea-receptor (SUR1), to form the ATP-sensitive potassium channel K_{ATP}. This K_{ATP} channel is expressed in the pancreas, skeletal muscle, smooth muscle and brain. The channel is particularly important in β cells of the pancreas, where the channel links the metabolic state of the cell to insulin secretion. In response to increased glucose levels, K_{ATP} channels close, since the higher ATP levels associated with glucose block K_{ATP} channels. This in turn causes membrane depolarization, an influx of Ca^{2+} and then insulin release. There are two types of mutations of Kir6.2 [21, 22, 43–57]: some mutations are inactivating (such as those in the pore region), but many other mutations are activating (located in the N and C termini) (Fig. 8.1.1). The inactivating mutations cause persistent hyperinsulinaemic hypoglycaemia of infancy (PHHI), a disorder that can be both autosomal dominant and autosomal recessive. The disorder is characterized by persistent insulin release even in the presence of hypoglycaemia, because the pancreatic β cells are persistently depolarized via the inactive potassium channels. PHHI usually presents during the first few hours after birth and the hypoglycaemia can be controlled by glucose infusion or a high carbohydrate diet. In addition to inactivating mutations of Kir6.2, PHHI can also be caused by mutations in SUR1. In contrast to the mutations in PHHI, the mutations in Kir6.2 leading to permanent neonatal diabetes are of the activating nature. Some of these Kir6.2 mutations in permanent neonatal diabetes introduce a decrease in sensitivity of the Kir6.2 channel to the blocking effect of ATP, and other mutations increased the probability of the channel being open. In both cases this leads to less depolarization of β cells and so impairs insulin release, with the consequent hyperglycaemia that is characteristic of diabetes.

8.1.2.2   Voltage-activated Potassium Channels

8.1.2.2.1   Kv1.1
The Kv1.1 channel belongs to the six-transmembrane voltage-activated potassium channel family (Fig. 8.1.1) and, like the other members of this family, displays

outward rectifying potassium currents. The gene corresponding to the Kv1.1 potassium channel is KCNA1, and is located on chromosome 12. Kv1.1 usually functions to keep action potentials short by depolarizing the membrane at the end of the action potential, through an influx of potassium ions. The channel is a neuronal potassium channel that is widely distributed in the brain, particularly in the hippocampus and cerebellum, where the channel regulates firing rates of neurons. Mutations in the KCNA1 gene are responsible for episodic ataxia type 1 (EA1) [1, 9–11, 20, 21, 58–66], which is an autosomal dominant inherited disorder. This is characterized by symptoms of paroxysmal cerebellar ataxia that are often associated with myokymia. Attacks of uncoordinated movements are normally visible in the arms, legs and head, and are usually short lasting and triggered by exertion, stress or startle. Myokymia is not always clinically evident and is sometimes only detectable with electromyography. Mutations affect the highly conserved residues of the Kv channel family, particularly in the intracellular loops and also the S1 region (Fig. 8.1.1). The effects of the mutations generally cause loss of function, although effects vary substantially, ranging from a small slowing of channel activation rate, depolarizing shifts in the voltage-dependence of activation, to completely non-functional channels. Mutations are mainly missense, and there is also one nonsense mutation at the start of the C terminus. As none of the mutations are found in the intracellular N terminal region (a region important for channel assembly), Kv1.1 mutant subunits are able to co-assemble with wild-type subunits and suppress their function, explaining the autosomal dominant basis of the disorder. The overall effect of the various non-functional mutations of the Kv1.1 channel is a reduction in the repolarizing effect of the channel. Unsurprisingly, there is a correlation in the reduction of potassium flux and the severity of the effect on the phenotype of the patient. The prolonged repolarization may lead to repetitive firing in motor axons, causing the myokymia; and may also lead to prolongation of neurotransmitter release, causing seizures and fits. The ataxia is of cerebellar origin, but it is not clear whether patients with EA1 have an increased or decreased activity from the cerebellum.

8.1.2.2.2 KCNQ1

The gene for the KvLQT1 potassium channel is KCNQ1 and it is located on chromosome 11. KvLQT1 is a slow delayed rectifying potassium channel that exhibits an outwardly rectifying current. The channel is expressed in cardiac myocytes as well as in many epithelia; in the heart the channel functions to repolarize the action potential. KvLQT1 is found combined with a small auxiliary subunit, MinK (also known as IsK), the corresponding gene for this auxiliary subunit being KCNE1. When KvLQT1 is co-expressed with MinK, the time course of activation is slower and the outward current is augmented (as compared with KvLQT1 alone). This augmented current underlies I_{Ks}, which is the slow delayed rectifier current in the heart. One of the channelopathies associated with mutations in the KCNQ1 gene is long QT syndrome, type 1 (LQT1) [3, 11, 21, 67–73], an autosomal dominant disorder. As for other types of long QT syndrome, the lengthening of the QT interval confers susceptibility to dangerous cardiac arrhythmias. A typical

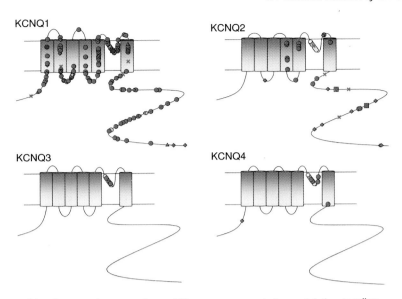

Mutations: ○ missense ◇ frameshift ✕ nonsense △ in frame deletion ▢ splice

Fig. 8.1.2 Mutations in KCNQ1 (red symbols, long QT type 1; green symbols Jervell and Lange-Nielson syndrome; blue symbols, atrial fibrillations), KCNQ2 and KCNQ3 (benign familial neonatal convulsions), KCNQ4 (deafness autosomal dominant type 2). Types of mutation are shown in the key.

example of the ventricular arrhythmias generated in long QT disorders generally are *torsade de pointes*, which are shown on an ECG recording by the QRS complex twisting around the isoelectric axis. Some of the mutations in the KCNQ1 gene that cause LQT1 are located in the pore region and helices S4, S5 and S6 (Fig. 8.1.2), but a number are also found in intracellular loops and the C terminus. KvLQT1 mutants cause loss of function, as indeed would be expected for instance for the pore mutants. The corresponding decrease in potassium current causes a prolongation of myocardial repolarization of the action potential, underlying the long QT interval. The autosomal dominant nature of the disorder is due to the association of mutant KCNQ1 subunits with wild-type subunits, suppressing their function. In the rare cases that an individual is homozygous for a KCNQ1 mutation, an additional inherited disorder is found, Jervell and Lange-Nielson syndrome (JLN) (Fig. 8.1.2). This recessive long QT disorder is accompanied by bilateral deafness. The deafness associated with this disorder can be explained by a decrease in production of endolymph via almost complete dysfunction of KvLQT1 channels in the stria vascularis of the inner ear.

Mutations of the associated auxiliary subunit, MinK (gene KCNE1) also occur. These mutations can hinder the assembly with KvLQT1, so reducing the potassium currents. This again leads to a long QT syndrome, known as LQT5 via its effect in the heart, and can also lead to JLN via effects in the ear.

Two mutations in KvLQT1 have also been reported that cause gain of function, and underlie atrial fibrillation (AF) syndromes [42, 74, 75]. The mutations are located in the membrane-spanning part of the channel (S1 and P regions) (Fig. 8.1.2). The gain of function mutations would be expected to cause shortening of action potential duration, and hence shortening of the QT interval. However, shortening of QT was only observed for one mutant, although both mutants were associated with AF.

8.1.2.2.3 **KCNQ2 and KCNQ3**

Mutations are also found in genes KCNQ2 on chromosome 20 and KCNQ3 on chromosome 8. When expressed as homotetramers, they both convey small currents, but when the two channels are co-expressed, as occurs *in situ*, the combination gives rise to much larger currents. Physiologically, this neuronal current is inhibited by muscarinic receptors and the current passed by this combination of subunits is known as the M current. The role of this M current is to dampen the tendency for repetitive firing of neurons and is therefore important in the regulation of the subthreshold of excitability of neurons and their firing patterns. Mutations in either the KCNQ2 or KCNQ3 genes can lead to the autosomal dominant epileptic channelopathy benign familial neonatal convulsions (BFNC) [2, 9–11, 20, 21, 76–89]. This condition is characterized by neonatal convulsions that clear spontaneously after a few weeks. The seizures associated with BFNC are usually frequent and of mixed type with tonic posturing, ocular signs and automatisms. Numerous different mutations in the KCNQ2 gene cause BFNC, and a few mutations in KCNQ3 (Fig. 8.1.2). For KCNQ2, mutations are distributed throughout the whole of the channel structure, although many mutations are found in the C terminus. The mutations in KCNQ3 are all found in the pore region. The result of the mutations in both genes is either variable loss of function of the channel or altered biophysical properties, generally leading to reductions in M currents. This in turn removes the dampening of repetitive firing, and the associated hyperexcitability may cause the convulsions characteristic of the disorder, BFNC.

8.1.2.2.4 **KCNQ4**

The gene for a fourth member of this family, KCNQ4, is located on chromosome 1. This potassium channel is found in the ear in the outer hair cells of the cochlea and is also expressed in the brain. Mutations in this channel, associated with loss of function, are responsible for an autosomal dominant deafness disorder, DFNA2 [21, 70, 90, 91]. This form of deafness is distinguished from that due to KCNQ1 mutations, being due to the outer hair cells rather than the endolymph. Mutant subunits combine with normal subunits to cause loss of function in a dominant negative manner, consistent with the dominant nature of the DFNA2 disorder. The mutations are located in the pore and S6 regions of the channel, as well as a nonsense mutation in the N terminus (Fig. 8.1.2), all of which would be expected to give non-functional channels, so interfering with normal hair cell function, and hence leading to deafness.

8.1.2.2.5 **Herg**

The herg potassium channel belongs to the human ether-a-go-go related family. A unique feature of the herg potassium channel is slow activation and deactivation with rapid inactivation. The corresponding gene (KCNH2) is found on chromosome 7 and is highly expressed in the heart. The herg channel combines with an auxiliary subunit, MiRP1 (KCNE2 gene), and forms a channel complex that underlies a major potassium current in the heart, I_{Kr}, which participates in the repolarization of the cardiac action potential. The channelopathy caused by mutations in the KCNH2 gene is another cardiac long QT syndrome: LQT2 [3, 11, 21, 68, 92–121]. As with other forms of LQT syndrome, patients are prone to ventricular fibrillations and heart arrhythmias. Mutations are found throughout the channel, and most are missense (Fig. 8.1.3). Mutations cause loss of function, either by deficiency in trafficking to the membrane, or by loss or altered function when trafficking is normal. Usually, mutations in the N-terminus accelerate deactivation, mutations in the pore region may carry more serious risk of arrhythmias, while mutations in the C terminus may affect trafficking or gating properties. The diminished function of this channel leads to a decrease in depolarizing I_{Kr} potassium current and a resultant lack of protection from premature firing of the cardiac action potential, leading in turn to after-depolarizations, *torsade de pointes* and arrhythmias. The dominant nature of the disorder can occur by a dominant-negative effect of mutant subunit on normal subunits within a channel tetramer. Since many of the mutations lead to trafficking disorders, much recent interest has focused on pharmacologically "rescuing" the mutant channel by various com-

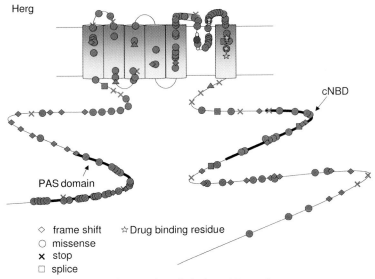

Fig. 8.1.3 Mutations in herg (red symbols, long QT type 2; blue symbol, short QT). The figure also shows the residues involved in drug binding to herg (green symbols).

pounds, so as to restore trafficking. For example, herg channel blocking reagents such as quinidine, or other non-blocking reagents such as thapsigargin, restore trafficking for some, but not all, mutants.

As already mentioned, the I_{Kr} current of the herg potassium channel is contributed to by the combination of herg with the MiRP1 auxiliary subunit. Mutations in MiRP1 can also hinder the ability of the channel to function properly and can also result in another type of long QT syndrome: LQT6, again with susceptibility to arrhythmias.

As for KCNJ2 and KCNQ1 mentioned above, a gain of function mutation has been reported for herg [42, 122, 123], leading, as before, to a short QT syndrome, arrhythmias and sudden death. The mutation is located on the helix between S5 and P regions (Fig. 8.1.3). Also, a gain of function mutation in the auxiliary subunit, MiRP1, with similar clinical effects, has been reported [124].

Finally, notably, the herg channel is important, even notorious, because many drugs can act on herg to trigger arrhythmias, a major unwanted effect. A whole range of drugs (e.g., some antihistamines, antibiotics and antipsychotics) can bind to sites on the pore and S6 region (Fig. 8.1.3) to block this channel, so causing prolongation of the QT interval, hence resulting in susceptibility to cardiac arrhythmias [125–131].

8.1.3
Non-selective Cation Channels

8.1.3.1 CNG Channels
Cyclic nucleotide-gated (CNG) channels have a similar membrane topology to the six-transmembrane potassium channel. The main characteristics of these channels is that they are cation conducting (sodium and potassium), and they are activated by cyclic nucleotides (cGMP and cAMP) via the binding of the latter to a domain on the C terminal region of the channel. These channels play roles in sensory perception throughout the nervous system, and are particularly important in the rods and cones of the eye. Mutations of CNGA1 and CNGB1 subunits in the rods, which co-assemble to form a tetramer, underlie an autosomal recessive type of retinitis pigmentosa (RP) [132–139]. This disorder includes night blindness, loss of peripheral vision and eventually blindness. Mutations in these subunits are mainly truncating (Fig. 8.1.4), implicating loss of channel function. An interesting truncating mutation is located near the end of the C terminus of CGNA1; this prevents association with CGNB1 and disrupts channel trafficking to the membrane. In the presence of light, cGMP levels are reduced and CNG channels close, leading to hyperpolarization of the photoreceptors and decreased neurotransmitter release. The mutant channels mimic wild-type photoreceptors exposed to constant light because mutant CNG channels will be constantly closed. However, the consequent events leading to cell death and blindness associated with RP are not understood.

Mutations in the CNGA3 and CNGB3 subunits in the cones of the eye underlie the autosomal recessive disorder achromatopsia [140–148], the main feature of

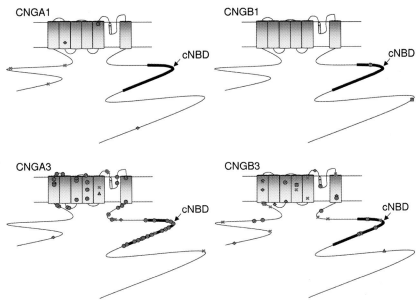

Mutations: ○ missense ◇ frameshift × nonsense △ in frame deletion □ splice

Fig. 8.1.4 Mutations in CNGA1 and CNGB1 (retinitis pigmentosa), CNGA3 and CNGB3 (achromatopsia).

this disorder being a total inability to distinguish colors. The CNGA3 and CNGB3 subunits co-assemble to form tetramers in the cone photoreceptors responsible for color vision. Mutations in both subunits are distributed throughout the channel protein; in CNGA3 these are predominantly missense whereas those in CNGB3 are mostly truncations (Fig. 8.1.4). Mutations are mainly loss of function, with trafficking defects or abnormal gating properties. The detailed mechanism of how these mutations cause the phenotype remains to be determined. Exceptionally, some missense mutations cause a gain of function; perhaps the phenotype in this case arises from calcium overload via excess calcium entry through the CNG channels.

8.1.3.2 HCN Channels

Hyperpolarization-activated cyclic nucleotide gated (HCN) six transmembrane channels are similar to CNG channels in also possessing a C-terminal cyclic nucleotide-binding domain, but by contrast these channels are activated by hyperpolarization. The HCN channels, which conduct sodium and potassium ions, are the molecular components of the pacemaker current (I_f) in the sinoatrial node (SAN) in the heart and are responsible for pacemaker activity. HCN4 is a predominantly expressed isoform of the HCN family in the SAN. Mutations in HCN4 cause sinus node dysfunction (SND), an inherited type of cardiac arrhythmia [149,

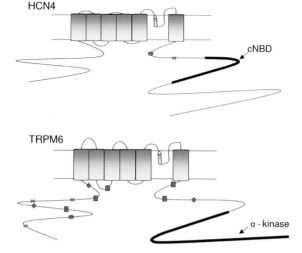

Fig. 8.1.5 Mutations in HCN4 (sinus node dysfunction) and TRPM6 (hypomagnesaemia with secondary hypocalcaemia).

Mutations: ○ missense ◇ frameshift × nonsense □ splice

150]. A nonsense mutation in the C terminus (Fig. 8.1.5) causes deletion of the cyclic nucleotide-binding domain and therefore loss of its response to cAMP. There is also a missense mutation (Fig. 8.1.5), again with loss of function. The disruption of normal pacemaking activity in some way leads to the observed arrhythmias characteristic of SND.

8.1.4
Transient Receptor Potential (TRP) Channels

The TRP channel family is also a six transmembrane family and has three sub-families of channels, TRPC, TRPV and TRPM. One interesting characteristic of these channels is that the S4 transmembrane domain lacks the full complement of charged residues and these channels are therefore only weakly voltage depen-dent. The TRPM6 channel is expressed in the epithelia of the colon and also in the kidney, and underlies the autosomal recessive disorder, hypomagnesaemia with secondary hypocalcaemia (HSH), a disorder involving low serum levels of these ions, as well as seizures and tetany [151–156]. The TRPM6 channel is also known as a "chanzyme" since it functions as both an ion channel and an en-zyme (α-kinase). The channel is essential in epithelial magnesium absorption since it is permeable to magnesium (and also calcium). Low magnesium serum levels are caused by both defective magnesium absorption in the intestine and excess renal wasting of magnesium. Mutations in TRPM6 causing HSH are all intracellular and, apart from one missense mutation, are truncating (Fig. 8.1.5). The expected loss of function caused by such truncating mutations would lead to interference with epithelial magnesium absorption, leading to the hypomagnesaemia of the disorder.

8.1.5
Voltage-gated Sodium Channels

8.1.5.1 **Nav1.4**

Several voltage-gated sodium channel subunits are involved in inherited disorders. The α subunit of the Nav1.4 channel (with corresponding gene SCN4A on chromosome 17) is expressed in skeletal muscle where the upstroke of the action potential is mediated by the opening of these sodium channels. More than one channelopathy is caused by mutations to this gene, but all the conditions are characterized by symptoms related to skeletal muscle, either myotonias (muscle stiffness disorders), or paralytic disorders that may also be accompanied by myotonia, or a myasthenic (muscle weakness) disorder [7, 10, 20, 157–179]. The myotonic disorders are potassium-aggravated myotonia (PAM) and paramyotonia congenita (PMC), and the paralytic disorders are hyperkalaemic periodic paralysis (HyperPP) and hypokalaemic periodic paralysis (HypoPP). These disorders are all autosomal dominant and to date all the mutations are missense. Not all of the mutations are distinct in the sense that there is overlap between disorders, particularly between HyperPP and PMC.

The symptoms of PAM are varied, featuring either continuous myotonia (muscle stiffness due to uncontrolled repetitive firing of action potentials) or attacks of muscle stiffness brought on during exercise. The condition is clinically distinguishable from other myotonias by its sensitivity to potassium; ingestion of potassium-rich foods can trigger an attack. The mutations causing the disorder are in the intracellular loops of the channel protein (Fig. 8.1.6). Paramyotonia congenita is a form of myotonia induced by cold temperatures and exercise, and can be followed by periods of muscle weakness or paralysis. Mutations causing PMC are predominantly found in domain IV of the channel protein (Fig. 8.1.6). Hyperkalaemic periodic paralysis is characterized by episodes of flaccid muscular weakness associated with hyperkalaemia (at the beginning of the attack) that last for some hours before spontaneous recovery. The attacks associated with this condition are usually precipitated by ingestion of potassium or by rest following vigorous exercise. The mutations are located mainly in the transmembrane regions, particularly in domain IV (Fig. 8.1.6). Hypokalaemic periodic paralysis is characterized by attacks of muscle paralysis that are triggered by exercise or carbohydrate-rich food and feature hypokalaemia during the attack (contrasting to HyperPP). Two of the mutations associated with HypoPP are in the S4 transmembrane region of domain II with the third mutation being intracellular (Fig. 8.1.6). The final disorder associated with this ion channel is a new myasthenic syndrome which is characterized by general weakness, by recurrent attacks of respiratory and bulbar paralysis, and by normal serum potassium levels. Two missense mutations have been identified with this disorder (Fig. 8.1.6).

The mutations in PAM, PMC and HyperPP appear to impair the fast inactivation of the Na channel. This gain of function results in impaired repolarization of the muscle fiber leading to uncontrolled repetitive firing of action potentials following an initial voluntary activation, leading to muscle stiffness. In contrast to

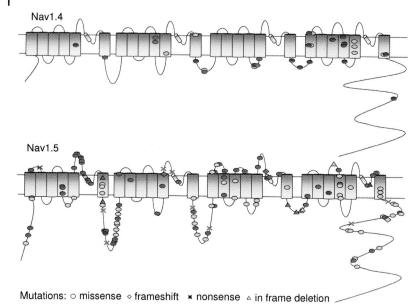

Mutations: ○ missense ◇ frameshift ✕ nonsense △ in frame deletion

Fig. 8.1.6 Mutations in Nav1.4 (green symbols, hyperkalaemic periodic paralysis; blue symbols, paramyotonia congenita; red symbols, potassium aggravated myotonia; magenta symbols, hypokalaemic periodic paralysis), and mutations in Nav1.5 (red symbols, Brugada's syndrome; green symbols, long QT type 3; blue symbols, isolated cardiac conduction defect).

these three disorders, HypoPP and the myasthenic syndrome involve loss of function defects. The mutations involved stabilize or enhance channel inactivation, although it is not understood how these mutations produce the long-lasting depolarizations associated with the muscle weakness of this condition.

8.1.5.2 Nav1.5

The Nav1.5 α subunit of the voltage-gated sodium channel (gene SCN5A, chromosome 3) is only expressed in heart muscle, where it plays a major role in cardiac action potentials, which in turn underlie normal cardiac rhythm. Several inherited disorders are associated with mutations in the SCN5A gene, long QT type3 (LQT3), Brugada syndrome, and isolated cardiac conduction defect (ICCD) [157, 178, 180–196]. LQT3 is a similar disorder to the types of long QT syndrome associated with potassium channels described above and has autosomal dominant inheritance. The disorder is again characterized by the prolongation of the QT interval of the ECG and patients are predisposed to ventricular tachyarrhythmias caused by unstable repolarization of the cardiac tissue. The mutations in the SCN5A gene associated with LQT3 are distributed throughout the channel protein (Fig. 8.1.6) and are predominantly missense with gain of function (usually by decreasing inactivation). The result of this is a small, persistent sodium current during the action potential plateau that delays myocyte repolarization and therefore evokes the long QT interval of this disorder.

Brugada syndrome is a form of idiopathic ventricular fibrillation characterized by elevated ST segments of the ECG and may involve right bundle branch block (RBBB), but there is no evidence of QT interval prolongation or structural heart disease. This is an autosomal dominant condition and the mutations in the SCN5A gene are again distributed throughout the channel protein and are mainly missense mutations with various truncating mutations (Fig. 8.1.6). These all lead to varying degrees of loss of function that reduce sodium channel function by either reducing functional channel expression or by increasing inactivation of the sodium channels, thus being the mirror image of LQT3. This reduction in the cardiac sodium current results in the transient outward potassium current being unopposed, leading to the ST elevation.

Isolated cardiac conduction defects are characterized by alteration of cardiac conduction through the His-Purkinje system, such that conduction between the atria and ventricles is slowed and the QRS complex of the ECG is widened. Conduction disease is distinct from LQT and Brugada syndromes since there is no tendency towards ventricular tachyarrhythmias and the electrophysiological basis of the disease is more complex than for the other two disorders. Mutations occur throughout the channel protein (Fig. 8.1.6). The cause of ICCD is complex since some mutations appear to cause gating defects to the Nav1.5 channel that are reminiscent of both LQT3 and Brugada's syndrome (i.e., some loss of function together with some decrease in inactivation). Other mutations seem to simply cause loss of function, such that other factors like allele penetrance or developmental factors may also influence the phenotype of the disorder.

8.1.5.3 Nav1.1

The Nav1.1 sodium channel (coded for by gene SCN1A on chromosome 2) is critical for the initiation and propagation of action potentials in the brain, where it is highly expressed. Unsurprisingly, given the fundamental role of this channel, mutations can cause epileptic disorders. Nav1.1 underlies a range of such disorders; these include generalized epilepsy with febrile seizures plus (GEFS+), intractable childhood epilepsy with generalized tonic-clonic seizures (ICEGTC), infantile spasms (IS), severe myoclonic epilepsy of infancy (SMEI), and borderline SMEI (SMEB) [10, 157, 197–212]. There is some overlap between these disorders in both level of severity and phenotype, with SMEI as the most severe form. In GEFS+, generally autosomal dominant, the mutations are all missense and are scattered throughout the membrane spanning regions (Fig. 8.1.7). For SMEI, there are over a hundred mutations, spread throughout the ion channel protein (Fig. 8.1.7); many of these mutations are *de novo* rather than inherited. The mechanism as to how these mutations cause the various epileptic disorders remains unclear, and other factors may be important in affecting the phenotype. Some of the mutations lead to gain of channel function (e.g., through decreased inactivation, as occurs in GEFS+), but the biophysical effects are complex. However, the

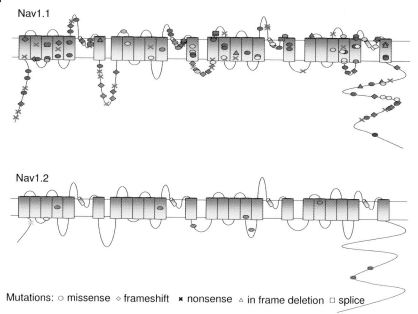

Mutations: ○ missense ◇ frameshift ✕ nonsense △ in frame deletion □ splice

Fig. 8.1.7 Mutations in Nav1.1 (red symbols, severe myoclonic epilepsy of infancy; green symbols, intractable childhood epilepsy with generalized tonic-clonic seizures; blue symbols, infantile spasms; magenta symbols, borderline severe myoclonic epilepsy of infancy; yellow symbols, generalized epilepsy with feb- rile seizures plus), and mutations in Nav1.2 (magenta symbols, benign familial neonatal infantile seizures; yellow symbols, febrile sei- zures associated with afebrile seizures; blue symbols, childhood absence epilepsy; red symbols, autism; green symbols, intractable epilepsy).

main result of most of the mutations is loss of function, and the mutations in the pore regions cause the most severe phenotypes. Many of the mutations for SMEI are truncating mutations (Fig. 8.1.7), leading to loss of function. Perhaps the (excitatory) epileptic disorders mainly arise from disruption of inhibitory pathways in the brain.

8.1.5.4 **Nav1.2**

The α subunit Nav1.2, with gene SCN2A on chromosome 2, is also critical in the initiation and propagation of action potentials in the brain. Mutations in this channel also underlie a range of epileptic disorders, including benign familial neonatal infantile seizures (BFNIS), febrile seizures associated with afebrile seizures, childhood absence epilepsy (CAE), intractable epilepsy, as well as autism (often associated with seizures) [10, 157, 212–220]. Mutations associated with these disorders are almost all missense, and are located throughout the channel protein (Fig. 8.1.7). The functional effects of the mutations in these disorders range from gain of function (via slowed inactivation) to complete loss of function, so that the phenotypic effects of these mutations remain difficult to explain.

8.1.5.5 β_1 Subunit

The auxiliary subunit of the sodium channel, β_1 (gene SCNB1, chromosome 19), is also involved in the autosomal dominant disorder of generalized febrile epilepsy plus (GEFS+) [221]. This subunit associates with voltage-gated sodium channels in both the brain and in skeletal muscle and normally has a major effect in accelerating the kinetics of the sodium current passed by these channels, particularly affecting the inactivation rate. When mutations causing GEFS+ are present in the SCNB1 gene, the rate of inactivation of the sodium channel becomes slower (as if the β_1 subunit were not present). These mutations therefore manifest themselves as a gain of function of the sodium channel, leading to GEFS+, as for some of the Nav1.1 mutants.

8.1.6
Nonvoltage-gated Sodium Channels

8.1.6.1 ENaC Channel

Another sodium ion channel that is associated with inherited disorders is the amiloride-sensitive epithelial sodium channel, ENaC. Unlike the voltage-dependent sodium channel family, the ENaC channel is composed of subunits with two membrane-spanning regions. The channel is formed from two α subunits, one β subunit and one γ subunit. The ENaC channel is expressed in epithelial cells of the kidney, the distal colon, and the airway. In the distal nephron of the kidney, the ENaC channel provides the primary mechanism for the reabsorption of Na^+ excretion and here allows the fine control of Na^+ balance, blood volume and therefore blood pressure under the hormonal control of aldosterone. There is a similar role for the ENaC channel in the distal colon where the channel functions to prevent excessive loss of Na^+ in the stools. A critical role is carried out by ENaC in the airways; the channel again promotes the absorption of Na^+, and consequently fluid, from the lungs – a process that is important both at birth and for subsequent control of lung fluid. There are two inherited disorders associated with mutations in ENaC channel subunits, pseudohypoaldosteronism type 1 (PHA1) and Liddle's syndrome [222–235]. In PHA1, the disorder is associated with loss of function mutations, and consequently is characterized by urinary loss of Na^+, despite an elevated level of aldosterone. The increased sodium load leads to increased potassium excretion as a result of increased activity of the sodium/potassium exchanger. A severe form of this disease is inherited as an autosomal recessive condition and results in sometimes lethal episodes of hypotension and shows alteration of Na^+ transport in several organs, the lungs also being affected. The less severe form of PHA1 is dominantly inherited and is usually only symptomatic during infancy, improving with age, this form of the condition not being associated with respiratory problems. The loss of function mutations in PHA1 occurs in α, β and γ subunits either at the extracellular region or the N terminus; many of these are nonsense, frameshift or missplice mutations, causing non-functional protein (Fig. 8.1.8). The other disorder of ENaC, Liddle's syndrome, is by contrast caused by gain of function mutations and an

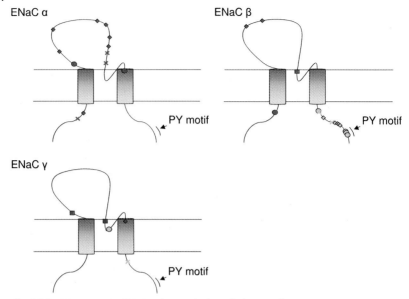

Fig. 8.1.8 Mutations in ENaC subunits (red symbols, pseudo-hypoaldosteronism type 1; green symbols, Liddle's syndrome).

autosomal dominant disorder. Overactivity of the ENaC channel in Liddle's syndrome leads to retention of sodium, plasma volume expansion, and hypertension, usually accompanied by metabolic alkalosis and hypokalaemia. These symptoms occur even though aldosterone levels are low. The mutations associated with Liddle's syndrome are all found within the β and γ subunits of the channel and all (except for one) are clustered in the C termini of these subunits (Fig. 8.1.8), leading to deletion or alteration of a PPPxY sequence ("PY motif") contained in the C terminus. The PY motif interacts with Nedd4, a protein that controls the number of active ENaC channels at the cell surface via ubiquitination and consequent downregulation. Disruptions in the PY motif prevent the interaction with Nedd4 and so there is an overexpression of channels at the cell surface since downregulation is impaired. This overactivity of ENaC leads to Liddle's syndrome. One mutation in Liddle's syndrome is located in the pore region of the γ subunit (Fig. 8.1.8) and, instead of the above mechanism, seems to cause increased activity of the channel by an increase in channel open probability, rather than by an increase in expression.

8.1.7
Calcium Channels

8.1.7.1 **Cav1.1**

The Cav1.1 (α_{1S}) calcium channel, together with its associated auxiliary subunits (α_2-δ, β and γ), forms a member of the L-type calcium channel family. This high voltage-activated channel is almost exclusively expressed in skeletal muscle. The α-subunit that forms the channel pore corresponds to the CACNA1S gene on chromosome 1. In skeletal muscle the Cav1.1 calcium channel is involved in excitation–contraction coupling, i.e., the coupling of electrical excitation of the muscle to the release of calcium from the sarcoplasmic reticulum (SR). For this, the Cav1.1 channel interacts with the ryanodine receptor located on the SR; the conformational change of Cav1.1 upon depolarization activates the ryanodine receptor, which in turn releases calcium, leading to the muscle contraction. Two autosomal dominant channelopathies are associated with this channel: hypokalaemic periodic paralysis (HypoPP) and malignant hyperthermia [3, 8, 14, 15, 23, 236–239]. As for channelopathies of Nav1.4, where the disorder HypoPP also occurs, the symptoms of HypoPP are carbohydrate-induced paralysis with low plasma potassium concentration. The muscle weakness associated with HypoPP may persist after the return of normal potassium levels and the plasma potassium concentration may not fall during the attack of paralysis. This indicates that the hypokalaemia associated with this disorder is somehow a result of the attack of weakness itself rather than being its cause. The mutations in the CACNA1S gene that have been identified as being causative for HypoPP are located in the S4 voltage sensing regions of domains II and IV of the α_{1S} subunit (Fig. 8.1.9). Although effects of these mutations on channel function have been reported, the mechanism leading to this disorder of Cav1.1 is unknown. During an attack of weakness, muscle fibers show long-lasting depolarization with resultant inactivation of sodium channels and hence paralysis. The other disorder associated with the CACNA1S gene is malignant hyperthermia susceptibility, which is a potentially fatal condition triggered by general anesthetics such as halothane or depolarizing muscle relaxants such as suxamethonium. Symptoms include severe hyperthermia, tachycardia and muscle rigidity, which must be treated rapidly with the muscle relaxant dantrolene. A mutation underlying this disorder is found (Fig. 8.1.9) in the intracellular III-IV linker, and the mutation appears to activate the ryanodine receptor with massive calcium release from the sarcoplasmic reticulum, although the mechanism is unclear since the calcium channel interacts with the ryanodine receptor at another site (II-III linker). Possibly, as Cav1.1 serves as a voltage sensor for the ryanodine receptor, the mutations may alter the voltage dependence between the two channels. Most of the malignant hyperthermia mutations are found in the ryanodine receptor itself.

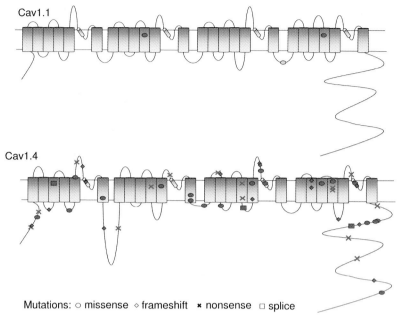

Cav1.1

Cav1.4

Mutations: ○ missense ◇ frameshift × nonsense □ splice

Fig. 8.1.9 Mutations in Cav1.1 (red symbols, hypokalaemic periodic paralysis; green symbols, malignant hyperthermia), and mutations in Cav1.4 (red symbols, congenital stationary night blindness type 2).

8.1.7.2 **Cav1.2**

The L-type Cav1.2 channel (α_{1C}) is coded for by the CACNA1C gene on chromosome 12. This voltage-dependent calcium channel is widely expressed, particularly in the heart and brain but also in smooth muscle, the gastrointestinal system, immune system and in developing fingers and teeth in foetal tissues. The function of Cav1.2 in these other tissues is unknown but in the heart and brain the channel functions to increase intracellular calcium via inward Ca^{2+} currents upon channel opening. The disorder associated with this channel is Timothy syndrome, and is characterised by symptoms of delayed cardiac repolarisation leading to long QT (type 8), webbing of the digits (syndactyly), immune system deficiency and autism. A single mutation in the CACNA1C gene, G406R, has been identified as the cause of the disease. This missense mutation is located at the C terminal end of the S6 segment of domain I of the channel and is a gain of function mutation since it disrupts voltage-dependent inactivation of Ca^{2+} current. It is this loss of inactivation that leads to the prolongation of inward Ca^{2+} current and a delay of cardiomyocyte repolarisation, visible as the elongation of the QT interval on an ECG. As discussed elsewhere, an elongation of the QT interval increases susceptibility to dangerous arrhythmias that are the primary cause of death associated with Timothy syndrome. Other characteristic defects of Timothy syndrome, such as syndactyly are developmental and are likely to be caused by Ca^{2+} induced cell death during development. The link between autism and the

Cav1.2 channel is also important as it implicates abnormal Ca^{2+} signalling as a possible cause of this condition [239a].

8.1.7.3 Cav1.4

Another channelopathy of L-type channels involves the α_{1F} subunit, with corresponding gene CACNA1F on the X chromosome. This channel is expressed almost exclusively in the retina, and mutations cause congenital stationary night blindness type 2 (CSNB2) [2, 14, 23, 237, 239–248], a recessive, non-progressive disorder of the retina. In addition to night blindness, the condition is characterized also by a decrease in visual acuity and myopia; electroretinograms indicate a reduction in both cone and rod synaptic function. Mutations are found throughout the channel protein (Fig. 8.1.9); most of the CSNB2 mutations cause protein truncation (nonsense, frameshift and splice mutants), leading to loss of function. This would impair the influx of calcium required for the release of glutamate from the photoreceptor presynaptic terminals in darkness, so underlying the disorder. Missense mutations are generally also loss of function (although a gain of function mutation also exists [249]).

8.1.7.4 Cav2.1

The Cav2.1 channel (α_{1A}) is coded for by the CACNA1A gene on chromosome 19. This high-voltage activating calcium channel is also known as P/Q-type because of its expression in Purkinje cells and granule cells; the channel is responsible for the Ca^{2+} influx that triggers neurotransmitter release. There are three channelopathies associated with mutations in this gene, familial hemiplegic migraine (FHM), episodic ataxia type 2 (EA2), and spinocerebellar ataxia type 6 (SCA6) [1, 9, 10, 14, 23, 239, 250–271], all autosomal dominant traits. Familial hemiplegic migraine is characterized by paralysis affecting one side of the body during attacks of migraine. Mutations in FHM are generally missense, mainly located in S4, S5, pore and S6 regions of all four domains (Fig. 8.1.10). For many mutations there are shifts in the IV curve to more negative potentials, so causing increased currents for small voltage steps. Although the way these mutations cause migraine is not clear, it may be due to increased neurotransmitter release (such as a consequence of the increased calcium entry into the neuron). Episodic ataxia type 2 is characterized by recurrent attacks of asymmetric limb movements and severe postural and gait ataxia. About half of the mutations of CACNA1A that cause EA2 lead to truncated channel proteins, which generally are incapable of forming functional channels. Many mutations are missense, located mainly at or near pore regions, and appear to cause channels with reduced or complete lack of function. Again the mechanism underlying the disorder is not clear, although a possibility may be a disturbance in GABA and glutamate release. Spinocerebellar ataxia, type 6 (SCA6) is similar to EA2 when studied phenotypically, although the main distinguishing features between the two disorders is the larger proportion of progressive ataxia among SCA6 patients (onset at 40 to 50 years of

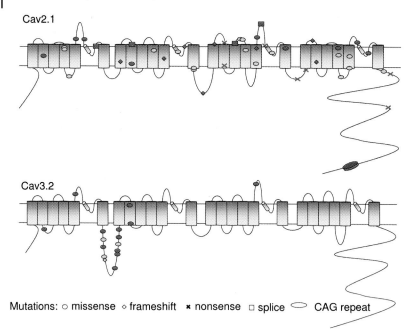

Cav2.1

Cav3.2

Mutations: ○ missense ◇ frameshift ✕ nonsense □ splice ⌒ CAG repeat

Fig. 8.1.10 Mutations in Cav2.1 (green symbols, familial hemiplegic migraine; red symbols, episodic ataxia type 2; blue symbols, spinocerebellar ataxia type 6), and mutations in Cav3.2 (red symbols, childhood absence epilepsy; green symbols, idiopathic generalized epilepsy).

age), and the presence of extracerebellar signs in these patients. The mutation in the CACNA1A gene underlying SCA6 is an expansion of the CAG repeat sequence in the coding region of the C terminus of the channel protein (Fig. 8.1.10), manifested as a polyglutamate extension from the normal 4–16 glutamate residues to some 21–28 residues. The more the repeat is expanded, the greater the disease severity and the earlier the age of onset. Expansion of the polyglutamine repeat affects P/Q channel activity by introducing a hyperpolarizing shift in the voltage dependence of channel inactivation, leading to reduced calcium influx into cells; details of how this leads to the disorder are again unclear.

A mutation in the CACNB4 gene has been associated with juvenile myoclonic epilepsy [272]. This gene codes for the β_4 subunit, which modulates the function and trafficking of the Cav2.1 calcium channel. The mutation responsible causes a truncation of the β_4 subunit protein with impaired calcium channel function; the reduction in the Ca^{2+} current in some way leads to the epileptic phenotype.

A further condition that is connected to the Cav2.1 channel, though not caused by mutations in the CACNA1A gene, is the Lambert-Eaton myasthenic syndrome (LEMS) [23, 24, 273, 274]. This is an autoimmune disorder in which the P/Q-type calcium channels are targeted and down-regulated by autoantibodies. The disorder is characterized by muscle weakness, autonomic neuropathies and, in many cases, small cell lung carcinoma. In this condition, the reduced number of presynaptic P/Q-type channels leads to a decrease in the release of acetylcholine at the neuromuscular junction, and therefore impairs neuromuscular transmission from nerve to muscle, leading to muscle weakness.

8.1.7.5 Cav3.2 Channels

The Cav3.2 calcium channel (α_{1H}) with the corresponding gene CACNA1H on chromosome 16 is a low voltage activated calcium channel, known as a T-type channel. T-type channels are expressed in the heart and in the brain, particularly in the thalamus and neocortex where they play an important role in oscillatory behavior and in neuronal firing of bursts, and have been implicated in the generation of idiopathic generalized epilepsies, including childhood absence epilepsy [275–280]. Most of the mutations underlying these disorders are missense (Fig. 8.1.10) and are found in the I-II linker (between the first and second domains). The channel mutations have relatively minor biophysical effects, with a range of effects. A common feature of some of the mutations is a gain of channel function, which would be expected to increase the number of spikes triggered per burst and alter spontaneous neuronal oscillations in the thalamus and neocortex, leading to epileptic seizures, although the exact mechanism of this is again unclear. The mild effects of the mutations in this gene led to the suggestion that the gene confers a susceptibility to epileptic attack as part of a multifactorial disorder.

8.1.8
Chloride Channels

8.1.8.1 ClC-1

There are nine different chloride channels that belong to the CLC gene family and they are structurally unrelated (Fig. 8.1.11) to other channels and transport systems. The channel is composed of two identical subunits, each with its own pore, so forming a two-pore channel. The skeletal muscle chloride channel ClC-1 (corresponding gene CLCN1 on chromosome 7) underlies the large resting Cl⁻ ion conductance of skeletal muscle. This conductance serves to stabilize the membrane potential at the resting potential level and it inhibits repetitive muscle action potentials after a single nerve impulse. The mutations in the CLCN1 gene lead to myotonia congenita, a condition that can be either dominant (Thomsen's disease) or recessive (Becker's syndrome) [7, 15, 25–27, 281–291]. The Thomsen's form of the disorder is usually present from birth whereas the Becker form develops during the first or second decade of life; however, the symptoms are more severe in the more common Becker form. The myotonia associated with these two conditions is characterized by muscle stiffness similar to the myotonia caused by mutations in the sodium channel gene, SCN4A. A distinctive feature of the chloride channelopathy is that weakness and myotonia usually appear during exercise, rather than during rest (after exercise) as occurs for the sodium channel mutations. For both Thomsen's disease and Becker's syndrome, mutations cause loss of function and are distributed throughout the channel protein; missense muta-

CLCN1

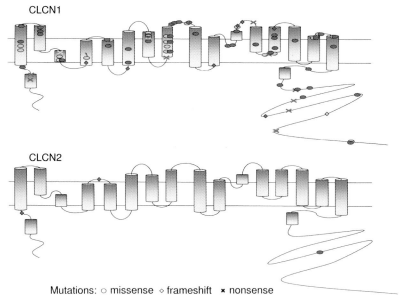

CLCN2

Mutations: ○ missense   ◇ frameshift   ✕ nonsense

Fig. 8.1.11 Mutations in CLCN1 (green symbols, Thomsen's disease; red symbols, Becker's syndrome), and mutations in CLCN2 (red symbols, autosomal dominant epileptic disorders).

tions are more common than nonsense and frameshift mutations (Fig. 8.1.11). The myotonia can be explained by a reduction in the membrane Cl^- conductance so that the muscle is rendered hyperexcitable. In turn, this leads to a single nerve impulse causing the repetitive firing of muscle action potentials, which creates the muscle stiffness. How the dominant and recessive forms of the condition have this effect is worth noting. The recessive form (Becker's) is due to impaired gating of each of the two conducting pores, while the dominant mutations in Thomsen's disease seem to cause their effect by acting on a common gating mechanism that simultaneously controls both pores in this dimeric channel.

8.1.8.2 ClC-2

The ClC-2 channel, with corresponding gene CLCN2 on chromosome 3, is activated by hyperpolarization, cell swelling and weakly acidic extracellular pH. Although the channel is almost ubiquitously expressed, it is particularly prevalent in neurons that are inhibited by GABA, where the channels function to maintain a high transmembrane Cl^- gradient necessary for the inhibitory GABA response. This inhibitory response normally counteracts the neuronal excitation that would be responsible for the generation of epileptic seizures. Mutations in the CLCN2 gene lead to autosomal dominant epileptic disorders, particularly juvenile myoclonic epilepsy, juvenile absence epilepsy, and epilepsy with grand mal seizures on awakening [25–27, 281, 283, 292, 293]. Two mutations are frameshifts, introducing truncated protein (Fig. 8.1.11), and cause loss of function as the mutant channels do not reach the membrane. This lowers the transmembrane chloride gradient that is essential for GABA inhibition, leading to an excitatory epileptic response. A third mutation is missense (Fig. 8.1.11), but it is not yet fully clear how this causes the disorder.

8.1.8.3 ClC-5

The ClC-5 channel with corresponding gene CLCN5 is primarily located in the kidney and intestine where it is expressed on endosome membranes. These intracellular compartments are responsible for internalizing molecules bound on the external surface of the cell. The ClC-5 channel appears to be essential for normal function of the endosomes, probably by allowing charge neutralization of H^+ ions that are pumped into the endosomes by H^+-ATPase. The physiological importance of the ClC-5 channel is illustrated by its dysfunction in Dent's disease, an X chromosome linked disorder of the renal system [25–27, 281, 283, 294–300]. The disorder is characterized by low molecular weight proteinuria and hypercalciuria, which in due course lead to the development of kidney stones. Renal failure during later life further typifies the disorder. In Dent's disease, many truncating mutations (frameshifts, splice site) are found (Fig. 8.1.12), introducing loss of function of the chloride channel. In addition, missense mutations are found that also cause loss of function (partial or complete); the missense mutations are clustered at the interface between the two subunits of the channel dimer, indicating a cru-

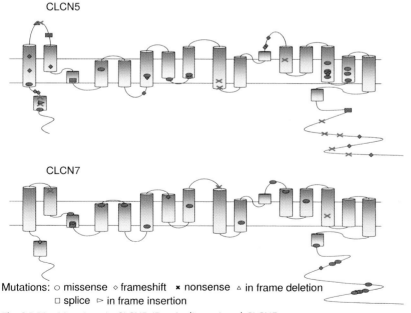

Fig. 8.1.12 Mutations in CLCN5 (Dent's disease) and CLCN7 (infantile malignant osteopetrosis).

cial role for the interaction between the two subunits at this interface. The result of the mutation is that endosome function is reduced; the proteinuria arises from the reduced small protein absorption in the kidney and consequent excretion in the urine. The hypercalciuria is likely an indirect effect of altered reabsorption of calcium-regulating hormones.

8.1.8.4 **CLC-7**

Another member of this ion channel family that is nearly ubiquitously expressed is ClC-7, with the corresponding gene CLCN7 (chromosome 16). Similar to ClC-5 channels, ClC-7 channels are localized in the membranes of intracellular compartments, particularly residing on late endosomes and lysosomes. This chloride channel is also found in the ruffled border (part of the plasma membrane) of osteoclasts, the cells involved in the absorption and removal of bone. In this location, the chloride channels function to provide the chloride conductance required for the efficient proton pumping by the H^+-ATPase of the ruffled membrane. In turn, the H^+-ATPase itself functions to maintain the acidic environment required for the efficient functioning of the lysosomal enzymes that degrade the bone matrix. Mutations in the CLCN7 gene underlie the autosomal dominant condition infantile malignant osteopetrosis [25–27, 281, 283, 301–305]. This is characterized by dense, fragile bones that are devoid of bone marrow and is caused by the disruption to the acidic environment of osteoclasts and resultant interference with lysosomal enzyme function. Mutations (mainly missense) occur throughout the channel protein (Fig. 8.1.12), apparently leading to loss of function, such that the chloride conductance does not allow maintenance of the acidic environment of the osteoclasts.

8.1.8.5 **ClC-Ka, ClC-Kb",4>**

The ClC-Ka and ClC-Kb channels (corresponding genes CLCN-KA and CLCN-KB on chromosome 1) are both predominantly expressed in the kidney and also in the inner ear. In the kidney, both channels are expressed mainly in the ascending limb of the loop of Henle and in the collecting duct, and are a crucial element in salt reabsorption. In both the kidney and the inner ear, the ClC-K channels always co-localize with the β-subunit barttin, which is required for the functional expression of both channels in the cell membrane. Mutations in CLCN-K genes are responsible for causing the renal transport defect, Bartter syndrome [25–27, 281, 283, 306–312]. This autosomal recessive inherited disorder is characterized by salt loss, which occurs through a reduction of NaCl reabsorption in the ascending limb of the loop of Henle. Most of the mutations in the ClC-KB channel (type 3 Bartter syndrome) are missense (Fig. 8.1.13), although complete deletions of the gene are prevalent. Lack of function, or reduced function, of these mutants would cause the disorder by its interference with salt reabsorption in the kidney. Interestingly, mutations in barttin are also causative of a more severe form of Bartter syndrome (type 4) because the lack of barttin disrupts the function of both ClC-K

CLCNKA

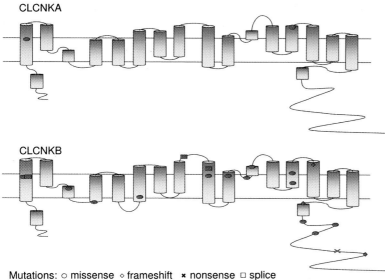

CLCNKB

Mutations: ○ missense ◇ frameshift × nonsense □ splice

Fig. 8.1.13 Mutations in CLCNKA and CLCNKB (Bartter's syndrome).

channels. Also, barttin mutations causing Bartter syndrome type 4 additionally lead to congenital deafness since they inhibit the functioning of both the ClC-K channels in the inner ear such that there is a secretory defect of the stria vascularis. Additionally, mutations in one patient affected both CLC-K genes simultaneously (Fig. 8.1.13 for ClC-KA mutation), with features mimicking those of Bartter syndrome type 4.

8.1.8.6 **CFTR**

Mutations in the CFTR chloride channel cause cystic fibrosis [281, 313–321], an autosomal recessive disorder that affects 1 in every 2500 live-births. The gene concerned is the "cystic fibrosis transmembrane conductance regulator" (CFTR) on chromosome 7, and the channel is a member of the ABC transporter supergene family (Fig. 8.1.14). The channel is expressed in many epithelial tissues, including the lungs, colon, pancreas and kidney. This chloride channel plays a critical role in the lung airways and in the pancreas where it functions in the regulation of fluid secretion and other absorptive processes. Secondary chronic bacterial pulmonary infection, facilitated by the dense mucous caused by defective CFTR, is the cause of death. The correct functioning of the CFTR channel is necessary for production of normal mucous that can be easily cleared by ciliated epithelial cells. There are more than 1000 mutations in the CFTR that cause cystic fibrosis and these are classified into five groups according to the functional effect of the mutation. Class I mutations are those that introduce a premature stop codon into the sequence of the channel such that no functional CFTR channels are synthe-

CFTR

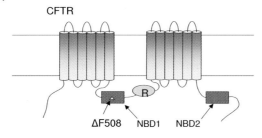

△ in frame deletion

Fig. 8.1.14 There are more than a thousand mutations in CFTR; the figure shows the most prevalent mutation, a deletion of a single phenylalanine at residue 508 (ΔF508). NBD1/2: nucleotide binding domains. R: regulatory domain.

sized. Aminoglycoside drugs can sometimes be used in the treatment of class I cystic fibrosis since these can cause the skipping of a stop codon during protein translation. Class II mutations are those that lead to defective processing or trafficking of the channel protein and therefore result in a reduction in channel expression in the plasma membrane. The most prevalent mutation in cystic fibrosis, ΔF508, belongs to this class (Fig. 8.1.14). Compounds (such as glycerol, anthracycline and butyrates) that aid the transportation of chloride channels to the membrane are effective in treating this form of the condition. Mutations in class III lead to channels that reach the cell membrane but are not able to be activated. Class IV mutations are those where the channel reaches the cell membrane and can be activated, but with decreased chloride conductance. Class V mutations result in reduced amounts of functional protein being expressed (rather than no protein expression at all as in class I) as a result of abnormal or alternative splicing.

8.1.9
Ligand-gated Channels

8.1.9.1 Muscle AChRs
Muscle acetylcholine receptors (AChRs) are concentrated on the postsynaptic folds at the neuromuscular junction. These ligand-gated ion channels conduct sodium and potassium ions, and they function to cause skeletal muscle depolarization when they are activated by acetylcholine released from the presynaptic nerve terminal. The channels are formed from five subunits, each with four transmembrane domains, and have the structure $\alpha_2\beta\gamma\delta$ in fetal muscle and $\alpha_2\beta\epsilon\delta$ in adult muscle. These channels briefly open in response to the binding of two acetylcholine molecules, one at the $\alpha\delta$ site and one at the $\alpha\epsilon$ site. Very different clinical syndromes can be caused by different mutations to the subunit genes, leading to various postsynaptic congenital myasthenic syndromes (CMS), i.e., inherited muscle weakness disorders [10, 20, 322–344].

One of the most prevalent CMS is acetylcholine receptor deficiency, which is due to a reduction in AChRs in the postsynaptic membrane. This is a recessive muscle weakness disorder with onset in infancy; however, the condition is not

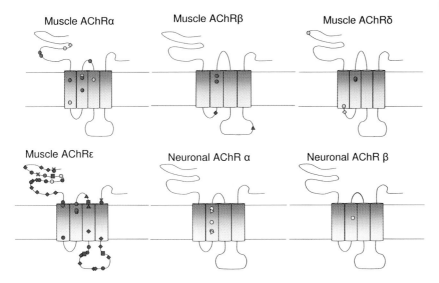

Mutations: ○ missense ◇ frameshift × nonsense △ in frame deletion
□ splice ▷ in frame insertion

Fig. 8.1.15 Mutations in muscle AChR subunits (red symbols, slow channel syndrome; green symbols, fast channel syndrome; blue symbols, AChR deficiency), and mutations in neuronal AChR subunits (yellow symbols, autosomal dominant nocturnal frontal lobe epilepsy).

progressive. Most of the mutations underlying this disorder are located throughout the ε subunit gene (Fig. 8.1.15). Mutations are mainly truncating types (frameshifts, splice mutations and nonsense mutations), so producing inactive channel proteins. These channel proteins frequently do not traffic to the cell surface properly and are retained in the endoplasmic reticulum. There are also mutations in the promoter region that reduce transcription and hence expression. The effect of these reductions in the number of functional AChRs, is a decrease in end-plate potential amplitude, which in turn leads to defective firing of action potentials, and hence characteristic muscle weakness.

Another CMS disorder is the autosomal dominant slow channel syndrome, which again is characterized by muscle weakness. This condition is a progressive disorder and can be distinguished from other CMS disorders in presenting in childhood, adolescence or adult life. This inherited disorder is associated with mutations in the α, β, δ and ε subunits of the AChR (Fig. 8.1.15), generally missense mutations. Many of the mutations are present in the M2 transmembrane domain that lines the channel pore and these mutations disrupt channel function, increasing channel open times. Other mutations in the extracellular N terminus increase AChR affinity for acetylcholine (which binds in this region) causing repetitive channel opening whilst ACh remains bound to the receptor. In addition, mutations in the M1 transmembrane region, surprisingly, also affect ACh binding. The prolonged ion channel openings (together with prolonged end-plate potentials) are accompanied by damage at the muscle synapse, probably caused by toxic overactivity due to calcium overload at the end-plate region.

Fast channel syndrome is a third CMS disorder associated with AChR mutations and is an autosomal recessive disorder. The disorder is characterized by muscle weakness and the onset of this condition is at birth. The mutations causing this disorder are located in the α, δ and ϵ subunits of the AChR (Fig. 8.1.15), and cause fewer and shorter channel openings. In some cases there is also some reduction in expression of the AChR at the end-plate. The reduced response of the AChR to acetylcholine is likely to be caused by the combined effect of both of these defects, and underlies the muscle weakness in this disorder.

8.1.9.2 Neuronal AChRs

The neuronal AChR is homologous with the AChR at the neuromuscular junction and has similar biophysical properties. Its structure is similar in that it is a pentameric protein; however, the subunit composition is different as the channels consist of α and β subunits only. The neuronal ligand-gated AChR also conducts sodium and potassium ions and is important in regulating sleep and arousal. The α_4 and β_2 subunits co-assemble to form the most common heteromeric receptor subtype and these receptors have a high affinity for acetylcholine and nicotine. The inherited disorder associated with mutations in both the α_4 and β_2 subunits is autosomal dominant nocturnal frontal lobe epilepsy (ADNFLE) [9, 12, 19, 344–358], which is characterized by clusters of seizures occurring during sleep, especially on falling asleep or waking. The mutations in the α_4 and β_2 subunits occur in or near the M2 transmembrane domain (Fig. 8.1.15), which lines the pore of this channel. Several distinct effects have been reported for the different mutations, including a decrease in maximal current amplitude, an increase or decrease in desensitization rate and both increases and decreases in acetylcholine affinity, and it is therefore unclear whether ADNFLE can be thought of as arising from a loss or gain of function of the AChR. Also unclear is how the seizures associated with this condition are localized to the frontal lobes of the brain since both the α_4 and β_2 subunits are expressed in all brain tissues without specificity to the frontal lobe. One explanation for this is the possibility of improved compensation for the effect of the mutated α_4/β_2 receptor in parts of the brain other than the frontal lobe.

8.1.9.3 Glycine Receptors

The ligand-gated glycine receptor channel is a pentameric protein consisting of α and β subunits (genes GLRA and GLRB), each with four transmembrane domains. The α-subunit exists as four forms α_1, α_2, α_3 and α_4, although only one β subunit has been reported. This channel conducts chloride ions so that glycine receptors mediate inhibition, especially of motor neurones in the brainstem and spinal cord. The inherited disorder associated with mutations in the glycine recep-

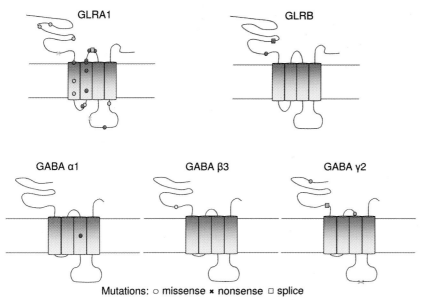

Mutations: ○ missense × nonsense □ splice

Fig. 8.1.16 Mutations in GLRA1 and GLRB (red symbols, dominant hyperekplexia; green symbols, recessive hyperekplexia), and mutations in GABA α1 (blue symbols, juvenile myoclonic epilepsy), GABA β3 (yellow symbols, insomnia), and GABA γ2 (magenta symbols, epilepsy).

tor is hyperekplexia [281, 344, 359–375], a condition that is characterized by an excessive startle response and exists in both major and minor forms. In the major form onset is in the newborn and motor development is delayed although cognitive development is normal. The major form has symptoms of stiffness, and brisk tendon reflexes in addition to the excessive startle response whereas the minor form is characterized by excessive startle alone. Hyperekplexia can be inherited in either an autosomal dominant or recessive manner, and various mutations in the α_1 and β subunits of the glycine receptor have been described (Fig. 8.1.16). Mutations causing the dominant and recessive forms of the disorder have different inhibitory effects on the glycine receptor. The dominant mutations (predominantly present in the M2 transmembrane domain) usually reduce both single channel chloride conductance and glycine sensitivity. The recessive mutations (dispersed throughout the channel protein) enhance the desensitization rate of the channels in addition to a decreased current magnitude and glycine sensitivity. There are, to date, two mutations associated with the β subunit, a missense mutation and a splice mutation. The missense mutation again affects the glycine sensitivity whereas the splice mutation prevents expression of functional β subunits.

8.1.9.4 **GABA$_A$ Receptors**

The main inhibitory transmitter in the brain is γ-aminobutyric acid (GABA), which acts on the heteropentameric GABA$_A$ receptor. Most GABA$_A$ receptors contain two α subunits, two β subunits and a γ subunit. Although there are six types of α subunits, four types of β subunit and three types of γ subunits, the $\alpha_1\beta_2\gamma_2$ combination is the most abundant in almost all regions of the brain. The pentameric complex forms a chloride ion channel so that when GABA is released from nerve terminals, the chloride ion influx generates hyperpolarization and hence inhibition. Only a few of the subunits are so far involved in inherited disorders [376–389]; these are α_1, α_2, β_3 and γ_2. A missense mutation in the third transmembrane region (Fig. 8.1.16) in α_1 leads to autosomal dominant juvenile myoclonic epilepsy as a result of a reduced response to GABA. The α_2 subunit is associated with susceptibility to alcohol dependence, perhaps by modulating the level of neural excitation since it also affects certain frequencies in the electroencephalogram. Mutations associated with the α_2 subunit are not present in the coding region of the subunit, but noncoding mutations probably affect the expression of this subunit. A chronic form of insomnia seems to be associated with a single missense mutation in the β_3 subunit at the N terminus (Fig. 8.1.16); the mutation may decrease the GABA-mediated inhibition. Various forms of epilepsy are associated with mutations in the γ_2 subunit; these include GEFS+ and CAE with considerable overlap between the conditions. There are two missense and two truncating mutations (Fig. 8.1.16), all giving varying degrees of inhibition of GABA$_A$ receptors. One interesting missense mutation in this subunit caused loss of benzodiazepine sensitivity, leading to the suggestion of an endogenous benzodiazepine receptor ligand that may reduce channel function *in vivo*.

8.1.10
Conclusions

Hopefully this wide-ranging overview of channelopathies will have given a useful basis in beginning to understand, and indeed simplifying, the wealth of information currently available on these ion channel disorders. With the exception of ryanodine receptor mutations in skeletal muscle and heart (not described here), we tried to cover all channelopathies that are currently known to us; within the space available we obviously could not do justice to each channelopathy by providing detailed analysis of each disorder. In particular, we have only briefly alluded to the potentially exciting area of drug treatment of some of the disorders, for instance by drugs designed to increase trafficking in cases where mutants are trafficking deficient. Another area of current therapeutic interest is by the utilization of gene therapy to attempt to replace defective genes by normal ones; so far this approach is still under development. Clearly, there is exciting scope for future research, not only in understanding better the functional basis of mutations in each disorder, but also in the search for therapeutic treatments of these important channelopathies.

Acknowledgements

Work supported by Biotechnology and Biological Sciences Research Council.

References

1. Baloh, R.W., Jen, J.C. (**2002**) *Ann. New York Acad. Sci.*, 956, 338–345.
2. Celesia, G.G. (**2001**) *Clin. Neurophysiol.*, 112, 2–18.
3. Felix, R. (**2000**) *J. Med. Genet.*, 37, 729–740.
4. George, A.L. (**2004**) *Arch. Neurol.*, 61, 473–478.
5. George, A.L. (**2004**) *Epilepsy Curr.*, 4(2), 65–70.
6. Hoffman, E.P. (**1995**). *Annu. Rev. Med.*, 46, 431–441.
7. Jurkat-Rott, K., Lerche, H., Lehmann-Horn, F. (**2002**) *J. Neurol.*, 249(11), 1493–1502.
8. Jurkat-Rott, K., Lehmann-Horn, F. (**2001**) *Curr. Opin. Pharmacol.*, 1, 280–287.
9. Kullmann, D.M. (**2002**) *Brain*, 125(6), 1177–1195.
10. Kullmann, D.M., Hanna, M.G. (**2002**) *Lancet Neurol.*, 1(3), 157–166.
11. Lehmann-Horn, F., Jurkat-Rott, K. (**1999**) *Physiol. Rev.*, 70(4), 1317–1372.
12. Lerche, H., Jurkat-Rott, K., Lehmann-Horn, F. (**2001**) *Am. J. Med. Genet.*, 106(2), 146–159.
13. Lossin, C., Wang, D.W., Rhodes, T.H., Vanoye, C.G., George, A.L. (**2002**) *Neuron*, 34(6), 877–884.
14. Lorenzon, N.M., Beam, K.G. (**2000**) *Kidney Int.*, 57, 794–802.
15. Meola, G., Sansone, V. (**2000**) *Neurol. Sci.*, 21, S953–S961.
16. Moulard, B., Picard, F., le Hellard, S., Agulhon, C., Weiland, S., Favre, I., Bertrand, S., Malafosse, A., Bertrand, D. (**2001**) *Brain Res. Rev.*, 36, 275–284.
17. Paulussen, A.D., Gilissen, R.A., Armstrong, M., Doevendans, P.A., Verhasselt, P., Smeets, H.J., Schulze-Bahr, E., Haverkamp, W., Breithardt, G., Cohen, N., Aerssens J. (**2004**) *J. Mol. Med.*, 82(3), 182–188.
18. Sanguinetti, M.C., Spector, P.S. (**1997**) *Neuropharmacology*, 36(6), 755–762.
19. Steinlein, O.K. (**2002**). *Eur. J. Pain*, 6 (Sup. A), S27–S34.
20. Surtees, R. (**2000**) *Eur. J. Pediatr.*, 159 (Sup.3), S199–S203.
21. Wray, D. (**2001**) *Pharmaceut. News*, 8(2), 12–17.
22. Abraham, M.R., Jahangir, A., Alekseev, A.E., Terzic, A. (**1999**) *FASEB J.*, 13(14), 1901–1910.
23. Benatar, M.G. (**1999**) *Q. J. Med.*, 92(3), 133–141.
24. Flink, M.T., Atchison, W.D. (**2003**) *J. Bioenerg. Biomembr.*, 35(6), 697–718.
25. Fahlke, C. (**2000**) *Kidney Int.*, 57(3), 780–786.
26. Jentsch, T.J., Poët, M., Fuhrmann, J.C., Zdebik, A.A. (**2005**) *Annu. Rev. Physiol.*, 67, 779–807.
27. Jentsch, T.J., Stein, V., Weinreich, F., Zdebik, A.A. (**2002**) *Physiol. Rev.*, 82(2), 503–568.
28. Flagg, T.P., Yoo, D., Sciortino, C.M., Tate, M., Romero, M.F., Welling, P.A. (**2002**) *J. Physiol.*, 544(2), 351–362.
29. Ai, T., Fujiwara, Y., Tsuji, K., Otani, H., Nakano, S., Kubo, Y, Horie, M. (**2002**) *Circulation*, 105(22), 2592–2594.
30. Andelfinger, G., Tapper, A.R., Welch, R.C., Vanoye, C.G., George, A.L., Benson, D.W. (**2002**) *Am. J. Hum. Genet.*, 71(3), 663–668.
31. Fodstad, H., Swan, H., Auberson, M., Gautschi, I., Loffing, J., Schild, L., Kontula, K. (**2004**) *J. Mol. Cell. Cardiol.*, 37(2), 593–602.
32. Hosaka, Y., Hanawa, H., Washizuka, T., Chinushi, M., Yamashita, F., Yoshida, T., Komura, S., Watanabe, H., Aizawa, Y. (**2003**) *J. Mol. Cell. Cardiol.*, 35(4), 409–415.
33. Plaster, N.M., Tawil, R., Tristani-Firouzi, M., Canun, S., Bendahhou, S., Tsunoda, A., Donaldson, M.R., Iannaccone, S.T., Brunt, E., Barohn, R., Clark, J., Deymeer, F., George, A.L., Fish, F.A., Hahn, A., Nitu, A., Ozdemir, C., Serdaroglu, P., Subramony, S. H., Wolfe, G., Fu, Y.H., Ptacek L.J. (**2001**) *Cell*, 105(4), 511–519.

34. Tristani-Firouzi, M., Jensen, J.L., Donaldson, M.R., Sansone, V., Meola, G., Hahn, A., Bendahhou, S., Kwiecinski, H., Fidzianska, A., Plaster, N., Fu, Y.H., Ptacek, L.J., Tawil, R. (**2002**) *J. Clin. Invest.*, 110(3), 381–388.

35. Bendahhou, S., Donaldson, M.R., Plaster, N.M., Tristani-Firouzi, M., Fu. Y.H., Ptácek. L.J. (**2003**) *J. Biol. Chem.*, 278(51), 51 779–51 785.

36. Bendahhou, S., Fournier, E., Sternberg, D., Bassez, G., Furby, A., Sereni, C., Donaldson, M.R., Larroque, M.M., Fontaine, B., Barhanin, J. (**2005**) *J. Physiol.*, 565, 731–741.

37. Chun, T.U., Epstein, M.R., Dick, M., Andelfinger, G., Ballester, L., Vanoye, C.G., George, A.L., Benson, D.W. (**2004**) *Heart Rhythm*, 1(2), 235–241.

38. Takahashi, T., Tandai, S., Toki, T., Sato, T., Eto, S., Sato, A., Ueda, T., Sato, S., Ichinose, K., Ito, E., Yonesaka, S. (**2005**) *Pediatr. Int.*, 47(2), 220–223.

39. Zhang, L., Benson, D.W., Tristani-Firouzi, M., Ptacek, L.J., Tawil, R., Schwartz, P.J., George, A.L., Horie, M., Andelfinger, G., Snow, G.L., Fu, Y.H., Ackerman, M.J., Vincent, G.M. (**2005**) *Circulation*, 111(21), 2720–2726.

40. Priori, S.G., Pandit, S.V., Rivolta, I., Berenfeld, O., Ronchetti, E., Dhamoon, A., Napolitano, C., Anumonwo, J., di Barletta, M.R., Gudapakkam, S., Bosi, G., Stramba-Badiale, M., Jalife, J. (**2005**) *Circ. Res.*, 96(7), 800–807.

41. Xia, M., Jin, Q., Bendahhou, S., He, Y., Larroque, M.M., Chen, Y., Zhou, Q., Yang, Y., Liu, Y., Liu, B., Zhu, Q., Zhou, Y., Lin, J., Liang, B., Li, L., Dong, X., Pan, Z., Wang, R., Wan, H., Qiu, W., Xu, W., Eurlings, P., Barhanin, J., Chen, Y. (**2005**) *Biochem. Biophys. Res. Commun.*, 332, 1012–1019.

42. Schimpf, R., Wolpert, C., Gaita, F., Giustetto, C., Borggrefe, M. (**2005**) *Cardiovasc. Res.*, 67, 357–366.

43. Cartier, E.A., Conti, L.R., Vandenberg, C.A., Shyng, S.L. (**2001**) *Proc. Natl. Acad. Sci. U.S.A.*, 98(5), 2882–2887.

44. Edghill, E.L., Gloyn, A.L., Gillespie, K.M., Lambert, A.P., Raymond, N.T., Swift, P.G., Ellard, S., Gale, E.A., Hattersley, A.T. (**2004**) *Diabetes*, 53(11), 2998–3001.

45. Eftychi, C., Howson, J.M., Barratt, B.J., Vella, A., Payne, F., Smyth, D.J., Twells, R.C., Walker, N.M., Rance, H.E., Tuomilehto-Wolf, E., Tuomilehto, J., Undlien, D.E., Rønningen, K.S., Guja, C., Ionescu-Tirgoviste, C., Savage, D.A., Todd, J.A. (**2004**) *Diabetes*, 53(3), 870–873.

46. Gloyn, A.L., Reimann, F., Girard, C., Edghill, E.L., Proks, P., Pearson, E.R., Temple, I.K., Mackay, D.J., Shield, J.P., Freedenberg, D., Noyes, K., Ellard, S., Ashcroft, F.M., Gribble, F.M., Hattersley, A.T. (**2005**) *Hum. Mol. Genet.*, 14(7), 925–934.

47. Gloyn, A.L., Pearson, E.R., Antcliff, J.F., Proks, P., Bruining, G.J., Slingerland, A.S., Howard, N., Srinivasan, S., Silva, J.M., Molnes, J., Edghill, E.L., Frayling, T.M., Temple, I.K., Mackay, D., Shield, J.P., Sumnik, Z., van Rhijn, A., Wales, J.K., Clark, P., Gorman, S., Aisenberg, J., Ellard, S., Njolstad, P.R., Ashcroft, F.M., Hattersley, A.T. (**2004**) *N. Engl. J. Med.*, 350(18), 1838–1849.

48. Huopio, H., Shyng, S.L., Otonkoski, T., Nichols, C.G. (**2002**) *Am. J. Physiol. Endocrinol. Metab.*, 283(2), E207–E216.

49. John, S.A., Weiss, J.N., Xie, L.H., Ribalet, B. (**2003**) *J. Physiol.*, 552(1), 23–34.

50. Massa, O., Iafusco, D., D'Amato, E., Gloyn, A.L., Hattersley, A.T., Pasquino, B., Tonini, G., Dammacco, F., Zanette, G., Meschi, F., Porzio, O., Bottazzo, G., Crino, A., Lorini, R., Cerutti, F., Vanelli, M., Barbetti, F. (**2005**) *Hum. Mut.*, 25(1), 22–27.

51. McCarthy, M.I. (**2004**) *Hum. Mol. Genet.*, 13, R33–R41.

52. Ohkubo, K., Nagashima, M., Naito, Y., Taguchi, T., Suita, S., Okamoto, N., Fujinaga, H., Tsumura, K., Kikuchi, K., Ono, J. (**2005**) *Clin. Endocrinol.*, 62(4), 458–465.

53. Proks, P., Antcliff, J.F., Lippiat, J., Gloyn, A.L., Hattersley, A.T., Ashcroft, F.M. (**2004**) *Proc. Natl. Acad. Sci. U.S.A.*, 101(50), 17 539–17 544.

54. Reimann, F., Tucker, S.J., Proks, P., Ashcroft, F.M. (**1999**) *J. Physiol.*, 518(2), 325–336.

55. Sagen, J.V., Ræder, H., Hathout, E., Shehadeh, N., Gudmundsson, K., Bævre, H., Abuelo, D., Phornphutkul, C., Molnes, J., Bell, G.I., Gloyn, A.L., Hattersley, A.T.,

Molven, A., Søvik, O., Njølstad, P.R. (**2004**) *Diabetes*, 53(10), 2713–2718.

56. Tornovsky, S., Crane, A., Cosgrove, K.E., Hussain, K., Lavie, J., Heyman, M., Nesher, Y., Kuchinski, N., Ben-Shushan, E., Shatz, O., Nahari, E., Potikha, T., Zangen, D., Tenenbaum-Rakover, Y., de Vries, L., Argente, J., Gracia, R., Landau, H., Eliakim, A., Lindley, K., Dunne, M.J., Aguilar-Bryan, L., Glaser, B. (**2004**) *J. Clin. Endocrinol. Metab.*, 89(12), 6224–6234.

57. Yorifuji, T., Nagashima, K., Kurokawa, K., Kawai, M., Oishi, M., Akazawa, Y., Hosokawa, M., Yamada, Y., Inagaki, N., Nakahata, T. (**2005**) *J. Clin. Endocrinol. Metab.*, 90(6), 3174–3178.

58. Boland, L.M., Price, D.L., Jackson, K.A. (**1999**) *Neuroscience*, 91(4), 1557–1564.

59. Cusimano, A., D'Adamo, M.C., Pessia, M. (**2004**) *FEBS Lett.*, 576(1–2), 237–244.

60. D'Adamo, M.C., Imbrici, P., Sponcichetti, F., Pessia, M. (**1999**) *FASEB J.*, 13(11), 1335–1345.

61. Eunson, L.H., Rea, R., Zuberi, S.M., Youroukos, S., Panayiotopoulos, C.P., Liguori, R., Avoni, P., McWilliam, R.C., Stephenson, J.B., Hanna, M.G., Kullmann, D.M., Spauschus, A. (**2000**) *Ann. Neurol.*, 48(4), 647–656.

62. Imbrici, P., Cusimano, A., D'Adamo, M.C., De Curtis, A., Pessia, M. (**2003**) *Pflügers Arch.*, 446(3), 373–379.

63. Klein, A., Boltshauser, E., Jen, J., Baloh, R.W. (**2004**) *Neuropediatrics*, 35, 147–149.

64. Lee, H., Wang, H., Jen, J.C., Sabatti, C., Baloh, R.W., Nelson, S.F. (**2004**) *Hum. Mut.*, 24(6), 536–542.

65. Manganas, L.N., Akhtar, S., Antonucci, D.E., Campomanes, C.R., Dolly, J.O., Trimmer, J.S. (**2001**) *J. Biol. Chem.*, 276(52), 49 427–49 434.

66. Scheffer, H., Brunt, E.R., Mol, G.J., van der Vlies, P., Stulp, R.P., Verlind, E., Mantel, G., Averyanov, Y.N., Hofstra, R.M., Buys, C.H. (**1998**) *Hum. Genet.*, 102(4), 464–466.

67. Gouas, L., Bellocq, C., Berthet, M., Potet, F., Demolombe, S., Forhan, A., Lescasse, R., Simon, F., Balkau, B., Denjoy, I., Hainque, B., Baro, I., Guicheney, P. (**2004**) *Cardiovasc. Res.*, 63(1), 60–68.

68. Liu, W., Yang, J., Hu, D., Kang, C., Li, C., Zhang, S., Li, P., Chen, Z., Qin, X., Ying, K., Li, Y., Li, Y., Li, Z., Cheng, X., Li, L., Qi, Y., Chen, S., Wang, Q. (**2002**) *Hum. Mut.*, 20(6), 475–476.

69. Park, K.H., Piron, J., Dahimene, S., Merot, J., Baro, I., Escande, D., Loussouarn, G. (**2005**) *Circ. Res.*, 96(7), 730–739.

70. Robbins, J. (**2001**) *Pharmacol. Ther.*, 90(1), 1–19.

71. Sharma, D., Glatter, K.A., Timofeyev, V., Tuteja, D., Zhang, Z., Rodriguez, J., Tester, D.J., Low, R., Scheinman, M.M., Ackerman, M.J., Chiamvimonvat, N. (**2004**) *J. Mol. Cell. Cardiol.*, 37(1), 79–89.

72. Yang, W.P., Levesque, P.C., Little, W.A., Conder, M.L., Shalaby, F.Y., Blanar, M.A. (**1997**) *Proc. Natl. Acad. Sci. U.S.A.*, 94(8), 4017–4021.

73. Zehelein, J., Thomas, D., Khalil, M., Wimmer, A.B., Koenen, M., Licka, M., Wu, K., Kiehn, J., Brockmeier, K., Kreye, V.A., Karle, C.A., Katus, H.A., Ulmer, H.E., Schoels, W. (**2004**) *Biochim. Biophys. Acta*, 1690(3), 185–192.

74. Chen, Y.H., Xu, S.J., Bendahhou, S., Wang, X.L., Wang, Y., Xu, W.Y., Jin, H.W., Sun, H., Su, X.Y., Zhuang, Q.N., Yang, Y.Q., Li, Y.B., Liu, Y., Xu, H.J., Li, X.F., Ma, N., Mou, C.P., Chen. Z., Barhanin. J., Huang, W. (**2003**) *Science*, 299(5604), 251–254.

75. Bellocq, C., van Ginneken, A.C., Bezzina, C.R., Alders, M., Escande, D., Mannens, M.M., Baro, I., Wilde, A.A. (**2004**) *Circulation*, 109(20), 2394–2397.

76. Biervert, C., Steinlein, O.K. (**1999**) *Hum. Genet.*, 104(3), 234–240.

77. Biervert, C., Schroeder, B.C., Kubisch, C., Berkovic, S.F., Propping, P., Jentsch, T.J., Steinlein, O.K. (**1998**) *Science*, 279(5349), 403–406.

78. Borgatti, R., Zucca, C., Cavallini, A., Ferrario, M., Panzeri, C., Castaldo, P., Soldovieri, M.V., Baschirotto, C., Bresolin, N., Dalla Bernardina, B., Taglialatela, M., Bassi, M.T. (**2004**) *Neurology*, 63(1), 57–65.

79. Castaldo, P., del Giudice, E.M., Coppola, G., Pascotto, A., Annunziato, L., Taglialatela, M. (**2002**) *J. Neurosci.*, 22(2), RC199.

80. Dedek, K., Fusco, L., Teloy, N., Steinlein, O.K. (**2003**) *Epilepsy Res.*, 54(1), 21–27.

81. Dedek, K., Kunath, B., Kananura, C., Reuner, U., Jentsch, T.J., Steinlein, O.K. (**2001**) *Proc. Natl. Acad. Sci. U.S.A.*, 98(21), 12 272–12 277.

82. del Giudice, E.M., Coppola, G., Scuccimarra, G., Cirillo, G., Bellini, G., Pascotto, A. (**2000**) *Eur. J. Hum. Genet.*, 8(12), 994–997.

83. Lee, W.L., Biervert, C., Hallmann, K., Tay, A., Dean, J.C., Steinlein, O.K. (**2000**) *Neuropediatrics*, 31(1), 9–12.

84. Pereira, S., Roll, P., Krizova, J., Genton, P., Brazdil, M., Kuba, R., Cau, P., Rektor, I., Szepetowski, P. (**2004**) *Epilepsia*, 45(4), 384–390.

85. Richards, M.C., Heron, S.E., Spendlove, H.E., Scheffer, I.E., Grinton, B., Berkovic, S.F., Mulley, J.C., Davy, A. (**2004**) *J. Med. Genet.*, 41(3), e35.

86. Rogawski, M.A. (**2000**) *Trends Neurosci.*, 23(9), 393–398.

87. Schroeder, B.C., Kubisch, C., Stein, V., Jentsch, T.J. (**1998**) *Nature*, 396(6712), 687–690.

88. Singh, N.A., Westenskow, P., Charlier, C., Pappas, C., Leslie, J., Dillon, J., Anderson, V.E., Sanguinetti, M.C., Leppert, M.F. (**2003**) *Brain*, 126(12), 2726–2737.

89. Tang, B., Li, H., Xia, K., Jiang, H., Pan, Q., Shen, L., Long, Z., Zhao, G., Cai, F. (**2004**) *J. Neurol. Sci.*, 221(1), 31–34.

90. Kubisch, C., Schroeder, B.C., Friedrich, T., Lutjohann, B., El-Amraoui, A., Marlin, S., Petit, C., Jentsch, T.J. (**1999**) *Cell*, 96(3), 437–446.

91. Van Camp, G., Coucke, P.J., Akita, J., Fransen, E., Abe, S., De Leenheer, E.M., Huygen, P.L., Cremers. C.W., Usami, S. (**2002**) *Hum. Mut.*, 20(1), 15–19.

92. Chen, J., Zou, A., Splawski, I., Keating, M.T., Sanguinetti, M.C. (**1999**) *J. Biol. Chem.*, 274(15), 10 113–10 118.

93. Christiansen, M., Tønder, N., Larsen, L.A., Andersen, P.S., Simonsen, H., Øyen, N., Kanters, J.K., Jacobsen, J.R., Fosdal, I., Wettrell, G., Kjeldsen K. (**2005**) *Am. J. Cardiol.*, 95(3), 433–434.

94. Clancy, C.E., Rudy, Y. (**2001**) *Cardiovasc. Res.*, 50(2), 301–313.

95. Curran, M.E., Splawski, I., Timothy, K.W., Vincent, G.M., Green, E.D., Keating, M.T. (**1995**) *Cell*, 80(5), 795–803.

96. Delisle, B.P., Anson, B.D., Rajamani, S., January, C.T. (**2004**) *Circ. Res.*, 94(11), 1418–1428.

97. Delisle, B.P., Anderson, C.L., Balijepalli, R.C., Anson, B.D., Kamp, T.J., January, C.T. (**2003**) *J. Biol. Chem.*, 278(37), 35 749–35 754.

98. Furutani, M., Trudeau, M.C., Hagiwara, N., Seki, A., Gong, Q., Zhou, Z., Imamura, S., Nagashima, H., Kasanuki, H., Takao, A., Momma, K., January, C.T., Robertson, G.A., Matsuoka, R. (**1999**) *Circulation*, 99(17), 2290–2294.

99. Hayashi, K., Shimizu, M., Ino, H., Yamaguchi, M., Terai, H., Hoshi, N., Higashida, H., Terashima, N., Uno, Y., Kanaya, H., Mabuchi, H. (**2004**) *Clin. Sci.*, 107(2), 175–182.

100. Hayashi, K., Shimizu, M., Ino, H., Yamaguchi, M., Mabuchi, H., Hoshi, N., Higashida, H. (**2002**) *Cardiovasc. Res.*, 54(1), 67–76.

101. Huang, F.D., Chen, J., Lin, M., Keating, M.T., Sanguinetti, M.C. (**2001**) *Circulation*, 104(9), 1071–1075.

102. Jahr, S., Lewalter, T., Hesch, R. D., Luderitz, B., Englisch, S. (**2000**) *Hum. Mut.*, 15(6), 584.

103. Johnson, W.H., Yang, P., Yang, T., Lau, Y.R., Mostella, B.A., Wolff, D.J., Roden, D.M., Benson, D.W. (**2003**) *Pediatr. Res.*, 53(5), 744–748.

104. Jongbloed, R.J., Wilde, A.A., Geelen, J.L., Doevendans, P., Schaap, C., Van Langen, I., van Tintelen, J.P., Cobben, J.M., Beaufort-Krol, G.C., Geraedts, J.P., Smeets, H.J. (**1999**) *Hum. Mut.*, 13(4), 301–310.

105. Kagan, A., Yu, Z., Fishman, G.I., McDonald, T.V. (**2000**) *J. Biol. Chem.*, 275(15), 11 241–11 248.

106. Laitinen, P., Fodstad, H., Piippo, K., Swan, H., Toivonen, L., Viitasalo, M., Kaprio, J., Kontula, K. (**2000**) *Hum. Mut.*, 15(6), 580–581.

107. Lees-Miller, J.P., Duan, Y., Teng, G.Q., Thorstad, K., Duff, H.J. (**2000**) *Circ. Res.*, 86(5), 507–153.

108. Moss, A.J., Zareba, W., Kaufman, E.S., Gartman, E., Peterson, D.R., Benhorin, J., Towbin, J.A., Keating, M.T., Priori, S.G., Schwartz, P.J., Vincent, G.M., Robinson, J.L., Andrews, M.L., Feng, C.,

Hall, W.J., Medina, A., Zhang, L., Wang, Z. (**2002**) *Circulation*, 105(7), 794–799.

109. Nakajima, T., Kurabayashi, M., Ohyama, Y., Kaneko, Y., Furukawa, T., Itoh, T., Taniguchi, Y., Tanaka, T., Nakamura, Y., Hiraoka, M., Nagai, R. (**2000**) *FEBS Lett.*, 481(2), 197–203.

110. Nakajima, T., Furukawa, T., Hirano, Y., Tanaka, T., Sakurada, H., Takahashi, T., Nagai, R., Itoh, T., Katayama, Y., Nakamura, Y., Hiraoka, M. (**1999**) *Cardiovasc. Res.*, 44(2), 283–293.

111. Paulussen, A., Raes, A., Matthijs, G., Snyders, D.J., Cohen, N., Aerssens, J. (**2002**) *J. Biol. Chem.*, 277(50), 48 610–48 616.

112. Paulussen, A., Yang, P., Pangalos, M., Verhasselt, P., Marrannes, R., Verfaille, C., Vandenberk, I., Crabbe, R., Konings, F., Luyten, W., Armstrong, M. (**2000**) *Hum. Mut.*, 15(5), 483.

113. Piippo, K., Laitinen, P., Swan, H., Toivonen, L., Viitasalo, M., Pasternack, M., Paavonen, K., Chapman, H., Wann, K.T., Hirvela, E., Sajantila, A., Kontula, K. (**2000**) *J. Am. Coll. Cardiol.*, 35(7), 1919–1925.

114. Rajamani, S., Anderson, C.L., Anson, B.D., January, C.T. (**2002**) *Circulation*, 105(24), 2830–2835.

115. Rossenbacker, T., Mubagwa, K., Jongbloed, R.J., Vereecke, J., Devriendt, K., Gewillig, M., Carmeliet, E., Collen, D., Heidbuchel, H., Carmeliet, P. (**2005**) *Circulation*, 111(8), 961–968.

116. Roti Roti, E.C., Myers, C.D., Ayers, R.A., Boatman, D.E., Delfosse, S.A., Chan, E.K., Ackerman, M.J., January, C.T., Robertson, G.A. (**2002**) *J. Biol. Chem.*, 277(49), 47 779–47 785.

117. Sun, Z., Milos, P.M., Thompson, J.F., Lloyd, D.B., Mank-Seymour, A., Richmond, J., Cordes, J.S., Zhou, J. (**2004**) *J. Mol. Cell. Cardiol.*, 37(5), 1031–1039.

118. Teng, S., Ma, L., Dong, Y., Lin, C., Ye, J., Bahring, R., Vardanyan, V., Yang, Y., Lin, Z., Pongs, O., Hui, R. (**2004**) *J. Mol. Med.*, 82(3), 189–196.

119. Thomas, D., Kiehn, J., Katus, H.A., Karle, C.A. (**2003**) *Cardiovasc. Res.*, 60(2), 235–241.

120. Zhang, L., Vincent, G.M., Baralle, M., Baralle, F.E., Anson, B.D., Benson, D.W., Whiting, B., Timothy, K.W., Carlquist, J., January, C.T., Keating, M.T., Splawski, I.

(**2004**) *J. Am. Coll. Cardiol.*, 44(6), 1283–1291.

121. Zhou, Z., Gong, Q., January, C.T. (**1999**) *J. Biol. Chem.*, 274(44), 31 123–31 126.

122. Hong, K., Bjerregaard, P., Gussak, I., Brugada, R. (**2005**) *J. Cardiovasc. Electrophysiol.*, 16(4), 394–396.

123. Brugada, R., Hong, K., Dumaine, R., Cordeiro, J., Gaita, F., Borggrefe, M., Menendez, T.M., Brugada, J., Pollevick, G.D., Wolpert, C., Burashnikov, E., Matsuo, K., Wu, Y.S., Guerchicoff, A., Bianchi, F., Giustetto, C., Schimpf, R., Brugada, P., Antzelevitch, C. (**2004**) *Circulation*, 109(1), 30–35.

124. Yang, Y., Xia, M., Jin, Q., Bendahhou, S., Shi, J., Chen, Y., Liang, B., Lin, J., Liu, Y., Liu, B., Zhou, Q., Zhang, D., Wang, R., Ma, N., Su, X., Niu, K., Pei, Y., Xu, W., Chen, Z., Wan, H., Cui, J., Barhanin, J., Chen, Y. (**2004**) *Am. J. Hum. Genet.*, 75(5), 899–905.

125. Fernandez, D., Ghanta, A., Kauffman, G.W., Sanguinetti, M.C. (**2004**) *J. Biol. Chem.*, 279(11), 10 120–10 127.

126. Ficker, E., Obejero-Paz, C.A., Zhao, S., Brown, A.M. (**2002**) *J. Biol. Chem.*, 277(7), 4989–4998.

127. Gessner, G., Zacharias, M., Bechstedt, S., Schonherr, R., Heinemann, S.H. (**2004**) *Mol. Pharmacol.*, 65(5), 1120–1129.

128. Ishii, K., Kondo, K., Takahashi, M., Kimura, M., Endoh, M. (**2001**) *FEBS Lett.*, 506(3), 191–195.

129. Perry, M., de Groot, M.J., Helliwell, R., Leishman, D., Tristani-Firouzi, M., Sanguinetti, M.C., Mitcheson, J. (**2004**) *Mol. Pharmacol.*, 66(2), 240–249.

130. Ridley, J.M., Milnes, J.T., Witchel, H.J., Hancox, J.C. (**2004**) *Biochem. Biophys. Res. Commun.*, 325(3), 883–891.

131. Scholz, E.P., Zitron, E., Kiesecker, C., Lueck, S., Kathofer, S., Thomas, D., Weretka, S., Peth, S., Kreye, V.A., Schoels, W., Katus, H.A., Kiehn, J., Karle C.A. (**2003**) *Naunyn Schmied. Arch. Pharmacol.*, 368(5), 404–414.

132. Bareil, C., Hamel, C.P., Delague, V., Arnaud, B., Demaille, J., Claustres, M. (**2001**) *Hum. Genet.*, 108(4), 328–334.

133. Dryja, T.P., Finn, J.T., Peng, Y.W., McGee, T.L., Berson, E.L., Yau, K.W. (**1995**) *Proc. Natl. Acad. Sci. U.S.A.*, 92(22), 10 177–10 181.

134. Farrar, G.J., Kenna, P.F., Humphries, P. (**2002**) *EMBO J.*, 21(5), 857–864.

135. Kondo, H., Qin, M., Mizota, A., Kondo, M., Hayashi, H., Hayashi, K., Oshima, K., Tahira, T., Hayashi, K. (**2004**) *Invest. Ophthalmol. Vis. Sci.*, 45(12), 4433–4439.

136. Paloma, E., Martínez-Mir, A., García-Sandoval, B., Ayuso, C., Vilageliu, L., Gonzàlez-Duarte, R., Balcells, S. (**2002**) *J. Med. Genet.*, 39(10), E66.

137. Trudeau, M.C., Zagotta, W.N. (**2002**) *Neuron*, 34(2), 197–207.

138. Wang, D.Y., Chan, W.M., Tam, P.O., Baum, L., Lam, D.S., Chong, K.K., Fan, B.J., Pang, C.P. (**2005**) *Clin. Chim. Acta*, 351(1–2), 5–16.

139. Zhang, Q., Zulfiqar, F., Riazuddin, S.A., Xiao, X., Ahmad, Z., Riazuddin, S., Hejtmancik, J.F. (**2004**) *Mol. Vis.*, 10, 884–889.

140. Kohl, S., Varsanyi, B., Antunes, G.A., Baumann, B., Hoyng, C.B., Jägle, H., Rosenberg, T., Kellner, U., Lorenz, B., Salati, R., Jurklies, B., Farkas, A., Andreasson, S., Weleber, R.G., Jacobson, S.G., Rudolph, G., Castellan, C., Dollfus, H., Legius, E., Anastasi, M., Bitoun, P., Lev, D., Sieving, P.A., Munier, F.L., Zrenner, E., Sharpe, L.T., Cremers, F.P., Wissinger, B. (**2005**) *Eur. J. Hum. Genet.*, 13(3), 302–308.

141. Kohl, S., Baumann, B., Broghammer, M., Jägle, H., Sieving, P., Kellner, U., Spegal, R., Anastasi, M., Zrenner, E., Sharpe, L.T., Wissinger, B. (**2000**) *Hum. Mol. Genet.*, 9(14), 2107–2116.

142. Kohl, S., Marx, T., Giddings, I., Jägle, H., Jacobson, S.G., Apfelstedt-Sylla, E., Zrenner, E., Sharpe, L.T., Wissinger, B. (**1998**) *Nat. Genet.*, 19(3), 257–259.

143. Liu, C., Varnum, M.D. (**2005**) *Am. J. Physiol. Cell Physiol.*, 289, C 187–198.

144. Michaelides, M., Aligianis, I.A., Ainsworth, J.R., Good, P., Mollon, J.D., Maher, E.R., Moore, A.T., Hunt, D.M. (**2004**) *Invest. Ophthalmol. Vis. Sci.*, 45(6), 1975–1982.

145. Nishiguchi, K.M., Sandberg, M.A., Gorji, N., Berson, E.L., Dryja, T.P. (**2005**) *Hum. Mut.*, 25(3), 248–258.

146. Rojas, C.V., María, L.S., Santos, J.L., Cortés, F., Alliende, M.A. (**2002**) *Eur. J. Hum. Genet.*, 10(10), 638–642.

147. Sundin, O.H., Yang, J.M., Li, Y., Zhu, D., Hurd, J.N., Mitchell, T.N., Silva, E.D., Maumenee, I.H. (**2000**) *Nat. Genet.*, 25(3), 289–293.

148. Wissinger, B., Gamer, D., Jägle, H., Giorda, R., Marx, T., Mayer, S., Tippmann, S., Broghammer, M., Jurklies, B., Rosenberg, T., Jacobson, S.G., Sener, E.C., Tatlipinar, S., Hoyng, C.B., Castellan, C., Bitoun, P., Andreasson, S., Rudolph, G., Kellner, U., Lorenz, B., Wolff, G., Verellen-Dumoulin, C., Schwartz, M., Cremers, F.P., Apfelstedt-Sylla, E., Zrenner, E., Salati, R., Sharpe, L.T., Kohl, S. (**2001**) *Am. J. Hum. Genet.*, 69(4), 722–737.

149. Schulze-Bahr, E., Neu, A., Friederich, P., Kaupp, U.B., Breithardt, G., Pongs, O., Isbrandt, D. (**2003**) *J. Clin. Invest.*, 111(10), 1537–1545.

150. Ueda, K., Nakamura, K., Hayashi, T., Inagaki, N., Takahashi, M., Arimura, T., Morita, H., Higashiuesato, Y., Hirano, Y., Yasunami, M., Takishita, S., Yamashina, A., Ohe, T., Sunamori, M., Hiraoka, M., Kimura, A. (**2004**) *J. Biol. Chem.*, 279(26), 27 194–27 198.

151. Hoenderop, J.G., Bindels, R.J. (**2005**) *J. Am. Soc. Nephrol.*, 16(1), 15–26.

152. Konrad, M., Schlingmann, K.P., Gudermann, T. (**2004**) *Am. J. Physiol. Renal Physiol.*, 286(4), F599–F605.

153. Montell, C. (**2003**) *Curr. Biol.*, 13(20), R799–R801.

154. Schlingmann, K.P., Gudermann, T. (**2005**) *J. Physiol.*, 566, 301–308.

155. Schlingmann, K.P., Weber, S., Peters, M., Niemann Nejsum, L., Vitzthum, H., Klingel, K., Kratz, M., Haddad, E., Ristoff, E., Dinour, D., Syrrou, M., Nielsen, S., Sassen, M., Waldegger, S., Seyberth, H.W., Konrad, M. (**2002**) *Nat. Genet.*, 31(2), 166–170.

156. Walder, R.Y., Landau, D., Meyer, P., Shalev, H., Tsolia, M., Borochowitz, Z., Boettger, M.B., Beck, G.E., Englehardt, R.K., Carmi, R., Sheffield, V.C. (**2002**) *Nat. Genet.*, 31(2), 171–174.

157. Lehmann-Horn, F., Jurkat-Rott, K. (**2001**) *Pharmaceut. News*, 8(2), 29–36.

158. Bendahhou, S., Cummins, T.R., Griggs, R.C., Fu, Y.H., Ptáček, L.J. (**2001**) *Ann. Neurol.*, 50(3), 417–420.

159. Bendahhou, S., Cummins, T.R., Tawil, R., Waxman, S.G., Ptácek, L.J. (**1999**) *J. Neurosci.*, 19(12), 4762–4771.

160. Bouhours, M., Sternberg, D., Davoine, C.S., Ferrer, X., Willer, J.C., Fontaine, B., Tabti, N. (**2004**) *J. Physiol.*, 554(3), 635–647.

161. Cannon, S.C. (**1997**) *Neuromuscul. Disord.*, 7(4), 241–249.

162. Davies, N.P., Eunson, L.H., Gregory, R.P., Mills, K.R., Morrison, P.J., Hanna, M.G. (**2000**) *J. Neurol., Neurosurg. Psychiatry*, 68(4), 504–507.

163. Hanna, M.G., Stewart, J., Schapira, A.H., Wood, N.W., Morgan-Hughes, J.A., Murray, N.M. (**1998**) *J. Neurol., Neurosurg. Psychiatry*, 65(2), 248–250.

164. Heine, R., Pika, U., Lehmann-Horn, F. (**1993**) *Hum. Mol. Genet.*, 2(9), 1349–1353.

165. Jurkat-Rott, K., Mitrovic, N., Hang, C., Kouzmekine, A., Iaizzo, P., Herzog, J., Lerche, H., Nicole, S., Vale-Santos, J., Chauveau, D., Fontaine, B., Lehmann-Horn, F. (**2000**) *Proc. Natl. Acad. Sci. U.S.A.*, 97(17), 9549–9554.

166. Kelly, P., Yang, W.S., Costigan, D., Farrell, M.A., Murphy, S., Hardiman, O. (**1997**) *Neuromuscul. Disord.*, 7(2), 105–111.

167. Marchant, C.L., Ellis, F.R., Halsall, P.J., Hopkins, P.M., Robinson, R.L. (**2004**) *Muscle Nerve*, 30, 114–117.

168. Okuda, S., Kanda, F., Nishimoto, K., Sasaki, R., Chihara, K. (**2001**) *J. Neurol.*, 248(11), 1003–1004.

169. Orrell, R.W., Jurkat-Rott, K., Lehmann-Horn, F., Lane, R.J. (**1998**) *J. Neurol., Neurosurg. Psychiatry*, 65(4), 569–572.

170. Rojas, C.V., Neely, A., Velasco-Loyden, G., Palma, V., Kukuljan, M. (**1999**) *Am. J. Physiol. Cell Physiol.*, 276(1), C259–C266.

171. Sasaki, R., Takano, H., Kamakura, K., Kaida, K., Hirata, A., Saito, M., Tanaka, H., Kuzuhara, S., Tsuji S. (**1999**) *Arch. Neurol.*, 56(6), 692–696.

172. Sternberg, D., Maisonobe, T., Jurkat-Rott, K., Nicole, S., Launay, E., Chauveau, D., Tabti, N., Lehmann-Horn, F., Hainque, B., Fontaine, B. (**2001**) *Brain*, 124(6), 1091–1099.

173. Struyk, A.F., Scoggan, K.A., Bulman, D.E., Cannon, S.C. (**2000**) *J. Neurosci.*, 20(23), 8610–8617.

174. Sugiura, Y., Makita, N., Li, L., Noble, P.J., Kimura, J., Kumagai, Y., Soeda, T., Yamamoto, T. (**2003**) *Neurology*, 61(7), 914–918.

175. Tahmoush, A.J., Schaller, K.L., Zhang, P., Hyslop, T., Heiman-Patterson, T., Caldwell, J.H. (**1994**) *Neuromuscul. Disord.*, 4(5/6), 447–454.

176. Tsujino, A., Maertens, C., Ohno, K., Shen, X.M., Fukuda, T., Harper, C.M., Cannon, S.C., Engel, A.G. (**2003**) *Proc. Natl. Acad. Sci. U.S.A.*, 100(12), 7377–7382.

177. Vicart, S., Sternberg, D., Fournier, E., Ochsner, F., Laforet, P., Kuntzer, T., Eymard, B., Hainque, B., Fontaine, B. (**2004**) *Neurology*, 63(11), 2120–2127.

178. Viswanathan, P.C., Balser, J.R. (**2004**) *Trends Cardiovasc. Med.*, 14(1), 28–35.

179. Wu, F.F., Gordon, E., Hoffman, E.P., Cannon, S.C. (**2005**) *J. Physiol.*, 565, 371–380.

180. Ackerman, M.J., Siu, B.L., Sturner, W.Q., Tester, D.J., Valdivia, C.R., Makielski, J.C., Towbin, J.A. (**2001**) *J. Am. Med. Assoc.*, 286(18), 2264–2269.

181. Benson, D.W., Wang, D.W., Dyment, M., Knilans, T.K., Fish, F.A., Strieper, M.J., Rhodes, T.H., George, A.L. (**2003**) *J. Clin. Invest.*, 112(7), 1019–1028.

182. Bezzina, C., Veldkamp, M.W., van den Berg, M.P., Postma, A.V., Rook, M.B., Viersma, J.W., van Langen, I.M., Tan-Sindhunata, G., Bink-Boelkens, M., van der Hout, A.H., Mannens, M.M.A.M., Widle, A.A.M. (**1999**) *Circ. Res.*, 85, 1206–1213.

183. Chen, Q., Kirsch, G.E., Zhang, D., Brugada, R., Brugada, J., Brugada, P., Potenza, D., Moya, A., Borggrefe, M., Breithardt, G., Ortiz-Lopez, R., Wang, Z., Antzelevitch, C., O'Brien, R.E., Schulze-Bahr, E., Keating, M.T., Towbin, J.A., Wang, Q. (**1998**) *Nature*, 392(6673), 293–296.

184. Chen, T., Inoue, M., Sheets, M.F. (**2005**) *Am. J. Physiol. Heart Circ. Physiol.*, 288(6), H2666–H2676.

185. Itoh, H., Shimizu, M., Mabuchi, H., Imoto, K. (**2005**) *J. Cardiovasc. Electrophysiol.*, 16, 378–383.

186. Kyndt, F., Probst, V., Potet, F., Demolombe, S., Chevallier, J.C., Baro, I., Moisan, J.P., Boisseau, P., Schott, J.J., Es-

cande, D., Le Marec, H. (**2001**) *Circulation*, 104(25), 3081–3086.

187. Mohler, P.J., Rivolta, I., Napolitano, C., LeMaillet, G., Lambert, S., Priori, S.G., Bennett, V. (**2004**) *Proc. Natl. Acad. Sci. U.S.A.*, 101(50), 17 533–17 538.

188. Moric-Janiszewska, E., Herbert, E., Cholewa, K., Filipecki, A., Trusz-Gluza, M., Wilczok, T. (**2004**) *J. Appl. Genet.*, 45(3), 383–390.

189. Rossenbacker, T., Carroll, S.J., Liu, H., Kuipéri, C., de Ravel, T.J.L., Devriendt, K., Carmeliet, P., Kass, R.S., Heidbüchel, H. (**2004**) *Heart Rhythm*, 1, 610–615.

190. Schulze-Bahr, E., Eckardt, L., Breithardt, G., Seidl, K., Wichter, T., Wolpert, C., Borggrefe, M., Haverkamp, W. (**2003**) *Hum. Mut.*, 21(6), 651–652.

191. Tester, D.J., Will, M.L., Haglund, C.M., Ackerman, M.J. (**2005**) *Heart Rhythm*, 2, 507–517.

192. Todd, S.J., Campbell, M.J., Roden, D.M., Kannankeril, P.J. (**2005**) *Heart Rhythm*, 2(5), 540–543.

193. Wang, Q., Chen, S., Chen, Q., Wan, X., Shen, J., Hoeltge, G.A., Timur, A.A., Keating, M.T., Kirsch, G.E. (**2004**) *J. Med. Genet.*, 41(5), e66.

194. Weiss, R., Barmada, M.M., Nguyen, T., Seibel, J.S., Cavlovich, D., Kornblit, C.A., Angelilli, A., Villanueva, F., McNamara, D.M., London, B. (**2002**) *Circulation*, 105(6), 707–713.

195. Yang, P., Kanki, H., Drolet, B., Yang, T., Wei, J., Viswanathan, P.C., Hohnloser, S.H., Shimizu, W., Schwartz, P.J., Stanton, M., Murray, K.T., Norris, K., George, A.L., Roden, D.M. (**2002**) *Circulation*, 105(16), 1943–1948.

196. Yokoi, H., Makita, N., Sasaki, K., Takagi, Y., Okumura, Y., Nishino, T., Makiyama, T., Kitabatake, A., Horie, M., Watanabe, I., Tsutsui, H. (**2005**) *Heart Rhythm*, 2(3), 285–292.

197. Abou-Khalil, B., Ge, Q., Desai, R., Ryther, R., Bazyk, A., Bailey, R., Haines, J.L., Sutcliffe, J.S., George, A.L. (**2001**) *Neurology*, 57(12), 2265–2272.

198. Escayg, A., MacDonald, B.T., Meisler, M.H., Baulac, S., Huberfeld, G., An-Gourfinkel, I., Brice, A., LeGuern, E., Moulard, B., Chaigne, D., Buresi, C., Malafosse A. (**2000**) *Nat. Genet.*, 24(4), 343–345.

199. Fujiwara, T., Sugawara, T., Mazaki-Miyazaki, E., Takahashi, Y., Fukushima, K., Watanabe, M., Hara, K., Morikawa, T., Yagi, K., Yamakawa, K., Inoue, Y. (**2003**) *Brain*, 126(3), 531–546.

200. Gennaro, E., Veggiotti, P., Malacarne, M., Madia, F., Cecconi, M., Cardinali, S., Cassetti, A., Cecconi, I., Bertini, E., Bianchi, A., Gobbi, G., Zara, F. (**2003**) *Epileptic Disord.*, 5(1), 21–25.

201. Kanai, K., Hirose, S., Oguni, H., Fukuma, G., Shirasaka, Y., Miyajima, T., Wada, K., Iwasa, H., Yasumoto, S., Matsuo, M., Ito, M., Mitsudome, A., Kaneko, S. (**2004**) *Neurology*, 63(2), 329–334.

202. Lossin, C., Rhodes, T.H., Desai, R.R., Vanoye, C.G., Wang, D., Carniciu, S., Devinsky, O., George, A.L. (**2003**) *J. Neurosci.*, 23(36), 11 289–11 295.

203. Mulley, J.C., Scheffer, I.E., Petrou, S., Dibbens, L.M., Berkovic, S.F., Harkin, L.A. (**2005**) *Hum. Mut.*, 25(6), 535–542.

204. Nabbout, R., Gennaro, E., Dalla Bernardina, B., Dulac, O., Madia, F., Bertini, E., Capovilla, G., Chiron, C., Cristofori, G., Elia, M., Fontana, E., Gaggero, R., Granata, T., Guerrini, R., Loi, M., La Selva, L., Lispi, M.L., Matricardi, A., Romeo, A., Tzolas, V., Valseriati, D., Veggiotti, P., Vigevano, F., Vallee, L., Dagna Bricarelli, F., Bianchi, A., Zara, F. (**2003**) *Neurology*, 60(12), 1961–1967.

205. Nagao, Y., Mazaki-Miyazaki, E., Okamura, N., Takagi, M., Igarashi, T., Yamakawa, K. (**2005**) *Epilepsy Res.*, 63(2/3), 151–156.

206. Ohmori, I., Ouchida, M., Ohtsuka, Y., Oka, E., Shimizu, K. (**2002**) *Biochem. Biophys. Res. Commun.*, 295(1), 17–23.

207. Pineda-Trujillo, N., Carrizosa, J., Cornejo, W., Arias, W., Franco, C., Cabrera, D., Bedoya, G., Ruiz-Linares, A. (**2005**) *Seizure*, 14(2), 123–128.

208. Rhodes, T.H., Lossin, C., Vanoye, C.G., Wang, D.W., George, A.L. (**2004**) *Proc. Natl. Acad. Sci. U.S.A.*, 101(30), 11 147–11 152.

209. Spampanato, J., Kearney, J.A., de Haan, G., McEwen, D.P., Escayg, A., Aradi, I., MacDonald, B.T., Levin, S.I., Soltesz, I., Benna, P., Montalenti, E., Isom, L.L., Goldin, A.L., Meisler, M.H. (**2004**) *J. Neurosci.*, 24(44), 10 022–10 034.

210. Spampanato, J., Escayg, A., Meisler, M.H., Goldin, A.L. (2003) *Neuroscience*, 116(1), 37–48.

211. Wallace, R.H., Hodgson, B.L., Grinton, B.E., Gardiner, R.M., Robinson, R., Rodriguez-Casero, V., Sadleir, L., Morgan, J., Harkin, L.A., Dibbens, L.M., Yamamoto, T., Andermann, E., Mulley, J.C., Berkovic, S.F., Scheffer, I.E. (2003) *Neurology*, 61(6), 765–769.

212. Weiss, L.A., Escayg, A., Kearney, J.A., Trudeau, M., MacDonald, B.T., Mori, M., Reichert, J., Buxbaum, J.D., Meisler, M.H. (2003) *Mol. Psychiatry*, 8(2), 186–194.

213. Berkovic, S.F., Heron, S.E., Giordano, L., Marini, C., Guerrini, R., Kaplan, R.E., Gambardella, A., Steinlein, O.K., Grinton, B.E., Dean, J.T., Bordo, L., Hodgson, B.L., Yamamoto, T., Mulley, J. C., Zara, F., Scheffer, I.E. (2004) *Ann. Neurol.*, 55(4), 550–557.

214. Haug, K., Hallmann, K., Rebstock, J., Dullinger, J., Muth, S., Haverkamp, F., Pfeiffer, H., Rau, B., Elger, C.E., Propping, P., Heils, A. (2001) *Epilepsy Res.*, 47(3), 243–246.

215. Heron, S.E., Crossland, K.M., Andermann, E., Phillips, H.A., Hall, A.J., Bleasel, A., Shevell, M., Mercho, S., Seni, M.H., Guiot, M.C., Mulley, J.C., Berkovic, S.F., Scheffer, I.E. (2002) *Lancet*, 360(9336), 851–852.

216. Ito, M., Shirasaka, Y., Hirose, S., Sugawara, T., Yamakawa, K. (2004) *Pediatr. Neurol.*, 31(2), 150–152.

217. Kamiya, K., Kaneda, M., Sugawara, T., Mazaki, E., Okamura, N., Montal, M., Makita, N., Tanaka, M., Fukushima, K., Fujiwara, T., Inoue, Y., Yamakawa, K. (2004) *J. Neurosci.*, 24(11), 2690–2698.

218. Kearney, J.A., Plummer, N.W., Smith, M.R., Kapur, J., Cummins, T.R., Waxman, S.G., Goldin, A.L., Meisler, M.H. (2001) *Neuroscience*, 102(2), 307–317.

219. Malacarne, M., Gennaro, E., Madia, F., Pozzi, S., Vacca, D., Barone, B., dalla Bernardina, B., Bianchi, A., Bonanni, P., De Marco, P., Gambardella, A., Giordano, L., Lispi, M.L., Romeo, A., Santorum, E., Vanadia, F., Vecchi, M., Veggiotti, P., Vigevano, F., Viri, F., Bricarelli, F.D., Zara, F. (2001) *Am. J. Hum. Genet.*, 68(6), 1521–1526.

220. Sugawara, T., Tsurubuchi, Y., Agarwala, K.L., Ito, M., Fukuma, G., Mazaki-Miyazaki, E., Nagafuji, H., Noda, M., Imoto, K., Wada, K., Mitsudome, A., Kaneko, S., Montal, M., Nagata, K., Hirose, S., Yamakawa, K. (2001) *Proc. Natl. Acad. Sci. U.S.A.*, 98(11), 6384–6389.

221. Wallace, R.H., Wang, D.W., Singh, R., Scheffer, I.E., George, A.L., Phillips, H.A., Saar, K., Reis, A., Johnson, E.W., Sutherland, G.R., Berkovic, S.F., Mulley, J.C. (1998) *Nat. Genet.*, 19(4), 366–370.

222. Abriel, H., Loffing, J., Rebhun, J.F., Pratt, J.H., Schild, L., Horisberger, J.D., Rotin, D., Staub, O. (1999) *J. Clin. Invest.*, 103(5), 667–673.

223. Horisberger, J.D. (2001) *Pharmaceut. News*, 8(2), 23–28.

224. Barbry, P., Hofman, P. (1997) *Am. J. Physiol. Gastrointest. Liver Physiol.*, 273, G571–G585.

225. Firsov, D., Schild, L., Gautschi, I., Merillat, A.M., Schneeberger, E., Rossier, B.C. (1996) *Proc. Natl. Acad. Sci. U.S.A.*, 93(26), 15 370–15 375.

226. Freundlich, M., Ludwig, M. (2005) *Pediatr. Nephrol.*, 20(4), 512–515.

227. Furuhashi, M., Kitamura, K., Adachi, M., Miyoshi, T., Wakida, N., Ura, N., Shikano, Y., Shinshi, Y., Sakamoto, K., Hayashi, M., Satoh, N., Nishitani, T., Tomita, K., Shimamoto, K. (2005) *J. Clin. Endocrinol. Metab.*, 90(1), 340–344.

228. Gormley, K., Dong, Y., Sagnella, G.A. (2003) *Biochem. J.*, 371(1), 1–14.

229. Gründer, S., Firsov, D., Chang, S.S., Jaeger, N.F., Gautschi, I., Schild, L., Lifton, R.P., Rossier, B.C. (1997) *EMBO J.*, 16(5), 899–907.

230. Hiltunen, T.P., Hannila-Handelberg, T., Petajaniemi, N., Kantola, I., Tikkanen, I., Virtamo, J., Gautschi, I., Schild, L., Kontula, K. (2002) *J. Hypertens.*, 20(12), 2383–2390.

231. Hummler, E., Horisberger, J.D. (1999) *Am. J. Physiol. Gastrointest. Liver Physiol.*, 276, G567–G571.

232. Kellenberger, S., Gautschi, I., Rossier, B.C., Schild, L. (1998) *J. Clin. Invest.*, 101(12), 2741–2750.

233. Pradervand, S., Vandewalle, A., Bens, M., Gautschi, I., Loffing, J., Hummler, E., Schild, L., Rossier, B.C. (2003) *J. Am. Soc. Nephrol.*, 14(9), 2219–2228.

234. Rossier, B.C., Pradervand, S., Schild, L., Hummler, E. (**2002**) *Annu. Rev. Physiol.*, 64, 877–897.

235. Tamura, H., Schild, L., Enomoto, N., Matsui, N., Marumo, F., Rossier, B.C. (**1996**) *J. Clin. Invest.*, 97(7), 1780–1784.

236. Kim, J.B., Lee, K.Y., Hur J.K. (**2005**) *J. Korean Med. Sci.*, 20(1), 162–165.

237. Striessnig, J., Hoda, J.C., Koschak, A., Zaghetto, F., Mullner, C., Sinnegger-Brauns, M.J., Wild, C., Watschinger, K., Trockenbacher, A., Pelster, G. (**2004**) *Biochem. Biophys. Res. Commun.*, 322(4), 1341–1346.

238. Wang, Q., Liu, M., Xu, C., Tang, Z., Liao, Y., Du, R., Li, W., Wu, X., Wang, X., Liu, P., Zhang, X., Zhu, J., Ren, X., Ke, T., Wang, Q., Yang, J. (**2005**) *J. Mol. Med.*, 83(3), 203–208.

239. van den Maagdenberg, A.M.J.M., Frants, R.R. (**2001**) *Pharmaceut. News*, 8(2), 37–44.

239a. Splawski, I., Timothy, K.W., Sharpe, L.M., Decher, N., Kumar, P., Bloise, R., Napolitano, C., Schwartz, P.J., Joseph, R.M., Condouris, K., Tager-Flusberg, H., Piori, S.G., Sanguinetti, M.C., Keating, M.T. (**2004**) *Cell*, 119(1), 19–31.

240. Allen, L.E., Zito, I., Bradshaw, K., Patel, R.J., Bird, A.C., Fitzke, F., Yates, J.R., Trump, D., Hardcastle, A.J., Moore, A.T. (**2003**) *Br. J. Ophthalmol.*, 87(11), 1413–1420.

241. Boycott, K.M., Maybaum, T.A., Naylor, M.J., Weleber, R.G., Robitaille, J., Miyake, Y., Bergen, A.A., Pierpont, M.E., Pearce, W.G., Bech-Hansen, N.T. (**2001**) *Hum. Genet.*, 108(2), 91–97.

242. Hoda, J.C., Zaghetto, F., Koschak, A., Striessnig, J. (**2005**) *J. Neurosci.*, 25(1), 252–259.

243. Hope, C.I., Sharp, D.M., Hemara-Wahanui, A., Sissingh, J.I., Lundon, P., Mitchell, E.A., Maw, M.A., Clover, G.M. (**2005**) *Clin. Exp. Ophthalmol.*, 33(2), 129–136.

244. Jacobi, F.K., Hamel, C.P., Arnaud, B., Blin, N., Broghammer, M., Jacobi, P.C., Apfelstedt-Sylla, E., Pusch, C.M. (**2003**) *Am. J. Ophthalmol.*, 135(5), 733–736.

245. McRory, J.E., Hamid, J., Doering, C.J., Garcia, E., Parker, R., Hamming, K., Chen, L., Hildebrand, M., Beedle, A.M., Feldcamp, L., Zamponi, G.W., Snutch, T.P. (**2004**) *J. Neurosci.*, 24(7), 1707–1718.

246. Wutz, K., Sauer, C., Zrenner, E., Lorenz, B., Alitalo, T., Broghammer, M., Hergersberg, M,, de la Chapelle, A., Weber, B.H., Wissinger, B., Meindl, A., Pusch, C.M. (**2002**) *Eur. J. Hum. Genet.*, 10(8), 449–456.

247. Zeitz, C., Minotti, R., Feil, S., Matyas, G., Cremers, F.P., Hoyng, C.B., Berger, W. (**2005**) *Mol. Vis.*, 11, 179–183.

248. Zito, I., Allen, L.E., Patel, R.J., Meindl, A., Bradshaw, K., Yates, J.R., Bird, A.C., Erskine, L., Cheetham, M.E., Webster, A.R., Poopalasundaram, S., Moore, A.T., Trump, D., Hardcastle, A.J. (**2003**) *Hum. Mut.*, 21(2), 169.

249. Hemara-Wahanui, A., Berjukow, S., Hope, C.I., Dearden, P.K., Wu, S.B., Wilson-Wheeler, J., Sharp, D.M., Lundon-Treweek, P., Clover, G.M., Hoda, J.C., Striessnig, J., Marksteiner, R., Hering, S., Maw, M.A. (**2005**) *Proc. Natl. Acad. Sci. U.S.A.*, 102(21), 7553–7558.

250. Alonso, I., Barros, J., Tuna, A., Seixas, A., Coutinho, P., Sequeiros, J., Silveira, I. (**2004**) *Clin. Genet.*, 65(1), 70–72.

251. Alonso, I., Barros, J., Tuna, A., Coelho, J., Sequeiros, J., Silveira, I., Coutinho, P. (**2003**) *Arch. Neurol.*, 60(4), 610–614.

252. Carrera, P., Stenirri, S., Ferrari, M., Battistini, S. (**2001**) *Brain Res. Bull.*, 56(3/4), 239–241.

253. Craig, K., Keers, S.M., Archibald, K., Curtis, A., Chinnery, P.F. (**2004**) *Ann. Neurol.*, 55(5), 752–755.

254. Denier, C., Ducros, A., Durr, A., Eymard, B., Chassande, B., Tournier-Lasserve, E. (**2001**) *Arch. Neurol.*, 58(2), 292–295.

255. Ducros, A., Denier, C., Joutel, A., Cecillon, M., Lescoat, C., Vahedi, K., Darcel, F., Vicaut, E., Bousser, M.G., Tournier-Lasserve, E. (**2001**) *N. Engl. J. Med.*, 345(1), 17–24.

256. Ducros, A., Denier, C., Joutel, A., Vahedi, K., Michel, A., Darcel, F., Madigand, M., Guerouaou, D., Tison, F., Julien, J., Hirsch, E., Chedru, F., Bisgard, C., Lucotte, G., Despres, P., Billard, C., Barthez, M.A., Ponsot, G., Bousser, M.G., Tournier-Lasserve, E. (**1999**) *Am. J. Hum. Genet.*, 64(1), 89–98.

257. Frontali, M. (**2001**) *Brain Res. Bull.*, 56(3/4), 227–231.

258. Guida, S., Trettel, F., Pagnutti, S., Mantuano, E., Tottene, A., Veneziano, L.,

Fellin, T., Spadaro, M., Stauderman, K., Williams, M., Volsen, S., Ophoff, R., Frants, R., Jodice, C., Frontali, M., Pietrobon, D. (**2001**) *Am. J. Hum. Genet.*, 68(3), 759–764.

259. Imbrici, P., Jaffe, S.L., Eunson, L.H., Davies, N.P., Herd, C., Robertson, R., Kullmann, D.M., Hanna, M.G. (**2004**) *Brain*, 127(12), 2682–2692.

260. Jodice, C., Mantuano, E., Veneziano, L., Trettel, F., Sabbadini, G., Calandriello, L., Francia, A., Spadaro, M., Pierelli, F., Salvi, F., Ophoff, R.A., Frants, R.R., Frontali, M. (**1997**) *Hum. Mol. Genet.*, 6(11), 1973–1978.

261. Kaunisto, M.A., Harno, H., Kallela, M., Somer, H., Sallinen, R., Hämäläinen, E., Miettinen, P.J., Vesa, J., Orpana, A., Palotie, A., Färkkilä, M., Wessman, M. (**2004**) *Neurogenetics*, 5(1), 69–73.

262. Kraus, R.L., Sinnegger, M.J., Koschak, A., Glossmann, H., Stenirri, S., Carrera, P., Striessnig, J. (**2000**) *J. Biol. Chem.*, 275(13), 9239–9243.

263. van den Maagdenberg, A.M., Kors, E.E., Brunt, E.R., van Paesschen, W., Pascual, J., Ravine, D., Keeling, S., Vanmolkot, K.R., Vermeulen, F.L., Terwindt, G.M., Haan, J., Frants, R.R., Ferrari, M.D. (**2002**) *J. Neurol.*, 249(11), 1515–1519.

264. Matsuyama, Z., Kawakami, H., Maruyama, H., Izumi, Y., Komure, O., Udaka, F., Kameyama, M., Nishio, T., Kuroda, Y., Nishimura, M., Nakamura, S. (**1997**) *Hum. Mol. Genet.*, 6(8), 1283–1287.

265. Scoggan, K.A., Chandra, T., Nelson, R., Hahn, A.F., Bulman, D.E. (**2001**) *J. Med. Genet.*, 38(4), 249–253.

266. Shizuka, M., Watanabe, M., Ikeda, Y., Mizushima, K., Okamoto, K., Shoji, M. (**1998**) *J. Neurol. Sci.*, 161(1), 85–87.

267. Spacey, S.D., Materek, L.A., Szczygielski, B.I., Bird, T.D. (**2005**) *Arch. Neurol.*, 62(2), 314–316.

268. Spacey, S.D., Hildebrand, M.E., Materek, L.A., Bird, T.D., Snutch, T.P. (**2004**) *Ann. Neurol.*, 56(2), 213–220.

269. Subramony, S.H., Schott, K., Raike, R.S., Callahan, J., Langford, L.R., Christova, P.S., Anderson, J.H., Gomez, C.M. (**2003**) *Ann. Neurol.*, 54(6), 725–731.

270. Terwindt, G.M., Ophoff, R.A., Haan, J., Sandkuijl, L.A., Frants, R.R., Ferrari, M.D. (**1998**) *Eur. J. Hum. Genet.*, 6(4), 297–307.

271. Tottene, A., Pivotto, F., Fellin, T., Cesetti, T., van den Maagdenberg, A.M., Pietrobon, D. (**2005**) *J. Biol. Chem.*, 280(18), 17 678–17 686.

272. Escayg, A., De Waard, M., Lee, D.D., Bichet, D., Wolf, P., Mayer, T., Johnston, J., Baloh, R., Sander, T., Meisler, M.H. (**2002**) *Am. J. Hum. Genet.*, 66, 1531–1539.

273. Wray, D. (1990) The Lambert-Eaton myasthenic syndrome in *Neuromuscular Transmisson: Basic and Applied Aspects*, Vincent, A, Wray, D. (Eds.), Manchester University Press, Manchester, UK.

274. Lang, B., Newsom-Davis, J., Wray, D., Vincent, A., Murray, N. (**1981**) *Lancet*, 2(8240), 224–226.

275. Chen, Y., Lu, J., Pan, H., Zhang, Y., Wu, H., Xu, K., Liu, X., Jiang, Y., Bao, X., Yao, Z., Ding, K., Lo, W.H., Qiang, B., Chan, P., Shen, Y., Wu, X. (**2003**) *Ann. Neurol.*, 54(2), 239–243.

276. Heron, S.E., Phillips, H.A., Mulley, J.C., Mazarib, A., Neufeld, M.Y., Berkovic, S.F., Scheffer, I.E. (**2004**) *Ann. Neurol.*, 55(4), 595–596.

277. Khosravani, H., Bladen, C., Parker, D.B., Snutch, T.P., McRory, J.E., Zamponi, G.W. (**2005**) *Ann. Neurol.*, 57(5), 745–749.

278. Khosravani, H., Altier, C., Simms, B., Hamming, K.S., Snutch, T.P., Mezeyova, J., McRory, J.E., Zamponi, G.W. (**2004**) *J. Biol. Chem.*, 279(11), 9681–9684.

279. Perez-Reyes, E. (**2003**) *Physiol. Rev.*, 83(1), 117–161.

280. Vitko, I., Chen, Y., Arias, J.M., Shen, Y., Wu, X.R., Perez-Reyes, E. (**2005**) *J. Neurosci.*, 25(19), 4844–4855.

281. Pusch, M. (**2001**) *Pharmaceut. News*, 8(2), 45–51.

282. Chen, L., Schaerer, M., Lu, Z.H., Lang, D., Joncourt, F., Weis, J., Fritschi, J., Kappeler, L., Gallati, S., Sigel, E., Burgunder, J.M. (**2004**) *Muscle Nerve*, 29(5), 670–676.

283. Dunø, M., Colding-Jørgensen, E., Grunnet, M., Jespersen, T., Vissing, J., Schwartz, M. (**2004**) *Eur. J. Hum. Genet.*, 12(9), 738–743.

284. Dutzler, R., Campbell, E.B., Cadene, M., Chait, B.T., MacKinnon, R. (**2002**) *Nature*, 415(6869), 287–294.

285. Grunnet, M., Jespersen, T., Colding-Jørgensen, E., Schwartz, M., Klaerke, D.A.,

Vissing, J., Olesen, S.P., Dunø, M. (**2003**) *Muscle Nerve*, 28(6), 722–732.

286. Jou, S.B., Chang, L.I., Pan, H., Chen, P.R., Hsiao, K.M. (**2004**) *J. Neurol.*, 251(6), 666–670.

287. Meyer-Kleine, C., Steinmeyer, K., Ricker, K., Jentsch, T.J., Koch, M.C. (**1995**) *Am. J. Hum. Genet.*, 57(6), 1325–1334.

288. Pusch, M. (**2002**) *Hum. Mut.*, 19(4), 423–434.

289. Simpson, B.J., Height, T.A., Rychkov, G.Y., Nowak, K.J., Laing, N.G., Hughes, B.P., Bretag, A.H. (**2004**) *Hum. Mut.*, 24(2), 185.

290. Sun, C., Tranebjærg, L., Torbergsen, T., Holmgren, G., Van Ghelue, M. (**2001**) *Eur. J. Hum. Genet.*, 9(12), 903–909.

291. Wu, F.F., Ryan, A., Devaney, J., Warnstedt, M., Korade-Mirnics, Z., Poser, B., Escriva, M.J., Pegoraro, E., Yee, A. S., Felice, K.J., Giuliani, M.J., Mayer, R.F., Mongini, T., Palmucci, L., Marino, M., Rüdel, R., Hoffman, E.P., Fahlke, C. (**2002**) *Brain*, 125(11), 2392–2407.

292. Haug, K., Warnstedt, M., Alekov, A.K., Sander, T., Ramírez, A., Poser, B., Maljevic, S., Hebeisen, S., Kubisch, C., Rebstock, J., Horvath, S., Hallmann, K., Dullinger, J.S., Rau, B., Haverkamp, F., Beyenburg, S., Schulz, H., Janz, D., Giese, B., Müller-Newen, G., Propping, P., Elger, C.E., Fahlke, C., Lerche, H., Heils, A. (**2003**) *Nat. Genet.*, 33(4), 527–532.

293. Niemeyer, M.I., Yusef, Y.R., Cornejo, I., Flores, C.A., Sepúlveda, F.V., Cid, L.P. (**2004**) *Physiol. Genom.*, 19(1), 74–83.

294. Carballo-Trujillo, I., Garcia-Nieto, V., Moya-Angeler, F.J., Anton-Gamero, M., Loris, C., Mendez-Alvarez, S., Claverie-Martin, F. (**2003**) *Nephrol. Dial. Transplant.*, 18(4), 717–723.

295. Cox, J.P., Yamamoto, K., Christie, P.T., Wooding, C., Feest, T., Flinter, F.A., Goodyer, P.R., Leumann, E., Neuhaus, T., Reid, C., Williams, P.F., Wrong, O., Thakker, R.V. (**1999**) *J. Bone Miner. Res.*, 14(9), 1536–1542.

296. Günther, W., Lüchow, A., Cluzeaud, F., Vandewalle, A., Jentsch, T.J. (**1998**) *Proc. Natl. Acad. Sci. U.S.A.*, 95(14), 8075–8080.

297. Igarashi, T., Inatomi, J., Ohara, T., Kuwahara, T., Shimadzu, M., Thakker, R.V. (**2000**) *Kidney Int.*, 58(2), 520–527.

298. Rebelo, M.A., Tostes, V., Araújo, N.C., Martini, S.V., Botelho, B.F., Guggino, W.B., Morales, M.M. (**2005**) *An. Acad. Bras. Ciênc.*, 77(1), 95–101.

299. Wu, F., Roche, P., Christie, P.T., Loh, N.Y., Reed, A.A., Esnouf, R.M., Thakker, R.V. (**2003**) *Kidney Int.*, 63(4), 1426–1432.

300. Yamamoto, K., Cox, J.P., Friedrich, T., Christie, P.T., Bald, M., Houtman, P.N., Lapsley, M.J., Patzer, L., Tsimaratos, M., Van'T Hoff, W.G., Yamaoka, K., Jentsch, T.J., Thakker, R.V. (**2000**) *J. Am. Soc. Nephrol.*, 11(8), 1460–1468.

301. Campos-Xavier, A.B., Saraiva, J.M., Ribeiro, L.M., Munnich, A., Cormier-Daire, V. (**2003**) *Hum. Genet.*, 112(2), 186–189.

302. Cleiren, E., Bénichou, O., Van Hul, E., Gram, J., Bollerslev, J., Singer, F.R., Beaverson, K., Aledo, A., Whyte, M.P., Yoneyama, T., deVernejoul, M.C., Van Hul, W. (**2001**) *Hum. Mol. Genet.*, 10(25), 2861–2867.

303. Henriksen, K., Gram, J., Schaller, S., Dahl, B.H., Dziegiel, M.H., Bollerslev, J., Karsdal, M.A. (**2004**) *Am. J. Pathol.*, 164(5), 1537–1545.

304. Kasper, D., Planells-Cases, R., Fuhrmann, J.C., Scheel, O., Zeitz, O., Ruether, K., Schmitt, A., Poet, M., Steinfeld, R., Schweizer, M., Kornak, U., Jentsch, T.J. (**2005**) *EMBO J.*, 24(5), 1079–1091.

305. Letizia, C., Taranta, A., Migliaccio, S., Caliumi, C., Diacinti, D., Delfini, E., D'Erasmo, E., Iacobini, M., Roggini, M., Albagha, O.M., Ralston, S.H., Teti, A. (**2004**) *Calcif. Tissue Int.*, 74(1), 42–46.

306. Fukuyama, S., Hiramatsu, M., Akagi, M., Higa, M., Ohta, T. (**2004**) *J. Clin. Endocrinol. Metab.*, 89(11), 5847–5850.

307. Károlyi, L., Koch, M.C., Grzeschik, K.H., Seyberth, H.W. (**1998**) *J. Mol. Med.*, 76(5), 317–325.

308. Kieferle, S., Fong, P., Bens, M., Vandewalle, A., Jentsch, T.J. (**1994**) *Proc. Natl. Acad. Sci. U.S.A.*, 91(15), 6943–6947.

309. Konrad, M., Vollmer, M., Lemmink, H.H., van den Heuvel, L.P., Jeck, N., Vargas-Poussou, R., Lakings, A., Ruf, R., Deschenes, G., Antignac, C., Guay-Woodford, L., Knoers, N.V., Seyberth, H.W., Feldmann, D., Hildebrandt, F. (**2000**) *J. Am. Soc. Nephrol.*, 11(8), 1449–1459.

310. Schlingmann, K.P., Konrad, M., Jeck, N., Waldegger, P., Reinalter, S.C., Holder, M., Seyberth, H.W., Waldegger, S. (**2004**) *N. Engl. J. Med.*, 350(13), 1314–1319.

311. Simon, D.B., Bindra, R.S., Mansfield, T.A., Nelson-Williams, C., Mendonca, E., Stone, R., Schurman, S., Nayir, A., Alpay, H., Bakkaloglu, A., Rodriguez-Soriano, J., Morales, J.M., Sanjad, S.A., Taylor, C.M., Pilz, D., Brem, A., Trachtman, H., Griswold, W., Richard, G.A., John, E., Lifton, R.P. (**1997**) *Nat. Genet.*, 17(2), 171–178.

312. Zelikovic, I., Szargel, R., Hawash, A., Labay, V., Hatib, I., Cohen, N., Nakhoul, F. (**2003**) *Kidney Int.*, 63(1), 24–32.

313. Dorwart, M., Thibodeau, P., Thomas, P. (**2004**) *J. Cyst. Fibros.*, 3 (Sup. 2), 91–94.

314. Gibson, R.L., Burns, J.L., Ramsey, B.W. (**2003**) *Am. J. Respir. Crit. Care Med.*, 168(8), 918–951.

315. Greger, R., Mall, M., Bleich, M., Ecke, D., Warth, R., Riedemann, N., Kunzelmann, K. (**1996**) *J. Mol. Med.*, 74(9), 527–534.

316. Guggino, W.B., Banks-Schlegel, S.P. (**2004**) *Am. J. Respir. Crit. Care Med.*, 170(7), 815–820.

317. Kidd, J.F., Kogan, I., Bear, C.E. (**2004**) *Curr. Top. Dev. Biol.*, 60, 215–249.

318. Kopito, R.R. (**1999**) *Physiol. Rev.*, 79(1), S167–S173.

319. Mehta, A. (**2005**) *Pediatr. Pulmonol.*, 39(4), 292–298.

320. Skach, W.R. (**2000**) *Kidney Int.*, 57(3), 825–831.

321. Welsh, M.J., Smith, A.E. (**1993**) *Cell*, 73(7), 1251–1254.

322. Beeson, D. (**2001**) *Pharmaceut. News*, 8(2), 18–22.

323. Oosterhuis, H.J.G.H., Newsom-Davis, J., Wokke, J.H.J., Molenaar, P.C., Weerden, T.V., Oen, B.S., Jennekens, F.G.I, Veldman. H., Vincent, A., Wray, D.W., Prior, C., Murray, N.M.F. (**1987**) *Brain*, 110, 1061–1079.

324. Beeson, D., Webster, R., Ealing, J., Croxen, R., Brownlow, S., Brydson, M., Newsom-Davis, J., Slater, C., Hatton, C., Shelley, C., Colquhoun, D., Vincent, A. (**2003**) *Ann. New York Acad. Sci.*, 998, 114–124.

325. Brownlow, S., Webster, R., Croxen, R., Brydson, M., Neville, B., Lin, J.P., Vincent, A., Newsom-Davis, J., Beeson, D. (**2001**) *J. Clin. Invest.*, 108(1), 125–130.

326. Burke, G., Cossins, J., Maxwell, S., Robb, S., Nicolle, M., Vincent, A., Newsom-Davis, J., Palace, J., Beeson, D. (**2004**) *Neuromuscul. Disord.*, 14(6), 356–364.

327. Croxen, R., Newland, C., Beeson, D., Oosterhuis, H., Chauplannaz, G., Vincent, A., Newsom-Davis, J. (**1997**) *Hum. Mol. Genet.*, 6(5), 767–774.

328. Engel, A.G., Ohno, K., Shen, X.M., Sine, S.M. (**2003**) *Ann. New York Acad. Sci.*, 998, 138–160.

329. Engel, A.G., Ohno, K., Sine, S.M. (**1999**) *Arch. Neurol.*, 56(2), 163–167.

330. Engel, A.G., Ohno, K., Milone, M., Wang, H.L., Nakano, S., Bouzat, C., Pruitt, J.N., Hutchinson, D.O., Brengman, J.M., Bren, N., Sieb, J.P., Sine, S.M. (**1996**) *Hum. Mol. Genet.*, 5(9), 1217–1227.

331. Engel, A.G., Ohno, K., Bouzat, C., Sine, S.M., Griggs, R.C. (**1996**) *Ann. Neurol.*, 40(5), 810–817.

332. Gomez, C.M., Maselli, R.A., Vohra, B.P., Navedo, M., Stiles, J.R., Charnet, P., Schott, K., Rojas, L., Keesey, J., Verity, A., Wollmann, R.W., Lasalde-Dominicci, J. (**2002**) *Ann. Neurol.*, 51(1), 102–112.

333. Hatton, C.J., Shelley, C., Brydson, M., Beeson, D., Colquhoun, D. (**2003**) *J. Physiol.*, 547(3), 729–760.

334. Lindstrom, J.M. (**2000**) *Muscle Nerve*, 23(4), 453–477.

335. Morar, B., Gresham, D., Angelicheva, D., Tournev, I., Gooding, R., Guergueltcheva, V., Schmidt, C., Abicht, A., Lochmüller, H., Tordai, A., Kalmár, L., Nagy, M., Karcagi, V., Jeanpierre, M., Herczegfalvi, A., Beeson, D., Venkataraman, V., Warwick Carter, K., Reeve, J., de Pablo, R., Kučinskas, V., Kalaydjieva, L. (**2004**) *Am. J. Hum. Genet.*, 75(4), 596–609.

336. Nichols, P., Croxen, R., Vincent, A., Rutter, R., Hutchinson, M., Newsom-Davis, J., Beeson, D. (**1999**) *Ann. Neurol.*, 45(4), 439–443.

337. Ohno, K., Quiram, P. A., Milone, M., Wang, H.L., Harper, M.C., Pruitt, J.N., Brengman, J.M., Pao, L., Fischbeck, K.H., Crawford, T.O., Sine, S.M., Engel, A.G. (**1997**) *Hum. Mol. Genet.*, 6(5), 753–766.

338. Ohno, K., Hutchinson, D.O., Milone, M., Brengman, J.M., Bouzat, C., Sine, S.M., Engel, A.G. (**1995**) *Proc. Natl. Acad. Sci. U.S.A.*, 92(3), 758–762.

339. Quiram, P.A., Ohno, K., Milone, M., Patterson, M.C., Pruitt, N.J., Brengman, J.M., Sine, S.M., Engel, A.G. (**1999**) *J. Clin. Invest.*, 104(10), 1403–1410.

340. Scola, R.H., Werneck, L.C., Iwamoto, F.M., Comerlato, E.A., Kay, C.K. (**2000**) *Muscle Nerve*, 23(10), 1582–1585.

341. Shelley, C., Colquhoun, D. (**2005**) *J. Physiol.*, 564(2), 377–396.

342. Shen, X.M., Ohno, K., Tsujino, A., Brengman, J.M., Gingold, M., Sine, S.M., Engel, A.G. (**2003**) *J. Clin. Invest.*, 111(4), 497–505.

343. Sieb, J.P., Kraner, S., Rauch, M., Steinlein, O.K. (**2000**) *Hum. Genet.*, 107(2), 160–164.

344. Surtees, R. (**1999**) *J. Inherit. Metab. Dis.*, 22(4), 374–380.

345. Sutor, B., Zolles, G. (**2001**) *Pflügers Arch.*, 442(5), 642–651.

346. Bertrand, D., Picard, F., Le Hellard, S., Weiland, S., Favre, I., Phillips, H., Bertrand, S., Berkovic, S.F., Malafosse, A., Mulley, J. (**2002**) *Epilepsia*, 43(Sup. 5), 112–122.

347. Cho, Y.W., Motamedi, G.K., Laufenberg, I., Sohn, S.I., Lim, J.G., Lee, H., Yi, S.D., Lee, J.H., Kim, D.K., Reba, R., Gaillard, W.D., Theodore, W.H., Lesser, R.P., Steinlein, O.K. (**2003**) *Arch. Neurol.*, 60(11), 1625–1632.

348. Combi, R., Dalprà, L., Tenchini, M.L., Ferini-Strambi, L. (**2004**) *J. Neurol.*, 251(8), 923–934.

349. di Corcia, G., Blasetti, A., De Simone, M., Verrotti, A., Chiarelli, F. (**2005**) *Eur. J. Paediatr. Neurol.*, 9(2), 59–66.

350. Duga, S., Asselta, R., Bonati, M.T., Malcovati, M., Dalprà, L., Oldani, A., Zucconi, M., Ferini-Strambi, L., Tenchini, M.L. (**2002**) *Epilepsia*, 43(4), 362–364.

351. Figl, A., Viseshakul, N., Shafaee, N., Forsayeth, J., Cohen, B.N. (**1998**) *J. Physiol.*, 513(3), 655–670.

352. Itier, V., Bertrand, D. (**2002**) *Neurophysiol. Clin.*, 32(2), 99–107.

353. Kuryatov, A., Gerzanich, V., Nelson, M., Olale, F., Lindstrom, J. (**1997**) *J. Neurosci.*, 17(23), 9035–9047.

354. Phillips, H.A., Favre, I., Kirkpatrick, M., Zuberi, S.M., Goudie, D., Heron, S.E., Scheffer, I.E., Sutherland, G.R., Berkovic, S.F., Bertrand, D., Mulley, J.C. (**2001**) *Am. J. Hum. Genet.*, 68(1), 225–231.

355. Rodrigues-Pinguet, N., Jia, L., Li, M., Figl, A., Klaassen, A., Truong, A., Lester, H.A., Cohen, B.N. (**2003**) *J. Physiol.*, 550(1), 11–26.

356. Rozycka, A., Trzeciak, W.H. (**2003**) *J. Appl. Genet.*, 44(2), 197–207.

357. Rozycka, A., Skorupska, E., Kostyrko, A., Trzeciak, W.H. (**2003**) *Epilepsia*, 44(8), 1113–1117.

358. Steinlein, O.K., Magnusson, A., Stoodt, J., Bertrand, S., Weiland, S., Berkovic, S.F., Nakken, K.O., Propping, P., Bertrand, D. (**1997**) *Hum. Mol. Genet.*, 6(6), 943–947.

359. Breitinger, H.G., Lanig, H., Vohwinkel, C., Grewer, C., Breitinger, U., Clark, T., Becker, C.M. (**2004**) *Chem. Biol.*, 11(10), 1339–1350.

360. Breitinger, H.G., Becker, C.M. (**2002**) *Chembiochem*, 3(11), 1042–1052.

361. Breitinger, H.G., Becker, C.M. (**2002**) *Neurosci. Lett.*, 331(1), 21–24.

362. Breitinger, H.G., Villmann, C., Becker, K., Becker, C.M. (**2001**) *J. Biol. Chem.*, 276(32), 29 657–29 663.

363. Cascio, M. (**2004**) *J. Biol. Chem.*, 279(19), 19 383–19 386.

364. del Giudice, E.M., Coppola, G., Bellini, G., Ledaal, P., Hertz, J.M., Pascotto, A. (**2003**) *J. Med. Genet.*, 40(5), e71.

365. del Giudice, E.M., Coppola, G., Bellini, G., Cirillo, G., Scuccimarra, G., Pascotto, A. (**2001**) *Eur. J. Hum. Genet.*, 9(11), 873–876.

366. Humeny, A., Bonk, T., Becker, K., Jafari-Boroujerdi, M., Stephani, U., Reuter, K., Becker, C.M. (**2002**) *Eur. J. Hum. Genet.*, 10(3), 188–196.

367. Kwok, J.B., Raskin, S., Morgan, G., Antoniuk, S.A., Bruk, I., Schofield, P.R. (**2001**) *J. Med. Genet.*, 38(6), e17.

368. Lewis, T.M., Schofield, P.R. (**1999**) *Ann. New York Acad. Sci.*, 868, 681–684.

369. Lewis, T.M., Sivilotti, L.G., Colquhoun, D., Gardiner, R.M., Schoepfer, R., Rees, M. (**1998**) *J. Physiol.*, 507(1), 25–40.

370. Lynch, J.W. (**2004**) *Physiol. Rev.*, 84(4), 1051–1095.

371. Rea, R., Tijssen, M.A., Herd, C., Frants, R.R., Kullmann, D.M. (**2002**) *Eur. J. Neurosci.*, 16(2), 186–196.

372. Rees, M.I., Lewis, T.M., Kwok, J.B., Mortier, G.R., Govaert, P., Snell, R.G., Schofield, P.R., Owen, M.J. (**2002**) *Hum. Mol. Genet.*, 11(7), 853–860.

373. Rees, M.I., Lewis, T.M., Vafa, B., Ferrie, C., Corry, P., Muntoni, F., Jungbluth, H., Stephenson, J.B., Kerr, M., Snell, R.G., Schofield, P.R., Owen, M.J. (**2001**) *Hum. Genet.*, 109(3), 267–270.

374. Saul, B., Kuner, T., Sobetzko, D., Brune, W., Hanefeld, F., Meinck, H.M., Becker, C.M. (**1999**) *J. Neurosci.*, 19(3), 869–877.

375. Tijssen, M.A., Vergouwe, M.N., van Dijk, J.G., Rees, M., Frants, R.R., Brown, P. (**2002**) *Mov. Disord.*, 17(4), 826–830.

376. Baulac, S., Huberfeld, G., Gourfinkel-An, I., Mitropoulou, G., Beranger, A., Prud'homme, J.F., Baulac, M., Brice, A., Bruzzone, R., LeGuern, E. (**2001**) *Nat. Genet.*, 28(1), 46–48.

377. Bonanni, P., Malcarne, M., Moro, F., Veggiotti, P., Buti, D., Ferrari, A.R., Parrini, E., Mei, D., Volzone, A., Zara, F., Heron, S.E., Bordo, L., Marini, C., Guerrini, R. (**2004**) *Epilepsia*, 45(2), 149–158.

378. Buhr, A., Bianchi, M.T., Baur, R., Courtet, P., Pignay, V., Boulenger, J.P., Gallati, S., Hinkle, D.J., Macdonald, R.L., Sigel, E. (**2002**) *Hum. Genet.*, 111(2), 154–160.

379. Buxbaum, J.D., Silverman, J.M., Smith, C.J., Greenberg, D.A., Kilifarski, M., Reichert, J., Cook, E.H., Fang, Y., Song, C.Y., Vitale, R. (**2002**) *Mol. Psychiatry*, 7(3), 311–316.

380. Cossette, P., Liu, L., Brisebois, K., Dong, H., Lortie, A., Vanasse, M., Saint-Hilaire, J.M., Carmant, L., Verner, A., Lu, W.Y., Wang, Y.T., Rouleau, G.A. (**2002**) *Nat. Genet.*, 31(2), 184–189.

381. Covault, J., Gelernter, J., Hesselbrock, V., Nellissery, M., Kranzler, H.R. (**2004**) *Am. J. Med. Genet. Part B Neuropsychiatr. Genet.*, 129(1), 104–109.

382. Edenberg, H.J., Dick, D.M., Xuei, X., Tian, H., Almasy, L., Bauer, L.O., Crowe, R.R., Goate, A., Hesselbrock, V., Jones, K., Kwon, J., Li, T.K., Nurnberger, J.I., O'Connor, S.J., Reich, T., Rice, J., Schuckit, M.A., Porjesz, B., Foroud, T., Begleiter, H. (**2004**) *Am. J. Hum. Genet.*, 74(4), 705–714.

383. Feusner, J., Ritchie, T., Lawford, B., Young, R.M., Kann, B., Noble, E.P. (**2001**) *Psychiatry Res.*, 104(2), 109–117.

384. Glatt, K., Glatt, H., Lalande, M. (**1997**) *Genomics*, 41(1), 63–69.

385. Harkin, L.A., Bowser, D.N., Dibbens, L.M., Singh, R., Phillips, F., Wallace, R.H., Richards, M.C., Williams, D.A., Mulley, J.C., Berkovic, S.F., Scheffer, I.E., Petrou, S. (**2002**) *Am. J. Hum. Genet.*, 70(2), 530–536.

386. Kananura, C., Haug, K., Sander, T., Runge, U., Gu, W., Hallmann, K., Rebstock, J., Heils, A., Steinlein, O.K. (**2002**) *Arch. Neurol.*, 59(7), 1137–1141.

387. Marini, C., Harkin, L.A., Wallace, R.H., Mulley, J.C., Scheffer, I.E., Berkovic, S.F. (**2003**) *Brain*, 126(1), 230–240.

388. Song, J., Koller, D.L., Foroud, T., Carr, K., Zhao, J., Rice, J., Nurnberger, J.I., Begleiter, H., Porjesz, B., Smith, T.L., Schuckit, M.A., Edenberg, H.J. (**2003**) *Am. J. Med. Genet. Part B Neuropsychiatr. Genet.*, 117(1), 39–45.

389. Wallace, R.H., Marini, C., Petrou, S., Harkin, L.A., Bowser, D.N., Panchal, R.G., Williams, D.A., Sutherland, G.R., Mulley, J.C., Scheffer, I.E., Berkovic, S.F. (**2001**) *Nat. Genet.*, 28(1), 49–52.

8.2
Structural and Ligand-based Models for HERG and their Application in Medicinal Chemistry

Yi Li, Giovanni Cianchetta, and Roy J. Vaz

8.2.1
Introduction: Perspective on the Necessity of Models

Blockage of human ether-a-go-go (hERG) potassium channels is a well-known mechanism for drug-induced cardiac arrhythmias and potential incidents of drug-induced sudden cardiac death [1–4]. For safety considerations, pharmaceutical developers have thus instituted programs to routinely screen out hERG channel blockers at the early stage of drug discovery using available *in vitro* methods [5] such as the Rb-efflux functional test [6, 7], radiolabeled binding assays [8, 9], or the whole cell patch-clamp assay [10–12]. Meanwhile, *in silico* methods using computational models to predict binding affinities of hERG channel blockers have been devised to design compounds devoid of any hERG inhibitory activities [13, 14]. In this chapter, we summarize the current status of structural and ligand-based models for hERG channel blockers and examples of applying these models in medicinal chemistry.

8.2.1.1 Availability of Structural Aspects of Models
The hERG channel, like other voltage-dependent K^+ channels, is a membrane protein with a topology of six membrane-spanning helices. The functional channel is an assembly of tetramers, and each monomer consists of six transmembrane (S1-S6) regions and the K^+ selectivity region (P region) that bridges the 5th (S5) and 6th (S6) transmembrane regions to form the pore of the channel. Voltage-dependent gating, or opening and closing of the pore, is controlled presumably by the arginine-rich fourth membrane spanning stretch (the S4 voltage sensor) present in all members of the voltage-dependent cation channel family [15]. The large intracellular C-terminal region of the hERG channel contains a cyclic nucleotide binding domain. The structure of the N-terminal domain, which has been resolved by X-ray crystallography, is structurally similar to the Per-Arnt-Sim (PAS) domains that are frequently involved in signal transduction [16]. The 3D structure of other regions of the hERG channel has not been resolved. However, the crystal structure of the KcsA channel [17], a simple channel consisting of merely the pore region (the S5, the P region and the S6 transmembrane sequences), together with the structure of more complicated channels and mutagenesis work provided the basis for understanding the structure–activity relationships of hERG and other channels. The crystal structures of a ligand-gated (Ca^{2+} as the ligand) K^+ channel, the Mthk channel [18, 19], and a voltage-gated K^+ channel Kvap have provided a snapshot into the open form of K^+ channel pore [20, 21]. The structure of an inward rectifying K^+ channel, KirBac1.1 [22], shows the same structural architecture as those of other solved structures. Recent work on the

shaker channel suggested an alternative mechanism of gating, including a smaller pore and possibly a unique hinge or swivel point [23]. The structural aspects of the hERG channel have been reviewed recently in terms of structural information of other K^+ channels [3].

8.2.1.2 What has been the Best Data to Model?

Since experimental data have been reported using various *in vitro* methods, one of the challenges to understanding hERG channel structure–activity relationships is to determine what *in vitro* biological activity best measures compound interactions with the channel. There are two variables that play a role in the measurement of biological activity, the biological system in which hERG is expressed and the method used for activity measurement [24]. Typically, the *in vitro* biological system consists of the hERG channel expressed either in Chinese Hamster Ovary (CHO) cells, Human Embryonic Kidney (HEK) cells or *Xenopus* oocytes. Of these, the stably transfected CHO and HEK cells seem to be the most reliable, since measurements in *Xenopus* oocytes lead to a significant underestimation of hERG potency due to partitioning of the compound into the yolk. Cell toxicity arising from high expression levels has led to alternatives to cell-based screening being investigated. One such example uses ion-sensitive dyes on a solid surface coated by a membrane with the transmembrane proteins at a predefined target density and lipid composition. In terms of the system of measurement, it can be binding, membrane potential or ion-flux. The binding assays are based on a competitive radiolabeled ligand binding assay and provide little information on hERG channel function. Moreover, many molecules that do not affect the binding but affect channel function go undetected. Surface plasmon resonance is also being investigated towards a high-throughput hERG binding assay. Membrane potential assays include assays using voltage-sensitive dyes and dye pairs via fluorescence resonance energy transfer (FRET). These methods largely suffer from a lack of sensitivity. Cell-based ion-flux assays include radiolabeled (^{86}Rb) as well as detection using atomic absorption spectroscopy of the Rb ion. Unfortunately, these assays can only pick up very potent hERG channel modulators. The most reliable method is an ion-flux assay, the patch clamp assay [25], which provides data measuring drug interactions with the hERG potassium channels. This assay uses electronic negative-feedback circuits to control transcellular membrane potential and measure ionic flux (current) through open ion channels. This high information content measurement provides valuable data on inhibitors as well as activators. Several emerging technologies aiming to automate the voltage-clamp method appear promising [12].

8.2.2
Structural Aspect of hERG Models

8.2.2.1 Mutation Data and Structural Mapping of Binding Sites
Mutagenesis studies have provided valuable information to the understanding of the structure–activity relationships (SAR) of the hERG channel blockers. Mutagenesis work on the hERG channel was first carried out with an alanine scanning mutagenesis of amino acids located in the S6 transmembrane domain [26]. The binding interactions of three compounds, MK-499, cisapride and terfenadine, were tested against each mutant. The measured binding data were interpreted based upon a homology model of hERG channel constructed from the crystal structure of KcsA. The residues implicated in the binding of all three drugs were Y652 and F656, whereas residues V625 and G648 only affected the binding of MK-499. Subsequent work [27] repositioned Y652 and F656 along the S6 α-helical axis. Mutations of these residues also reduced blocking sensitivity by cisapride, providing evidence that they were essential to interact with hERG channel blockers.

Further mutation study at only Y652 and its binding of chloroquine suggested that this compound only blocked the channel in the open form by cation-π and π-stacking interactions with Y652 and F656 respectively [28]. Systematic mutations at Y652 and F656 with various natural amino acids [29] and measuring the effects on cisapride, MK-499 and terfenadine confirmed these interactions. In comparing the sequence of the S6 domain of hERG and other K$^+$ channels [16], it was proposed that in hERG, the sequence at 655–657, PVP was replaced by IFG, making the hERG channel larger due to the lack of the second proline residues. Replacement of the IFG sequence with PVP [29] also resulted in a channel less sensitive to blockage by MK-499, cisapride and terfenadine. All this mutational work has aided tremendously in delineating key determinants of drug binding to the hERG channel.

A recent attempt at modeling hERG channel structure [30] seems to be interesting. The study models the channel in various states, starting from the closed form of KcsA, having an intermediate open state and finally an open state represented by the Mthk structure. Various ligands were docked and based on the interaction energies – some ligands were shown to prefer the open form and some the intermediate form of the channel. The outliers that could not be represented by either model probably favored a still further intermediate that was not reported.

8.2.3
Ligand-based Chemometric Models

Several computational models have been reported recently for hERG channel blockers. Most of these models were derived from a ligand-based approach using a relatively small experimental data set containing molecules of positively charged functional group(s). Additionally, these models pertain to inhibitors and blockers that block the channel from the intracellular direction. There are known blockers, namely peptides and large molecules, that block the channel from the extracellular direction, which are not the focus in this review [30, 31].

Molecular models reported in the literature about the hERG channel and its blockers were developed for intended uses, often with limited scope. The first type of model arises out of efforts trying to identify potential molecules that block the hERG channel. These models would be used as a first pass to select molecules as candidates for an *in vitro* assay [32]. These models could also be used for library design and synthesis [33]. Most 2D physicochemical and chemometric models fall in this category.

8.2.3.1 2D Physicochemical Models

A decision-tree-based approach based on a binary classification model was proposed using just three simple calculated physicochemical properties (Clog$P$, CMR and pK_a) for classification of hERG channel blockers [34]. These variables were found statistically significant in describing a hERG blocker with Clog$P$ (calculated log$P$) $\pm$ 3.7, a size descriptor CMR (calculated molar refractivity) in the range of 110 to 176, and the molecule having a minimum pK_a of 7.3 (molecule with a basic group). These physicochemical properties are often associated with molecular permeability into membranes. Thus it can be viewed as describing the first necessary step for a molecule to reach the binding site intracellularly, rather than requirements by the hERG channel for the recognition of a blocker.

8.2.3.2 Three-dimensional Pharmacophore Models

Similarly, chemometric approaches have been developed for binary classifications to discriminate hERG channel blockers from a library of compounds. Training of an artificial neural network with a set of 244 compounds and over one thousand calculated descriptors, based on 2D substructures as well as other descriptors (including 3D-Volsurf descriptors), led to a prediction scheme used as a general *in silico* filter [35]. In this study, patch-clamp data was obtained from mammalian cells transfected with hERG channel. No specific descriptors were pointed out as being critical or statistically deterministic for hERG binders among the molecules used in the study. Analysis of a training set of 85 active hERG channel blocker plus 329 inactive compounds led to an elaborate scheme in a "veto" format to predict hERG blockers using a combined pharmacophore fingerprint descriptor based model and topological 2D similarity nearest neighbors type classification [36]. A hologram QSAR (HQSAR) model including 3D-Volsurf descriptors was constructed from an initial 13 compound training set and claimed to be highly predictive [37].

Ekins has proposed a pharmacophore model generated by Catalyst using a training set of 15 molecules. This model consisted of four hydrophobic features and one positive ionizable feature [38, 39] as the essential elements for the recognition by the hERG channel. Conversely, another model [40] found that hydrophobic surface area, the number of hydrogen bond acceptors and pK_a of a piperidine moiety that was common to all structures used in the study turned out to be important descriptors. Thus, it suggested that initial partition into the cell membrane was a key step in hERG inhibition.

8.2.3.3 GRIND-based Model

It is still a challenge to make a quantitative *in silico* prediction on how a drug influences the hERG channel function. There is ample evidence to suggest that binding interactions of drug molecules at the pore cavity depend upon the conducting state of the channel [41, 42]. Thus, a reliable model by chemometric approach requires a large data set using a consistent experimental measurement.

Using a large dataset, we have developed a general method employing multiple models for the prediction of hERG channel blockers [43]. The dataset consisted of 882 compounds with experimentally measured IC_{50}s for hERG channel inhibition using the standard whole-cell patch clamp electrophysiology method [10]. The compounds included both molecules with ionizable basic nitrogen atoms and those with non-ionizable nitrogen atoms such as amide, aromatic and aniline nitrogen atoms.

The hERG IC_{50}s were analyzed using GRIND descriptors and the program ALMOND [43]. The complete set of molecules was divided into two groups, one including only those containing one or more basic protonated nitrogen atoms and the other containing no basic nitrogen. The first group consisted of 518 molecules in the learning set with 26 in the test set. The second group consisted of 322 compounds with 16 in the test set. The descriptors with the highest coefficient described a pair of pharmacophores. Thus, GRIND based pharmacophore models (Fig. 8.2.1) were generated respectively for the molecules with or without a basic nitrogen atom. Table 8.2.1 lists the descriptors in the two models.

Of the seven descriptors that seemed to weigh heavily in the two models, four were almost identical. Of the other three descriptors that were not the same, two

Tab. 8.2.1 Salient GRIND descriptors in the two PLS models.

GRIND descriptor	GRID MIFs	Subset I		Subset II	
		Step	Å	Step	Å
11	Dry-dry	36	18	36	18
44	Tip-tip	49	24.5	58	29
12	Dry-Hbond donor	28	14	41	20.5
13	Dry-Hbond acceptor	42	21	40	20
14	Dry-tip	46	23	47	23.5
24	Hbond donor-tip	46	23	52	26
34	Hbond acceptor-tip	45	22.5	49	24.5

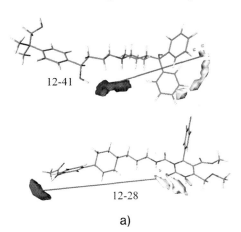

Fig. 8.2.1 (a) The common H-bond donor and hydrophobic area pharmacophoric element identified by the ALMOND models using structures [43] – one with a basic nitrogen and the other without a basic nitrogen. (b) GRIND descriptors related to the presence of a H-bond acceptor group for the same molecules as (a) – hydrophobic area (yellow), field generated by a H-bond acceptor group (blue) and TIP field (green).

a)

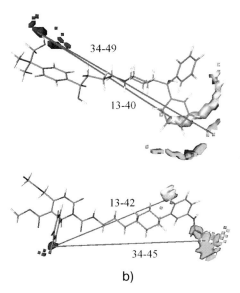

b)

included descriptors involving an H-bond donor that corresponded to the protonated basic nitrogen, and the last corresponded to a 5 Å difference in the dimension of the molecule. Of the four descriptors that were the same, one corresponded to a hydrophobic–hydrophobic distance, two corresponded to a descriptor involving an H-bond acceptor and the last corresponded to an edge distance from the hydrophobic center. The H-bond acceptor is a pharmacophoric element that was not previously noted as being important in any of the datasets previously analyzed. The study also suggested a single or similar binding site for both protonated and non-positively charged molecules. An interaction analysis of a compound without a basic nitrogen with the various mutant sequences as was done by Fernandez and coworkers [29] would be interesting to validate or invalidate the statement.

8.2.4
Ligand-based QSAR Models

8.2.4.1 **CoMSiA/CoMFA Models**

The next two models could be used in the lead optimization phase of a drug discovery program. The first, developed by Cavalli and coworkers [44], was a Comparative Molecular Field Analysis (CoMFA) 3D-QSAR model constructed from literature data. This pharmacophore model was derived from CoMFA analysis of a set of 31 QT-prolonging drugs, all of which involved a positively charged tertiary amine flanked by three aromatic or hydrophobic centers. Care was taken in culling the literature; *Xenopus* oocyte derived data was not used. In the study, all highly active molecules contained a basic nitrogen and thus a basic nitrogen was included as part of the proposed pharmacophore. The pharmacophore also consisted of three hydrophobic pharmacophoric elements. The pharmacophore was further supported using CoMFA. The steric coefficients obtained from the CoMFA could be used in a design to eliminate hERG activity from the molecule after deciphering how a particular active hERG molecule aligns with the pharmacophore.

We proposed a "drain plug" model from a combined homology modeling of the hERG channel pore region and a ligand-based 3D-QSAR Comparative Molecular Similarity Analysis (CoMSiA) of an experimentally consistent dataset [45]. It was based on a series of molecules derived from sertindole, together with a series of other drugs, most of which were tested in the same laboratory. The corresponding IC_{50}s were used to derive a CoMSiA model that is also a 3D-QSAR method that gives similar information as the previously discussed CoMFA method. The model was further validated in the following manner. First a homology model of the open form of the channel was constructed from the Mthk crystal structure. Then three molecules, MK499, terfenadine and cisapride, as aligned in the CoMSiA model (Fig. 8.2.2), were docked into the homology model. The previously described mutations [26] were found to agreed with the docked model in terms of affecting activity. Figure 8.2.3 shows the docked model. Again our CoMSiA model can be, and has been, used to design molecules with reduced hERG affinity. If a molecule with an IC_{50} obtained from our laboratories showed activity beyond a "safe" range, then appropriate changes can be made as suggested by the coefficient plots obtained from the model. Examples of using this model to some drug discovery programs are described below.

8.2.5
Application of Models to Improve Selectivity: Case Studies

There are several recently reported examples that successfully remove binding interactions of compounds with the hERG channel during the lead optimization phase of drug discovery. Friesen and coworkers [46] reported optimization of a series of phosphodiesterase-4 (PDE4) inhibitors. Initial optimization of compound CDP-840 (1) with respect to PDE4 inhibition, metabolism issues and pharmaco-

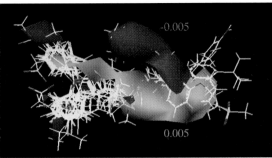

Fig. 8.2.2 (Top) CoMSiA steric coefficient plots, showing regions (in red) where substitution would enhance selectivity against the HERG channel. (Bottom) CoMSiA electrostatic coefficient plots, showing regions where substitution by groups that increased negative potential (in orange) would increase selectivity against the HERG channel.

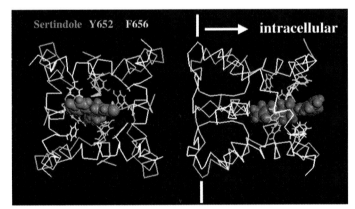

Fig. 8.2.3 Orthogonal views of sertindole docked into the homology model of HERG obtained from the crystal structure of Mthk, showing interactions with Y652 and F656.

kinetic properties led to the identification of tertiary alcohol (**2**) shown in Fig. 8.2.4. However, this compound exhibited significant prolongation of the QT interval following intravenous administration in anesthetized dogs. Further medicinal chemistry efforts were directed at optimizing structure with respect to the binding affinity to the hERG channel, in addition to other pharmacological properties. By

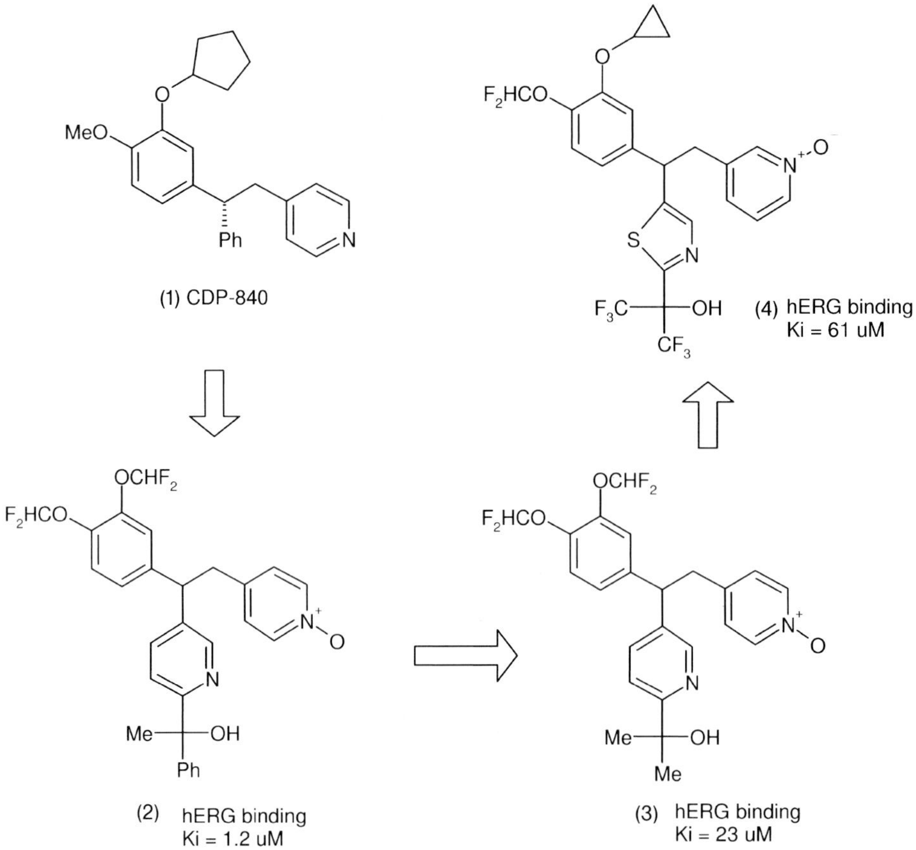

Fig. 8.2.4 Lead optimization and alleviation of binding interactions to the HERG channel for phosphodiesterase-4 inhibitors based upon CDP-840 (1).

simply replacing the tertiary alcohol phenyl group with a methyl or trifluoromethyl group, they were able to arrive at compounds that exhibited lower hERG binding affinity, similar *in vitro* profile and improved pharmacokinetics. Interestingly, no binding data were given for the original lead CDP-840, which lacks the tertiary alcohol phenyl moiety. In this case, the activity measurements were from a radiolabeled ligand displacement assay.

Another example is demonstrated by the medicinal chemistry efforts devoted to minimize affinity for the hERG channel and to improve the pharmacokinetic profile of neuropeptide Y, Y5-receptor antagonists (Fig. 8.2.5) [47]. By attaching hydrophilic functionalities to the hydrophobic center, Blum and coworkers discovered it was possible to modulate binding interactions of Y5 antagonists with the hERG channel. An inherently acidic acyl-sulfonamide substituent was effective in removing the hERG activity, but adversely affected the compound's potency

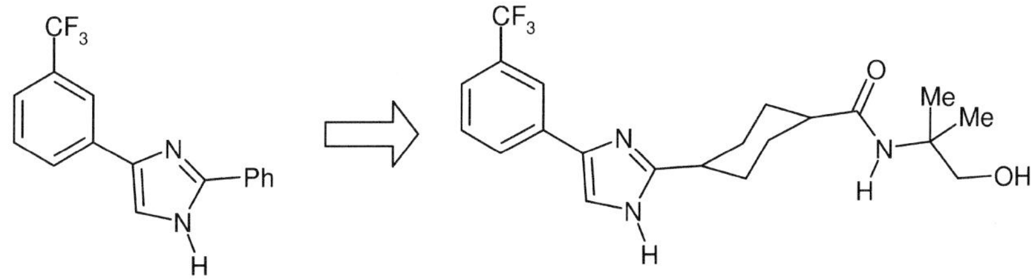

(5)   (6) weaker hERG binder

Fig. 8.2.5 Lead optimization and alleviation of binding interactions to the HERG channel for neuropeptide Y Y5 receptor
antagonists.

at the Y5 receptor. These workers were able ultimately to develop a potent and
selective Y5-receptor antagonist with favorably weak hERG activity.

Incorporation of an acidic group near the basic nitrogen is also an effective way
to remove hERG activity. Figure 8.2.6 shows an example from a report by Fraley et
al. at Merck during the optimization of the indolyl quinolinone class of KDR
kinase inhibitors [48]. Using the alignment depicted in Fig. 8.2.7, our CoMSiA
model was able to predict that the carboxylic acid derivative **8** is a weaker
hERG blocker by about three-fold, which is consistent with the experimental observations. The hERG activity was measured in these last two examples by an in-
flection point in the patch-clamp measurements.

We have applied our predictive CoMSiA model successfully to alleviate hERG
blockage of drug candidates during lead optimization. The general strategy is
to make structural modifications at or around the pharmacophoric centers of

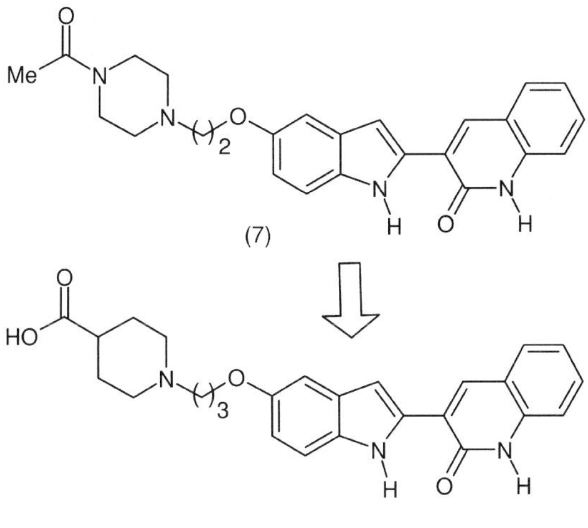

(7)

(8) weaker hERG binder

Fig. 8.2.6 Lead optimization
and alleviation of binding interactions to the HERG channel for
KDR kinase inhibitors.

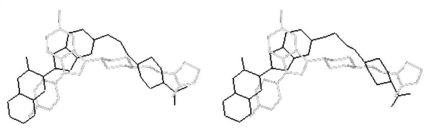

Fig. 8.2.7 Alignment of **7** (left) and **8** (right) with sertindole (gray stick) for the prediction by the CoMSiA model to rank the hERG activity.

hERG blockers with minimal impact on the targeted biological activity. These pharmacophoric centers correspond to the pharmacophore elements of aromatic centroids and basic nitrogen atom used for superimposing compounds onto sertindole in our CoMSiA model.

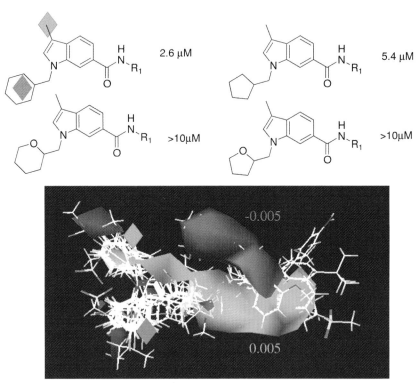

Fig. 8.2.8 An example of the successful application of the CoMSiA model to improve selectivity against HERG. Diamonds indicate the key points of alignment on the CoMSiA model. Decreasing the hydrophobicity of the aliphatic group corresponding to the primary hydrophobic group in the pharmacophore led to improved selectivity against HERG. Activity towards the HERG channel is listed and is obtained from the same assay as described in the text. R_1 is kept constant in this series only.

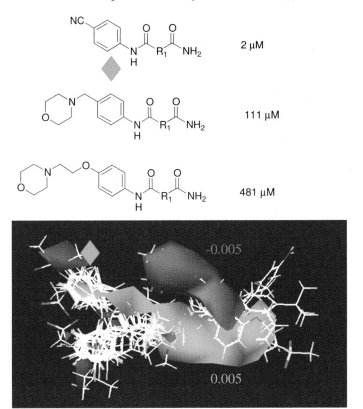

Fig. 8.2.9 An example of the successful application of the CoMSiA to improve selectivity against HERG. The diamond indicates the key point of alignment on the CoMSiA model. Subsequent substitution led to improved selectivity against HERG. Activity towards the HERG channel is listed and is obtained from the same assay as described in the text. R_1 is kept constant in this series only.

This approach is clearly justified from the CoMSiA coefficient plots. The first strategy would involve modification of the basic or positively charged nitrogen, if it exists in the molecule, and if this modification is tolerated by the targeted activity of the molecule. Electrostatic coefficient plots validate this suggestion. Most antihistamine compounds and those targeted to the CNS have a basic nitrogen, which is essential for activity towards the target, and hence this strategy might not work. The secondary strategy is based on the assumption that decreased hydrophobicity around aromatic centers by means of heterocycle replacement or elimination of aromatic ring would lead to diminished hERG blockage. Figure 8.2.8 shows an example where this strategy has been used. Here, in a series of compounds lacking a positively charged or basic nitrogen, increasing the polar nature of substituents that corresponded to the primary aromatic group in sertindole decreased hERG IC_{50} as measured by patch-clamping. In a different example, judicious addition of polar groups in a position corresponding to the para

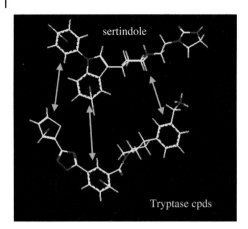

Fig. 8.2.10 Optimization of the HERG activity of tryptase inhibitors.

position of the benzene ring in sertindole would also decrease hERG activity. This can be seen in the example provided in Fig. 8.2.9 where, again in a series lacking a basic nitrogen, appropriate substitution decreased the patch-clamping IC_{50}s. In both cases, represented in Figs. 8.2.8 and 8.2.9, activities towards the targets in question were not adversely affected and some of the physical and pharmacokinetic properties were affected in a positive manner.

Another example of the use of the model to improve selectivity is shown in the published example of β-tryptase inhibitors [49]. The initial inhibitor had an IC_{50} of 0.8 μM in the hERG patch clamp assay. Figure 8.2.10 shows the alignment of the lead molecules with sertindole. Based on the negative steric coefficient at around the 7 position of the indole ring, a substitution was suggested. This substitution was also suggested to improve selectivity towards CYP2C9, another liability associated with this series. The exact nature of the substitution was later suggested to be a carboxamide since this was suggested to form a virtual dimer because the active form of β-tryptase is the tetrameric form of the enzyme. This substitution was shown not only to improve selectivity towards hERG, but also improve selectivity towards CYP2C9 while maintaining potent activity towards β-tryptase by forming a virtual dimer.

8.2.6
Conclusions

Trying to achieve agreement between the ligand-based models and the models based on structure of the channel would tremendously impact our understanding of hERG blockage. However, during the last five years, tremendous progress has been made towards utilizing computational, biological and chemical tools in understanding the interactions with the hERG K^+ channel. This progress will continue, especially with the use of these tools in achieving compounds in discovery that have better selectivity against the hERG channel and, hopefully, better cardiac safety.

References

1. Sanguinetti, M. C., C. Jiang, M. E. Curran, M. T. Keating. A mechanistic link between an inherited and an acquired cardiac arrhythmia: HERG encodes the IKr potassium channel. *Cell* **1995**, *81*, 299–307.

2. De Ponti, F., E. Poluzzi, A. Cavalli, M. Recanatini, N. Montanaro. Safety of non-antiarrhythmic drugs that prolong the QT interval or induce torsade de pointes: an overview. *Drug Saf.* **2002**, *25*, 263–286.

3. Mitcheson, J. S., M. D. Perry. Molecular determinants of high-affinity drug binding to HERG channels. *Curr. Opin. Drug Discov. Devel.* **2003**, *6*, 667–674.

4. Pearlstein, R., R. Vaz, D. Rampe. Understanding the structure-activity relationship of the human ether-a-go-go-related gene cardiac K^+ channel. A model for bad behavior. *J. Med. Chem.* **2003**, *46*, 2017–2022.

5. Netzer, R., U. Bischoff, A. Ebneth. HTS techniques to investigate the potential effects of compounds on cardiac ion channels at early-stages of drug discovery. *Curr. Opin. Drug Discov. Devel.* **2003**, *6*, 462–469.

6. Tang, W., et al. Development and evaluation of high throughput functional assay methods for HERG potassium channel. *J. Biomol. Screen* **2001**, *6*, 325–331.

7. Cheng, C. S., et al. A high-throughput HERG potassium channel function assay: an old assay with a new look. *Drug Dev. Ind. Pharm.* **2002**, *28*, 177–191.

8. Finlayson, K., L. Turnbull, C. T. January, J. Sharkey, J. S. Kelly. [3H]dofetilide binding to HERG transfected membranes: a potential high throughput preclinical screen. *Eur. J. Pharmacol.* **2001**, *430*, 147–148.

9. Chiu, P. J., et al. Validation of a [3H]astemizole binding assay in HEK293 cells expressing HERG K+ channels. *J. Pharmacol. Sci.* **2004**, *95*, 311–319.

10. Kang, J., L. Wang, F. Cai, D. Rampe. High affinity blockade of the HERG cardiac $K(+)$ channel by the neuroleptic pimozide. *Eur. J. Pharmacol.* **2000**, *392*, 137–140.

11. Zou, A., M. E. Curran, M. T. Keating, M. C. Sanguinetti. Single HERG delayed rectifier K^+ channels expressed in Xenopus oocytes. *Am. J. Physiol.* **1997**, *272*, H1309–1314.

12. Kiss, L., P. B. Bennett, V. N. Uebele, K. S. Koblan, S. A. Kane, B. Neagle, K. Schroeder. High throughput ion-channel pharmacology: planar-array-based voltage clamp. *Assay Drug Dev. Technol.* **2003**, *1*, 127–135.

13. Recanatini, M., E. Poluzzi, M. Masetti, A. Cavalli, F. De Ponti. QT prolongation through hERG $K(+)$ channel blockade: Current knowledge and strategies for the early prediction during drug development. *Med. Res. Rev.*, **2005**, *25*, 133–166.

14. Aronov, A. M. Predictive in silico modeling for hERG channel blockers. *Drug Discov. Today* **2005**, *10*, 149–155.

15. Morais Cabral, J. H., A. Lee, S. L. Cohen, B. T. Chait, M. Li, R. Mackinnon. Crystal structure and functional analysis of the HERG potassium channel N terminus: a eukaryotic PAS domain. *Cell* **1998**, *95*, 649–655.

16. Keating, M. T., M. C. Sanguinetti. Molecular and cellular mechanisms of cardiac arrhythmias. *Cell* **2001**, *104*, 569–580.

17. Doyle, D. A., et al. The structure of the potassium channel: molecular basis of K^+ conduction and selectivity. *Science* **1998**, *280*, 69–77.

18. Jiang, Y., A. Lee, J. Chen, M. Cadene, B. T. Chait, R. MacKinnon. Crystal structure and mechanism of a calcium-gated potassium channel. *Nature* **2002**, *417*, 515–522.

19. Jiang, Y., A. Lee, J. Chen, M. Cadene, B. T. Chait, R. MacKinnon. The open pore conformation of potassium channels. *Nature* **2002**, *417*, 523–526.

20. Jiang, Y., V. Ruta, J. Chen, A. Lee, R. MacKinnon. The principle of gating charge movement in a voltage-dependent K^+ channel. *Nature* **2003**, *423*, 42–48.

21. Jiang, Y., A. Lee, J. Chen, V. Ruta, M. Cadene, B. T. Chait, R. MacKinnon. X-ray structure of a voltage-dependent K^+ channel. *Nature* **2003**, *423*, 33–41.

22. Kuo, A., et al. Crystal structure of the potassium channel KirBac1.1 in the closed state. *Science* **2003**, *300*, 1922–1926.

23. Webster, S. M., D. Del Camino, J. P. Dekker, G. Yellen. Intracellular gate opening in Shaker K⁺ channels defined by high-affinity metal bridges. *Nature* **2004**, *428*, 864–868.

24. Bennett, P. B., H. R. Guthrie. Trends in ion channel drug discovery: advances in screening technologies. *Trends Biotechnol.* **2003**, *21*, 563–569.

25. Molleman, A. *Patch Clamping - An Introductory Guide to Patch Clamp Electrophysiology*, Wiley, Chichester, **2003**.

26. Mitcheson, J. S., J. Chen, M. Lin, C. Culberson, M. C. Sanguinetti. A structural basis for drug-induced long QT syndrome. *Proc. Natl. Acad. Sci. U.S.A.* **2000**, *97*, 12 329–12 333.

27. Chen, J., G. Seebohm, M. C. Sanguinetti. Position of aromatic residues in the S6 domain, not inactivation, dictates cisapride sensitivity of HERG and eag potassium channels. *Proc. Natl. Acad. Sci. U.S.A.* **2002**, *99*, 12 461–12 466.

28. Sanchez-Chapula, J. A., R. A. Navarro-Polanco, C. Culberson, J. Chen, M. C. Sanguinetti. Molecular determinants of voltage-dependent human ether-a-go-go related gene (HERG) K⁺ channel block. *J. Biol. Chem.* **2002**, *277*, 23 587–23 595.

29. Fernandez, D., A. Ghanta, G. W. Kauffman, M. C. Sanguinetti. Physicochemical features of the HERG channel drug binding site. *J. Biol. Chem.* **2004**, *279*, 10 120–10 127.

30. Korolkova, Y. V., et al. New binding site on common molecular scaffold provides HERG channel specificity of scorpion toxin BeKm-1. *J. Biol. Chem.* **2002**, *277*, 43 104–43 109.

31. Pardo-Lopez, L., M. Zhang, J. Liu, M. Jiang, L. D. Possani, G. N. Tseng. Mapping the binding site of a human ether-a-go-go-related gene-specific peptide toxin (ErgTx) to the channel's outer vestibule. *J. Biol. Chem.* **2002**, *277*, 16 403–16 411.

32. Biller, S. A., et al. In *Pharmaceutical Profiling in Drug Discovery for Lead Selection*, Borchardt, R. T., Kerns, E. H., Lipinski, C. A., Thakker, D. R. and Wang, B. (eds.). AAPS Press, Virginia, **2004**.

33. Beresford, A. P., M. Segall, M. H. Tarbit. In silico prediction of ADME properties: are we making progress? *Curr. Opin. Drug Discov. Devel.* **2004**, *7*, 36–42.

34. Buyck, C., J. Tollenaere, M. Engels, F. De Clerck. 2002. An in silico model for detecting potential hERG blocking, *The 14th European Symposium on Quantitative Structure-Activity Relationships*, Bournemouth, UK.

35. Roche, O., G. Trube, J. Zuegge, P. Pflimlin, A. Alanine, G. Schneider. A virtual screening method for prediction of the HERG potassium channel liability of compound libraries. *Chembiochem* **2002**, *3*, 455–459.

36. Aronov, A. M., B. B. Goldman. A model for identifying HERG K+ channel blockers. *Bioorg. Med. Chem.* **2004**, *12*, 2307–2315.

37. Keseru, G. M. Prediction of hERG potassium channel affinity by traditional and hologram qSAR methods. *Bioorg. Med. Chem. Lett.* **2003**, *13*, 2773–2775.

38. Ekins, S., W. J. Crumb, R. D. Sarazan, J. H. Wikel, S. A. Wrighton. Three-dimensional quantitative structure-activity relationship for inhibition of human ether-a-go-go-related gene potassium channel. *J. Pharmacol. Exp. Ther.* **2002**, *301*, 427–434.

39. Ekins, S. Predicting undesirable drug interactions with promiscuous proteins in silico. *Drug Discov Today* **2004**, *9*, 276–285.

40. Fischer, R. personal communication.

41. Kamiya, K., J. S. Mitcheson, K. Yasui, I. Kodama, M. C. Sanguinetti. Open channel block of HERG K(+) channels by vesnarinone. *Mol. Pharmacol.* **2001**, *60*, 244–253.

42. Witchel, H. J., C. E. Dempsey, R. B. Sessions, M. Perry, J. T. Milnes, J. C. Hancox, J. S. Mitcheson. The low-potency, voltage-dependent HERG blocker propafenone–molecular determinants and drug trapping. *Mol. Pharmacol.* **2004**, *66*, 1201–1212.

43. Cianchetta, G., Y. Li, J. Kang, D. Rampe, G. Cruciani, A. Fravolini, R. J. Vaz. Predictive models for HERG potassium channel blockers. *Bioorg. Med. Chem. Lett.*, **2005**, *15*, 3637–3642.

44. Cavalli, A., E. Poluzzi, F. De Ponti, M. Recanatini. Toward a pharmacophore for drugs inducing the long QT syndrome: insights from a CoMFA study of HERG K(+) channel blockers. *J. Med. Chem.* **2002**, *45*, 3844–3853.

45. Pearlstein, R. A., et al. Characterization of HERG potassium channel inhibition using CoMSiA 3D QSAR and homology modeling approaches. *Bioorg. Med. Chem. Lett.* **2003**, *13*, 1829–1835.

46. Friesen, R. W., et al. Optimization of a tertiary alcohol series of phosphodiesterase-4 (PDE4) inhibitors: structure-activity relationship related to PDE4 inhibition and human ether-a-go-go related gene potassium channel binding affinity. *J. Med. Chem.* **2003**, *46*, 2413–2426.

47. Blum, C. A., X. Zheng, S. De Lombaert. Design, synthesis, and biological evaluation of substituted 2-cyclohexyl-4-phenyl-1H-imidazoles: potent and selective neuropeptide Y Y5-receptor antagonists. *J. Med. Chem.* **2004**, *47*, 2318–2325.

48. Fraley, M. E., et al. Optimization of the indolyl quinolinone class of KDR (VEGFR-2) kinase inhibitors: effects of 5-amido- and 5-sulphonamido-indolyl groups on pharmacokinetics and hERG binding. *Bioorg. Med. Chem. Lett.* **2004**, *14*, 351–355.

49. Vaz, R. J., et al. Design of bivalent ligands using hydrogen bond linkers: synthesis and evaluation of inhibitors for human beta-tryptase. *Bioorg. Med. Chem. Lett.* **2004**, *14*, 6053–6056.

8.3
Ion Channel Safety Issues in Drug Development

Armando A. Lagrutta and Joseph J. Salata

8.3.1
Introduction

During recent years, there has been heightened concern about safety issues in drug development that are related to ion channels. These issues arise not only in programs where identified ion channel targets or suspected ion channelopathies underlie a therapeutic need. Rather, they affect all drug development programs. Areas where ion channels are therapeutic targets pose a particular set of problems, but these are only a small subset of the issues currently addressed under the discipline known as *Safety Pharmacology*. Acquired cardiac channelopathies, in particular long QT syndrome (LQTS) and, with few exceptions, LQT involving the hERG potassium channel, have been identified as a major safety concern by regulatory agencies and pharmaceutical industries worldwide. We will discuss the steps researchers are taking to provide an "integrated risk assessment" of long QT, and to assess cardiac risk beyond long QT, and we will conclude with safety issues that have been the specific concern of drug development programs targeting ion channels.

8.3.2
Regulatory–Industry Relationship (ICH); Safety Pharmacology

Safety issues in drug development are paramount concerns of the pharmaceutical industry and regulatory agencies responsible for their oversight. The advent of globalization, and a common desire for an increased efficiency in the processes of bringing safe medicines to a global marketplace, is leading to the harmonization of national and regional business practices and regulations. The long-term objective is ultimately the implementation of safety testing standards recognized and accepted worldwide.

The International Conference on Harmonization of Technical Requirements for Registration of Pharmaceuticals for Human Use (ICH) was created in 1990 as a project uniting the regulatory agencies of Europe, Japan and the United States, and experts from the pharmaceutical industry in the three regions, to discuss scientific and technical aspects of product registration. The stated goal of the ICH is a more economical use of human, animal and material resources, and the elimination of unnecessary delays in the global development and availability of new medicines, while maintaining safeguards on quality, safety, and efficacy, and regulatory obligations to protect public health [1]. There are six parties directly involved in ICH decisions, representing the regulatory agencies and pharmaceutical industry in Europe, Japan and the United States: the European Union – European Commission – European Medicines Agency (EMEA); the European Federation of Pharmaceutical Industries and Associations (EFPIA);

the Ministry of Health, Welfare and Labor, Japan (MHLW); the Japan Pharmaceutical Manufacturers Association (JPMA); the United States Food and Drug Administration (FDA), and the Pharmaceutical Research and Manufacturers of America (PhRMA).

The mechanism for harmonization among the six parties involved in ICH decisions is a stepwise process of consensus building on a draft guideline or recommendation, followed by the initiation of consultation and regulatory action within each of the three regions, arriving at a tripartite harmonized text. At the consultation step, feedback is received, from regulatory organizations and industry associates, both within the ICH and outside. When a harmonized document is fully adopted, regulatory implementation according to normal mechanisms within each of the three regions ensues.

ICH guidelines, categorized into quality, safety, efficacy and multidisciplinary topics, move through the steps of harmonization, and become the instrument of useful discussion and debate, until their eventual recognition as generally applicable and enforceable standards and procedures throughout the pharmaceutical industry. Quality topics relate to chemical and pharmaceutical Quality Assurance; safety topics, to *in vitro* and *in vivo* preclinical studies; efficacy topics, to clinical studies in human subjects; multidisciplinary topics, to cross-cutting issues not neatly defined as any of the other three categories. Figures 8.3.1 and 8.3.2 briefly summarize the ICH structure and the ICH guideline harmonization process. For additional information, readers are referred to the ICH web site and related links [1].

While the assessment of safety in animal models prior to registration of investigational drugs for human use has long been an important activity in the pharmaceutical industry worldwide, the recognition of *safety pharmacology* as a discipline in its own right must be attributed in no small part to the implementation, in Europe, Japan, and the United States, of one of the safety guidelines harmonized at the ICH level. Designated as ICH guideline S7A, on *Safety Pharmacology Studies for Human Pharmaceuticals*, this working document reached step 4 of the ICH harmonization process in November 2000. Its implementation by the EU, MHLW, and FDA quickly followed, making it an important component of regulatory submissions since that time. Safety Pharmacology studies are defined in the

---

International Conference on Harmonization

Regulatory Agencies:
- EMEA (EU – EC)
- MHWL (Japan)
- FDA (USA)

Pharmaceutical Industry:
- EFPIA (EU – EC)
- JPMA (Japan)
- PhRMA (USA)

Fig. 8.3.1 Organization of the International Conference on Harmonization of Technical Requirements for Registration of Pharmaceuticals for Human Use (ICH). The ICH has representatives from the pharmaceutical industry and regulatory agencies of three geographical regions/markets: the European Union (EU), Japan, and the United States of America.

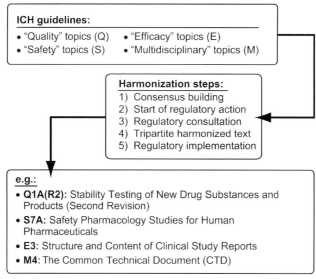

ICH guidelines:

- "Quality" topics (Q)
- "Safety" topics (S)
- "Efficacy" topics (E)
- "Multidisciplinary" topics (M)

Harmonization steps:
1) Consensus building
2) Start of regulatory action
3) Regulatory consultation
4) Tripartite harmonized text
5) Regulatory implementation

e.g.:
- **Q1A(R2):** Stability Testing of New Drug Substances and Products (Second Revision)
- **S7A:** Safety Pharmacology Studies for Human Pharmaceuticals
- **E3:** Structure and Content of Clinical Study Reports
- **M4:** The Common Technical Document (CTD)

Fig. 8.3.2 Description of the stepwise process of harmonization within the ICH. Guidelines are drafted by experts, and subjected to consultation with the regulatory agencies, the scientific community, the industry, and the public at large. The text of ICH guidelines can be revised during this consultation step, in response to feedback. The guidelines are adopted by the respective regions, either as regulations or as "guidances" based on existing regulations. The object of harmonization is that adopted guidelines, having been reviewed and accepted by all parties involved, become the basis for standard procedures in all three regions.

S7A document as "those studies that investigate the potential undesirable pharmacodynamic effects of a substance on physiological functions in relation to exposure in the therapeutic range and above" [2]. The view of safety pharmacology as a distinct scientific discipline integrating best practices from other disciplines "to further the discovery, development and safe use of biologically active chemical entities by the identification, monitoring and characterization of potentially undesirable pharmacodynamic activities in non-clinical studies" is now well established [3] and has been recently reviewed [4].

The role played by the ICH as an opinion-forming body regarding safety pharmacology and related issues cannot be overstated. It is therefore relevant to review some of the initiatives that have either been adopted or are being actively studied and debated. Here, we summarize the ICH S7A safety guideline, a related guideline termed ICH S7B, and an efficacy proposal for safety in clinical studies (ICH E14). These three initiatives inform current thinking on safety issues in drug development.

8.3.2.1 ICH Guidelines on Safety Pharmacology Studies

While the ICH S7A guideline achieved regulatory implementation (step 5) in November 2000, a parallel guideline, designated ICH S7B, on *The Nonclinical Evaluation of the Potential for Delayed Ventricular Repolarization (QT Interval Prolongation) by Human Pharmaceuticals*, did so only recently (May 2005), after much scrutiny, input and revision. Debate centered on the preclinical testing requirements needed to make an "integrated risk assessment" regarding the potential for QT interval prolongation and proarrhythmia, and on the knowledge base necessary to make such a risk assessment. An indisputable point at the core of the S7B initiative is the general recognition that the electrical activity of the heart and the ion channels underlying this electrical activity constitute a safety area that needs to be closely assessed for risk during preclinical pharmaceutical development.

8.3.2.1.1 ICH Guideline on *Safety Pharmacology Studies for Human Pharmaceuticals* (S7A). Safety Pharmacology Studies. Core Battery: Cardiovascular, Respiratory, and Central Nervous System. Follow-up & Supplemental Studies

The ICH S7A guideline places detailed emphasis on a core battery of *in vivo* observations, investigating the cardiovascular, respiratory and nervous systems, preferably in non-anesthetized animals chronically instrumented for telemetry measurements. Core cardiovascular studies mentioned in the document include measurements of blood pressure, heart rate, and the electrocardiogram. Core respiratory measurements include respiratory rate and other measurements of respiratory function, such as tidal volume and hemoglobin oxygen saturation. Core nervous system measurements include motor activity, coordination, behavioral changes, sensory/motor reflex responses and body temperature, using commonly accepted methodologies, such as the functional observation battery (Ref. [1] and documents therein).

In addition to the core battery on vital functions, S7A mentions the possible need of follow-up studies within each of the core systems, and for supplemental studies on other systems, such as the renal/urinary, gastrointestinal or other organ systems. The document is not meant to be an exhaustive or prescriptive list of Safety Pharmacology studies but very simply, yet thoroughly, is a reminder of the importance and usefulness of traditional pharmacology on well accepted and tested animal models.

8.3.2.1.2 ICH Guideline on *Safety Pharmacology Studies for Assessing the Potential for Delayed Ventricular Repolarization (QT interval prolongation) by Human Pharmaceuticals* (S7B)

The ICH S7B guideline addresses cardiac safety, and specifically a safety issue clearly identified and connected with ion "channelopathies": long QT syndrome (LQTS). Prolongation of the QT interval in the electrocardiogram has been recognized as a marker for abnormal electrical activity, which may lead to ventricular fibrillation and cardiac death. At the end of the last century, many cases of "drug acquired" LQTS were documented, resulting in the withdrawal of a number of major drugs from the market (e.g., terfenadine, astemizole, grepafloxacin, and cisapride), the imposition of availability restrictions on several others, and a general desire for an improvement in the preclinical determination of this risk for new investigational drugs. In 1997, the Committee for Proprietary Medicinal Products (CPMP) of the EMEA, EU, issued a "points to consider" document outlining a number of recommendations for *in vitro* and *in vivo* safety studies to address the potential for QT prolongation [5], which was later harmonized in the ICH draft guideline.

ICH S7B has undergone extensive review and modifications since its initial release for consultation in early 2002. The first version contained extensive detail about various *in vivo*, *in vitro* and *ex vivo* safety assays (e.g., action potential measurements in isolated cardiac myocytes or multicellular cardiac preparations), and offered rather specific guidance regarding the interpretation of the various assays to assess risk. The final version emphasizes *in vivo* studies of long QT prolongation in animal models and *in vitro* studies of I_{Kr}/hERG current, and refers to an "integrated risk assessment" in more general terms.

A primary driver behind the extensive review of S7B is the complexity of LQTS. The I_{Kr}/hERG cardiac channel has been identified as a molecular target implicated in most cases of inherited and drug acquired LQTS. Yet, the correlation between drug effects on the hERG channel and on the electrocardiogram is imperfect at best, because drugs that affect the hERG channel may affect other cardiac ion channels, with attenuating or aggravating consequences. A large body of evidence establishing the merits of supporting preclinical assays (e.g., *ex vivo* action potential measurements, studies of arrhythmia in animal models) makes the case for their inclusion in any integrated risk assessment. However, the predictive value of *in vitro* supporting assays, as well as *in vivo* studies of QT, is still debated. No animal model completely captures the clinical reality, and some models are not quantitative predictors of human risk. Physiological differences between various animal models and human exist as a consequence of differences in the type and distribution of various ion channels in cardiac tissues. Nevertheless, *in vitro* and *in vivo* preclinical tests can help predict risk of QT prolongation, and provide guidance in the design of clinical trials. Their importance, as reviewed below, cannot be overstated.

8.3.2.2 ICH Project to Develop Guidance on *Clinical Evaluation of QT/QTc Interval Prolongation and Proarrhythmic Potential for Non-Antiarrhythmic Drugs* (E14)

ICH E14 is a project to develop guidance on the clinical evaluation of long QT and the proarrhythmic potential for all drugs. This document outlines the conduct of a "thorough QT/QTc study" during clinical trials (QTc refers to the QT interval corrected for changes in heart rate). The main consideration driving this proposal is the question of whether or not preclinical testing can sufficiently exclude a clinical risk for the prolongation of QT and proarrhythmia. Since this is a highly controversial subject (see above), ICH E14 envisions that the conduct of the "thorough QT/QTc study" would be a requirement of all clinical trials. It is recognized that implementing such a proposal would have a profound impact on all drug development programs. Furthermore, the ability of such a clinical study to determine unequivocally the risk of *torsade de pointes* is not a foregone conclusion. As reviewed below, additional factors can affect this risk. Judging from the circumstances leading to the withdrawal of drugs from the market due to their torsadogenic potential, the incidence of a clinically identifiable risk is extremely low. The number of subjects studied during clinical trials might not be sufficient to predict such a risk, regardless of how "thorough" a study is conducted. At issue is also the study of normal subjects versus a diverse population with idiosyncratic predisposition to LQT. After long debate, both the E14 and S7B draft guidelines have recently advanced to regulatory implementation (May 2005).

8.3.3
Safety Issues Specific to the hERG Channel

We now turn to some of the specific findings that have underscored the need for global safety initiatives. The numerous case studies reporting drug-induced QT prolongation have focused interest on drug interactions with the hERG channel. This "acquired channelopathy" is a major concern of industry and regulators alike. Our review of the recent literature will focus on any structure–activity relations (SAR) derived from the study of specific chemical classes, and general lessons about drug-hERG channel interactions.

8.3.3.1 Antihistamines, Quinolone Antibacterials, Antipsychotics, Macrolide Antibiotics, & other hERG Channel Blockers: No Evidence for Chemical Class Effects

The first notable examples of drugs removed from the market due to documented cases of *torsade de pointes* (*TdP*, a potentially fatal polymorphic ventricular arrhythmia) are the antihistamines terfenadine and astemizole (marketed, respectively, as *Seldane*™ and *Hismanal*™). Since the 1990s and the early 2000s, considerable effort has been spent researching and reviewing these two examples, to help assess safety issues in the piperidine chemical class of H_1 receptor antagonists, and among antihistamines in general [6–14 and references therein]. For both terfenadine and astemizole, the proarrhythmic potential is clearly due to hERG inhibition resulting in long QT, and precipitating into *TdP*. Both drugs inhibit the hERG channel very potently, at low nanomolar concentrations. Because of extensive first-pass metabolic transformation before entering the systemic circulation, the drugs do not normally accumulate to nanomolar levels in plasma, and the concentrations that inhibit hERG channels are deemed "supratherapeutic." Clinical cases of *TdP* have been generally instances of changes in the normal drug metabolism, such as the reported increase in unmetabolized terfenadine upon concomitant ketoconazole administration [6]. Most of these alterations are now understood to be the result of inhibition of cytochrome P450 isoform CYP3A4.

This finding has led to the development of identified metabolites possessing therapeutic activity but less hERG channel inhibitory activity. This is exemplified by fexofenadine [13] and desloratadine [14], the major metabolites of terfenadine and loratadine. An important consideration in pursuing such a drug development strategy is the hERG inhibitory activity of the metabolites themselves, activity that factors into the adverse cardiac outcome. In a clinical case reported in the mid-1990s, proarrhythmic effects were correlated to undetectable plasma concentrations of astemizole, and "therapeutic" concentrations of its first-pass metabolite, desmethylastemizole. Close examination of hERG inhibition revealed that desmethylastemizole inhibits hERG with nanomolar potency similar to astemizole [7, 8]. Newer members of the piperidine chemical class of H_1 receptor antagonists, such as cetirizine, have shown lower hERG inhibitory potency [9]. It has been suggested that substituting groups attached to the tertiary amine of the molecule, and not the presence of the piperidine ring [10] or the aromatic ring structures involved in H_1 antagonistic activity [11], are major structural determinants of proarrhythmic effects in second-generation antihistamines. Figure 8.3.3 summarizes all relevant SAR connected to the piperidine class of H_1 receptor antagonists reviewed here.

Fluoroquinolone antibacterials are more recent examples of hERG inhibition resulting in QT interval prolongation and instances of *TdP*. Complicating the safety profile for the chemical class as a whole is the finding of various adverse effects in the clinic, such as phototoxicity and liver damage, in addition to proarrhythmic potential, among members of the class. There have been, however, encouraging lessons regarding adverse cardiac effects. Substitution at position 5 may have a role in hERG inhibition and QTc prolongation, with little effect on their potency against Gram-positive, Gram-negative or atypical bacteria [15–17]. This suggestion is based on comparison of drugs associated with proarrhythmic potential (e.g., grepafloxacin, sparfloxacin), and those deemed safer (e.g., ciprofloxacin, ofloxacin). Figure 8.3.4 summarizes all relevant SAR connected to fluoroquinolones reviewed here.

Antipsychotic drugs represent a chemically diverse group with similar actions on D_2 or $5HT_{2A}$ receptors. The proarrhythmic potential of several members of this large therapeutic class, including phenothiazines (e.g., thioridazine), butyrophenones (e.g., haloperidol), and various other chemical series, is undeniable. Efforts to address and review proarrhythmic effects on the therapeutic class in some kind of systematic fashion center on: activity on therapeutic targets (D_2 and $5HT_{2A}$ receptors) versus hERG [18]; comparisons of QTc changes, hERG inhibition and plasma concentrations [19]; and comparisons of myocardium to plasma concentrations [20]. This last type of study suggests another area complicating safety concerns, namely the potential for drug accumulation in cardiac tissues at plasma levels deemed to be "safe" [21].

The $5HT_4$ receptor partial agonist cisapride, a benzamide developed as a gastroprokinetic agent, is another example of proarrhythmic liability involving hERG inhibition [22]. No extensive studies on the benzamide series exist. One report compares the activity of mosapride and cisapride on rabbit I_{Kr} [23], and a more recent

TERFENADINE
hERG (or I_{Kr}) IC_{50} μM: 0.02 – 0.2
ETPC unbound nM: 0.1 – 0.29

                                   1.2 – 9 (P450 inh)

FEXOFENADINE
hERG (or I_{Kr}) IC_{50} μM: 5 – 23
ETPC unbound nM: 345

ASTEMIZOLE
hERG (or I_{Kr}) IC_{50} μM: 0.0009 – 0.026
ETPC unbound nM: 0.2 – 0.26

DESMETHYLASTEMIZOLE
hERG (or I_{Kr}) IC_{50} nM: 0.0009
ETPC unbound nM: ~30x astemiz.

LORATADINE
hERG (or I_{Kr}) IC_{50} μM: 0.173
ETPC unbound nM: 0.1 – 0.45

DESLORATADINE
hERG (or I_{Kr}) IC_{50} μM: >>10
Cmax nM: 200

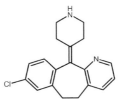

Fig. 8.3.3 Structures and relevant activities of anti-histamines belonging to the piperidine chemical class of H_1 receptor antagonists. Most HERG/I_{Kr} IC_{50} values and ETPC (effective therapeutic plasma concentrations) unbound were obtained from a recent review by Redfern et al. [46]. Data for desmethylastemizole were obtained from Ref. [8]. Data for desloratadine were obtained from Ref. [14].

CETIRIZINE
hERG (or I_{Kr}) IC_{50} μM: 108 – 300
ETPC unbound nM: 56

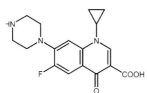

GREPAFLOXACIN
hERG (or I$_{Kr}$) IC$_{50}$ µM: 27 – 104
ETPC unbound nM: 1669 – 2087

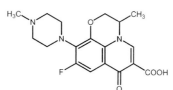

SPARFLOXACIN
hERG (or I$_{Kr}$) IC$_{50}$ µM: 23 – 34.4
ETPC unbound nM: 196 – 1766

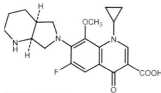

CIPROFLOXACIN
hERG (or I$_{Kr}$) IC$_{50}$ µM: >100 – 966
ETPC unbound nM: 2408 – 5281

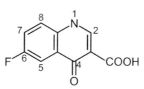

OFLOXACIN
hERG (or I$_{Kr}$) IC$_{50}$ µM: 1420
Peak free plasma conc. nM: 8700 – 14000

MOXIFLOXACIN
hERG (or I$_{Kr}$) IC$_{50}$ µM: 129
Peak free plasma conc. nM: 5900

FLUOROQUINOLONE

Fig. 8.3.4 Structures and relevant activities of fluoroquinolone antibacterials. The basic fluoroquinolone structure is also shown. Substitutions at position 5 have been suggested to play an important role in hERG channel inhibition and QT prolongation liability. Most one compares hERG inhibition by cisapride, mosapride, prucalopride, and renzapride [24]. Conclusions drawn by the first study, suggesting that the fluoro-substituted phenyl ring and the distance to the tertiary amine are important structural determinants of I$_{Kr}$ inhibition by cisapride, do not seem to be supported by the second study. Figure 8.3.5 summarizes all relevant SAR connected to the benzamide series of 5HT$_4$ agonists reviewed here. Newer 5HT$_4$ agonists developed as gastroprokinetic agents, such as tegaserod, belong to a separate chemical series, and seem to be devoid of adverse QT prolongation [25].

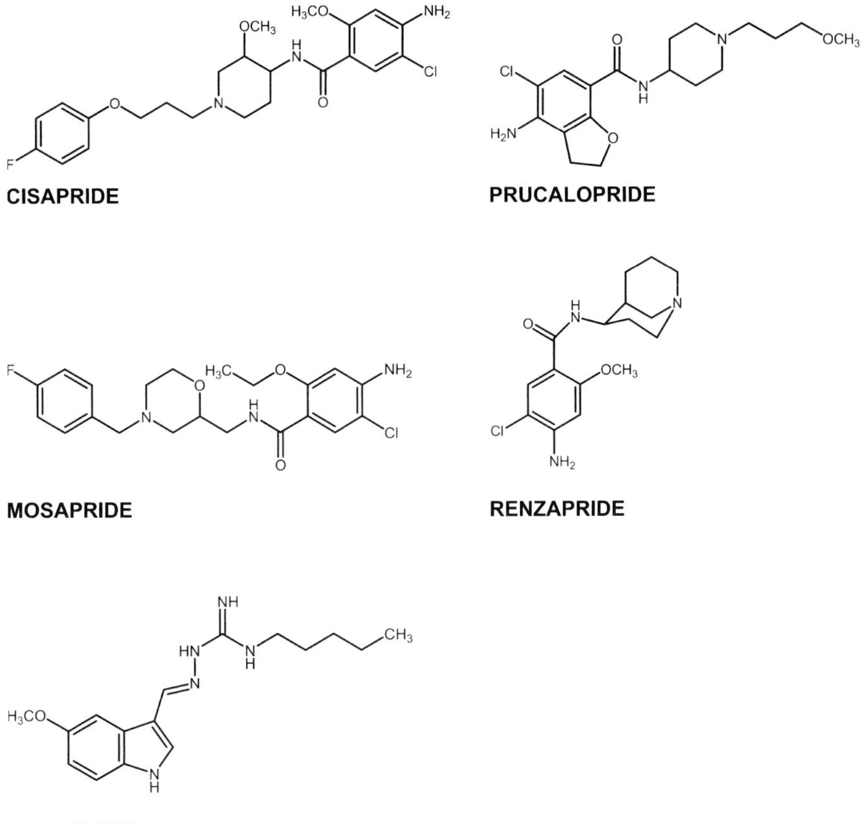

CISAPRIDE

PRUCALOPRIDE

MOSAPRIDE

RENZAPRIDE

TEGASEROD

Fig. 8.3.5 Structures of 5HT$_4$ receptor agonists developed as gastroprokinetic drugs. HERG or I$_{Kr}$ IC$_{50}$ values reported in the literature as results of direct comparison are: cisapride (9 nM), mosapride (4 μM), compared in rabbit I$_{Kr}$ (23); cisapride (240 nM), mosapride (>30 μM), prucalopride (5.7 μM), renzapride (1.8 μM) compared in cloned hERG (24). Redfern et al. [46], summarizing published data, report a IC$_{50}$ range for cisapride on hERG or I$_{Kr}$ of 0.002–0.045 μM, and an ETPC unbound range of 2.6–4.9 nM. For mosapride, a similar ETPC unbound range, of 2–5 nM, has been reported [23].

Macrolide antibiotics (e.g., erythromycin, clarithromycin) have been implicated in drug–drug interactions that, by inhibition of cytochrome P450 enzymes (e.g., 3A4, 2D6), increase the accumulation of other drugs or their metabolites with proarrhythmic potential. In addition, these antibiotics have been shown to be hERG channel inhibitors themselves, although the correlation between therapeutic concentrations and hERG inhibition is not straightforward [26]. Other antibiotics, such as ketoconozale, are similarly implicated in LQTS liabilities because of indirect and direct effects on hERG activity [27].

8.3.3.2 HERG Trafficking Defects and their Rescue by Pharmacological Chaperones

Recently, there have been reports in the literature about drug-induced LQTS by inhibition of hERG channel trafficking to the plasma membrane, as opposed to block of the mature functional channel. Inhibition of hERG channel trafficking by arsenic trioxide has been presented to explain its effects on QT in the clinic, and on QT and action potential duration in a guinea pig model [28, 29]. The description of hERG trafficking effects for arsenic trioxide, a widely used chemotherapeutic agent for the treatment of leukemias, and for heat shock protein inhibitors geldanamycin and radicicol [30], warrants close monitoring. Ironically, the rescue of hERG channel subunits defective in trafficking by known hERG channel blockers [31, 32] has been exploited as the basis of a biochemical assay identifying hERG inhibitors [33].

8.3.3.3 Verapamil, Clomiphene & other HERG Blockers that do not Produce LQT

Detailed description of hERG channel inhibitors that, paradoxically, do not produce LQTS is of great interest to those looking to establish more precise preclinical correlates for *TdP*, and to those formulating possible safety strategies in the face of an established hERG liability (see Sections 8.3.4 and 8.3.5). Two examples, verapamil and clomiphene, block hERG with potencies close to their therapeutic plasma concentrations, but seem to offset the potential QT prolongation effect by also blocking cardiac calcium channels [34, 35]. It is not known whether all "safe" hERG blockers possess a multi-channel action, or specifically action on cardiac calcium channels.

8.3.3.4 Search for Predictable Pharmacophores

One of the most challenging tasks concerning the development of new drugs devoid of hERG inhibitory activity is to predict the characteristics of ligands interacting with the hERG channel. This topic is reviewed in detail in Chapter 8.2, and has been discussed recently in the published literature [36]. *In silico* models of hERG inhibition are based on the structure of the hERG channel, and on ligand-based quantitative SAR. The hERG channel structure has been modeled, starting with sequence homology of the hERG channel subunit with subunits of crystallized bacterial potassium channels. Details about specific residues involved in hERG inhibition are modeled from results of site-directed mutagenesis experiments, testing changes in the effect of well-characterized hERG inhibitors [37]. These data suggest that the hERG channel possesses an unusually large pore, compared to other potassium channels; in addition, some of the residues lining the hERG pore (e.g., Y652, F656) are unique among other potassium channels, possessing aromatic side chains capable of interacting with pharmacophores with high affinity, by means of π-cationic or π-stacking interactions. More recent data suggest that these residues are not the major structural determinants of all high-affinity interactions [38]. Ligand-based models begin with, and are limited by, the quality and quantity of available biological data [39–41].

8.3.4
"Integrated Risk Assessment" of Delayed Ventricular Repolarization

There are no examples of chemical classes where hERG inhibition is universal for the whole class, or where the hERG SAR coincides completely with the therapeutic SAR. These findings warrant the current strategy by pharmaceutical industries to monitor, as early as possible in the drug discovery and development process, activity on the hERG channel (e.g., displacement binding assays, fluorescence-based depolarization assays, traditional or automated patch clamp platforms). Among all existing hERG assays, patch clamp is the "gold standard", and the advent of automated platforms has facilitated the throughput needed for drug development. Human channel subunits cloned in mammalian cell lines are the preferred test system for drug effects (as opposed to *Xenopus* oocytes injected with hERG channel subunit mRNA, which systematically yield underestimates of drug potency, attributed to drug–oocyte yolk interactions). Homotetramers of hERG channel alpha subunits seem to reconstitute native I_{Kr} channel activity, although the need for heterotetramers with two different types of hERG alpha subunits for more precise reconstitution [42] and the existence of accessory hERG beta subunits clearly implicated in inherited LQTS [43] have been reported. While reports of hERG inhibition by mechanisms other than channel block perhaps suggest a need for different kinds of hERG-based assays [33], reports of compounds that clearly block hERG channel without producing an adverse effect on the electrocardiogram [34, 35] warrant a multi-component assessment of risk.

There is a consensus among regulatory agencies and pharmaceutical industries regarding this need for an "integrated risk assessment" beyond hERG channel liability. Several appraisals and recommendations have been published recently [44–49] on the elements comprising an "integrated" assessment of QT prolongation risk. The overriding concern in all these is the predictive value of preclinical *in vivo* and *in vitro* data with regards to *TdP* in humans.

With respect to *in vivo* studies, there is agreement on the need to obtain a surface ECG in animals. There is convergence on the use of some test systems, such as the dog, which seem to be good preclinical indicators of QT liability in humans, but also increasing awareness about their limitations [50]. Because there is greater awareness about the potentially confounding effects of anesthetics on cardiac function, *in vivo* assays in conscious animals are favored. There is no agreement on the method used to correct QT for changes in heart rate, although there is a general recognition about the shortcomings of some of these corrections (e.g., QTc Bazett on conscious dogs). Alternatives that capitalize on the capabilities of automated data collection and telemetry have been suggested; for example, using variable rate adjustments, such as an analysis of covariance based on pretest or control data [51, 52] or dynamic beat-to-beat QT-RR interval measurements [53]. Finally, there is a recognition of problems in obtaining reliable and meaningful measurements of drug-induced changes of the QT interval in preclinical and clinical studies (interindividual and intraindividual variability in QTc values and metabolic capacity for a given drug, QT interval measurement variability, pharmacokinetics issues, data analysis and interpretation issues; reviewed in [47]). In response to these problems, there is an increasing desire for standardization of procedures for regulatory purposes (see ICH guidelines, above).

With respect to *in vitro* studies supporting hERG assays and surface ECG results in animals, tests of drug effects on cell lines expressing other cardiac ion channels implicated in inherited LQTS, such as I_{Ks} (KvLQT1/minK) or Na (hH1), are suggested when no evidence of hERG inhibition but LQTS are detected. Action potential measurements in isolated myocytes from appropriate species, in cardiac tissues (e.g., Purkinje fibers, papillary muscle, perfused wedge preparations of ventricular myocardium) or in perfused or isolated heart, represent integrative strategies bridging the molecular and organismal ends of the preclinical risk assessment spectrum (reviewed in [49]).

A careful review of each of the approaches that make part of an integrated risk assessment of the potential for torsadogenesis in humans reveals assay limitations, and many paradoxical findings that defy easy interpretation (e.g., lack of effect of terfenadine in canine Purkinje action potential measurements despite its potent effect on hERG, guinea pig myocyte I_{Kr} and action potential duration (ADP), rabbit Purkinje fibers; [10, 46, 54]). Continuing standardization in methodologies, with opportunities for discussion, review, and harmonization, is warranted.

8.3.4.1 Determination of "Provisional Safety Margins"

Recently, a representative sample of drugs with varying degrees of QT liability in human was categorized according to their torsadogenic propensity (five classes: repolarization prolonging antiarrhythmics, drugs with unacceptable *TdP* risk removed from the market, drugs with widely reported *TdP* risk, drugs with isolated reports of *TdP* risk, and drugs with no reports of *TdP* risk). The drugs were then evaluated with regards to their "safety margins" [46]: the ratios of adverse effect activity (IC_{50} on hERG/I_{Kr} or on the consequent lengthening of action potential duration) divided by the effective therapeutic plasma concentrations (ETPC). This summary of a large body of data not collected systematically, and limited by those data, nonetheless revealed a trend: drugs like antiarrhythmics, known to block hERG, tend to exhibit low safety margins (TdP in humans tend to exhibit high safety margins (>30-fold). Despite the finding of several "outliers", a practical "safe" margin of 30-fold seemed to apply to the whole set. This "30-fold" margin has been extrapolated, and is generally accepted as a provisional benchmark against which preclinical activity data (hERG/I_{Kr} IC_{50}/ETPC) can be measured. Details about each of the outliers ("false positives", such as amiodarone, and "false negatives", such as verapamil) should continue to be explored and discussed.

8.3.4.2 Concept of "Reduced Repolarization Reserve"

From the clinical perspective, there is an increasing recognition of variable risk with regards to acquired LQTS. Multiple clinical risk factors for drug-induced *TdP* have been identified: female gender, electrolyte imbalances, bradycardia, base-line QT prolongation, and others (reviewed in [55]). A unifying framework, "reduced repolarization reserve" has been used to explain the variability [55, 56]. Briefly, the suggestion is that physiologic mechanisms that maintain normal cardiac repolarization vary among individuals, but the differences are not apparent in a basal state. The idea of a repolarization reserve is based on the redundancy of physiological mechanisms underlying ventricular depolarization. An example of this in terms of ion channels is the partially overlapping roles played by I_{Kr} (hERG) and I_{Ks} (KvLQT1/minK) in the cardiac action potential. Recently, experimental support has been presented for the suggestion that drugs that inhibit hERG but show no QT liability (e.g., verapamil), or drugs that display a bell-shaped concentration response in action potential duration assays (e.g., haloperidol, cisapride), mitigate the potentially adverse effects caused by hERG inhibition by targeting other cardiac ion channels [57]. A prevalent opinion, however, is that one should take no comfort in the possibility of "safe" non-antiarrhythmic candidate drugs with mixed channel activity [46].

8.3.5
Beyond QT Prolongation

An important effort in preclinical drug research has been the search for a more precise connection between inhibition of repolarizing currents in the cardiac ventricle and *TdP*. The general consensus is that animal models of arrhythmia, such as the methoxamine-treated rabbit [58], generally offer accurate qualitative but no quantitative assessments of drug safety. It has been recognized that *in vitro* hERG channel inhibition and *in vivo* QT/QTc measurements are the closest tools available to what could be termed an ideal "biomarker" for *TdP* [45]. "Integrated risk assessment", in this context, is a necessity, but not a panacea. Many researchers are actively pursuing better "biomarkers" – signals that may more reliably and accurately reflect an early stage in the progression to *TdP*, and which are mechanistically linked to *TdP*.

The SCREENIT system, based on rabbit Langendorff-perfused hearts, was developed as an *in vitro* model with sufficient stability to conduct long-lasting, systematic, computerized measurements of action potential properties at various cycle lengths, and with increasing drug concentrations [59]. Using this system, blinded tests of proarrhythmic and antiarrhythmic drugs have been conducted [60, 61]. Based on these tests, it has been proposed that prolongation of action potential duration, in itself, is antiarrhythmic; other parameters that may or may not accompany prolongation, such as instability, triangulation (rendering the action potential more triangular in shape), and reverse use dependence, are proarrhythmic [60]. The extent to which these conclusions apply to species other than the rabbit is not known.

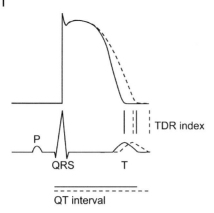

Fig. 8.3.6 Schematic diagram of action potential duration (APD) and its correlation with the electrocardiogram (ECG). Features described in the text are highlighted, such as triangulation of the action potential, QT interval prolongation, and the ECG index for transmural dispersion of repolarization (TDR).

In agreement with the notion that action potential prolongation in itself is not proarrhythmic, it has been proposed that early afterdepolarizations (oscillations in membrane potential that arise during the repolarization phase of the action potential) and concomitant ectopic beats detected in the electrocardiogram are the trigger for *TdP*. It is argued that the substrate of *TdP* is an increase in the "dispersion of repolarization" in the ventricle [62]. This refers to an abnormal increase in the differences in action potential duration across the ventricular wall (transmural dispersion of repolarization or TDR), between the left and right ventricles, or between the base and apex of the heart. Specifically for TDR, it has been suggested that its index in the electrocardiogram, measured as the distance from the peak to the end of the T wave, can be a more reliable marker of proarrhythmia than QT prolongation, not only in animal models but also in humans [63, 64].

Another proarrhythmic marker suggested as a more reliable measurement than QT prolongation in the ECG is based on beat-to-beat alternations of the cardiac monophasic action potential ("electrical alternans"), and the way proarrhythmic drugs affect them [65]. This *in vivo* assay, conducted in guinea pigs, seems to detect proarrhythmic effects at therapeutically relevant concentrations of drug. It is proposed to be a useful preclinical assay, predictive of clinical outcomes.

A recent review compares several surrogate markers of *TdP* derived from the electrocardiogram, but which depart from the traditional heart-rate corrected QT interval measurement. These include: QT dispersion (QTd), a method widely used, based on measurements of interlead variability of QT interval in the standard 12-lead surface ECG; TDR index, measured from the T wave, as described above; T wave changes (and a related technique known as principal component analysis, or PCA, quantitatively describing the shape of the T wave as a sum of several components); dynamic analysis of QT interval rate dependence (QT/RR slope), discussed above as an alternative to QT heart-rate correction formulas. The main conclusions are highly critical of QT dispersion. QTd is not necessarily a good predictor of dispersion of repolarization in the ventricle; unsurprisingly, among all the alternatives, it appears to be the worst predictor of proarrhythmic risk. The other alternatives listed, based on T wave morphology or addressing QT correction formula limitations, seem to be more promising [66]. Figure 8.3.6 summarizes, in schematic fashion, the correlation between action potential measurements and the electrocardiogram, and some of the markers discussed here.

8.3.6
Issues Specific to Ion Channel Targets

In addition to acquired channelopathies on the hERG channel, connected to LQTS, pharmaceutical research is concerned with inadvertent inhibition of any of the various cardiovascular or neuronal ion channels. For many of these targets, the first line of safety screens is simple pharmacological assays, such as displacement binding assays. The core battery of *in vivo* safety tests outlined in the ICH S7A guideline complements this approach. Specific follow-up tests can be conducted when warranted.

One last, but not least important, issue concerns drug development when the intended therapeutic target is an ion channel. Pharmacogenetic paradigms are aimed at correcting defects known to be caused by inherited channelopathies. In these, the risk assessment rests on the precise identification of the defect in the patient, and is not a true drug development issue.

A more traditional paradigm of drug development is aimed at "therapeutic targets", which, upon modulation, are predicted to yield measurable benefit to a number of closely related defects or needs. Sometimes, the therapeutic need might not be completely understood at the outset of the drug discovery and development process, or the "target" might be a physiological parameter, rather than a molecular entity. In many cases, retrospective examination can reveal that a given agent is effective on several targets at its therapeutic concentrations, and safety issues may be associated with these findings. In this traditional paradigm, a large number of ion channels have been targeted for a number of unmet medical needs, and there are substantial numbers of published reports on safety issues related to their discovery and development.

Class I (blocking sodium channels) and Class III (prolonging cardiac action potential) antiarrhythmics represent the best example of a calculated risk in drug development in the face of a paradoxical medical situation. The genesis, current use and prospects for antiarrhythmic drug development have been reviewed [67]. Briefly, Class I antiarrhythmics were the first identified treatment for atrial fibrillation and for potentially lethal ventricular tachyarrhythmias associated with acute myocardial infarction. Clinical trials developing newer class I antiarrhythmics in the late 1980s established a correlation between mortality and this class of drugs. Drug discovery and development efforts shifted to Class III agents, which act predominantly by prolonging action potential duration. However, the association of action potential prolongation with LQTS, *TdP*, and ventricular fibrillation, described extensively here, has warranted a closer look. Several "safety-minded" strategies have been pursued, based on target choice and drug properties. Regarding target choice, a recent example is the study of I_{Ks} versus I_{Kr} as proposed treatment for ventricular tachyarrhythmias. While the two major depolarizing currents in the human ventricle, I_{Kr} and I_{Ks}, have been implicated in inherited LQTS, the incidence of LQT2 (affecting hERG/I_{Kr}) is much higher than the incidence of LQT1 or LQT6 (affecting KvLQT1/minK). Furthermore, selective block may provide a profile of effects imparting improved antiarrhythmic efficacy (e.g., frequency dependence, homogeneity of effect, sustained activity during sympathetic stimulation; [68] and references therein). Regarding drug properties, a recent example is the study of dronedarone and other derivatives of amiodarone, aimed at removing unwanted side effects while preserving or improving the relatively safe profile of the parent structure, with regards to *TdP* [69].

The therapeutic area of neuropathic pain presents other examples of "safety-minded" drug development involving ion channels. A snail conopeptide ω-conotoxin and its synthetic derivative, ziconotide, were developed as antinociceptive agents with a safety strategy intending to reduce a documented adverse effect by excluding drug access from systemic circulation. The sympathetic vasoconstrictor effects of ω-conotoxin on the N-type Ca^{2+} channels in the vasculature, and the consequent orthostatic hypotension that ensues, are well documented [70]. A strategy aiming to target the N-type Ca^{2+} channels in dorsal root ganglia for the treatment of opioid-resistant pain sought to by-pass this adverse effect by intrathecal injection. Despite numerous preclinical and clinical studies documenting the relative safety of this route of administration, there has been one report in the literature documenting adverse effects in the central nervous system (CNS) in patients ([71], references therein). Strategies to circumvent these issues include the search for novel N-type calcium channel blockers [72], which might present a better safety profile than ziconitide, and the search for drugs on other targets in dorsal root ganglia, such as sodium channel subunits [73].

Drug development centered on the voltage-dependent potassium channel Kv1.3, targeted for immunosuppression and autoimmune disease therapies, exemplifies the difficulties in finding selective blockers and drugs with restricted tissue distribution (e.g., systemic, but no CNS accumulation). The Kv1.3 channel subunit is a major player in controlling membrane potential in T lymphocytes. Structurally, it has some unique features in its pore region, but also exhibits many of the highly conserved amino acid residues present in all members of the Kv1.x family. Regarding its role and tissue distribution, the Kv1.3 channel plays a major role in membrane potential control in T lymphocytes, but it is also involved in repolarization in the CNS. As a whole, Kv1.x channels are widely distributed, regulating membrane potential and Ca^{2+} signaling in various cell types. For these reasons, the search for drugs targeting Kv1.3 channels devoid of adverse acute toxicity has not been easy [74–76].

Voltage-gated and ligand-gated ion channels are critically important in excitability of cardiac, skeletal and smooth muscle, neuronal excitability, endocrine secretion, and immunomodulation. In most instances, their quaternary structure is achieved by the assembly of several α subunits and accessory subunits with fairly unrestricted distribution throughout tissues, organs, and systems. As evidenced by the examples presented here, major safety concerns relate to the precise identity of the target and the precise characterization of the drug action. "Safety-minded" decisions include targeting tissue-restricted isoforms of the channel subunits or associated regulatory molecules, designing drugs with state-dependent activity to maximize therapeutic effects and minimize potential adverse effects, and understanding in sufficient detail drug metabolism and drug tissue distribution, especially in organs and systems of safety concern.

References

1. http://www.ich.org and related links.
2. M.M. Dotzel, *Federal Register* 66, 36791, **2001**. Docket 00D-1407. Full text at http://www.fda.gov/cder/guidance/4461fnl.htm.
3. http://www.safetypharmacology.org.
4. L.B. Kinter, J.P.Valentin. Safety pharmacology and risk assessment. *Fundament. Clin. Pharmacol.* (**2002**) 16, 175–182.
5. Committee for Proprietary Medicinal Products (CPMP). Points to consider: the assessment of the potential for QT interval prolongation by non-cardiovascular medicinal products. The European Agency for the Evaluation of Medicinal Products, London. Human Medicines Evaluation Unit (**1997**). Full text at http://www.emea.eu.int.
6. P.K. Honig, D.C. Wortham, K. Zamani, D.P Connor, J.C. Mulin, L.R. Cantilena. Terfenadine-ketoconazole interaction. *J. Am. Med. Assoc.* (**1993**) 269, 1513–1518.
7. V. R. Vorperian, Z. Zhou, S. Mohammad, T.J. Hoon, C. Studenik, C.T. January. Torsade de Pointes with an antihistamine metabolite: potassium channel blockade with desmethylastemizole. *J. Am. Coll. Cadiol.* (**1996**) 28, 1556–1561.
8. Z. Zhou, V.R. Vorperian, Q. Gong, S. Zhang, C.T. January. Block of HERG potassium channels by the antihistamine astemizole and its metabolites desmethylastemizole and norastemizole. *J. Cardiovasc. Electrophys.* (**1999**) 10, 836–843.
9. M. Taglialatela, A. Pannacione, P. Castaldo, G. Giorgio, Z. Zhou, C.T. January, A.Genovese, G. Marone, L. Annunziato. Molecular basis for the lack of hERG channel block-related cardiotoxicity by the H_1 receptor blocker cetirizine compared to other second-generation anti-

10. J.J. Salata, N.K. Jurkiewicz, A.A. Wallace, R.F. Stupienski, P.J. Guinoso, J.J. Lynch Jr. Cardiac electrophysiological actions of the histamine H_1 receptor antagonists astemizole and terfenadine compared to chlorpheniramine and pyrilamine. *Circ. Res.* (**1995**) 76, 110–119.
11. M.-Q. Zhang. Chemistry underlying the cardiotoxicity of antihistamines. *Curr. Med. Chem.* (**1997**) 4, 187–200.
12. R.L. Woosley. Cardiac actions of antihistamines. *Annu. Rev. Pharmacol. Toxicol.* (**1996**) 36, 233–252.
13. C. Pratt, A.M. Brown, D. Rampe. Cardiovascular safety of fexofenadine HCl. *Clin. Exp. Allergy* (**1999**) 29, 212–216.
14. D.K. Agrawal. Pharmacology and clinical efficacy of desloratadine as an anti-allergic and anti-inflammatory drug. *Exp. Opin. Invest. Drugs* (**2001**) 10, 547–560.
15. B.A. Lipsky, C.A. Baker. Fluoroquinolone toxicity profiles: a review focusing on newer agents. *Clin. Infectious Diseases* (**1999**) 28, 352–364.
16. J. Kang, L. Wang, X.-L. Chen, D.J. Triggle, D. Rampe. Interactions of a series of fluoroquinolone antibacterial drugs with the human cardiac K^+ channel HERG. *Mol. Pharmacol.* (**2001**) 59, 122–126.
17. M.I. Andersson, A.P. MacGowan. Development of the quinolones. *J. Antimicrob. Chemother.* (**2003**) 51, Suppl. S1, 1–11.
18. D.M. Taylor. Antipsychotics and QT prolongation. *Acta Psychiatr. Scand.* (**2003**) 107, 85–95.
19. S. Kongsamut, J. Kang, X.-L. Chen, J. Roehr, D. Rampe. A comparison of the receptor binding and hERG channel affinities for a series of antipsychotic

drugs. *Eur. J. Pharmacol.* (**2002**) 450, 37–41.

20. K. Titier, M. Canal, E. Deridet, A. Abouelfath, S. Gromb, M. Molimard, N. Moore. Determination of myocardium to plasma concentration ratios of five antipsychotic drugs: comparison with their ability to induce arrhythmia and sudden death in clinical practice. *Toxicol. Appl. Pharmacol.* (**2004**) 199, 52–60.

21. W. Crumb, I. Cavero. QT interval prolongation by non-cardiovascular drugs: issues and solutions for novel drug development. *Pharmaceutical Science and Technology Today* (**1999**) 2, 270–280.

22. D. Rampe, M.-L. Roy, A. Dennis, A.M. Brown. A mechanism for the proarrhythmic effects of cisapride (Propulsid): high affinity blockade of the human cardiac potassium channel HERG. *FEBS Lett.* (**1997**) 417, 28–32.

23. L. Carlsson, G.J. Amos, B. Andersson, L. Drews, G. Duker, G. Wadstedt. Electrophysiological characterization of the prokinetic agents cisapride and mosapride *in vivo* and *in vitro*: implications for proarrhythmic potential? *J. Pharmacol. Exp. Therap.* (**1997**) 282, 220–227.

24. F. Potet, T. Bouyssou, D. Escande, I. Baro. Gastrointestinal prokinetic drugs have different affinity for the human cardiac human ether-a-gogo K$^+$ channel. *J. Pharmacol. Exp. Therap.* (**2001**) 299, 1007–1012.

25. M. Corsetti, J. Tack. Tegaserod: a new 5-HT4 agonist in the treatment of irritable bowel syndrome. *Exp. Opin. Pharmacother.* (**2002**) 3, 1211–1218.

26. W.A. Volberg, B.J. Koci, W. Su, J. Lin, J. Zhou. Blockade of human cardiac potassium channel human *ether-a-go-go*-related gene (*HERG*) by macrolide antibiotics. *J. Pharmacol. Exp. Ther.* (**2002**) 302, 320–327.

27. R. Dumaine, M.-L. Roy, A.M. Brown. Blockade of HERG and Kv1.5 by ketoconazole. *J. Pharmacol. Exp. Therap.* (**1998**) 286, 727–735.

28. C.-E. Chiang, H.-N. Luk, T.-M. Wang, P.Y.-A. Ding. Prolongation of cardiac repolarization by arsenic trioxide. *Blood* (**2002**) 100, 2249–2252.

29. E. Ficker, Y.A. Kuryshev, A.T. Dennis, C. Obejero-Paz, L. Wang, P. Hawryluk, B.A. Wible, A.M. Brown. Mechanisms of arsenic-induced prolongation of cardiac repolarization. *Mol. Pharmacol.* (**2004**) 66, 33–44.

30. E. Ficker, A.T. Dennis, L. Wang, A.M. Brown. Role of the cytosolic chaperones Hsp70 and Hsp90 in the maturation of the cardiac potassium channel hERG. *Circ. Res.* (**2003**) 92, E87–E100.

31. E. Ficker, C.A. Obejero-Paz, S. Zhao, A.M. Brown. The binding site for channel blockers that rescue misprocessed human long QT syndrome type 2 ether-a-go-go related gene (HERG) mutations. *J. Biol. Chem.* (**2002**) 277, 4989–4998.

32. Q. Gong, C.L. Anderson, C.T. January, Z. Zhou. Pharmacological rescue of trafficking defective HERG channels formed by coassembly of wild-type and long QT mutant N470D subunits. *Am. J. Physiol. Heart Circ. Physiol.* (**2004**) 287, H652–658.

33. A.M. Brown. Drugs, hERG and sudden death. *Cell Calcium* (**2004**) 35, 543–547.

34. S. Zhang, Z. Zhou, Q. Gong, J.C. Makielski, C.T. January. Mechanism of block and identification of the verapamil binding domain to HERG potassium channels. *Circ. Res.* (**1999**) 84, 989–998.

35. K.H. Yuill, J.J. Borg, J.M. Ridley, J.T. Milnes, H.J. Witchel, A.A. Paul, R. Z. Kozlowski, J.C. Hancox. Potent inhibition of human cardiac potassium (HERG) channels by the anti-estrogen agent clomiphene – without QT interval prolongation. *Biochem. Biophys. Res. Commun.* (**2004**) 318, 556–561.

36. R. Pearlstein, R. Vaz, D. Rampe. Understanding the structure-activity relationship of the human ether-a-go-go related gene cardiac K$^+$ channel. A model for bad behavior. *J. Med. Chem.* (**2003**) 46, 2017–2022.

37. J.S. Mitcheson, J. Chen, M. Lin, C. Culberson, M.C. Sanguinetti. A structural basis for drug-induced long QT syndrome. *Proc. Natl. Acad. Sci. U.S.A.* (**2000**) 97, 12329–12333.

38. J.M. Ridley, J.T. Milnes, H.J. Witchel, J.C. Hancox. High affinity HERG channel blockade by the antiarrhythmic agent dronedarone: resistance to mutations of the S6 residues Y652 and F656. (**2004**) 325, 883–891.

39. W. Bains, A. Basman, C. White. HERG binding specificity and binding site structure: evidence from a fragment-based evolutionary computing SAR study. *Prog. Biophys. Mol. Biol.* (**2004**) 86, 205–233.

40. A.M. Aronov, B.B. Goldman. A model for identifying HERG K channel blockers. *Bioorg. Med. Chem.* (**2004**) 12, 2307–2315.

41. R.A. Pearlstein, R.J. Vaz, J. Kang, X.-L. Chen, M. Preobrazhenskaya, A.E. Shchekotikhin, A.M. Korolev, L.N. Lysenkova, O.V. Miroshnikova, J. Hendrix, D. Rampe. Characterization of HERG potassium channel inhibition using CoMSiA 3D QSAR and homology modeling approaches. *Bioorg. Med. Chem. Lett.* (**2003**) 13, 1829–1835.

42. E.M.C. Jones, E.C. Roti Roti, J. Wang, S.A. Delfosse, G.A. Robertson. Cardiac I_{Kr} channels minimally comprise hERG 1a and 1b subunits. *J. Biol. Chem.* (**2004**) 279, 44 690–44 694.

43. Y. Lu, M.P. Mahaut-Smith, C.L.-H. Huang, J.L. Vanderberg. Mutant MiRP1 subunits modulate HERG K^+ channel gating: a mechanism for proarrhythmia in long QT syndrome type 6.

44. T.G. Hammond, L. Carlsson, A.S. Davis, W.G. Lynch, I. MacKenzie, W.S. Redfern, A.T. Sullivan, A.J. Camm. Methods of collecting and evaluating non-clinical cardiac electrophysiology data in the pharmaceutical industry: results of an international survey. *Cardiovasc. Res.* (**2001**) 49, 741–750.

45. L.B. Kinter, P.K.S. Siegl, A.S. Bass. New preclinical guidelines on drug effects on ventricular repolarizaton: Safety pharmacology comes of age. *J. Pharmacol. Toxicol. Method* (**2004**) 49, 153–158.

46. W.S. Redfern, L. Carlsson, A.S. Davis, W.G. Lynch, I. MacKenzie, S. Palethorpe, P.K.S. Siegl, I. Strang, A.T. Sullivan, R. Wallis, A.J. Camm, T.G. Hammond. Relationships between preclinical cardiac electrophysiology, clinical QT interval prolongation and torsade de pointes for a broad range of drugs: evidence for a provisional safety margin in drug development. *Cardiovasc. Res.* (**2003**) 58, 32–45.

47. F. De Ponti, E. Poluzzi, N. Montanaro. QT-interval prolongation by non-cardiac drugs: lessons to be learned from recent experience. *Eur. J. Clin. Pharmacol.* (**2000**) 56, 1–18.

48. B. Fermini, A. Fossa. The impact of drug-induced QT interval prolongation on drug discovery and development. *Nat. Rev.: Drug Discov.* (**2003**) 2, 439–447.

49. R. Fenichel, M. Malik, C. Antzelevitch, M. Sanguinetti, D.M. Roden, S.G. Priori, J.N. Ruskin, R.J. Lipicky, L.R. Cantilena. Drug-induced torsades de pointes and implications for drug development. *J. Cardiovasc. Electrophysiol.* (**2004**) 15, 475–495.

50. A.A. Fossa, M.J. DePasquale, D.L. Raunig, M.J. Avery, D.J. Leishman. The relationship of clinical QT prolongation to outcome in the conscious dog using a beat-to-beat QT-RR interval assessment. *J. Pharmacol. Exp. Therap.* (**2002**) 202, 828–833.

51. S. Spence, K. Soper, C. Hoe, J. Coleman. The heart rate-corrected QT interval of conscious beagle dogs: a formula based on analysis of covariance. *Toxicol. Sci.* (**1998**) 45, 247–258.

52. H. Miyazaki, M. Tagawa. Rate-correction technique for QT interval in long-term telemetry ECG recording in beagle dogs. *Exp. Anim.* (**2002**) 51, 465–475.

53. A.A. Fossa, T. Wisialowski, A. Magnano, E. Wolfgang, R. Winslow, W. Gorczyca, K. Crimin, D.L. Raunig. Dynamic beat-to-beat modeling of the QT-RR interval relationship: analysis of QT prolongation during alterations of autonomic state versus human ether a-go-go-related gene inhibition. *J. Pharmacol. Exp. Therap.* (**2005**) 312, 1–11.

54. G.A. Gintant, J.T. Limberis, J.S. McDermott, C.D. Wegner, B.F. Cox. The canine Purkinje fiber: an in vitro model system for acquired long QT syndrome and drug-induced arrhythmogenesis. *J. Cardiovasc. Pharmacol.* (**2001**) 37, 607–618.

55. D.M. Roden. Drug-induced prolongation of the QT interval. *N. England J Med* (**2004**) 350, 1013–1022.

56. D.M. Roden. Taking the "idio" out of "idiosyncratic": predicting torsades de pointes. *Pacing Clin. Electrophysiol.* (**1998**) 21, 1029–1034.

57. R.L. Martin, J.S. McDermott, H.J. Salmen, J. Palmatier, B.F. Cox, G.A. Gintant. The utility of hERG and repolarization assays in evaluating delayed cardiac repolarization: influence of multi-channel block. *J. Cardiovasc. Pharmacol.* (**2004**) 43, 369–379.

58. L. Carlsson, O. Almgren, G. Duker. QTU-prolongation and torsades de pointes induced by putative class III antiarrhythmic agents in the rabbit: etiology and interventions. *J. Cardiovasc. Pharmacol.* (**1990**) 16, 276–285.

59. L.M. Hondeghem. Computer aided development of antiarrhythmic agents with class IIIa properties. *J. Cardiovasc. Electrophysiol.* (**1994**) 5, 711–721.

60. L.M. Hondeghem, L. Carlsson, G. Duker. Instability and triangulation of the action potential predict serious proarrhythmia, but action potential duration prolongation is antiarrhythmic. *Circulation* (**2001**) 103, 2004–2013.

61. L.M. Hondeghem, P. Hoffman. Blinded test in isolated female rabbit heart reliably identifies action potential duration during prolongation and proarrhythmic drugs: importance of triangulation, reverse use dependence, and instability. *J. Cardiovasc. Pharmacol.* (**2003**) 41, 14–24.

62. L. Belardinelli, C. Antzelevitch, M.A.Vos. Assessing predictors of drug-induced torsade de pointes. *Trends Pharmacol. Sci.* (**2003**) 24, 619–625.

63. J.M. Di Diego, L. Belardinelli, C. Antzelevitch. Cisapride-induced transmural dispersion of repolarization and torsade de pointes in the canine left ventricular wedge preparation during epicardial stimulation. *Circulation* (**2003**) 108, 1027–1033.

64. N. Watanabe, Y. Kobayashi, K. Tanno, F. Miyoshi, T. Asano, M. Kawamura, Y. Mikami, T. Adachi, S. Ryu, A. Miyata, T. Katagiri. Transmural dispersion of repolarization and ventricular tachyarrhythmias. *J. Electrocard.* (**2004**) 37, 191–200.

65. T. Fossa, E. Wisialowski, E. Wolfgang, M.Wang, D. Avery, B. Raunig, B. Fermini. Differential effect of HERG blocking agents on cardiac electrical alternans in the guinea pig. *Eur. J. Pharmacol.* (**2004**) 486, 209–221.

66. R.R. Shah. Drug-induced QT dispersion: does it predict the risk of torsade de pointes? *J. Electrocard.* (**2005**) 38, 10–18.

67. S. Nattel, B.N. Singh. Evolution, mechanisms, and classification of antiarrhythmic drugs: focus on class III actions. *Am. J. Cardiol.* (**1999**) 84, 11R–19R.

68. J.J. Salata, H.G. Selnick, J.J. Lynch. Pharmacological modulation of I_{Ks}: potential for antiarrhythmic therapy. *Curr. Med. Chem.* (**2004**) 11, 29–44.

69. S.A. Doggrell, J.C. Hancox. Dronedarone: an amiodarone analogue. *Exp. Opin. Investig. Drugs* (**2004**) 13, 415–426.

70. C. Wright, A. Hawkes, J. Angus. Postural hypotension following N-type Ca^{2+} channel blockade is amplified in experimental hypertension. *J. Hypertension* (**2000**) 18, 65–73.

71. K.K. Jain. An evaluation of intrathecal ziconotide for the treatment of chronic pain. *Exp. Opin. Investig. Drugs* (**2000**) 9, 2403–2410.

72. E. Teodori, E. Baldi, S. Dei, F. Gualtieri, M.N. Romanelli, S. Scapecchi, C. Belluci, C. Ghelardini, R. Matucci. Design, synthesis, and preliminary pharmacological evaluation of 4-aminopyridine derivatives as N-type calcium channel blockers active on pain and neuropathic pain. *J. Med. Chem.* (**2004**) 47, 6070–6081.

73. B. Priest, M.L. Garcia, R.E. Middleton, R.M. Brochu, S. Clark, G. Dai, I.E. Dick, J.P. Felix, C.J. Liu, B.S. Reiseter, W. A. Schmalhofer, P. P. Shao, Y.S. Tang, M. Z. Chou, M.G. Kohler, M.M. Smith, V.A. Warren, B.S. Williams, C.J. Cohen, W.J. Martin, P.T. Meinke, W.H. Parsons, K.A. Wafford, G. J. Kaczorowski. A disubstituted succinamide is a potent sodium channel blocker with efficacy in a rat pain model. *Biochemistry* (**2004**) 43, 9866–9876.

74. J.P. Felix, R.M. Bugianesi, W.A. Schmalhofer, R. Borris, M.A. Goetz, O.D. Hensens, J. Bao, F. Kayser, W.H. Parsons, K. Rupprecht, M.L. Garcia, G. J. Kaczorowski, R.S. Slaughter. Identification and biochemical characterization of a novel nortriterpene inhibitor of the human lymphocyte voltage-gated potassium channel, Kv1.3. *Biochemistry* (**1999**) 38, 4922–4930.

75. W.A. Schmalhofer, J. Bao, O.B. McManus, B. Green, M. Matyskiela, D. Wunderler, R.M. Bugianesi, J.P. Felix, M. Hanner, A.R. Linde-Arias, C.G. Ponte, L.

Velasco, G. Koo, M.J. Staruch, S. Miao, W.H. Parson, K. Rupprecht, R.S. Slaughter, G.J. Kaczorowski, M.L. Garcia. *Biochemistry* (**2002**) 41, 7781–7794.

76. H. Wulff, C. Beeton, K.G. Chandy. Potassium channels as therapeutic targets for autoimmune disorders. *Curr. Opin. Drug Discov. Dev.* (**2003**) 6, 640–647.

Index

Voltage-Gated Ion Channels as Drug Targets. Edited by D. Triggle
Copyright © 2006 WILEY-VCH Verlag GmbH & Co. KGaA, Weinheim
ISBN 3-527-31258-7